Trophoblast Research
Volume 4

Trophoblast Invasion and Endometrial Receptivity

Novel Aspects of the Cell Biology of Embryo Implantation

Trophoblast Research

Series Editors

Richard K. Miller and Henry A. Thiede
University of Rochester Medical Center
Rochester, New York

Trophoblast Research
VOLUME 4

TROPHOBLAST INVASION AND ENDOMETRIAL RECEPTIVITY

Novel Aspects of the Cell Biology of Embryo Implantation

Edited by

Hans-Werner Denker

University Clinics
Essen, Federal Republic of Germany

and

John D. Aplin

St. Mary's Hospital
Manchester, United Kingdom

PLENUM MEDICAL BOOK COMPANY
New York and London

Library of Congress Cataloging in Publication Data

Trophoblast invasion and endometrial receptivity: novel aspects of the cell biology
of embryo implantation / edited by Hans-Werner Denker and John D. Aplin.
    p.   cm. – (Trophoblast research; v. 4)
    "Derived from a workshop . . . held during the Twenty–Fourth Annual Meeting of
the Cell, Tissue, and Organ Culture Study Group (C.T.O.C.) in 1986, at Heidelberg,
Federal Republic of Germany" – T.p. verso.
    Includes bibliographical references.
    ISBN 0-306-43520-9
    1. Human embryo – Transplantation – Congresses. 2. Trophoblast – Physiology –
Congresses. 3. Endometrium – Physiology – Congresses. 4. Surgery, Experimental –
Congresses. I. Denker, Hans-Werner, 1941–   . II. Aplin, John Dalzell, date. III.
Cell Tissue and Organ Culture Study Group. Meeting (24th: 1986: Heidelberg, Ger-
many) IV. Series.
    [DNLM: 1.   Embryo   Transfer – congresses.   2.   Endometrium – physiology –
congresses. 3. Fertilization in Vitro – congresses. 4. Models, Molecular – congresses. 5. Trophoblast – physiology – congresses. W1 TR877 v. 4 / WQ 205 T856 1986]
RG135.T76   1990
618.1′78059-dc20
DNLM/DLC   90-7197
for Library of Congress   CIP

Derived from a workshop on The Cell Biology of Trophoblast Invasion
In Vivo and In Vitro, held during the Twenty-Fourth Annual Meeting
of the Cell, Tissue, and Organ Culture Study Group (C.T.O.C.) in 1986,
at Heidelberg, Federal Republic of Germany

Plenum Medical Book Company is an imprint of
Plenum Publishing Corporation
233 Spring Street, New York, N.Y. 10013

Printed in the United States of America

TROPHOBLAST RESEARCH

Trophoblast Research publishes contributions concerning the placenta and the extraembryonic membranes as they relate to embryonic and fetal development and to trophoblastic neoplasia. Original articles, reviews, and reports are published in single bound volumes. All articles are peer-reviewed.

EDITORS

The Editorial Office for *Trophoblast Research*:
Department of Obstetrics and Gynecology
University of Rochester School of Medicine and Dentistry
601 Elmwood Avenue, Rochester, New York USA 14642
(716) 275-3638

PREFACE

Interest in mechanisms of embryo implantation is increasing, particularly with the realization that failure of implantation after in vitro fertilization and embryo transfer places significant limits on the success of treatment. In addition, there is a need to provide hypotheses, and ultimately mechanisms, for the high rates of embryonic loss in women in the population at large.

Traditionally, implantation research has concentrated on genetics and endocrinology without providing many therapeutic benefits. A new era is now beginning with the application of modern cellular and molecular approaches to the investigation of the relationship between trophoblast and endometrium. At the same time, older data can be reevaluated in the light of current research into cell-cell and cell-matrix interactions.

The feeling that new avenues of research are open was apparent when an international group of scientists came together at a workshop on "The Cell Biology of Trophoblast Invasion In Vivo and In Vitro" held during the XXIV Annual Meeting of the Cell, Tissue and Organ Culture Study Group (C.T.O.C.) at Heidelberg in 1986. What was unusual about this Conference was the interdisciplinary dialogue between implantation researchers and tumor biologists, highlighting aspects common to invasion of trophoblast and tumor cells. The nature of invasiveness is still poorly understood. Nevertheless, examination of the interactions of invasive cells with cell adhesion molecules in extracellular matrix and at other cell surfaces, and of the regulation of buildup and degradation of cell surface and matrix components suggest that there are specific characteristics associated with the invasive phenotype.

As far as the host tissue is concerned, it was at this meeting that a new concept of the mechanism of hormonally regulated endometrial receptivity, an unsolved cell biological paradox, was discussed for the first time before an international forum of reproductive biologists: the concept that a partial loss of elements of apico-basal polarity may render the uterine epithelium receptive to attachment of trophoblast.

It was decided to publish these ideas in a volume of expanded and updated papers based on selected presentations from the Heidelberg meeting, as well as some related invited papers. We are grateful to Drs. R.K. Miller and H.A. Thiede, as well as to the Editorial Board of Trophoblast Research for including this volume in the series, to the reviewers for their comments, to Mss. G. Mathieu, J. White, B. Witte, and J. Crombie for taking up the immense load of secretarial and computer work, and to Plenum Press for their cooperation in producing the book.

<table>
<tr><td>Hans-Werner Denker</td><td>Essen, Federal Republic of Germany</td></tr>
<tr><td>John Aplin</td><td>Manchester, England</td></tr>
</table>

CONTENTS

INTRODUCTION

MORPHOLOGY

EXPERIMENTAL MODELS

CELL BIOLOGY AND IMMUNOLOGY OF THE INVASIVE TROPHOBLAST

THE HOST TISSUE
Uterine Epithelium: Cell Biological Changes In Relation To Endometrial "Receptivity"

Contents

xi

Basement Membranes And Endometrial Stroma

INTRODUCTION

TROPHOBLAST - ENDOMETRIAL INTERACTIONS AT EMBRYO IMPLANTATION: A CELL BIOLOGICAL PARADOX

H.-W. Denker

Institut für Anatomie
Universitätsklinikum, Hufelandstr. 55
D-4300 Essen 1, Federal Republic of Germany

INTRODUCTION

Invasion of the trophoblast into the endometrium, which forms an essential element of embryo implantation in most mammalian species including the human, has long impressed investigators. It resembles invasion of malignant tumors in many respects, including host tissue destruction, blood vessel erosion, a certain degree of repair processes and neovascularization (cf. Denker, 1977, 1980, 1983). It presents an immunologic paradox appearing to violate transplantation laws: antigenically different cells being tolerated within a basically immuno-competent milieu (cf. Bulmer et al., 1990; Loke et al., 1990).

In addition it has recently been pointed out that this process also appears to disobey certain basic principles of cell biology: When implantation is being initiated, the trophoblast of the blastocyst attaches with its apical plasma membrane to the apical plasma membrane of the uterine epithelium. It is well established that a fundamental property of epithelia is to possess two distinct membrane domains: the apical plasma membrane which is non-adhesive (so that e. g., obliteration of the intestinal canal or of body cavities is avoided) and the basolateral domain which is adhesive due to the expression of adhesion molecules. The fact that at implantation initiation, the trophoblast and the uterine epithelium do establish their first contact via their respective apical cell membranes is, therefore, most astonishing and can be termed a cell biological paradox (Denker, 1986, 1988). Understanding the mechanisms underlying this apparent paradox should allow one to better understand the nature of trophoblast invasiveness and endometrial "receptivity", the two fundamental properties of the "graft" (the trophoblast) and the "host" (the endometrium). Recently, new insights have been gained into the cell biological basis of these two phenomena as will be discussed in detail below.

MORPHOLOGY OF TROPHOBLAST-ENDOMETRIUM INTERACTION IN THE INITIAL PHASE OF IMPLANTATION

Based on electron microscopic findings, penetration of the trophoblast through the uterine epithelium has been described to follow any of three different modes in different species having invasive implantation (Schlafke and Enders, 1975; see Figure 1).

In the *"displacement type"* (found e.g., in the rat and mouse), attachment of the trophoblast to the apical plasma membrane of the uterine epithelium is followed by the degeneration and sloughing of whole groups of uterine epithelial cells. These cells dissociate from neighboring epithelial cells and detach from their basal lamina so that the latter becomes exposed and accessible to the advancing trophoblast (Figure 1A). The subsequent penetration of the basal lamina is apparently not initiated by the trophoblast but by processes of decidual cells which erode the remains of the basal lamina from underneath (Schlafke et al., 1985).

In the *"fusion type"* of penetration (rabbit; basically the same, although limited to few uterine epithelial cells, applies to the ruminants, Wooding, 1984), the apical plasma membranes of certain parts of the trophoblast fuse with those of the uterine epithelium and form, strangely enough, a mixed syncytium containing nuclei of both embryonic and maternal origin (Figure 1B). This mixed syncytium then reveals in the rabbit its truly invasive behavior by penetrating through the basal lamina (or what is left of it, cf. Marx et al., 1990) and by eroding underlying blood vessels to form the typical hemochorial type of contact. In addition to the species just mentioned, fusion between trophoblast and uterine epithelium was also reported to occur in the human (stages studied: day 11 and 22; Larsen, 1970, 1974; Larsen and Knoth, 1971; Knoth and Larsen, 1972). However, these observations would obviously benefit from confirmation by studies of additional material since fusion was only seen in a few places, in the day 22 specimen, and only when the epithelium was degenerating. However, the same authors reported also on fusion between trophoblast and liver cells when (human) choriocarcinoma was transplanted into hamster liver. This important question concerning human implantation will be discussed later in this chapter.

In the *"intrusion type"* (ferret; perhaps also other carnivores, see Leiser, 1979, 1982), small tongues of cytoplasm of the syncytiotrophoblast are seen to penetrate between uterine epithelial cells. The latter may or may not fuse with each other simultaneously thus forming small syncytia, to a different degree depending on the species. The processes of syncytiotrophoblast obviously open intercellular junctions between uterine epithelial cells (Figure 1C), and immediately thereafter junctions are apparently re-formed between the trophoblast and the adjacent uterine epithelial cells. Thus, a complicated process of breaking and reformation of junctions must be dealt with. On the basis of published observations it cannot be excluded that a combination of these mechanisms occurs in at least some species (e.g., in the rhesus monkey and perhaps in the human, cf. Denker, 1983). In the former species morphology suggests that the trophoblast does intrude between uterine epithelial cells (Enders et al., 1983; Enders and Schlafke, 1986). However, the reported morphological details do not always allow for clearcut identification of adjacent cells as trophoblast or uterine epithelium, during the phase of epithelial penetration, so that a possible contribution of fusion events cannot be excluded. In the human, even light microscopy (Carnegie Collection specimens) suggests that fusion occurs at least in the uterine epithelium adjacent to the implantation site (cf. Denker, 1983). Larsen (1970, 1974) has reported on fusion between trophoblast and uterine epithelium; obviously supporting data are desirable as mentioned above. The question whether fusion or intrusion is typically found is relevant for the mechanisms involved, since fusion of trophoblast with uterine epithelial cells would eliminate the need to split intercellular junctions between uterine epithelial cells and then re-form new junctions with trophoblast. In any case, after the uterine epithelial cells have been overcome, the tongues of trophoblast cytoplasm are seen to pause for a while at (to attach to?) the basal lamina and then to penetrate it in order to proceed with their invasion (Schlafke and Enders, 1975).

Common to all three described modes of penetrating the uterine epithelium is that the process always starts with attachment of the apical plasma membrane of trophoblast to the apical plasma membrane of the uterine epithelium. This is a general phenomenon found not only in the described invasive types of implantation but also in the non-invasive epithelio–chorial type (pig, not included in Figure 1; Dantzer, 1985). As mentioned in the beginning, this is a most astonishing phenomenon since apical plasma membranes of epithelia are normally known to be non-adhesive so that this process needs to be explained on a molecular level.

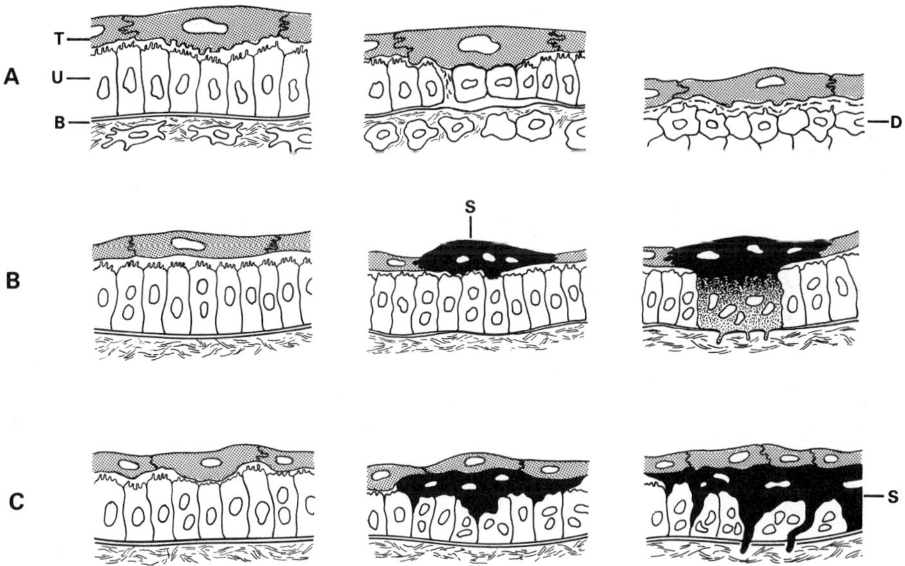

Figure 1. Sketch of the major morphological characteristics of trophoblast interactions with the uterine epithelium during the initial phase of implantation. Blastocyst coverings (zona pellucida and its equivalents) and the process of their shedding are omitted. A) Displacement penetration (rat, mouse); B) fusion penetration (rabbit); C) intrusion penetration (carnivores). T: trophoblast; U: uterine epithelium; B: basement membrane of the latter; S: syncytiotrophoblast; D: decidual cells. (Modified after Schlafke and Enders, 1975; Leiser, 1981).

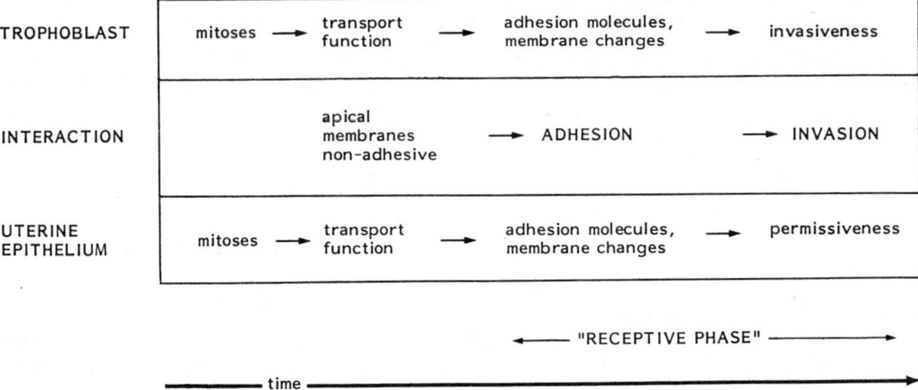

Figure 2. General terms in use to describe processes relevant for the transition from the pre-implantation phase to implantation initiation. A direct cellular interaction (adhesion, invasion) between trophoblast and uterine epithelium is possible only if the sequence of physiological states occurs in synchrony, in both partners, i.e., if the adhesive/invasive state of the trophoblast meets the endometrium in the "receptive phase" (for further discussion, see text).

CELL BIOLOGY OF THE INTERACTIONS BETWEEN TROPHOBLAST AND ENDOMETRIUM

The physiological processes that can be expected to go on in both partners, the trophoblast and the uterine epithelium, during implantation initiation may be listed as in Figure 2. During the preimplantation phase, both tissues start off with proliferation and differentiation to form typical epithelia, specialized for transport functions (uterine epithelium: transport of certain blood plasma constituents and of specific secretory products to the uterine cavity; trophoblast: transport of fluid and nutrients to the blastocyst cavity; Beier, 1974; Borland, 1977; Borland et al., 1977). During this phase, each apical plasma membrane is non-adhesive, and the blastocyst remains (more or less, depending on the mechanism of "grasp", Böving, 1963) mobile.

Implantation starts when these two adjacent membranes become adhesive to each other. During the subsequent phases starting with epithelial penetration, the trophoblast demonstrates its invasiveness, and the endometrium its permissiveness. As far as the latter is concerned, the period during which firm adhesion and invasion of the trophoblast is possible is usually termed the "receptive phase" of the endometrium. The concept that such a specific phase (the "implantation window", Psychoyos, 1988) exists in the physiology of the endometrium was developed, by inference, on the basis of two types of data: 1.) Results of embryo transfer experiments have often been cited as proof for the existence of such a phase, although this interpretation is equivocal since implantation *sensu strictu* has usually not been studied in those investigations (e.g., Chang, 1950) but rather the importance of synchrony between daily changes in the maternal milieu (e.g.,uterine secretions) and embryonic development during preimplantation stages. 2.) Endometrial "receptivity" for trophoblast attachment is more convincingly demonstrated by investigations on the hormonal control of periimplantation development and implantation (Psychoyos, 1973; Psychoyos and Casimiri, 1980), studies done basically in the rat and rabbit. The concept has recently been proposed to apply also to the human (Psychoyos, 1988).

The uterine epithelium appears to play the central role in "receptivity" or "non-receptivity": If the uterine epithelium is removed, blastocysts can implant completely independent of any hormonal control (Cowell, 1969). Outside the uterus, at ectopic sites, the trophoblast if transplanted interstitially is highly invasive no matter what the hormonal status of the host (even in males, Kirby, 1965, 1967, 1970; Porter, 1967). Even the trophoblast of pig blastocysts which never becomes invasive in utero was reported to show the ability to attach to fibroblasts in vitro (Kuzan and Wright, 1981) and invasive behavior when transplanted ectopically (Samuel and Perry, 1972). To express a phase of "receptivity" is a unique property of the uterine epithelium not seen in the tubal epithelium which the trophoblast cannot penetrate, at least not in animals (Tutton and Carr, 1984).

An obvious assumption is, therefore, that the apical plasma membranes of the trophoblast and/or the uterine epithelium behave like typical apical plasma membranes of other epithelia, during the preimplantation phase, i.e., they either do not express adhesion molecules or they express, in addition, molecules that impede adhesion. At implantation initiation, however, such adhesion molecules may be expressed at these apical plasma membranes (trophoblast and/or uterine epithelium). Consequently, a number of contributions in this volume ask the question whether such molecules can be identified on either surface during this phase, and how they can be classified (cell-cell adhesion molecules, matrix receptors, lectin-like molecules, glycosyltransferases, etc.) (Anderson et al., 1990; Aplin and Charlton, 1990; Bükers et al., 1990; Carson et al., 1990; Chávez, 1990; Fisher et al., 1990; Foidart et al., 1990; Hoffman et al., 1990; Morris and Potter, 1990). Another question is how this adhesive or non-adhesive phenotype is regulated, which will be discussed in the following sections.

Cell Surface Properties Of The Invasive Trophoblast

The trophoblast has been found to show changes in various cell surface parameters (such as cell surface charge, lectin binding properties, and glycoprotein expression), at late preimplantation and implantation stages (Chávez, 1986; Chávez and Enders, 1982; Enders and Schlafke, 1974; Guillomot et al., 1982; Holmes and Dickson, 1973; Jenkinson and Searle, 1977; Johnson and Calarco, 1980; Noeslund, 1979; Nilsson et al., 1975; Sherman et al., 1979). All these changes may be somehow involved in the expression of the invasive phenotype, but their exact role remains to be defined. It appears improbable that the cell surface charge is the major determinant; at least the observed degree of cell type specificity (trophoblast vs. various invasive tumor cells, Hohn et al., 1985) suggests that recognition phenomena are involved that would be difficult to explain solely on the basis of cell surface charge values. Lectin-glycoprotein or glycosyltransferase-glycoprotein interactions could provide for such a specificity, but experimental evidence for their involvement is still very limited (Chávez, 1990). An interesting possibility is a specific role of alpha galactosyl end groups (Foidart et al., 1990).

The existence and role of endogenous lectins (possibly serving as ligands for the respective carbohydrate groups) in this system remains to be proven.

Interesting data are becoming available about the expression of receptors for matrix molecules, on the invasive trophoblast (Aplin and Charlton, 1990; Carson et. al., 1990; Fisher et al., 1990; Foidart et al., 1990; Sutherland et al., 1988). It remains to be seen whether any of these molecules may be identical with any of the above-mentioned glycoproteins or lectin-like molecules. In a functional sense, one would expect such matrix receptors to play a role primarily during the phases of invasion after penetration through the uterine epithelium, i.e., when the trophoblast is confronted with various types of interstitial extracellular matrix. Such receptors are indeed expressed on the trophoblast at developmental stages equivalent to the phases of stromal invasion (Aplin and Charlton, 1990; Fisher et al., 1990; Foidart et al., 1990).

For tumor cell interaction with host tissues during invasion, a cycle of events including adhesion to matrix, localized matrix degradation (e.g., by proteinases, in case of penetration through the basal lamina type IV collagenase), migration, and again adhesion to non-degraded parts of matrix has been proposed to play a central role (Liotta et al., 1984, 1986). In order to demonstrate the involvement of these processes in implantation experiments are necessary not only on the interaction of trophoblast with extracellular matrix models but also with endometrium (in vitro or in vivo). In particular it should be interesting to investigate, in the latter system, the effects of low molecular weight peptides that are able to compete with receptor binding (like the RGD or YIGSR-containing peptides, Gehlsen et al., 1988; Fisher et al., 1990).

A finding that requires a functional explanation is that trophoblast and chorioncarcinoma cells actively produce matrix molecules, e.g., laminin and fibronectin (Peters et al., 1985; Queenan et al., 1987; Ulloa-Aguirre et al., 1987). Histochemistry suggests that, in vivo, the invasive trophoblast contains a particularly rich store of intracellular matrix molecules, e.g., fibronectin (Wartiovaara et al., 1979) and laminin (Strunck, in preparation); these and other matrix molecules appear to be also associated with its cell surface (cf. Carson et al., 1990). Similar findings have also been reported for invasive tumor cells (Castronovo et al., 1985, 1987). It should be interesting to know whether these molecules have been adsorbed and internalized from surrounding matrix, or whether they have been synthesized by the trophoblast, and what purpose this accumulation may serve.

Additional data are urgently needed concerning the cytochemical distribution of matrix receptor molecules (and perhaps of matrix molecules themselves, see above) on the plasma membrane of trophoblast cells as well as of the matrix molecules in the pericellular environment. Are these molecules indeed (maximally?) expressed at the trophoblast apical plasma membrane at implantation initiation, and the leading edge of isolated invading trophoblast cells during stromal invasion? Determination of the presence of such molecules by experiments on the interaction of the trophoblast with two-dimensional models of extracellular matrix (e.g., the blastocyst outgrowth model or experiments with isolated trophoblast cells plated on extracellular matrix-coated dishes) can be misleading in this respect: It remains uncertain which plasma membrane domain (apical or basolateral) is confronted with the extracellular matrix in such a system, morphology suggesting that during outgrowth of monolayers of trophoblast on matrix the apical plasma membrane faces the medium, not the matrix (Enders et al., 1981).

For the initial phase of implantation, i.e., for adhesion to the uterine epithelium, one would expect cell-cell adhesion molecules (CAMs) to be of particular importance. Knowledge of their distribution on the trophoblast and on the uterine epithelium is, however, very limited so far. In particular, data on the expression at the apical vs. the basolateral membrane domain are largely lacking. Cell CAM 120/80 (E-cadherin, uvomorulin), typically expressed on lateral membranes of polarized epithelia, was shown to play a role in cell-cell interactions during differentiation of the two primary cell lines in ontogenesis, the embryoblast vs. the primitive trophoblast (Johnson et al., 1986). There is no evidence so far that it may be involved in trophoblast attachment to tissue culture dishes (trophoblast outgrowth model) (Richa, 1986). This, however, would not be expected to be the case with this homotypic adhesion molecule. What is clearly needed are experiments on a possible role of uvomorulin and other CAMs in the interaction of the trophoblast with the uterine epithelium. It should be possible to perform such experiments with the model systems described in this volume (Morris and Potter, 1990; Hohn and Denker, 1990). Cell CAM 105 was found to be downregulated in the mural trophoblast of rat blastocysts (Svalander et al., 1978). This specific CAM may thus not be involved in trophoblast-uterine epithelium adhesion but in lateral contacts within the population of trophoblast cells, and down regulation may facilitate dissociation of trophoblast cells and may promote invasion of isolated cells.

A particularly interesting class of molecules may be heparan sulfate proteoglycan (HSPG) and its receptor, according to data discussed in detail by Carson et al. (1990) and by Morris and Potter (1990). It appears quite possible that this pair of molecules plays a central role in mediating the first attachment between the apical plasma membranes of the trophoblast and the uterine epithelium, prior to formation of intercellular junctions between both epithelia and subsequent penetration of the trophoblast through the uterine epithelium and into the endometrial stroma. As far as a functional classification is concerned, one would tend to place HSPG receptors closer to the matrix receptor molecules discussed above than to CAMs. Respect to their expression on the apical plasma membrane of the uterine epithelium, together with other parameters, in relation to the "receptive state" of the endometrium will be discussed below.

The Host Tissue: Cell Biological Aspects Of Endometrial "Receptivity"

As discussed above, the apical plasma membrane of the uterine epithelium is non-adhesive for the trophoblast in the non-pregnant state and during the preimplantation phase. In order to fulfill this "repellent" function all the uterine epithelium needs to do is to behave like any typical epithelium: to possess a non-adhesive apical plasma membrane domain in addition to the adhesive basolateral domain. In fact, invasive cells (e.g., tumor cells) cannot adhere to and invade through intact epithelia (except for mesothelia, the amniotic epithelium and endothelia) from the apical pole (although they can do so from the basal pole) (de Ridder et al., 1975). The same seems to hold true for tumor cells introduced

Table 1

Endometrial "Receptivity" For Blastocyst Implantation:
Partial Loss/Destabilization Of Apico-Basal Polarity Of The Uterine Luminal Epithelium

PROPERTIES OF PLASMA MEMBRANES

Apical: Loss of Marker Enzymes	Classen-Linke et al., 1987
Changes In Lectin Binding Properties	Chavez and Anderson, 1985; Nalbach, 1985; Anderson et al., 1986; Bükers, et al., 1989
Reduction Of Thickness Of Glycocalix/Cell Surface Charge	Enders and Schlafke, 1977; Anderson et al., 1986; Morris and Potter, 1984; 1989
Increased Density Of Intramembranous Protein Particles (~ basolateral membrane)	Murphy et al., 1982a; Winterhager, 1985; Winterhager et al., in press
Acquisition Of New Proteins/Glycoproteins	Lampelo et al., 1985; Anderson et al., 1988; Hoffman et al., 1989
Acquisition Of Receptors For Matrix/Cell Surface Molecules (e.g., HSPG)	Carson et al., 1989
Acquisition Of The Ability To Form Hemidesmosome-Like Junctions	Denker, 1977
Acquisition Of The Ability To Form "Reflexive" Gap Junctions	Murphy et al., 1982d
(Redistribution Of Proteins That Were Restricted To The Baso-lateral Membrane In the Pre-Receptive Phase?)	

Lateral:	Proliferation Of Tight Junction Strands Towards Basal Cell Pole	Murphy et al., 1982b; Murphy et al., 1982c; Winterhager and Kühnel, 1982
	Loss Of Subapical Maximum Of Desmoplakin Concentration, More Even Distribution On The Lateral Membrane	Classen-Linke and Denker, 1989
Basal:	Reduced Adhesion To Basal Lamina (Rat)	Tachi et al., 1970
	Defective Basal Lamina (Rabbit)	Marx et al., 1989

INTRACELLULAR/TRANSCELLULAR TRANSPORT

Changed Activity And Direction Of Endocytosis And Transepithelial Transport	Parr, 1980, 1982, 1983; Parr and Parr, 1977, 1978; Marengo et al., 1986
Changed Sorting of Membrane Precursors (Inferred From Membrane Changes, See Above)	

ORGANIZATION OF CYTOSKELETON

Polar Distribution Of Vimentin	Hochfeld et al., 1989

experimentally into the lumen of the non-receptive uterus, although results differ somewhat in different systems. In contrast, in the "receptive phase" of the endometrium, controlled by sex steroids, adhesion and invasion of tumor cells become possible (at least in rodents, not in the rabbit: Wilson, 1963; Wilson and Potts, 1970; Short and Yoshinaga, 1967; Hohn et al., 1985) as do trophoblast invasion and embryo implantation (in all species with the apparent exception of the pig).

Consequently, one has to assume that the uterine epithelium either loses adhesion inhibiting molecules or expresses adhesion molecules (in the broadest sense) at the apical plasma membrane, during this phase, in a fashion analogous to the invasive trophoblast (see above). Indeed, a reduction in the thickness of the glycocalyx and in cell surface charge has been observed, perhaps suggesting the reduction of an inhibitory influence (Anderson et al., 1986, 1990; Enders and Schlafke, 1977; Morris and Potter, 1984, 1990). The types of molecules involved need to be defined; adhesion mediating molecules may be of the cell-cell or cell-matrix adhesion molecule type, or they may be lectin-like or glycosyltransferases. These questions and relevant experimental data are discussed in detail in various contributions to this volume (Anderson et al., 1990; Bükers et al., 1990; Carson et al., 1990; Chávez, 1990; Hoffman et al., 1990; Seif and Aplin, 1990).

However, if one considers not only properties of the apical plasma membrane and not only adhesion molecules but also other parameters of the uterine epithelium it turns out that far more general changes in cellular organisation are observed during acquisition of "receptivity". In addition to the apical membrane, changes are seen in the lateral and basal membrane, in the association with the basal lamina, and even in the organization of the cytoskeleton (Table 1):

At the *apical plasma membrane*, one observes (for references, see Table 1) the expression of new proteins/glycoproteins some of which may be adhesion–mediating molecules (of the types discussed above), e.g., HSPG receptors. Changes in lectin binding properties are conspicuous but are not in every respect in agreement with predictions made previously suggesting, e.g., a positive correlation of the expression of galactosyl end groups with receptivity (Anderson, Chávez, see above): In the implantation chamber of the rabbit, such lectin binding sites were found to decrease rather than to increase in density at least at the placental folds (Bükers et al., 1990). Marker enzymes of the apical plasma membrane are lost in this phase so that this membrane looses part of its apical characteristics. In contrast, the apical plasma membrane gains the ability to form hemidesmosome-like junctions and gap junctions, a property otherwise found only at the basolateral membrane domain. The density of intra-membranous protein particles seen in freeze fracture is found to increase so that during the receptive phase the apical plasma membrane resembles in this respect the basolateral membrane (for references, see Table 1). One tends to assume, therefore, that these observations reflect the acquisition of adhesion molecules at the apical plasma membrane, molecules that are either specifically expressed here or molecules that are otherwise found only in the basolateral membrane domain but become redistributed.

In the *lateral plasma membrane* region, tight junctional strands were found to proliferate during this phase towards the basal cell pole. The position of these strands along the apico-basal axis is a marker of the functional polarity of epithelial cells, as shown in other systems (Chevalier et al., 1985; Kitajima et al., 1985). Desmoplakin is found (in rabbit uterine epithelium) to loose, during this phase, its typical subapical maximum and to become more evenly distributed on the lateral membrane (for references, see Table 1).

At the *basal cell pole* adhesion of the epithelial cells to their basal lamina becomes reduced (rat) or the basal lamina becomes defective (rabbit) (for references, see Table 1).

Intracellular/transcellular transport activities in the uterine epithelium change considerably during the implantation phase. This was traditionally considered to be an expression of stage-specific changes in endometrial secretory activity aimed at providing a daily changing optimal milieu for blastocyst development, but it may well be that it is at least in part also an expression of changes in sorting and intracellular transport of membrane precursors and breakdown products as a mechanism for the changes in membrane composition discussed here (for references, see Table 1).

The *cytoskeleton* of the uterine epithelium is also found to show surprising changes during the implantation period. This is one of the few epithelia in which, in addition to cytokeratins, vimentin is also expressed. Vimentin was found mainly in the basal cytoplasm of uterine epithelial cells, outside the "receptive phase", but in the implantation chamber of the rabbit it was seen to change it distribution finally to become maximally concentrated in the apical cytoplasm (for references, see Table 1).

All the parameters mentioned in Table 1 are characteristics of the apico-basal polarity of epithelia. With all these parameters, a general trend is observed when the "receptive phase" is approached, i.e., a reduction of the uneven distribution along the apico-basal axis. It is hypothesized, therefore, that *uterine epithelial cells lose, during acquisition of "receptivity", certain elements of apico-basal polarity* that are strongly expressed during the pre-receptive phases (Denker, 1986, 1988). This may mean a partial loss of some of the most typical epithelial characteristics, in preparation for trophoblast invasion, and even the acquisition of elements of an invasive phenotype (basal processes penetrating into the stroma in the rabbit, see Marx et al., 1990; penetration of the "epithelial plaque" into the stroma in the Rhesus monkey, Wislocki and Streeter, 1938; Rossman, 1940; Enders et al., 1983; Enders and Schlafke, 1986). The notion that parts of the uterine epithelium not only in the immediate vicinity of the invading trophoblast exhibit (in certain species) a "disorganized" morphology (Hill, 1898; Leiser, 1979) is also consistent with this view.

In Figure 3, a schematic illustration of this concept is given with respect to the distribution of membrane-bound receptor type molecules. Figure 3A shows a uterine epithelial cell as one expects it to be during the pre-receptive phase, with a distribution of receptor molecules as typical for normal epithelia: cell-cell adhesion molecules (like the adherens junctions molecule, uvomorulin) localized at the lateral membranes, and cell-matrix receptors at the basal membrane (Y, e.g., laminin receptors playing a role here in binding to the basal lamina). The apical plasma membrane, however, lacks these types of molecules but instead

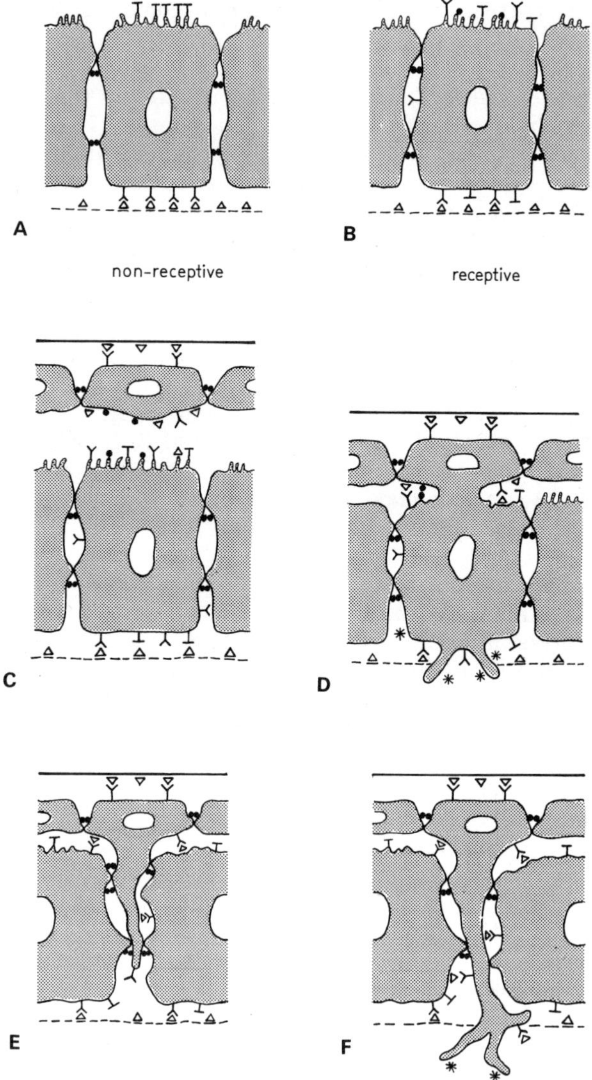

A
non-receptive
B
receptive
C
D
E
F

expresses other types of molecules which may have adhesion-inhibiting properties (T). In contrast, a cell of the receptive phase (Figure 3B) is postulated to show a redistribution of some or all of these molecules, so that the apical plasma membrane does now express cell-cell and/or cell-matrix adhesion molecules, whereas adhesion to the basal lamina may be reduced.

It must be pointed out that the details of this model are still quite hypothetical. What is evident so far is only that the apical plasma membrane loses some of its marker molecules while gaining certain new (mostly poorly defined) proteins and glycoconjugates and the ability to allow adhesion, that on the lateral membranes molecules become more evenly distributed along the apico-basal axis and that binding of the basal plasma membrane to the basal lamina changes (see Table 1 for references). However, precise data on the spatio-temporal behavior of adhesion molecules (e.g., cell-cell adhesion molecules like uvomorulin and matrix receptor molecules) at these sites are largely lacking. The only such molecule for which relevant data are available so far is the HSPG receptor. This molecule does indeed appear to relocate to the apical plasma membrane in the receptive phase (Carson et al., 1990) although an increase of binding sites could also be due to release from bound HSPG (Morris and Potter, 1990). It would be very exciting to have histochemical data supporting these findings.

Figure 3C/D and E/F give a hypothetical molecular model of events during trophoblast attachment to the uterine epithelium and invasion through it, for the fusion type and the intrusion type of penetration, respectively. This may, it is hoped, serve to stimulate thinking about experimental protocols to enable us to find out about the exact mechanisms involved.

Figure 3. A morphologist's view of changes in cell-cell and cell-matrix adhesion as related to the acquisition of endometrial "receptivity" (A, B) and to the interaction of trophoblast and uterine epithelium during the initial phases of implantation (C-F; C, D: fusion penetration as seen in the rabbit; E, F: intrusion penetration as seen in carnivores). This is a highly speculative scheme which intends to be thought-provoking but does not intend to be correct in detail. The basic idea is that cell-cell and cell-matrix receptor molecules change their polar distribution (relative abundance) at the basolateral versus the apical plasma membrane domain of uterine epithelial cells, during acquisition of endometrial "receptivity". Appearance of receptors at the (formerly non-adhesive) apical plasma membrane of the uterine epithelium (B) allows the trophoblast to attach here, followed by either fusion with (D) or by penetration between epithelial cells (E,F). Also indicated is the changing relationship between the uterine epithelium and its basal lamina. The type of adhesion molecules is quite hyphothetical and needs to be defined experimentally (see Denker 1986, 1988; Carson et al., 1990). T: apical type intramembraneous proteins (ectodomain non-adhesive); filled circles: cell-cell adhesion molecules (CAMs); Y: receptors for e. g., HSPG (or laminin, others); triangles: HSPG (laminin, other ligands); stars: integrin type receptors (binding sites for fibronectin, vitronectin, collagen, and others).

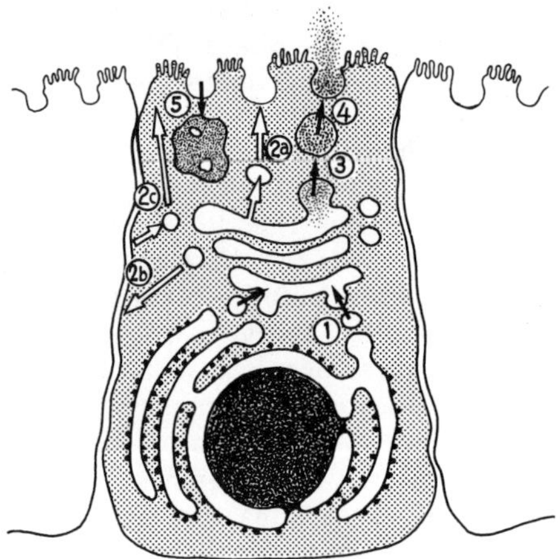

Figure 4. In order to acquire the state of "receptivity", uterine epithelial cells are postulated to undergo changes in intracellular sorting of membrane precursor material for the apical vs. the basolateral membrane domain. In the pre-receptive phase, sorting is expected to be as effective as in any typical epithelial cell, following separate routes to the two regions (1 - 2a or 1 - 2b), extrusion of secretion contributing to insertion of new membrane material (3, 4). Sorting may also be exerted through secondary relocation (2c). Removal of membrane material by endocytosis (5) or by membrane shedding (not shown) contributes to controlling the final composition of the membranes. It is proposed that appropriate changes in these pathways cause the appearance of adhesion molecules at the apical plasma membrane of the uterine epithelium at acquisition of "receptivity" for trophoblast attachment.

        Figure 4 illustrates how a typical polarized epithelial cell maintains its two distinct membrane domains (apical and basolateral) through sorting of membrane precursor material. Sorting may involve a priori differing routes of membrane precursor material to the apical or the basolateral region, or all this material may be inserted primarily unsorted and sorting may be achieved through secondary processes like selective removal of certain molecules during membrane internalization (endocytosis). Exocytosis as part of secretory activity also contributes to changes of membrane composition. For example, apical protrusions of uterine epithelial cells as seen in many species in the receptive phase may serve this purpose (pinching off leading to shedding of apical-type molecules from the apical plasma membrane domain) rather than (or in addition to) transport functions providing an appropiate intraluminal "milieu" for blastocyst development as previously suggested (Beier and Kühnel, 1973; Enders and Nelson, 1973; for discussion of occurrence cf. Parr and Parr, 1982).

        These processes of sorting must be disturbed during endometrial "receptivity". An interesting question is whether this change is of a global nature or specific for certain types of molecules. This will be an important point for future investigations. The data given in Table 1 suggest that the changes are in fact relatively far reaching. This leads to the question of whether the uterine epithelial cell may, during acquistion of "receptivity", partially give up its epithelial nature, and whether it may go back part of its way in development and regress to a more mesenchymal (or even semi-invasive) phenotype. It is known that such a phenotypic (morphologic and biochemical) change is a normal process during various phases of embryonic development (Hay, 1985). This is coupled to changes e. g., in the expression of cytoskeletal proteins (Greenburg and Hay, 1988). Vimentin expression is most common in motile cells, although there is no strict correlation between vimentin expression and motility. As mentioned above, the uterine luminal epithelium is one of the few epithelia that express vimentin in addition to cytokeratins; it is among these epithelia that the property of allowing invasive (tumor) cells to attach and invade from the apical cell pole appears to be more common (e. g., mesothelia). Thus, one may speculate that the genetic program of the uterine epithelium may have been geared only partially into the epithelial direction, during embryonic development, and that it may maintain mesenchymal potentials (including properties of the cytoskeleton as well as membrane characteristics) that can be increased by appropriate stimuli, hormonal, or others. Thus, the induction of the "receptive state" of the uterine epithelium by steroids may involve regression of the phenotype to a more primitive state, including partial loss of cell polarity.

In addition to systemic conditioning by steroid hormones, there must be local signaling from the blastocyst. This could involve e.g., matrix molecules. A switch from the epithelial (polar) to the mesenchymal (apolar) phenotype can be induced in certain cultured epithelial cells by extracellular matrix if the cells are in contact with these molecules on their whole surface (Hay, 1985; Greenburg and Hay, 1988). Some epithelia seem to be able to respond to matrix added to their apical surface by reversing polarity, even without dissociating (thyroid: Garbi et al., 1986). The invasive trophoblast appears to express certain matrix molecules on its surface, thus confronting the uterine epithelium with such molecules at (non-typically) the apical plasma membrane. These molecules (and perhap others "secreted" into the implantation chamber) may thus additionally stimulate loss of polarization of uterine epithelial cells which, however, was already pre-programmed through labilization of the apico-basal polarity due to conditioning by steroid hormones. It has been proposed, therefore, that the preparation of the uterine epithelium for embryo implantation ("receptivity") is achieved through a de-stabilization of the apico-basal polarity due to steroid hormone action, and that the loss of polarity parameters is stimulated further by locally acting blastocyst-derived signals (Denker, 1986).

CONCLUSION AND PERSPECTIVES

Scant evidence is yet available as to the nature of *trophoblast invasiveness*. For this peculiar state of the trophoblast, expression of certain molecules, e.g., matrix molecules or matrix receptors normally not found at the surface of epithelial cells may play an important role together with certain properties of the motility apparatus of the cell: high motility and lack of contact inhibition. However, the motility characteristics of trophoblast (at least choriocarcinoma) cells are complex, and it will be worthwile to expand on differential investigation of the various cell types emerging in culture and to relate this to the behavior and role of trophoblast subpopulations during embryo implantation/placentation in vivo (Aplin and Charlton, 1990). It should be fascinating to see whether the formation of these subtypes and their behavior are influenced by contact with endometrium, e.g., its cell surfaces and its matrix, experiments which could be done using the model systems described in this Volume. Data on the nature of the cell surface-bound molecules involved are becoming available and it can be expected that our understanding of trophoblast invasion will be much improved during the next years. In addition to adhesion molecules, expression/production of certain proteinases and perhaps other degradative enzymes is certainly involved in facilitating penetration of the trophoblast into the stroma. Much will have to be done, however, in order to elucidate how invasiveness (which, in the normal situation, lasts for only a limited time period) is regulated.

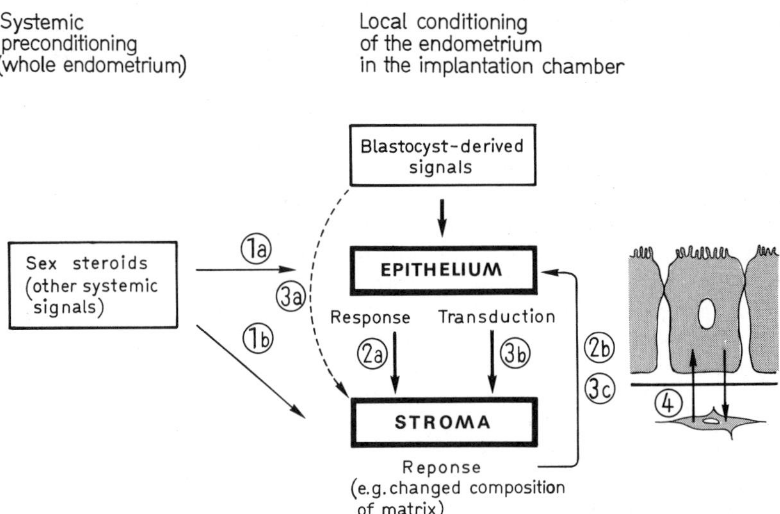

Figure 5. Signal exchange at the implantation site. Influences that the uterus exerts on the blastocyst are omitted. 1a: The epithelial cells respond directly to maternal steroid hormones. 1b: The stroma cells respond directly to maternal steroid hormones. 2a: The stroma cells respond to changes in the epithelium that have been induced as described in 1a. 2b: The epithelial cells respond to changes in the stroma that have been induced as described in 1b. 3a: Blastocyst-derived signals may reach stroma cells directly (being transported unchanged through the uterine epithelium), stroma cells respond. 3b: The uterine epithelium transduces blastocyst-derived signals, without otherwise reacting, to the stroma cells, which will respond. 3c: As a result, the epithelial cells respond to changes in the stroma. 4: Epithelium-stroma interactions are changed due to changed production rates or to redirected transport (not anymore to the interface of both partners as before) of second/third order signal substances playing a role in these interactions (possibly including matrix molecules).

As far as the *endometrium* is concerned, a new concept has been developed that may explain what is going on in the uterine epithelium when it acquires "receptivity". If it is correct that the uterine epithelium loses part of the typical apico-basal polarity during this phase, it should be possible to derive from this some new ideas on the nature of the primary response of endometrium to steroid hormones. For example, since intracellular sorting processes in epithelial cells are dependent on microtubules one would expect to be able to influence acquisition of "receptivity" by manipulation of the microtubular system. On the other hand, any mediator substance that can stimulate epithelial-mesenchymal transformation can be considered a very interesting candidate for a new class of molecules that can influence endometrial "receptivity". Based on this concept it should be possible to manipulate the process by non-steroidal means.

It is well known, as for example from experiments on various glandular tissues, that many properties of epithelia including their state of differentiation depend strongly on the stroma on which these cells sit. In mammary epithelial cells, the turnover of HSPG was found to depend on cell-substratum associations (rapid when cells were rounded and slow on collagen gels) (Bernfield et al., 1984). The observation that, on mouse uterine epithelial cells, HSPG turnover is increased at receptivity (Morris and Potter, 1990) may thus reflect the changed (loosened) association with its basal lamina and the stroma (see Table 1).

Steroid hormone-induced reactions of the endometrial stroma will certainly influence the behavior of the uterine epithelium (Figure 5; cf. Glasser, 1990, for discussion). In the implantation chamber, endometrial changes as a response to systemic maternal steroid hormone conditioning are probably complemented by locally acting blastocyst-derived signals (Figure 5). These are transduced to the stroma by the uterine epithelium which plays an important role in initiation of decidualization in rodents (Lejeune and Leroy, 1980). Local stimulation of loss of epithelial polarity may be involved in the mechanism of signal transduction, but may, alternatively, also be a consequence of it. Thus, we are dealing with a system which involves complex regulation with mutual interdependence at various levels. It is hoped that the concept of destabilization of the apico-basal polarity of uterine epithelial cells and its consequences for the composition of the apical plasma membrane and cell behavior in general during "receptivity" will help in elucidating details of this regulation by providing a testable hypothesis.

ACKNOWLEDGEMENTS

The author would like to thank cordially his coworkers and colleagues who have contributed substantially to the reviewed work: M. von Bentheim, Dr. I. Classen-Linke, A. Donner, J. Friedrich, G. Helm, E. Hölscher, Dr. H.-P. Hohn, M. Marx, Dr. B. Nalbach, PD Dr. E. Winterhager. Thanks are also due to Ms. G. Freise for typing the manuscript and to Mr. Stapper for drawing the diagrams. Our own investigations have been supported by the Deutsche Forschungsgemeinschaft and by the Minister für Wissenschaft und Forschung Nordrhein-Westfalen.

REFERENCES

Anderson, T.L., Olson, G.E., and Hoffman, L.H. (1986) Stage-specific alterations in the apical membrane glycoproteins of endometrial epithelial cells related to implantation in rabbits. *Biol. Reprod.* 34, 701-720.

Anderson, T.L., Sieg, S.M., and Hodgen, G.D. (1988) Membrane composition of the endometrial epithelium: Molecular markers of uterine receptivity to implantation. In: *Human Reproduction*, (ed.), R. Iizuka and K. Semm, Amsterdam-New York-Oxford: Excerpta Medica (Int'l. Congress Ser., No. 768), pp. 513-516.

Anderson, T.L., Simon, J.A., and Hodgen, G.D. (1990) Histochemical characteristics of the endometrial surface related temporally to implantation in the non-human primate (Macaca fascicularis). In: *Trophoblast Invasion And Endometrial Receptivity: Novel Aspects Of The Cell Biology Of Embryo Implantation, Trophoblast Research*, Vol. 4, (eds.), H.-W. Denker and J.D. Aplin, New York: Plenum Press, pp. 273-284.

Aplin, J.D. and Charlton, A.K. (1990) The role of matrix macromolecules in the invasion of decidua by trophoblast. Model studies using BeWo cells. In: *Trophoblast Invasion and Endometrial Receptivity: Novel Aspects of the Cell Biology of Embryo Implantation, Trophoblast Research*, Vol. 4, (eds.), H.-W. Denker and J.D. Aplin, New York: Plenum Press, pp. 139-158.

Beier, H.M. (1974) Oviducal and uterine fluids. *J. Reprod. Fert.* 37, 221-237.

Beier, H.M. and Kühnel, W. (1973) Pseudopregnancy in the rabbit after stimulation by human chorionic gonadotropin. *Horm. Res.* 4, 1-27.

Bernfield, M., Banerjee, S., Rapraeger, A., Jalkanen, M., Koda, J., Nguyen, H., and Kaznowski, C. (1984) Regulation of epithelial morphogenesis by matrix-cell interactions. *J. Cell Biol.* 99, 234a.

Böving, B.G. (1963) Implantation mechanisms, In: *Mechanisms Concerned With Conception*, (ed.), C.G. Hartman, Oxford-London-New York:Pergamon Press, pp. 321-396.

Borland, R.M. (1977) Transport processes in the mammalian blastocyst. In: *Development In Mammals*, Vol. 1, (ed.), M.H. Johnson, North-Holland Publishing Company Amsterdam, New York, Oxford, pp. 31-67.

Borland, R.M., Hazra, S., Biggers, J.D., and Lechene, C.P. (1977) The elemental composition of the environments of the gametes and preimplantation embryo during the initiation of pregnancy. *Biol. Reprod.* 16, 147-157.

Bükers, A., Friedrich, J., Nalbach, B.P., and Denker, H.-W. (1990) Changes in lectin binding patterns in rabbit endometrium during pseudopregnancy, early pregnancy and implantation. In: *Trophoblast Invasion and Endometrial Receptivity: Novel Aspects of the Cell Biology of Embryo Implantation, Trophoblast Research*, Vol. 4, (eds.), H.-W. Denker and J.D. Aplin, New York: Plenum Press, pp. 285-305.

Bulmer, J.N., Ritson, A., and Pace, D. (1990) Endometrial leucocytes in human pregnancy. In: *Trophoblast Invasion and Endometrial Receptivity: Novel Aspects of the Cell Biology of Embryo Implantation, Trophoblast Research*, Vol. 4, (eds.), H.-W. Denker and J.D. Aplin, New York: Plenum Press, pp. 431-451.

Carson, D.D., Wilson, O.F., and Dutt, A. (1990) Glycoconjugate expression and interactions at the cell surface of mouse uterine epithelial cells and periimplantation-stage embryos. In: *Trophoblast Invasion and Endometrial Receptivity: Novel Aspects of the Cell Biology of Embryo Implantation, Trophoblast Research*, Vol. 4, (eds.), H.-W. Denker and J.D. Aplin, New York: Plenum Press, pp. 211-241.

Castronovo, V., Bracke, M., Mareel, M., and Foidart, J.-M. (1985) An accumulation of laminin precedes invasion by transformed MO4 cells. In: *Basement Membranes*, (ed.), S. Shibata, Amsterdam, New York, Oxford: Elsevier Science Publishers, pp. 447-448.

Castronovo, V., Mahieu, P., Bracke, M., Mareel, M., Colin, C., Lambotte, R., and Foidart, J.-M. (1987) Role of laminin and alpha-galactosyl cell surface residues in tumor invasion and metastasis. *Eur. J. Cell Biol.* Suppl. 20, (Vol. 44), 3-4.

Chang, M.C. (1950) Development and fate of transferred rabbit ova or blastocyst in relation to the ovulation time of recipients. *J. Exp. Zool.* 114, 197-225.

Chávez, D.J. (1986) Cell surface of mouse blastocysts at the trophectoderm-uterine interface during the adhesive stage of implantation. *Amer. J. Anat.* 176, 153-158.

Chávez, D.J. (1990) Possible involvement of D–galactose in the implantation process. In: *Trophoblast Invasion and Endometrial Receptivity: Novel Aspects of the Cell Biology of Embryo Implantation, Trophoblast Research*, Vol. 4, (eds.), H.-W. Denker and J.D. Aplin, New York: Plenum Press, pp. 259-272.

Chávez, D.J. and Anderson, T.L. (1985) The glycocalyx of the mouse uterine luminal epithelium during estrus, early pregnancy, the peri-implantation period, and delayed implantation. I. Acquisition of Ricinus communis I binding sites during pregnancy. *Biol. Reprod.* 32, 1135–1142.

Chávez, D.J. and Enders, A.C. (1982) Lectin binding of mouse blastocysts: Appearance of Dolichos biflorus binding sites on the trophoblast during delayed implantation and their subsequent disappearance during implantation. *Biol. Reprod.* 26, 545-552.

Chevalier, J., Bourguet, J., and Pinto da Silva, P. (1985) Osmotic gradient reversal induces massive assembly of tight junction strands at the basal pole of toad bladder epithelial cells. *J. Cell Biol.* 101, 304a.

Classen-Linke, I. and Denker, H.-W. (1990) Preparation of uterine epithelium for trophoblast attachment: Histochemical changes in the apical and lateral membrane compartment. In: *Trophoblast Invasion and Endometrial Receptivity: Novel Aspects of the Cell Biology of Embryo Implantation, Trophoblast Research*, Vol. 4, (eds.), H.-W. Denker and J.D. Aplin, New York: Plenum Press, pp. 307-322.

Classen-Linke, I., Denker, H.-W., and Winterhager, E. (1987) Apical plasma-membrane-bound enzymes of rabbit uterine epithelium: Pattern changes during the periimplantation phase. *Histochem.* 87, 517-529.

Cowell, T.P. (1969) Implantation and development of mouse eggs transferred to the uteri of non-progestational mice. *J. Reprod. Fertil.* 19, 239-245.

Dantzer, V. (1985) Electron microscopy of the initial stages of placentation in the pig. *Anat. Embryol.* 172, 281-293.

Denker, H.-W. (1977) Implantation: The role of proteinases, and blockage of implantation by proteinase inhibitors. *Adv. Anat. Embryol. Cell Biol.*, Berlin-Heidelberg-New York: Springer-Verlag, 53, Part 5.

Denker, H.-W. (1980) Embryo implantation and trophoblast invasion. In: *Cell Movement And Neoplasia*, (ed.), M. De Brabander et al., Oxford-New York: Pergamon Press, pp. 151-162.

Denker, H.-W. (1983) Basic aspects of ovoimplantation. In: *Obstetrics And Gynecology Annual*, (ed.), R.M. Wynn, Vol. 12, Norwalk, Connecticut: Appleton-Century-Crofts, pp. 15-42.

Denker, H.-W. (1986) Epithel-Epithel-Interaktionen bei der Embryo-Implantation: Ansätze zur Lösung eines zellbiologischen Paradoxons. Verh. Anat. Ges. 80 (Prag 1985), *Anat. Anz. Suppl.* 160, 93-114.

Denker, H.-W. (1988) Implantation: Recent approaches to understand a cell biological paradox. In: *Human Reproduction.Current Status/Future Prospect*, (Proceed. VIth World Congr. on Human Reproduction, Tokyo 1987), (eds.), R. Iizuka and K. Semm, Amsterdam: Excerpta Medica, Elsevier Science Publ. (International Congress Ser., No. 768), pp. 237-240.

De Ridder, L., Mareel, M., and Vakaet, L. (1975) Adhesion of malignant and nonmalignant cells to cultured embryonic substrates. *Canc. Res.* 35, 3164-3171.

Enders, A.C., Chávez, D.J., and Schlafke, S. (1981) Comparison of implantation in utero and in vitro. In: *Cellular And Molecular Aspects Of Implantation*, (eds.), S.R. Glasser and D.W. Bullock, New York and London: Plenum Press, pp. 365-382.

Enders, A.C., Hendrickx, A.G., and Schlafke, S. (1983) Implantation in the Rhesus monkey: Initial penetration of endometrium. *Amer. J. Anat.* 167, 275-298.

Enders, A.C. and Nelson, D.M. (1973) Pinocytotic activity of the uterus of the rat. *Amer. J. Anat.* 138, 277-300.

Enders, A.C. and Schlafke, S. (1974) Surface coats of the mouse blastocyst and uterus during the preimplantation period. *Anat. Rec.* 180, 31-46.

Enders, A.C. and Schlafke, S. (1977) Alteration in uterine luminal surface at the implantation site. *J. Cell Biol.* 75, 70a.

Enders, A.C. and Schlafke, S. (1986) Implantation in nonhuman primates and in the human. *Comparative Primate Biology*, Vol. 3, *Reproduction and Development*, New York: Alan R. Liss, Inc, pp. 291-310.

Fisher, S.J., Sutherland, A., Moss, L., Hartman, L., Crowley, E., Bernfield, M., Calarco, P., and Damsky, C. (1990) Adhesive interactions of murine and human trophoblast cells. In: *Trophoblast Invasion and Endometrial Receptivity: Novel Aspects of the Cell Biology of Embryo Implantation, Trophoblast Research*, Vol. 4, (eds.), H.-W. Denker and J.D. Aplin, New York: Plenum Press, pp. 115-138.

Foidart, J.M., Christiane, Y., and Emonard, H. (1990) Interactions between human trophoblast cells and the extracellular matrix of the endometrium. Specific expression of alpha 1-3 galactose residues by invasive human trophoblastic cells. In: *Trophoblast Invasion and Endometrial Receptivity: Novel Aspects of the Cell Biology of Embryo Implantation, Trophoblast Research*, Vol. 4, (eds.), H.-W. Denker and J.D. Aplin, New York: Plenum Press, pp. 159-177.

Garbi, C., Tacchetti, C., and Wollman, S.H. (1986) Change of inverted thyroid follicle into a spheroid after embedding in a collagen gel. *Exp. Cell Res.* 163, 63-77.

Gehlsen, K.R., Argraves, W.S., Pierschbacher, M.D., and Ruoslahti, E. (1988) Inhibition of in vitro tumor cell invasion by Arg-Gly-Asp-containing synthetic peptides. *J. Cell Biol.* 106, 925-930.

Glasser, S.R. (1990) Biochemical and structural changes in uterine endometrial cell types following natural or artificial deciduogenic stimuli. In: *Trophoblast Invasion and Endometrial Receptivity: Novel Aspects of the Cell Biology of Embryo Implantation, Trophoblast Research*, Vol. 4, (eds.), H.-W. Denker and J.D. Aplin, New York: Plenum Press, pp. 377-416.

Greenburg, G. and Hay, E.D. (1988) Cytoskeleton and thyroglobulin expression change during transformation of thyroid epithelium to mesenchyme-like cells. *Develop.* 102, 605-622.

Guillomot, M., Fléchon, J.-E., and Wintenberger-Torres, S. (1982) Cytochemical studies of uterine and trophoblastic surface coats during blastocyst attachment in the ewe. *J. Reprod. Fert.* 65, 1-8.

Hay, E.D. (1985) Extracellular matrix, cell polarity and epithelial-mesenchymal transformation. In: *Molecular Determinants Of Animal Form*, UCLA Symp. Molec. Cell Biol., New Ser., (ed.), G.M. Edelman, Vol. 31, New York: Alan R. Liss, Inc., pp. 293-318.

Hill, J.P. (1898) The placentation of Perameles (Contributions to the embryology of the Marsupialia - I.). *Quart. J. Microsc. Sci.* (London) 40, 385-446 and Plates 29-33.

Hochfeld, A., Beier, H.M., and Denker, H.-W. (1990) Changes of intermediate filament protein localization in endometrial cells during early pregnancy. In: *Trophoblast Invasion and Endometrial Receptivity: Novel Aspects of the Cell Biology of Embryo Implantation, Trophoblast Research*, Vol. 4, (eds.), H.-W. Denker and J.D. Aplin, New York: Plenum Press, pp. 357-374.

Hoffman, L.H., Winfrey, V.P., Anderson, T.L., and Olson, G.E. (1990) Uterine receptivity to implantation in the rabbit: Evidence for a 42 kDa glycoprotein as a marker of receptivity. In: *Trophoblast Invasion and Endometrial Receptivity: Novel Aspects of the Cell Biology of Embryo Implantation, Trophoblast Research*, Vol. 4, (eds.), H.-W. Denker and J.D. Aplin, New York: Plenum Press, pp. 243-258.

Hohn, H.-P. and Denker, H.-W. (1990) A three-dimensional organ culture model for the study of implantation of rabbit blastocysts in vitro. In: *Trophoblast Invasion and Endometrial Receptivity: Novel Aspects of the Cell Biology of Embryo Implantation, Trophoblast Research*, Vol. 4, (eds.), H.-W. Denker and J.D. Aplin, New York: Plenum Press, pp. 71-95.

Hohn, H.-P., Donner, A., and Denker, H.-W. (1985) Evaluation of an in vitro model for embryo implantation: Selective receptivity of rabbit endometrium for trophoblast attachment. *Eur. J. Cell Biol.* Suppl. 12 (Vol. 39), 18.

Holmes, P.V. and Dickson, A.D. (1973) Estrogen-induced surface coat and enzyme changes in the implanting mouse blastocyst. *J. Embryol. Exp. Morph.* 29, 639-645.

Jenkinson, E.J. and Searle, R.F. (1977) Cell surface changes on the mouse blastocyst at implantation. *Exp. Cell Res.* 106, 386-390.

Johnson, L.V. and Calarco, P.G. (1980) Mammalian preimplantation development: The cell surface. *Anat. Rec.* 196, 201-219.

Johnson, M.H., Maro, B., and Takeichi, M. (1986) The role of cell adhesion in the synchronization and orientation of polarization in 8-cell mouse blastomeres. *J. Embryol. Exp. Morph.* 93, 239-255.

Kirby, D.R.S. (1965) The role of the uterus in the early stages of mouse development. In: *Ciba Foundation Symposium: Preimplantation Stages Of Pregnancy*, (eds.), G.E.W. Wolstenholme and M. O'Connor, London: Churchill.

Kirby, D.R.S. (1967) Ectopic autografts of blastocysts in mice maintained in delayed implantation. *J. Reprod. Fert.* 14, 512-517.

Kirby, D.R.S. (1970) The extra-uterine mouse egg as an experimental model. In: *Schering Symposium On Mechanisms Involved In Conception*, (ed.) G. Raspe, Oxford: Pergamon Press/Vieweg, pp. 255-273.

Kitajima, K., Yamashita, K., and Fujita, H. (1985) Fine structural aspects of the shift of zonula occludens and cytoorganelles during inversion of cell polarity in cultured porcine thyroid follicles. *Cell Tiss. Res.* 242, 221-224.

Knoth, M. and Larsen, J.F. (1972) Ultrastructure of a human implantation site. *Acta Obstet. Gynecol. Scand.* 51, 385.

Kuzan, F.B. and Wright, R.W., Jr. (1981) Attachment of porcine blastocysts to fibroblast monolayers in vitro. *Theriogenology* 16, 651-658.

Lampelo, S.A., Ricketts, A.P., and Bullock, D.W. (1985) Purification of rabbit endometrial plama membranes from receptive and non-receptive uteri. *J. Reprod. Fert.* 75, 475-484.

Larsen, J.F. (1970) Electron microscopy of nidation in the rabbit and observations on the human trophoblastic invasion. In: *Ovo-Implantation, Human Gonadotrophins And Prolactin*, (eds.), P.O. Hubinont, F. Leroy, C. Robyn, and P. Leleux, Basel, München, New York: S. Karger, pp. 38-51.

Larsen, J.F. (1974) Ultrastructural studies of the implantation process. *Basic Life Sciences* (New York) 4, 287-296.

Larsen, J.F. and Knoth, M. (1971) Ultrastructure of the anchoring villi and trophoblastic shell in the second week of placentation. *Acta Obstet. Gynecol. Scand.* 50, 117.

Leiser, R. (1979) Blastocystenimplantation bei der Hauskatze. Licht- und elektronenmikroskopische Untersuchung. *Zbl. Vet. Med. C, Anat. Histol. Embryol.* 8, 79-96.

Leiser, R. (1982) Development of the trophoblast in the early carnivore placenta of the cat. *Biblthca Anat.* 22, 93-107.

Leiser, R. (1981) Ultrastructural aspects of implantation. In: *Human Reproduction*, Proceedings of III World Congress, Berlin, March 22-26, (eds.), K. Semm and L. Mettler, Amsterdam-Oxford-Princeton: Excerpta Medica, pp. 378-382.

Lejeune, B. and Leroy, F. (1980) Role of the uterine epithelium in inducing the decidual cell reaction. *Prog. Reprod. Biol.* 7, 92-101.

Liotta, L.A., Rao, C.N., and Barsky, S.H. (1984) Tumor cell interaction with the extracellular matrix. In: *The Role Of Extracellular Matrix In Development*, (ed.), R.L. Trelstad, New York: Alan R. Liss, Inc., pp. 357-371

Liotta, L.A., Rao, C.N., and Wewer, U.M. (1986) Biochemical interactions of tumor cells with the basement membrane. *Ann. Rev. Biochem.* 55, 1037-1057.

Loke, Y.W., King, A., and Grabowska, A. (1990) Antigenic expression by migrating trophoblast and its relevance to implantation: A review. In: *Trophoblast Invasion and Endometrial Receptivity: Novel Aspects of the Cell Biology of Embryo Implantation, Trophoblast Research*, Vol. 4, (eds.), H.-W. Denker and J.D. Aplin, New York: Plenum Press, pp. 191-207.

Marengo, S.R., Bazer, F.W., Thatcher, W.W., Wilcox, C.J., and Wetteman, R.P. (1986) Prostaglandin F_2 as the luteolysin in swine: VI. Hormonal regulation of the movement of exogenous PGF_2 from the uterine lumen into the vasculature. *Biol. Reprod.* 34, 284-292.

Marx, M., Winterhager, E., and Denker, H.-W. (1990) Penetration of the basal lamina by processes of the uterine epithelial cells during implantation in the rabbit. In: *Trophoblast Invasion and Endometrial Receptivity: Novel Aspects of the Cell Biology of Embryo Implantation, Trophoblast Research*, Vol. 4, (eds.), H.-W. Denker and J.D. Aplin, New York: Plenum Press, pp. 417-430.

Morris, J.E. and Potter, S.W. (1984) A comparison of developmental changes in surface charge in mouse blastocysts and uterine epithelium using DEAE beads and dextran sulfate in vitro. *Develop. Biol.* 103, 190-199.

Morris. J.E. and Potter, S.W. (1990) An in vitro model for studying interactions between mouse trophoblast and uterine epithelial cells. A brief review of in vitro systems and observations on cell-surface changes during blastocyst attachment. In: *Trophoblast Invasion and Endometrial Receptivity: Novel Aspects of the Cell Biology of Embryo Implantation, Trophoblast Research*, Vol. 4, (eds.), H.-W. Denker and J.D. Aplin, New York: Plenum Press, pp. 51-69.

Murphy, C.R., Swift, J.G., Mukherjee, T.M., and Rogers, A.W. (1982a) Changes in the fine structure of the apical plasma membrane of endometrial epithelial cells during implantation in the rat. *J. Cell Sci.* 55, 1-12.

Murphy, C.R., Swift, J.G., Mukherjee, T.M., and Rogers, A.W. (1982b) The structure of tight junctions between uterine luminal epithelial cells at different stages of pregnancy in the rat. *Cell Tiss. Res.* 223, 281-286.

Murphy, C.R., Swift, J.G., Need, J.A., Mukherjee, T.M., and Rogers, A.W. (1982c) A freeze-fracture electron microscopic study of tight junctions of epithelial cells in the human uterus. *Anat. Embryol.* 163, 367-370.

Murphy, C.R., Swift, J.G., Mukherjee, T.M., and Rogers, A.W. (1982d) Reflexive gap junctions on uterine luminal epithelial cells. *Acta Anat.* 112, 92-96.

Noeslund, G. (1979) Growth control of the mouse blastocyst in vitro. Dissertation, Uppsala.

Nalbach, B.P. (1985) Lektinbindungsmuster in Uterus und Blastozyste des Kaninchens während der Präimplantationsphase und der frühen Implantation. Histochemie und Methodenkritik. Dissertation, Medizinische Fakultät der RWTH Aachen.

Nilsson, O., Lindqvist, I., and Ronquist, G. (1975) Blastocyst surface charge and implantation in the mouse. *Contraception* 11, 441-450.

Parr, M.B. (1980) Endocytosis at the basal and lateral membranes of rat uterine epithelial cells during early pregnancy. *J. Reprod. Fert.* 60, 95-99.

Parr, M. (1982) Apical vesicles in the rat uterine epithelium: A morphometric study. *Anat. Rec.* 202, 145A-146A.

Parr, M.B. (1983) Relationship of uterine closure to ovarian hormones and endocytosis in the rat. *J. Reprod. Fert.* 68, 185-188.

Parr, M.B. and Parr, E.L. (1977) Endocytosis in the uterine epithelium of the mouse. *J. Reprod. Fert.* 50, 151-153.

Parr, M.B. and Parr, E.L. (1978) Uptake and fate of ferritin in the uterine epithelium of the rat during early pregnancy. *J. Reprod. Fert.* 52, 183-188.

Parr, M.B. and Parr, E.L. (1982) Relationship of apical domes in the rabbit uterine epithelium during the peri-implantation period to endocytosis, apocrine secretion and fixation. *J. Reprod. Fert.* 66, 739-744.

Peters, B.P., Hartle, R.J., Krzesicki, R.F., Kroll, T.G., Terini, F., Balun, J.E., Goldstein, I.J., and Ruddon, R.W. (1985) The biosynthesis, processing, and secretion of laminin by human choriocarcinoma cells. *J. Biol. Chem.* 260, 14732-14742.

Porter, D.G. (1967) Observations on the development of mouse blastocysts transferred to the testis and kidney. *Amer. J. Anat.* 121, 73-85.

Psychoyos, A. (1973) Endocrine control of egg implantation. In: *Handbook Of Physiology*, Sect. 7 (Endocrinology) Vol. II (Female Reproductive System) Part 2, (ed.) R.O. Greep, Washington: American Physiological Society, pp. 187-215.

Psychoyos, A. (1988) The "implantation window": Can it be enlarged or displaced ? In: *Human Reproduction. Current Status/Future Prospect*, (ed.) R. Iizuka and K. Semm, Amsterdam-New York-Oxford:Excerpta Medica (International Congress Ser., No. 768), pp. 231-232.

Psychoyos, A. and Casimiri, V. (1980) Factors involved in uterine receptivity and refractoriness. In: *Blastocyst-Endometrium Relationships* (Prog. Reprod. Biol., Vol. 7), (eds.), F. Leroy, C.A. Finn, A. Psychoyos, and P.O. Hubinont, Basel München, Paris, London, New York, Sydney: S. Karger, pp. 143-157.

Queenan, J.T. Jr., Kao, L.-C., Arboleda, C.E., Ulloa-Aguirre, A., Golos, T.G., Cines, D.B., and Strauss III, J.F. (1987) Regulation of urokinase-type plasminogen activator production by cultured human cytotrophoblasts. *J. Biol. Chem.* 262, 10903-10906.

Richa, J. (1986) Cell surface glycoproteins and cellular interactions in the preimplantation mouse embryo. Diss., Univ. of Pennsylvania (*Diss. Abstr. Int.* 47B, 1378).

Rossman, I. (1940) The deciduomal reaction in the rhesus monkey (Macaca mulatta). I. The epithelial proliferation. *Amer. J. Anat.* 66, 277-365.

Samuel, C.A. and Perry, J.S. (1972) The ultrastructure of pig trophoblast transplanted to an ectopic site in the uterine wall. *J. Anat.* (London) 113, 139-149.

Schlafke S. and Enders, A.C. (1975) Cellular basis of interaction between trophoblast and uterus at implantation. *Biol. Reprod.* 12, 41-65.

Schlafke, S., Welsh, A.O., and Enders, A.C. (1985) Penetration of the basal lamina of the uterine luminal epithelium during implantation in the rat. *Anat. Rec.* 212, 47-56.

Sherman, M.I., Shalgi, R., Rizzino, A., Sellens, M.H., Gay, St., and Gay, R. (1979) Changes in the surface of the mouse blastocyst at implantation. In: *Maternal Recognition Of Pregnancy*. Ciba Foundation Series (New Series) 64, 33-52.

Short, R.V. and Yoshinaga, K. (1967) Hormonal influences on tumour growth in the uterus of the rat. *J. Reprod. Fert.* 14, 287-293.

Strunck, B. (in preparation) Dissertation, Medizinische Fakultät der RWTH Aachen.

Sutherland, A.E., Calarco, P.G., and Damsky, C.H. (1988) Expression and function of cell surface extracellular matrix receptors in mouse blastocyst attachment and outgrowth. *J. Cell Biol.* 106, 1331-1348.

Svalander, P.C., Odin, P., Nilsson, B.O., and Obrink, B. (1987) Trophectoderm surface expression of the cell adhesion molecule cell-CAM 105 on rat blastocysts. *Develop.* 100, 653-660.

Tachi, S., Tachi, C., and Lindner, H.R. (1970) Ultrastructural features of blastocyst attachment and trophoblastic invasion in the rat. *J. Reprod. Fert.* 21, 37-56.

Tutton, D.A. and Carr, D.H. (1984) The fate of trophoblast retained within the oviduct in the mouse. *Gynecol. Obstet. Invest.* 17, 18-24.

Ulloa-Aguirre, A., August, A.M., Golos, T.G., Kao, L.-C., Sakuragi, N., Kliman, H.J., and Strauss III, J.F. (1987) 8-Bromo-adenosine 3,5'-monophosphate regulates expression of chorionic gonadotropin and fibronectin in human cytotrophoblasts. *J. Clin. Endocrinol. Metab.* 64, 1002-1009.

Wartiovaara, J., Leivo, I., and Vaheri, A. (1979) Expression of the cell surface-associated glycoprotein, fibronectin, in the early mouse embryo. *Develop. Biol.* 69, 247-257.

Wilson, I.B. (1963) A tumour tissue analogue of the implanting mouse embryo. *Proc. Zool. Soc. Lond.* 141, 137.

Wilson, I.B. and Potts, D.M. (1970) Melanoma invasion in the mouse uterus. *J. Reprod. Fert.* 22, 429-434.

Winterhager, E. (1985) Dynamik der Zellmembran: Modellstudien während der Implantationsreaktion beim Kaninchen. Habilitationsschrift, Med. Fak. der RWTH Aachen.

Winterhager, E., Classen-Linke, I. and Denker, H.-W. (in press) Strukturelle Veränderungen der apikalen Uterusepithelzellmembran in Vorbereitung auf die Embryoimplantation. *Verh. Anat. Ges.* 83.

Winterhager, E. and Kühnel, W. (1982) Alterations in intercellular junctions of the uterine epithelium during the preimplantation phase in the rabbit. *Cell Tissue Res.* 224, 517-526.

Wislocki, G.B. and Streeter, G.L. (1938) On the placentation of the macaque (Macaca mulatta), from the time of implantation until the formation of the definitive placenta. *Carnegie Inst. Wash. Contrib. Embryol.* 27, 1-66.

Wooding, F.B.P. (1984) Role of binucleate cells in fetomaternal cell fusion at implantation in the sheep. *Amer. J. Anat.* 170, 233-250.

MORPHOLOGY

TROPHOBLAST INVASION AND PLACENTATION IN THE HUMAN: MORPHOLOGICAL ASPECTS

Robert Pijnenborg

Department of Obstetrics and Gynecology
Katholieke Universiteit
Leuven, Belgium

INTRODUCTION

The evolution of viviparity in mammals has been possible because of the development of placentation, which means the apposition of two vascular systems, maternal and fetal to allow physiological exchanges between the two (Mossman, 1937). That trophoblast invasion is a key phenomenon in this regard has been recognized for many years (Grosser, 1927). The ultimate expression of this evolutionary tendency is found in species with hemochorial placentation which include the human being. It is obvious, however, that the invasive process needs to be restricted or modulated in order to allow a gradual succession of the different developmental steps. Different uterine tissue components need to be penetrated successively i.e., the uterine epithelium (during blastocyst implantation), endometrial stroma and vessel walls in order to reach the hemochorial condition (Pijnenborg et al., 1985). It may well be that for each step specialized trophoblastic cells have to be developed.

If one considers the wealth of information acquired in different areas of mammalian reproductive biology during the past decades, it is a depressing thought how little is known about early development in the human. With the elaboration of in vitro fertilization techniques, a considerable amount of information has been acquired about early human embryogenesis, but the very early stages of blastocyst implantation, early invasion and placentation are still poorly understood. A better insight into the early events is, however, essential because deficiencies in this early period may provide a reason for defects later in pregnancy.

One possible approach is the study of laboratory animals that are characterized by a hemochorial type of placentation. In a previous review such an approach was evaluated (Pijnenborg et al., 1981a). It is clear that certain ideas, mainly concerning the nature of the decidualization process and the preparatory changes in vessel walls before trophoblast invasion starts, have provided a new stimulus for the examination of human material. On the other hand, suitable human material has had to be accumulated, and some of this has been provided by the unique collection of more than 200 hysterectomy specimens from 8 weeks to 18 weeks of pregnancy by the late Professor G. Dixon at the University of Bristol (England). Additional investigations have been performed to fill some of the gaps in understanding early implantation (Pijnenborg et al., 1980; 1981a; 1981b; 1983). However, one is still left with a considerable gap from the day of implantation to about 8 weeks. It is exactly during the early weeks of gestation that trophoblast interacts with epithelium and decidua. It is the purpose of this contribution to present an updated review of trophoblast invasion in the human.

Initial Penetration Of The Uterine Epithelium

Studies of early implantation events in different mammals are illustrative of the variability that exists between species. According to the comparative studies by Schlafke and Enders (1975), three types of implantation can be distinguished: (a) intrusive implantation, in which trophoblast extensions penetrate between apparently intact epithelial cells, (b) displacement implantation, characterized by degeneration of uterine epithelium before the actual implantation starts, and (c) fusion implantation, where syncytial trophoblast is thought to fuse with maternal epithelial cells during the invasion process.

A detailed description of these very early events in the human is still lacking, as the youngest known specimen is 7 1/2 days of age, in which the epithelium is already well penetrated (Hertig, 1968). It is not at all sure whether a comparison with related primate species is really trustworthy. Early stages of implantation have been described for the Rhesus monkey by Heuser and Streeter (1941) at the light microscopical level. Their results have recently been refined and extended by the electron microscopical studies of Enders et al. (1983). In the latter publication one restricted area of uterine invasion was illustrated in a 9 1/2 day specimen. Extensions of syncytial trophoblast seem to play the active role in penetrating an intact looking epithelium, thus showing an intrusive type of implantation. At 10 days syncytial trophoblast is mainly distributed in a ring-like fashion at the margins of the implantation area, while at the center cytotrophoblast has started to proliferate. Invasion seems to be interrupted temporarily at the basal lamina, but at 10 1/2 days the uterine epithelium has disappeared in the central area of invasion, and a complex mixture of cytotrophoblast and syncytial trophoblast faces directly the endometrial connective tissues.

In a recent report, "implantation" in vitro of a human blastocyst on a monolayer of endometrial epithelial cells was described (Lindenberg et al., 1986). Just as in the Rhesus monkey the polar trophoblast, which is in contact with the inner cell mass of the blastocyst, is responsible for attachment and invasion. A partial peripheral outgrowth of cytotrophoblast seems to displace the epithelial cells and makes contact with the culture vessel. During this process ectoplasmatic extensions of cytotrophoblast are penetrating between the healthy looking endometrial cells, again suggesting an intrusive type of implantation. Indications of early syncytial formation are found in the central area of "implantation", where the trophoblast is in contact with the culture vessel. Of course such data have to be treated with caution as it is not known how representative such an artificial culture system is for the in vivo situation.

There is some controversy whether cytotrophoblast or syncytial trophoblast plays the active role in invasion. According to Enders (1976), there is a correlation between early syncytial formation and invasiveness of trophoblast. On the other hand, in primates early syncytial formation may not reflect the invasiveness of trophoblast as such, but rather represents a preliminary step in the development of the primitive lacunar system, the first stage in placental development.

Cytotrophoblast Cell Proliferation And Early Invasion

After the formation of the chorionic villi within the primitive syncytium, cytotrophoblast starts to proliferate and forms the so-called cytotrophoblastic shell during the first few weeks of pregnancy. This structure is thought to allow a rapid circumferential extension of the implantation site (Boyd and Hamilton, 1970). The result is a gradual encroachment of the decidua, but the mechanisms of growth and expansion, including the role of the decidua, are not understood. Besides the general increase in cytotrophoblastic shell mass, an active invasion of the decidua by individual trophoblastic cells takes place. Boyd and Hamilton comment upon the presence of syncytial streamers, which they think are the remnants of the primitive syncytium that extend into the surrounding decidua. However, the majority of the invading trophoblast consists of predominantly uninuclear cytotrophoblast, i.e., released from the trophoblastic shell and progressively and diffusely infiltrates the decidua, as is illustrated in a 4 1/2 week gestational specimen by Robertson et al. (1981).

In later stages, 4 to 6 weeks, the cytotrophoblastic shell starts to regress gradually, but foci of proliferating cytotrophoblast remain at the tips of the anchoring villi. Hamilton and Boyd (1960) and Harris and Ramsey (1966) have presented evidence that the developing conceptus starts to communicate with maternal blood via the maternal capillaries or venules. There is controversy whether these capillaries are tapped or alternatively become incorporated within the primitive syncytial lacunar system. In very early specimens there is no evidence for a direct communication with the endometrial spiral arteries. The latter is thought to be initiated only during the progressive expansion of the cytotrophoblastic shell, when spiral arterial walls are breached by the invading trophoblastic cells. Again, there is no clear evidence for a contributory role of the maternal tissues, e.g., whether a spontaneous decidual regression takes place at this level. From this period onwards, endovascular trophoblast, presumably derived from the cytotrophoblastic shell, appears within the lumina of the decidual spiral arteries. This type of invasion seems to begin during the first month of pregnancy, but the paucity of material has not allowed a careful timing of the events. The relation of endovascular trophoblast with the spiral vessel walls is also not clear in this early period. Boyd and Hamilton commented upon degenerative changes in the spiral vessel walls, i.e., endothelial hypertrophy, muscular retrogression, and the appearance of swollen cells that may represent part of the the general decidual reaction (Brettner, 1964), or alternatively are the consequence of the invasion of the wall by the endovascular trophoblastic cells. We do not think that enough material has been systematically studied in order to propose some definite sequence of cause and effect relationships between trophoblast invasion and vascular changes in this very early pregnancy period. In Figure 1, successive steps of trophoblastic invasion are schematically illustrated.

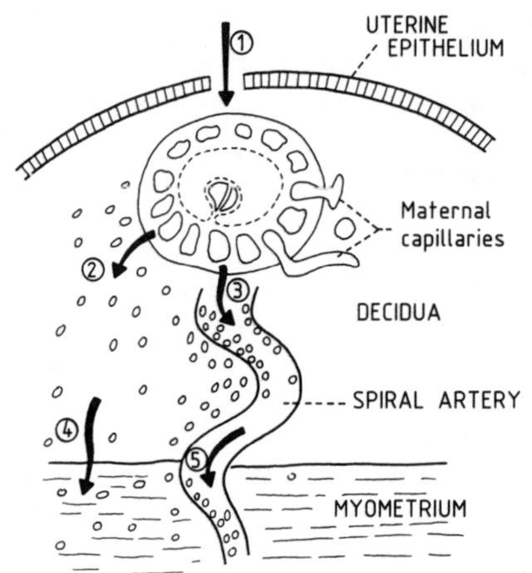

Figure 1. Diagram illustrating successive steps in trophoblast invasion during placentation in the human.

MATERIALS AND METHODS

The phenomena that were described in the introduction were all observed in a limited number of specimens in very early pregnancy. From the second month of pregnancy onwards, more pregnancy hysterectomy specimens have been available for examination. For the present study, the period of 8 to 18 weeks of pregnancy was emphasized. While at the level of the decidua, most of the significant events have already occurred by that time, the opportunity to follow step by step trophoblastic invasion and associated changes in the myometrium was presented. From the collection of pregnant uteri at the University of Bristol, 48 specimens were selected which were considered as representative of normal pregnancy. Specimens with clinical and pathological complications were excluded from this study. The material had been fixed in 4% formaldehyde saline. Slices of about 1 cm thickness were cut through the central portion of placenta and adjacent uterine wall, and these slices were further processed as a whole and embedded in paraffin wax. Step serial sections were prepared and stained according to standard (Hematoxylin and Eosin; Periodic Acid Schiff) techniques.

For morphometrical analysis, histology sections through whole placental bed areas were divided into different compartments, from lateral to central. Within each compartment, the presence of interstitial trophoblast was evaluated in three different layers of myometrium, starting at the superficial myometrium directly underneath the decidual-myometrial junction, and moving deeper according to fixed distances (Figure 2). Standard morphometrical techniques were used for this study, using a point-counting technique by projecting a square grid in the microscopic field (Dunnill, 1968). As a first step the volume density of cytotrophoblast was examined to determine if it is related to the position within the placental bed and if there is any change during the period of pregnancy under study. This information was needed as a background for the next step, when changes in the myometrial spiral arteries were studied and related to the presence of interstitial cytotrophoblast. In order to acquire quantitative data, compartments in the placental bed were grouped to demarcate three categories of myometrial tissue with: (a) no associated interstitial cytotrophoblast, (b) interstitial trophoblast of volume density of less than 5%, and (c) interstitial trophoblast of volume density of 5% or more (Figure 2). Additionally, spiral arteries were assessed in non-placental bed areas, for comparative purposes.

More extensive and detailed information about the morphometrical techniques and the statistical analysis are published in two previous articles (Pijnenborg et al., 1981b; 1983).

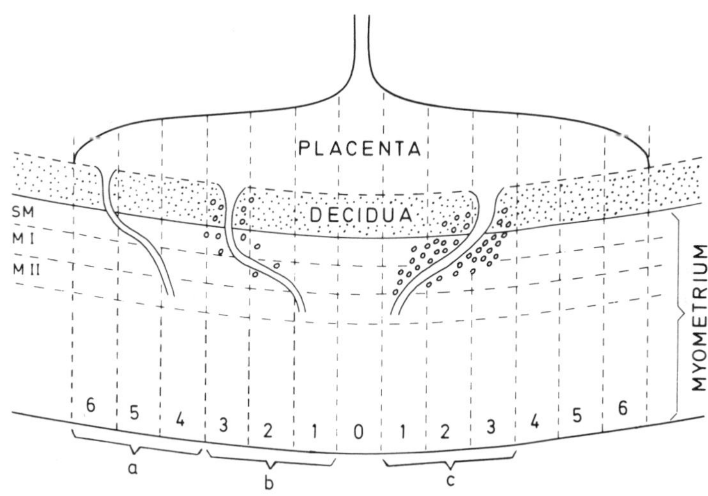

Figure 2.   Diagram illustrating arbitrary compartments and layers of the placental bed used for morphometrical analysis of the distribution of invading trophoblast. The open circles represent interstitial trophoblast related to spiral (uteroplacental) arteries: 'a' compartment with no interstitial trophoblast, 'b' compartment with less than 5 percent volume density of trophoblast, and 'c' compartment with greater than 5 percent volume density of trophoblast. SM = superficial myometrium; MI and MII deeper layers of myometrium. With kind permission of the Editor, *Placenta* 4, 397-414, Pijnenborg et al., 1983.

RESULTS AND DISCUSSION

Interstitial Invasion Of The Inner Myometrium

At 8 weeks, the invasion of the myometrium had just begun (Figure 1), and so this area of the placental bed has been studied extensively throughout the remainder of the first trimester and part of the second trimester (up to 18 weeks of pregnancy) (Pijnenborg et al., 1981b; 1983). From the proliferating tips of anchoring villi, streamers of cytotrophoblast can be followed through the decidua where they tend to be concentrated around spiral arteries. Subsequently they seem to fan out in the subjacent myometrium, where their association with the vessels is less obvious. Although invading cytotrophoblastic cells appear in the myometrium in huge numbers, their presence in this region is not mentioned by Boyd and Hamilton (1970), but it has been documented by Park (1971) and Robertson and Warner (1974).

According to the results of morphometrical analysis of the distribution of these cytotrophoblastic cells in different areas of myometrium there seems to occur a shift from a predominantly quadratic distribution (i.e., highest volume density in the center of the placental bed area) to a quartic distribution from about 10 weeks that suggests a ring-like maximal invasive zone around a less intensive center in later stages (Pijnenborg et al., 1981b). It is not known whether the increase in volume densities of invading cytotrophoblast in lateral compartments during later stages is due to a lateral migration of trophoblast from the center within the myometrium, or whether additional invasion occurs via the decidua, possibly from proliferating tips of more laterally situated anchoring villi (Figure 3). The cytotrophoblastic cells at the height of their invasion seem to become embedded in a matrix of dispersed connective tissue elements that represents the original smooth muscle tissue in an atrophic state (Figure 4). Because of this histological picture, a local effect of the invading cytotrophoblast on the myometrium may be postulated, possibly contributing to the increase in expansile capacity that occurs during the first trimester of pregnancy (Wood, 1964). This may involve changes in ground substance or the collagen network (Wood, 1972) that at least partly may be related to the presence of cytotrophoblast. There is also evidence for a role of the invading interstitial trophoblast in early changes in spiral arteries that will be presented in the next section.

The final question concerns the ultimate fate of the invading trophoblast. An obvious feature in the material that was studied was an increase in numbers of multinucleated trophoblastic giant cells after 10 weeks of gestation. In many instances some structures suggestive of the development of giant cells by fusion of cytotrophoblastic elements were observed (Figure 5). Giant cell formation seems to be the ultimate fate of the cytotrophoblast that in this way may become restricted in its invasive behavior. In a placental bed at term multinucleated giant cells are the only obvious interstitial trophoblastic elements scattered through decidua and myometrium.

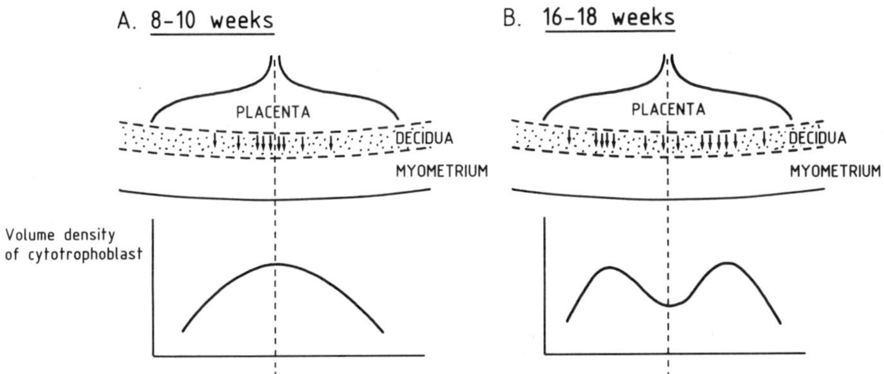

Figure 3. Diagrams illustrating the distribution of interstitial cytotrophoblast in the placental bed at 8-10 weeks (A) and 16-18 weeks (B).

Vascular Changes In The Myometrium And The Second Wave Of Endovascular Trophoblast Invasion

In the current collection of hysterectomy specimens an excellent opportunity was available to study week by week the changes in the myometrial segments of spiral arteries that are destined to develop into the uteroplacental arteries. From the beginning the regressive changes in the vessel wall that included endothelial swelling, vacuolation of the intima, appearance of hypertrophied basophilic cells and disruption of the media were striking (Figure 6). The question asked was whether these changes developed spontaneously, or if they were induced by local factor(s) including possible effects of invading trophoblast. Therefore, these features were studied in different spiral arteries in the placental bed as well as in non-placental areas. Additionally, these changes were related to the presence of perivascular trophoblast and the levels of interstitial trophoblast that were studied before.

The different steps in the statistical analysis were explained in Pijnenborg et al. (1983). From this analysis it was clear that intimal vacuolation was common to placental bed and non-placental bed arteries and increased with gestational age. Some other changes, endothelial swelling and appearance of basophilic cells, were noted also in spiral arteries of non-placental bed endometrium but to a considerably lesser extent than in the placental bed. There was a significant correlation between the volume density of interstitial cytotrophoblast in the myometrium and in particular the presence of perivascular trophoblast, with the morphological alterations in the spiral arteries. Thus, these data suggested that regressive changes might be induced by the presence of interstially invading cytotrophoblast. The next question then is, to what purpose?

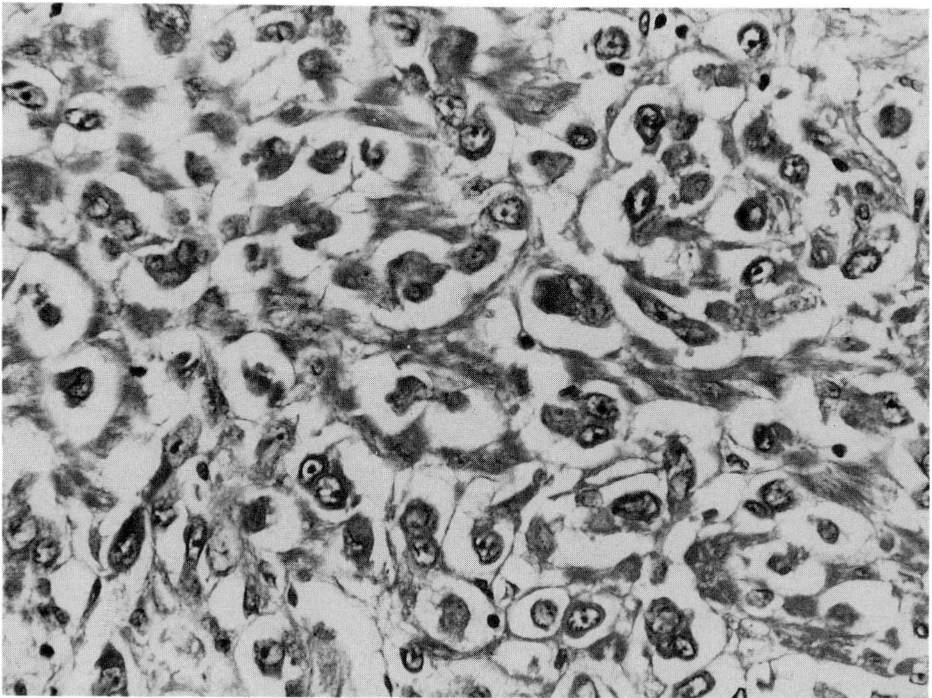

Figure 4. Cytotrophoblast in dispersed smooth muscle elements of the superficial myometrium in a central compartment of the placental bed at 11 weeks. Remnants of smooth muscle can be distinguished (X500). With kind permission of the Editor, *Placenta* 2, 303-316, Pijnenborg et al., 1981.

At this stage, attention was turned again to the endovascular trophoblast. As mentioned before, in the course of the first few months of pregnancy endovascular trophoblast is encountered in the decidual spiral arteries but not beyond the decidual-myometrial junction. In the 16-18 week specimens, however, endovascular trophoblast was noticed in myometrial segments of many spiral arteries, which means that a second wave of endovascular migration is triggered off quickly after a "resting phase" of several weeks (Figure 1). Additionally, endovascular trophoblast appeared in vessels only when there is severe disruption of the media (Figure 7). Following these analyses, it was proposed that non-villous migratory cytotrophoblast acts locally in the myometrium to induce changes in spiral arteries as a preparation for the second wave of endovascular trophoblast migration. According to the histological observations, this second wave must be viewed as an extension of the endovascular migration of the same trophoblast that contributes to the first wave, while there is no (histological) evidence of interstitial (perivascular) trophoblast penetrating through the arterial walls from outside the vessel. It is clear however that more studies using different techniques will be required in order to resolve the exact nature and mechanisms of trophoblast-maternal tissue interactions.

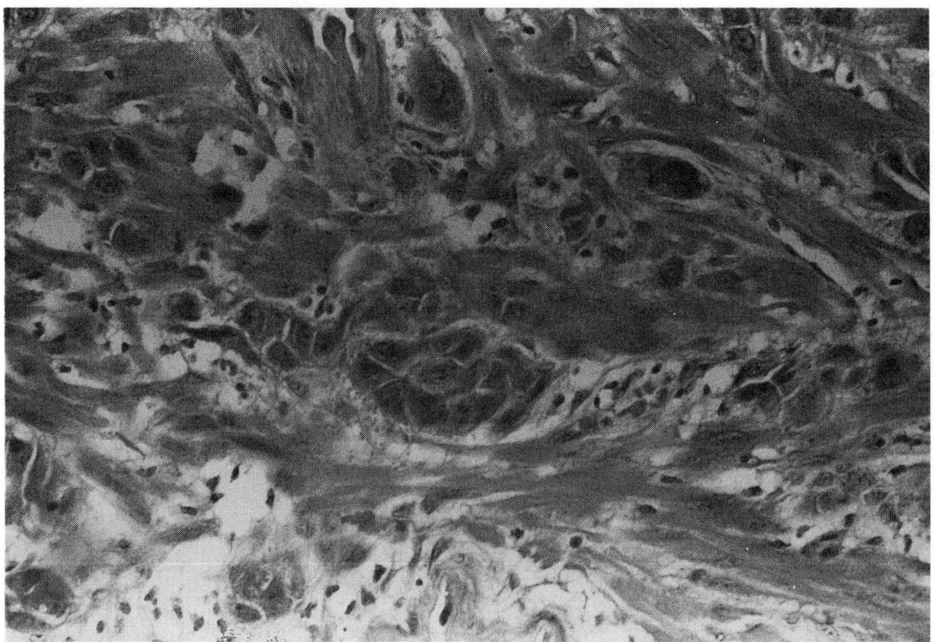

Figure 5. Cluster of cytotrophoblast in early stage of fusion in the superficial myometrium at 10 weeks (X250). With kind permission of the Editor, *Placenta* 2, 303-316, Pijnenborg et al., 1981.

Physiological Changes In Placental Bed Spiral Arteries

After the arrival of endovascular trophoblastic cells, they become gradually incorporated into the arterial walls, as can be appreciated by the appearance of PAS-positive material in between the cells in continuity with the vessel wall extracellular material (Figure 7). This process has been studied at the electron microscopic level by De Wolf et al. (1980). Trophoblast incorporation seems to take place in areas with disrupted endothelium. The trophoblast that does penetrate the vessel wall is characterized by the presence of dilated cisternae of rough endoplasmic reticulum and seems to secrete some unidentified substance(s) probably contributing to the so-called fibrinoid material. The end result is that the spiral arteries are converted into widened tubes in which no normal muscular and elastic elements can be identified. Instead, a wall is present containing amorphous fibrinoid material in which trophoblastic cells are embedded. The first systematic study of these arteries in normal pregnancy was published in 1967 by Brosens et al., who commented upon the "physiological" nature of the changes that otherwise would have been described as pathological. In this now classical paper they put forward the hypothesis that the big cells found in the spiral artery walls at the end of pregnancy were derived from the endovascular trophoblastic cells observed in the beginning of pregnancy. It is obvious that vessels in which the normal muscular architecture is so thoroughly altered (De Wolf et al., 1973; Sheppard and Bonnar, 1974) must be completely different from the hemodynamic point of view, as has been pointed out by Moll et al. (1975). It is equally clear that by the absence of the normal muscular elements, these vessels can no longer respond to vasoactive stimuli. Physiological changes must be regarded as an adaptation of the uteroplacental vascular system to ensure a continuous supply of maternal blood in sufficient quantities to allow proper fetal development. So it is seen how a peculiar invasive behavior of trophoblast in the beginning of pregnancy has major consequences for the well-being of the fetus throughout pregnancy.

The latter can be particularly appreciated if one examines some pathological events in human pregnancy. In cases of preeclampsia for example, it has been shown by Brosens et al. (1972) that physiological changes of spiral arteries

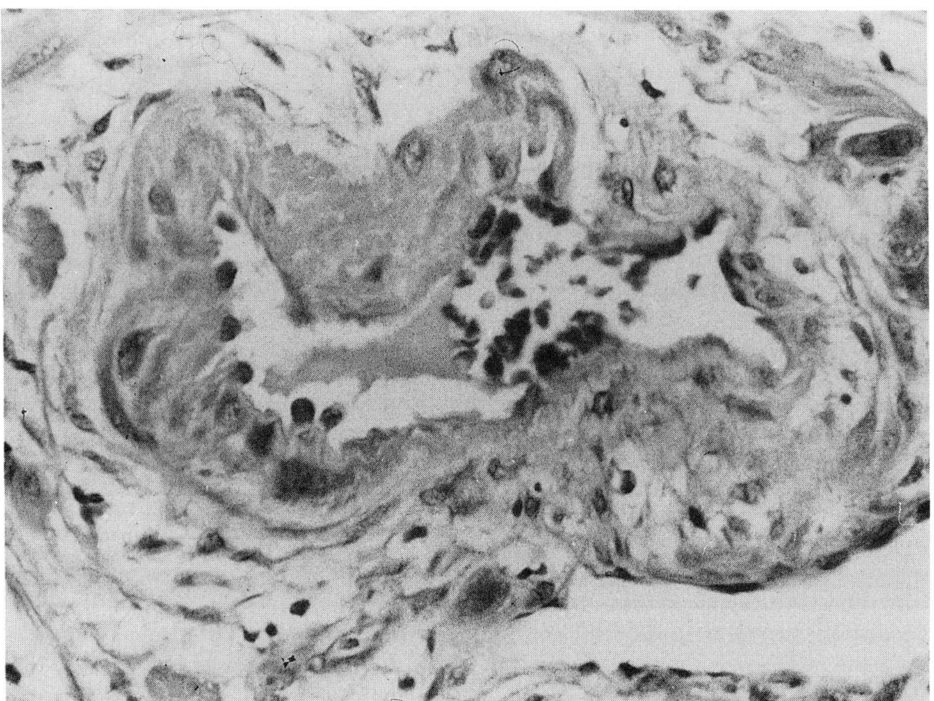

Figure 6. Placental bed spiral artery showing extreme disorganization of the vessel wall, before the second wave of endovascular trophoblastic invasion (X500).

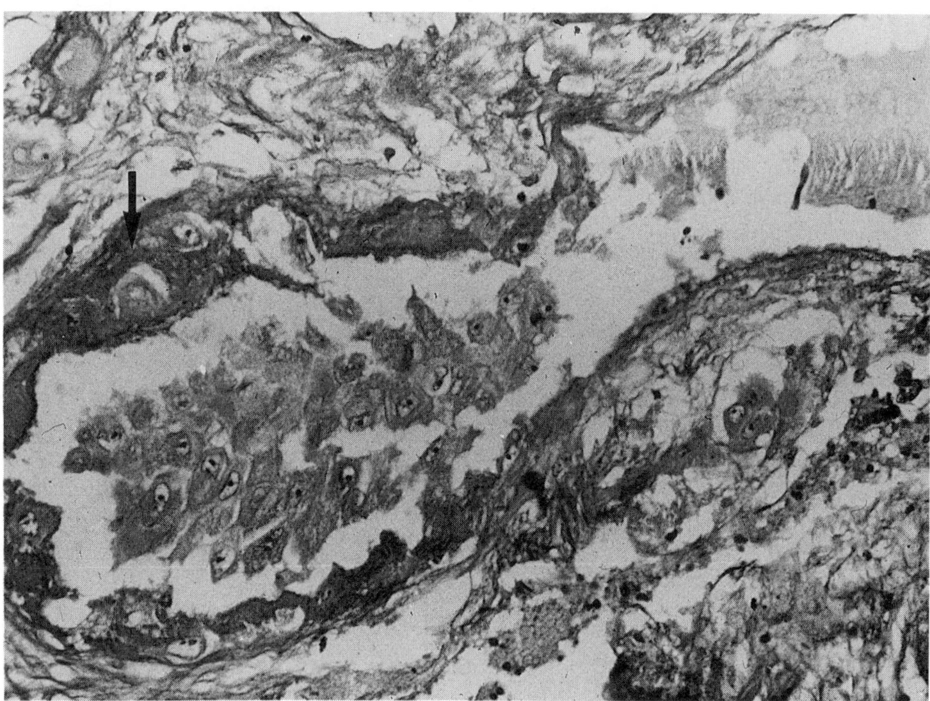

Figure 7. Distended, disorganized placental bed spiral artery at 16 weeks' gestation containing endovascular trophoblast that is partly embedded in the vessel wall (left, arrowed) (X500).

are limited to the decidual segments of the spiral arteries, which led Robertson to hypothesize that in such pathological conditions the second wave of endovascular trophoblast migration is inhibited. No mechanism is yet proposed for how this inhibition is produced. Some evidence has been offered in the past for a possible immune factor as contributing to the disease (Jenkins, 1976). On the other hand, other factors may be equally involved. For example, early hemodynamic disturbances may influence trophoblast endovascular migration and evidence for such an effect was found in experimental animals (Pijnenborg et al., 1975). This leaves, however, the reasons for aberrations in uterine blood flow patterns. In line with the current observations, one must consider the possibility that the priming action of interstitial cytotrophoblast on the myometrial spiral arteries is in some way disturbed. It is clear that the structural changes described in the myometrial vessels, which may be induced by interstitial cytotrophoblast, must have major hemodynamical consequences. Are spiral arteries less responsive to interstitial cytotrophoblast in patients that are destined to develop preeclampsia or, alternatively, is there a restriction or inhibition of interstitial cytotrophoblast invasion in such cases? Gerretsen et al. (1981; 1983) pointed to the fact that in preeclampsia higher numbers of multinucleated giant cells can be found around the spiral arteries at the end of pregnancy. This observation is not easy to interpret and although we think that this finding must be further substantiated, it again points to changed interrelationships between the interstitial trophoblast and the blood vessels. Finally, most obstetricians will be familiar with pathological changes of spiral arteries that develop in preeclampsia (fibrinoid necrosis, foam cells etc., Robertson et al., 1967; Brosens et al., 1972; De Wolf et al., 1975; Sheppard and Bonnar, 1981) but these may be considered as epiphenomena (Robertson et al., 1986). It is appropriate for a work about implantation and early trophoblast invasion to stress that understanding the early establishment of trophoblast-endometrium relationships is essential for understanding the later events and this is highlighted in the development of the uteroplacental blood supply and its defects.

SUMMARY

An updated review of trophoblast invasion in early human pregnancy is presented. There are hints that the earliest penetration of the uterine wall is affected by trophoblastic intrusion between healthy epithelial cells. During the first month, cells derived from the cytotrophoblastic shell start to invade the decidua. In this process, the maternal spiral arteries are breached, allowing the first wave of endovascular trophoblast invasion into the decidual segments of spiral arteries. From 8 weeks, interstitial cytotrophoblastic cells invade the myometrium. Initially they reach their highest concentrations in the central areas, but later on, i.e., after 10 weeks, they acquire a ring-like maximal distribution in the placental bed. A series of regressive changes occur in the myometrial segments of the spiral arteries and a correlation of some of them with the presence of interstitial cytotrophoblast was found. The second wave of endovascular trophoblast invasion starts at 14-15 weeks of pregnancy only after the initial regressive changes have developed in spiral artery walls. Understanding of invasive behavior of trophoblast in early pregnancy is essential for explaining pathological conditions that develop later in pregnancy, such as preeclampsia.

REFERENCES

Boyd, J.D. and Hamilton, W.J. (1970) *The Human Placenta*, Cambridge: W. Heffer, p. 365.

Brosens, I., Robertson, W.B., and Dixon, H.G. (1967) The physiological response of the vessels of the placental bed to normal pregnancy. *J. Path. Bact.* 93, 569-579.

Brosens, I., Robertson, W.B., and Dixon, H.G. (1972) The role of the spiral arteries in the pathogenesis of preeclampsia. In: *Obstet. Gynecol. Annual*, (ed.) R. Wynn, New York: Appleton-Century-Crofts, pp. 177-191.

Brettner, A. (1964) Zum Verhalten der Sekundaren Wand der Uteroplacentargefasse bei der Decidualen Reaktion. *Acta Anat.* 57, 367-376.

De Wolf, F., De Wolf-Peeters, C., and Brosens, I. (1973) Ultrastructure of the spiral arteries in the human placental bed at the end of normal pregnancy. *Am. J. Obstet. Gynecol.* 117, 833-848.

De Wolf, F., Robertson, W.B., and Brosens, I. (1975) The ultrastructure of acute atherosis in hypertensive pregnancy. *Am. J. Obstet. Gynecol.* 123, 164-174.

De Wolf, F., De Wolf-Peeters, C., Brosens, I., and Robertson, W.B. (1980) The human placental bed: Electron microscopic study of trophoblastic invasion of spiral arteries. *Am. J. Obstet. Gynecol.* 137, 58-70.

Dunnill, M.S. (1968) Quantitative methods in histology. In: *Recent Advances In Clinical Pathology*, (ed.) D Dyke, Edinburgh and London: Churchill Livingstone, pp. 401-416.

Enders, A.C. (1976) Anatomical aspects of implantation. *J. Reprod. Fert. Suppl.* 25, 1-15.

Enders, A.C., Hendrickx, A.G., and Schlafke, S. (1983) Implantation in the Rhesus monkey: Initial penetration of endometrium. *Am. J. Anat.* 167, 275-298.

Gerretsen, G., Huisjes, H.J., and Elema, J.D. (1981) Morphological changes of the spiral arteries in the placentalbed in relation to pre-eclampsia and fetal growth retardation. *Br. J. Obstet. Gynaecol.* 88, 876-881.

Gerretsen, G., Huisjes, H.J., Hardonk, M.J., and Elema, J.D. (1983) Trophoblast alterations in the placental bed in relation to physiological changes in spiral arteries. *Br. J. Obstet. Gynaecol.* 90, 34-39.

Grosser, O. (1927) *Fruhentwicklung, Eihautbildung und Placentation des Menschen und der Saugetiere*, Munchen: J.F. Bergmann Verlag, p. 454.

Hamilton, W.J. and Boyd, J.D. (1960) Development of the human placenta in the first three months of gestation. *J. Anat.* 94, 297-328.

Harris, J.W.S. and Ramsey, E.M. (1966) The morphology of human uteroplacental vasculature. *Contrib. Embryol.* 38, 43-58.

Hertig, A.T. (1968) *Human Trophoblast*, Springfield: Charles C. Thomas, p. 363.

Heuser, C.H. and Streeter, G.L. (1941) Development of the Macaque Embryo. *Contrib. Embryol.* 29, 15-55.

Jenkins, D.M. (1976) Pre-eclampsia/eclampsia (Gestosis) and other pregnancy complications with possible immunologic basis. In: *Immunology Of Human Reproduction*, (eds.) J.S. Scott and W.R. Jones, New York: Academic Press, pp. 297-328.

Lindenberg, S., Hyttel, P., Lenz, S., and Holmes, P.V. (1986) Ultrastructure of the early human implantation in vitro. *Human Reproduction* 1, 533-538.

Moll, W., Kunzel, W., and Herberger, J. (1975) Hemodynamic implications of hemochorial placentation. *Eur. J. Obstet. Gynaecol. Reprod. Biol.* 5, 67-74.

Mossman, H.W. (1937) Comparative morphogenesis of the fetal membranes and accessory uterine structures. *Contrib. Embryol.* 26, 129-246.

Park, W.W. (1971) *Choriocarcinoma: A Study Of Its Pathology*, London: Heinemann, pp. 13-27.

Pijnenborg, R., Robertson, W.B., and Brosens, I. (1975) The role of ovarian steroids in placental development and endovascular trophoblast migration in the golden hamster. *J. Reprod. Fert.* 44, 43-51.

Pijnenborg, R., Dixon, G., Robertson, W.B., and Brosens, I. (1980) Trophoblastic invasion of human decidua from 8 to 18 weeks of pregnancy. *Placenta* 1, 3-19.

Pijnenborg, R., Robertson, W.B., Brosens, I., and Dixon, G. (1981a) Review article: Trophoblast invasion and the establishment of haemochorial placentation in man and laboratory animals. *Placenta* 2, 71-92.

Pijnenborg, R., Bland, J.M., Robertson, W.B., Dixon, G., and Brosens, I. (1981b) The pattern of interstitial trophoblastic invasion of the myometrium in early human pregnancy. *Placenta* 2, 303-316.

Pijnenborg, R., Bland, J.M., Robertson, W.B., and Brosens, I. (1983) Uteroplacental arterial changes related to interstitial trophoblast migration in early human pregnancy. *Placenta* 4, 397-414.

Pijnenborg, R., Robertson, W.B., and Brosens, I. (1985) Morphological aspects of placental ontogeny and phylogeny. *Placenta* 6, 155-162.

Robertson, W.B., Brosens, I., and Dixon, H.G. (1967) The pathological response to the vessels of the placental bed to hypertensive pregnancy. *J. Path. Bact.* 93, 581-592.

Robertson, W.B. and Warner, B. (1974) The ultrastructure of the human placental bed. *J. Pathol.* 112, 203-211.

Robertson, W.B., Brosens, I.A., and Dixon, H.G. (1981) Maternal blood supply in fetal growth retardation. In: *Fetal Growth Retardation*, (eds.) F.A. Van Assche, W.B. Robertson, M. and Renaer, London: Churchill Livingstone, pp. 126-138.

Robertson, W.B., Khong, T.Y., Brosens, I., De Wolf, F., Sheppard, B.L., and Bonnar, J. (1986) The placental bed biopsy: Review from three European centers. *Am. J. Obstet. Gynecol.* 155, 401-412.

Schlafke, S. and Enders, A.C. (1975) Cellular basis of interaction between trophoblast and uterus at implantation. *Biol. Reprod.* 12, 41-65.

Sheppard, B.L. and Bonnar, J. (1974) The ultrastructure of the arterial supply of the human placenta in early and late pregnancy. *J. Obstet. Gynaecol. Br. Cwlth.* 81, 497-511.

Sheppard, B.L. and Bonnar, J. (1981) An ultrastructural study of uteroplacental spiral arteries in hypertensive and normotensive prenancy and fetal growth retardation. *Br. J. Obstet. Gynaecol.* 88, 695-705.

Wood, C. (1964) The expansile behaviour of the human uterus. *J. Obstet. Gynaecol. Br. Cwlth.* 71, 615-620.

Wood, C. (1972) Myometrial and tubal physiology. In: *Human Reproductive Physiology*, (ed.) R.P. Shearman, Oxford: Blackwell, pp. 324-375.

EXPERIMENTAL MODELS

AN IN VITRO MODEL FOR STUDYING INTERACTIONS BETWEEN MOUSE TROPHOBLAST AND UTERINE EPITHELIAL CELLS

A brief review of in vitro systems and observations on cell-surface changes during blastocyst attachment

John E. Morris and Sandra W. Potter

Department of Zoology
Oregon State University
Corvallis, Oregon 97331 USA

INTRODUCTION

Even a cursory examination of the published attempts to devise an in vitro implantation system for mammalian embryos reveals that there have been two distinct approaches (Table I). One approach involves the use of complex tissues to mimic as closely as possible the natural conditions for the embryo or invasive cells. The other approach uses single cells in order to identify specific cellular and molecular events.

The pioneering studies of in vitro implantation were those of Glenister (Glenister, 1961, 1966, 1967, 1971), in which he used strips of rabbit endometrium as the substratum. The strips were flattened on a nutrient agar surface in culture medium and seeded with blastocysts, which occasionally attached and developed to some extent. One problem with this system is that it was not clear whether the blastocysts were attaching and invading as would be expected in normal implantation, or whether they were attaching at sites of tissue damage. Attachment apparently occurred at any site and in any orientation, and there was clear evidence of tissue damage and necrosis. It is significant that, unlike the situation in vivo, neither progesterone nor estradiol alone or in combination had any influence on attachment of the embryo. Although he had success in getting rabbit blastocysts to attach to endometrial explants, Glenister was unsuccessful with other species (Glenister, 1966). The importance of his work is that it clearly demonstrated that the trophoblast can interact with uterine cells in vitro, and this has encouraged other attempts.

Grant and coworkers (Grant, 1973a, b; Grant et al., 1975), however, were able to demonstrate attachment by mouse blastocysts injected into cultured horns of immature or ovariectomized mature mice. Progesterone in the medium significantly enhanced attachment and invasion in parallel with an induced closure of the uterine lumen, the only histological response they observed. Decidualization in vitro only appeared in uteri from mice that had been pretreated with progesterone 16-25 hours before culture. Trophoblast invasion and embryonic differentiation were not associated with decidualization of the stroma (Grant et al.,

1975). The normal pattern of embryogenesis was disturbed in culture: the yolk cavity collapsed, the distal endoderm failed to complete its encirclement within the trophoblastic shell, and Reichert's membrane did not form. The systems of Glenister and Grant exemplify attempts to reproduce normal implantation as closely as possible in vitro. Another use of a complex system is that of Hohn and coworkers in Denker's laboratory (Hohn et al., 1987), in which they investigated the invasiveness of several tumor cell lines compared with blastocysts in a modification of Glenister's system and found that only blastocysts were able to penetrate the epithelium and invade the stroma.

The major drawback of whole tissue systems for studying the cellular basis of implantation is their complexity. This has led a number of laboratories to model implantation with monolayers of uterine cells in dishes upon which blastocysts are seeded. Among the earliest of such studies were those of Sherman and coworkers (Salomon and Sherman, 1975; Sherman and Wudl, 1976; Sherman, 1978), who observed attachment of mouse blastocysts to monolayers of uterine cells at about the stage normal attachment occurs in the mouse. Following attachment the cultured cells moved away to permit spreading of trophoblast on the dish surface, a process they defined as implantation in vitro. They observed no influence of hormones, and there was no difference between experiments in which the cells were mixed cell types or pure epithelial or stromal cells.

A serious weakness with techniques that attempt to understand normal blastocyst attachment by studying attachment to a cell layer is that the process of adhesion appears to be different from that occuring in vivo (Morris et al., 1983). Glass and coworkers (Glass et al., 1979, 1980, 1983) observed that monolayers of trophoblast cells are nonadhesive for other cells, at least on their apical or upper surfaces which are the surfaces of first contact in vivo. In contrast blastocysts, freed of their zonae pellucidae, are notoriously able to stick to glass and plastic (e.g., Böving, 1971), making it difficult at times to pipet or manipulate them with glass needles. After attachment to a dish there are occasional lateral focal sites where junctional complexes form between trophoblast and adjacent uterine cells (Chávez and van Blerkom, 1981). In most regions a clear "halo" is observed separating the outgrowing trophoblast and epithelium (Sherman and Wudl, 1976; Glass et al., 1979, 1983; Chávez and van Blerkom, 1981). Because only the apical surface of trophoblast cells is involved in the initial adhesive interactions in vivo, the major advantage of monolayer culture, therefore, is that it permits the investigator to ask questions about adhesive interactions that may influence the basal surfaces of the outgrowing trophoblast. For example, mouse embryos cultured on plastic synthesize matrix heparan sulfate proteoglycan (Heifetz et al., 1980; Dziadek et al., 1985), which may be directly involved in their adhesion to this surface (Farach et al., 1987). Cultures on hydrated collagen gels promote differentiation and more normal function of epithelial cells(Sengupta et al., 1986; Karst and Merker, 1988).

Table I

Comparison Of Some Systems For Investigating The In Vitro
Implantation Of Mammalian Blastocysts

Description	Species	System	Results	References
Blastocysts on opened and stretched strips of endometrium, pregnant donors	Rabbit	Complex	Over 50% invaded, non-specific site, hormones without effect	Glenister 1961,1966 1967,1971
Delayed blastocysts in isolated uterine horns, ovariectomized mature & immature donors	Mouse	Complex	Invasion of stroma, but no decidualization, progesterone (not estradiol) enhanced invasion and development	Grant 1973 Grant et al. 1975
Blastocysts or malignant cell lines on pieces of endometrium	Rabbit	Complex	Invasion only by trophoblast cells unless the epithelium was removed or damaged	Hohn et al. 1987
Blastocysts on monolayers of uterine epithelium and/or stroma	Mouse	Simple	Attach and spread on surface, cells displaced by trophoblast, hormones without effect	Salomon and Sherman 1975; Sherman and Wudl 1976 Sherman 1978
Blastocysts on layers of various cell types, or various cells on spreading trophoblast	Mouse	Simple	No adhesive interaction, cultured cells move away to permit trophoblast to adhere to dish	Glass et al. 1979, 1980
Blastocysts together with fragments of uterine epithelium in hanging drops	Mouse	Simple	Low frequency adhesion by apical surfaces, followed by disruption of the epithelium, hormones without effect	Morris et al. 1982, 1983

In view of these observations it is likely that the trophoblast, when confronted with a uterine cell monolayer in culture, may not adhere to the monolayer at all, at least not until after it is firmly attached to and is spreading on the dish. Surfaces of hatched blastocysts (or blastocysts with artificially removed zonae pellucidae) are covered with microvilli, which may insert between the underlying uterine cells in culture and be the initial sites of contact with the culture dish. In early studies Sherman and Salomon (1975) noted that attachment to plastic typically is 9 to 10 hours faster than attachment to uterine cell monolayers. Within this time period in culture the initial movement of underlying cells away from the blastocyst is seen. Following its attachment, spreading of the trophoblast involves extension of filopodia, lobopodia, and lamellipodia, all of which are very much more like typical tissue culture cells than trophoblast in vivo (Enders et al., 1981). Probably the most significant difference between attachment in vivo and in vitro is that initial adhesive interactions in vivo involve the apical surfaces of two epithelia (e.g., Schlafke and Enders, 1975), whereas in culture the adhesion and spreading of the trophoblast involves the basal surface (Wiley and Pedersen, 1977; Enders et al., 1981).

In an effort to circumvent the tendency of blastocysts to adhere to nonliving surfaces we have developed a culture method in which blastocysts and vesicles of uterine epithelium are placed together in hanging drop culture (Morris et al., 1982, 1983). The hanging drop culture method (see Figure 3 and text below), actually one of the oldest cell culture methods (Paul, 1975), offers several advantages for the study of implantation. The epithelium organizes with the former lumenal surface facing outward, and blastocysts go through a series of apical adhesive interactions in response to this epithelium similar to those in vivo.

Differentiation Of Blastocysts And Uterine Epithelium In Hanging Drop Culture

Mouse blastocysts attached to a culture dish are distorted by the trophoblastic outgrowth and clearly develop abnormally (Enders et al., 1981), but tissues do differentiate (Wiley and Pedersen, 1977) and recognizable organs form (Hsu, 1979). During the 6 days studied, the survival in hanging drops was essentially identical with that on culture dishes, and differentiation of the inner cell mass was similar up to at least the egg cylinder stage (Potter and Morris, 1985). Also, similar to cultures on plastic, differentiation was much slower than in vivo. The striking difference compared with blastocysts in vivo or on plastic was the manner of trophectoderm development. The cells migrated away from the inner cell mass, just as they do in embryos attached to plastic, but without a surface on which to spread they aggregated as a morular cluster at the abembryonic end (Figure 1). This clearly demonstrates that cells of the trophectoderm have different adhesive properties than those of the inner cell mass. Their attraction for each other is greater than their attraction to embryonic cells. The implication of this for undisturbed implantation in vivo is that the trophectoderm remains surrounding the inner cell mass, not because of its affinity for the inner cell mass, but because of its affinity for the tightly encasing uterine lumenal epithelial cells.

Vesicles of uterine epithelium were prepared by trypsinizing the uteri and then allowing them to recover for 20 minutes before subjecting them to shear (Morris et al., 1982, 1988a, 1988b; Morris and Potter, 1984). The released epithelial

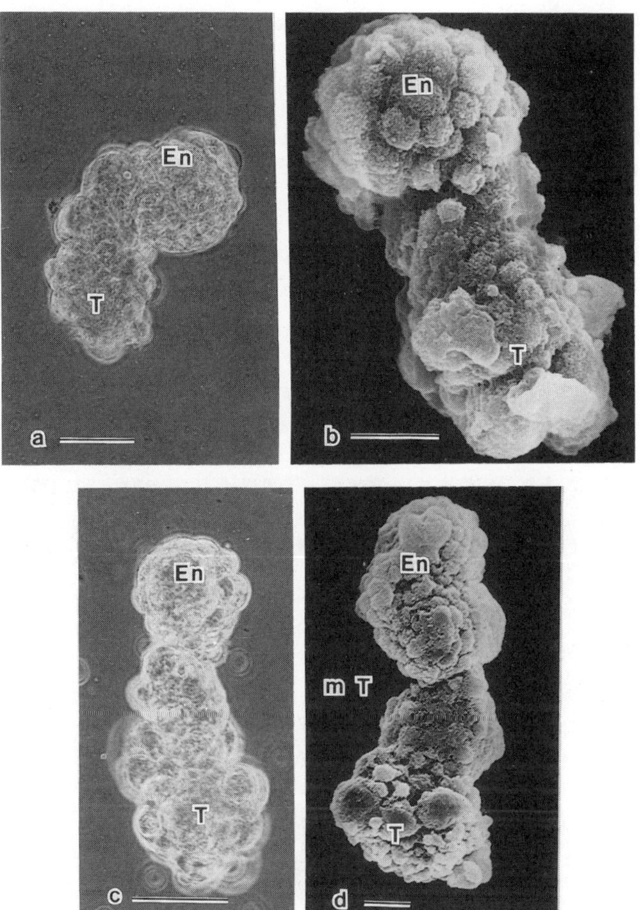

Figure 1. Differentiation of blastocysts in hanging drop culture. Blastocysts, 3.5 days post-coitum, were removed from their zonae pellucidae and cultured an additional four (a, b) or five (c, d) days in hanging drops. Shown here are living embryos photographed through the light microscope (a, c) and similar embryos following fixation and scanning electron microscopy (b, d). During this culture period, trophoblast cells clustered opposite the endoderm (En, as identified by its histological similarity to embryos in utero, Potter and Morris, 1984) and transformed into giant cells (T). Most polar trophectodermal cells migrated away from the endoderm to become clustered in an untransformed region in the middle (mT). Scale bar = 50 μm.

fragments were essentially free of basal lamina but otherwise in sheets of tightly cohering cells (Figure 2a). During overnight culture in hanging drops or on a gyrating shaker the fragments rolled up, always with the basal surface inward and the former lumenal surface outward (Figure 2b). By the next day many epithelial fragments had formed sealed vesicles that pumped medium and had swollen to become transparent (Figure 2c, d).

Interactions Between Blastocysts And Uterine Epithelial Vesicles

Single blastocysts and vesicles were paired in hanging drops on the cover of a petri dish (Figure 3a) and examined periodically for adhesion. In those instances where adhesion occurred (Figures 3b and 4a) all stages previously described for in vivo implantation in mice (Schlafke and Enders, 1975; Wimsatt, 1975) have been observed: intertwining of embryonic and maternal microvilli, close apposition of membranes (Figure 4d), and eventual engulfing of epithelial cells by trophectoderm (Morris et al., 1983). Judging by the formation of microextensions and lamellipodia by the trophectoderm, the blastocyst is the most active member of the pair (Figures 4c, d).

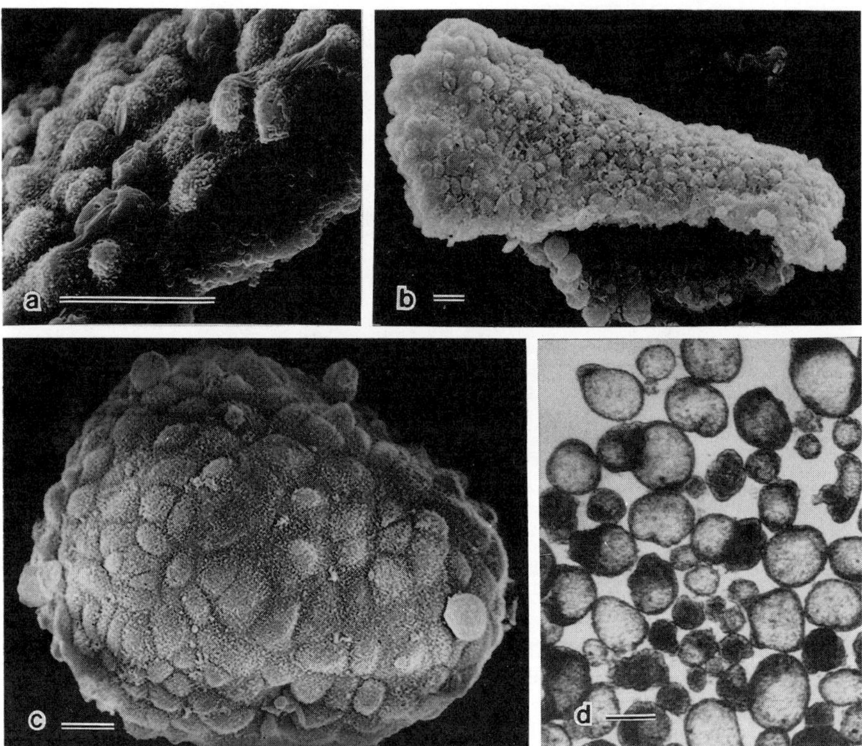

Figure 2. Development of vesicles from fragments of uterine epithelium. Epithelium was isolated from uteri of mice on day 4 of pregnancy (i.e., 3.5 days post-coitum) and cultured en masse on a shaker. Scanning electron micrographs were made initially (a), after 3 hours (b), and after 24 hours (c). Vesicles were selected for size and pooled (shown live in d). Note that the rolling of uterine fragments occurred with the apical surface facing outward and the basal surface inside the vesicle. Scale bar = 10 μm (a-c) 50 μm (d).

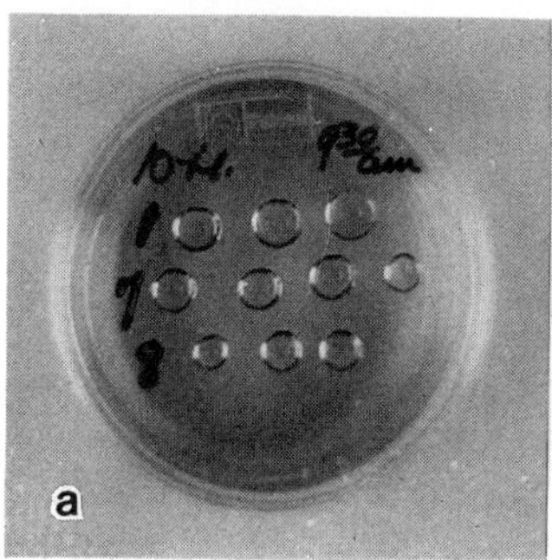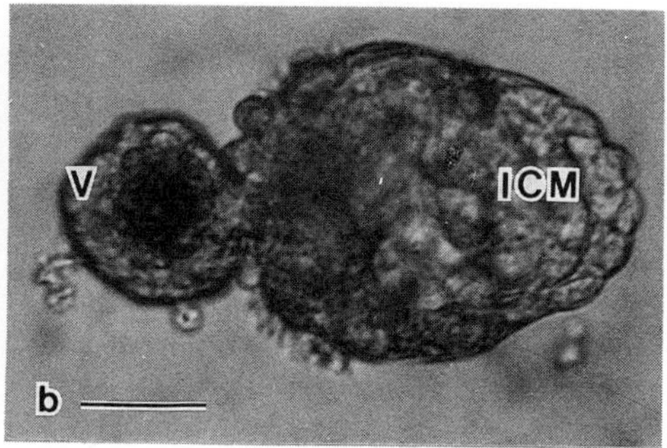

Figure 3. Combined blastocysts and epithelial vesicles in hanging drop culture. Fragments of uterine epithelium were allowed to form vesicles by culture overnight on a shaker as in Figure 2, while blastocysts from the same uteri were cultured on the surface of a bacteriological petri dish, a surface to which they do not readily attach. After this equilibration period single blastocysts and vesicles were paired in hanging drops on the under surface of the cover of a 35 mm petri dish (a). By this method individual pairs could be observed at successive time periods and scored for adhesion (b) simply by rapping the surface of the dish with a pencil and noting whether the agitated embryo and vesicle moved together. The frequency of adhesion increased on subsequent days as the epithelial vesicles began to deteriorate in the presence of the embryos. The site of adhesion appeared to be random, and in this case the position of the inner cell mass (ICM) away from the vesicle (V) indicates attachment to the abembryonic trophoblast . Scale bar = 50 µm (b).

Despite its promise as a system in which normal adhesion and initial invasiveness may be studied at the cellular level there is one major drawback to the hanging drop method that has not yet been overcome. Because the frequency of adhesion depends on the closeness of contact between blastocyst and the vesicle, such things as vibrations in the incubator, deposition of cell debris, and variations in vesicle shape all conspire to prevent the maintenance of close contact and, thus, reduce the frequency of adhesion (Morris et al., 1983). Overall frequencies of adhesion were approximately 10%, 25%, and 31% after 1, 2, or 3 days, respectively (Morris et al., 1983), but within these figures there was wide variability (0 to nearly 80%). The increase in adhesion frequency with time was in part because epithelium in the presence of blastocysts did not remain as vesicles but was in various degrees of collapse or fragmentation and was phagocytosed by the trophectoderm (Morris et al., 1983). The important conclusion from this work is that there is a natural resistance of the blastocyst to adhere to living cells and that the process of adhesion and implantation probably involves changes in the epithelium and/or blastocyst that permit close contact between plasma membranes of the adhering cells.

As an approach to determining the molecular mechanism of this resistance to blastocyst adhesion we have carried out an extensive series of experiments, in which hanging drop cultures were treated with hormones, inhibitors, and enzymes, in an effort to modify the frequency of adhesion.

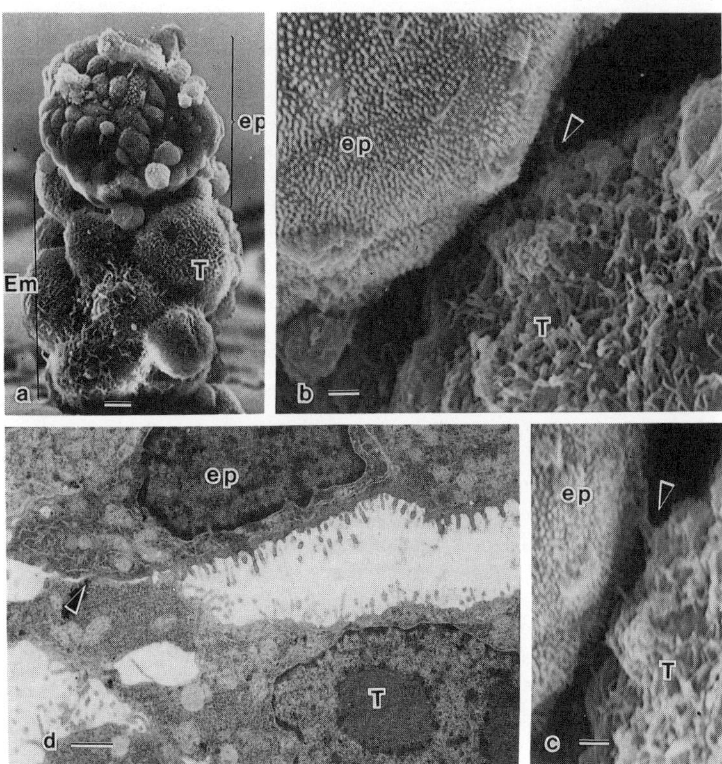

Figure 4. Details of the site of embryo-vesicle adhesion. A scanning electron micrograph shows an epithelial vesicle (ep) attached to a young cylinder-stage embryo (Em) after three days in culture (a). When the site of attachment is looked at with a higher magnification (b, c viewed from different angles) it is possible to see microvilli, microextensions, and lamellipodia (arrow) on the trophectoderm (T) surface making contact with the epithelium. In section (d) these protrusions from the trophectoderm surface typically were seen making contact with the epithelium over very limited areas (arrow). Scale bar = 10 µm, (a); 1 µm, (b, c, d).

MATERIALS AND METHODS

We have previously described the methods for preparation and culture of uterine epithelium (Morris and Potter, 1984), culture of blastocysts (Potter and Morris, 1985), and culture of vesicles of uterine epithelium together with blastocysts in hanging drops (Morris et al., 1982, 1983). Essentially, blastocysts were removed 3.5 days post-coitum (i.e., during day 4) and stripped of zonae by treatment with acid Tyrode's solution. Embryos were transferred through several drops of Spindle and Pedersen's basal medium (Spindle and Pedersen, 1973) containing 10% fetal bovine serum and placed singly in 10 to 15 µl drops of the same medium on the inside of a 35 mm petri dish cover. The cover was quickly flipped over and placed on the dish bottom to maintain humidity (Figure 3a). The cultures were prepared as rapidly as possible to prevent evaporation from the drops, which reached lethal levels within 3-5 minutes. The dishes were gassed with 5% CO_2/95%

air in sealed, humidified chambers (Billups-Rothenberg, Del Mar, California) for incubation.

The technique we found most reliable for preparing epithelial vesicles from pregnant mice involved cutting uteri into quarters and incubating them in crude pancreatin (25 mg/ml) supplemented with crystalline trypsin (2.5 mg/ml) in calcium- and magnesium-free Hanks' solution on a gyratory shaker at 50 rpm at 4°C for 40 minutes, similar to the method of McCormack and Glasser (1980). Following digestion the uteri were washed in complete serum-containing medium and incubated in this medium on the shaker at 37°C for 20 minutes (Morris and Potter, 1984). After incubation, they were drawn in and out of a wide-mouth (2-3 mm) pipet to shear off the epithelial fragments, which were then separated from contaminating cells by two or three 20-second washes with low-speed centrifugation. They were cultured overnight in a gyrating (50 rpm) Erlenmeyer flask to form vesicles. Pancreatin alone, but not trypsin alone, was nearly as effective as the combination of the two during the digestion step, suggesting that a minor component of the pancreatin was the active agent. Collagenase was not used because of the larger number of stromal cells released. The 20 minute incubation following digestion greatly enhanced the yield of epithelial fragments, possibly by inactivating residual proteases and allowing a partial recovery of tight junctions within the epithelium with only minimal recovery of attachment to the stroma. Longer incubation times reduced yields of epithelium, and shorter times released smaller epithelial fragments plus free cells, many of which were damaged.

Combined cultures usually were prepared the day following isolation of blastocysts and uterine epithelial fragments. The medium for cultures containing vesicles was the same as that for embryos alone except that it was supplemented with 1 mg/ml of extra glucose, 5 µg/ml of insulin, and 5 µg/ml DNase I. During overnight culture of epithelial fragments to form vesicles the blastocysts were maintained in bacteriological petri dishes, to which they either did not attach or attached only very loosely. Vesicles were washed in fresh medium and individually selected for size and transparency. The latter criterion helped assure the health of the epithelium and its freedom from contaminating stroma.

The experimental treatments used in this research involved, as indicated in Table II, the addition of a reagent to the culture medium or pretreatment of the uterine epithelial vesicles. In the latter case, vesicles were washed in bulk in balanced salt solution, and selected vesicles were incubated either in the inhibitor or enzyme. After the incubation period the vesicles were passaged through several drops of complete medium on the bottom of a dish and individually pipeted to hanging drops containing single blastocysts. Controls consisted of untreated epithelium in standard medium. Adhesion was monitored, as before (Morris et al., 1983), by direct observation of the hanging drops with a microscope and simply tapping the surface of the dish with a pencil, causing the drops to vibrate. If adhesion had occurred, the tissue and embryo were clearly seen to dance about together within the drop. Usually at least 10 drops were used for each experiment, with an equal number of control drops. The maximum number of drops per dish usually was about 15 to minimize the exposure time, and hence evaporation, during preparation. There was considerably more variability between experiments than within an experiment with respect to differences in adhesion frequency between experimental and control treatments. Therefore, adhesion frequency was normalized (Table II) as a percentage of adhesion in controls for that experiment. For example, an experiment in which 3/9 (33%) of the treated samples adhered and 5/10 (50%) of the control samples adhered is represented as 33/50, or 66% adhesions.

Surface charge was assayed by adhesion to DEAE-Sephadex (G-50) beads in complete medium with or without serum (Morris and Potter, 1984). The vesicles were prepared and treated as described above, and then 10 selected vesicles were placed in 0.25 ml of medium with equilibrated beads. Adhesion was scored when a vesicle-bead pair remained together during probing with a ball-tipped glass needle.

RESULTS

Test Of Potential Inhibitors And Promotors Of Adhesion

Table II summarizes the results of the experimental treatments of blastocysts and/or vesicles. They were of three types: (1) introduction of steroids and other serum factors into the medium on the assumption that these would leak transepithelially between the cells as the vesicles swelled and thus enter the cells normally at their basal surfaces; (2) culture in the presence of inhibitors of transcription (Actinomycin D), synthesis of aminosugar-containing oligosaccharides (6-diazo-5-oxo-L-norleucine, DON), or N-linked glycosylation of proteins (tunicamycin); or (3) pretreatment of the epithelial vesicles with enzymes to alter specifically the surface. Because of the wide variability inherent in this type of experiment, the data were partially normalized by expressing the adhesion frequency as a percent of the adhesion seen in the paired untreated controls (see Materials and Methods). Within limits dictated by the variability, our results suggest that (1) all enzymatic digestions attempted tended to inhibit adhesion, possibly because any loss of presumptive inhibitor molecules was counterbalanced by loss of adhesion sites. Digestion with chondroitinase resulted in apparent cell damage, so that by the second day the high frequency of attachment actually represented phagocytosis of damaged epithelial cells by the trophoblast. (2) Inhibitor treatments that resulted in increased frequency of adhesion, similarly were correlated with damage to the epithelium. (3) When steroids were added to the medium, there was an apparent enhancing effect that roughly doubled the average frequency of adhesion. We do not know whether the effect of steroids was directly on the epithelium, on the blastocyst, or indirectly through action on undetected contaminating stromal cells. There is evidence to indicate that stroma may have to be present as a mediator of the epithelial response to estradiol (e.g., Cunha et al., 1985), but recent work of Salamonsen et al. (1987) with sheep uterine epithelium in culture indicates that cells from estradiol-primed animals can respond to progesterone in culture by a significant increase in the synthesis of a specific protein. The same laboratory had previously demonstrated (Salamonsen et al., 1986) the synthesis of a different group of proteins in the epithelium in response to blastocyst-conditioned culture medium. The differences in response of sheep and mouse cells in culture may reflect their natural differences in hormonal requirements.

Table II

Influence Of Various Treatments On Adhesion Between
Blastocysts And Uterine Epithelial Vesicles

		Total number of:		Adhesion as a % of paired controls (Avg. % adhesion)	
Agent	Concentration	Expts	Embryos	1 day	2 days
1. Hydrocortisone	10^{-7}M	1	26	31 (20)	5 (4)[a]
Sephadex G-50	High MW	1	19	12 (5)	69 (53)
fraction of serum	Low MW	1	21	43 (19)[b,c]	89 (68)[b,c]
Unassayed serum: +progesterone and estradiol	10^{-7} M 10^{-8} M	6	119	220 (18)	130 (38)
Low progesterone serum[d] +progesterone and estradiol	10^{-7} M 10^{-8} M	7	96	180 (4)	58 (14)
Human chorionic gonadotropin	0.5 IU/ml	1	12	0	93 (25)
	2.5 IU/ml	3	63	240 (5)	290 (26)
2. Actinomycin D	0.5 μg/ml	2	17	270 (24)	0
	1.0 μg/ml	1	11	0[b]	
Diazoxynor-leucine	0.1 μg/ml	4	184	33 (0.4)	170 (20)
Tunicamycin	0.1 μg/ml	3	129	200 (3)	120 (15)
3. Chondroitinase ABC	0.5 unit/ml	1	15	0	290 (20)
Pronase	0.5 mg/ml	2	17	15 (6)	50 (3)
Trypsin	2.5 mg/ml, crystalline	4	57	72 (11)	65 (25)
Hyaluronidase, testicular (1 hr)	1 mg/ml	7	155	6 (7)	59 (18)
Neuraminidase Cl.perfringens	0.01 IU/ml	1	7	58 (25)[e]	0[b]
V. cholerae	0.02 IU/ml	4	32	78 (19)[e]	0[b]

The data express adhesion as a percentage of paired controls (actual percentage adhesion for the experiment in parentheses). Out of 26 experiments using 509 untreated controls, 16% (range 0 to 62%) had adhered by the first day and 30% (range 4-77%) by the second day. Controls often served for more than one of the experiments listed.

[a]Attachment accompanied by major phagocytosis
[b]Epithelium damaged or destroyed
[c]Blastocyst damaged or destroyed
[d]The serum contained 0.023 μg/ml of progesterone
[e]Single cells shed from epithelium attached in 1/2 to 2/3 of pairs

Examination Of Cell Surface Charge

A general impression from these and previous experiments (above and Morris et al., 1983) is that blastocysts preferentially attach to inert substrata and to epithelial vesicles that collapse and/or have begun to fragment into single cells. This suggests that the estradiol trigger for normal implantation in mice may involve the removal of a block to adhesion through the depression of synthetic activity, rather than by the induction of new adhesion molecules. This idea is not unique to us (Enders and Schlafke, 1979, and Chávez and Anderson, 1987), and it is supported by the observation that preimplantation uterine epithelium in several species has a thick, negatively charged coat which becomes thinner at the time of implantation (Enders and Schlafke, 1979; Hewitt et al., 1979; Enders et al., 1980), although the thinning at implantation is not as evident in rodents as in some other species (Enders and Schlafke, 1974, 1979). There may also be a negatively charged coat on the blastocyst that is reduced at the time of implantation (Nilsson et al., 1973; Jenkinson and Searle, 1977; Nilsson and Hjerten, 1982).

To examine charged groups that could directly interact with an adjacent tissue, binding of blastocysts and uterine epithelial vesicles to ion exchange beads (DEAE-Sephadex) was investigated. In previous work (Morris and Potter, 1984) increasing concentrations of the polyanion dextran sulfate competitively inhibited cellular attachment to the beads was, and thus an estimate of the relative charge differences was provided. Epithelial vesicles had approximately a four-fold higher charge density than blastocysts and the charge decreased at least 50% between 3.5 and 4.5 days of pregnancy, the time when implantation normally is initiated. In an attempt to identify the molecules responsible for the surface charge the cell surface was digested with specific enzymes that would remove charged groups from the cell surface. Only neuraminidase (sialidase) had an unambiguous ability to abolish binding between epithelial vesicles and DEAE beads, but it had virtually no effect on blastocysts (not shown here, see Morris and Potter, 1984), suggesting that they bound to the beads by mechanisms not requiring charge interaction. Hydrolases directed against glycosaminoglycans had no effect on the binding (Table III), supporting the observation that uterine epithelial proteoglycans are predominantly on the basolateral surface (Morris et al., 1988a, b). Proteases also inhibited binding, but the effect was inconsistent, suggesting that terminal sialic acids on glycolipids as well as those on glycoproteins need to be considered as possible major contributors to the net negative charge. This was also indicated by experiments (not shown) in which glutaraldehyde-fixed vesicles were tested for their ability to bind DEAE beads. Fixation either masked or immobilized charged groups so that binding was very weak compared with unfixed vesicles and was, therefore, impossible to quantify (only slight probing of beads with a needle was sufficient to dislodge a vesicle). This weak binding was not modified following digestion with neuraminidase, trypsin, or Pronase. On the other hand, extraction of the fixed vesicles with nonionic detergent (1% Nonidet P-40 in saline for 5 minutes) modified the cell surface sufficiently, apparently by exposing strongly anionic groups, that 100% of the vesicles bound tightly to the beads.

Table III

Influence Of Enzyme Digestion On Attachment Of
Epithelial Vesicles To DEAE-Sephadex Beads

	Without Serum	With 10% FBS
Control	9/9, 10/10, 10/10	9/9, 10/10, 10/10
Vibrio cholerae neuraminidase 0.02 IU/ml, 10 minutes	0/5, 0/11	1/8, 0/7
Chondroitinase ABC 0.01 IU/ml, 10 minutes	10/10	10/10
Heparitinase 12 mIU/ml, 10 minutes	10/10	10/10
Heparinase 12 mIU/ml, 10 minutes	10/10	10/10
Crystalline bovine trypsin 2.5 mg/ml, 5 minutes	5/5, 0/9	8/8, 0/6
Pronase 1.0 mg/ml, 10 minutes	1/8	0/7

Frequencies are shown as a fraction of the actual number of epithelial vesicles attached per total present for each experiment. Each pair of numbers is a separate experiment.

DISCUSSION

Enders and coworkers (Schlafke and Enders 1975; Enders et al., 1981) have emphasized that one important difference between blastocysts on cell monolayers and in vivo is that in the latter case they are surrounded on all sides by uterine epithelium that is very closely apposed to the blastocyst surface. It is also clear that blastocysts do not readily adhere to the apical surfaces of living cells but do adhere to a variety of artificial surfaces (Glass et al., 1980; Morris et al., 1983). Thus, there appear to be two requirements that must be met in any in vitro implantation system that seeks to understand cellular interactions: (1) close physical contact between the apical surface of trophoblast and uterine epithelium and (2) isolation from competing adhesive surfaces. Coculture of blastocysts and uterine epithelial cells in suspension on a gyrating shaker fulfills the second requirement, but apparently it does not fulfill the first, because under these conditions no cells adhere to blastocysts (although they aggregate with each other) unless they are forced to attach by cross-bridging agents (Morris et al., 1983). A compromise solution has been to culture blastocysts in hanging drops together with one or two vesicles of epithelium. The vesicles consist of fragments of epithelium that have rolled up to become little blisters with the former luminal surface facing outward. Under these conditions there is no competing substratum except for the possibility of denatured serum proteins at the air-liquid interface of the hanging drop (Giaever and Kesse, 1983). The requirement for juxtaposition of blastocyst and epithelial apical surfaces is only partially met, however, because the surfaces are in contact at only one point, and the amount of force is only that provided by the gentle curvature of the hanging drop. On the other hand, having only a single point of contact is an advantage for rapid and unambiguous determination of the site of adhesion, and the system enables simple monitoring of specific blastocyst-epithelial pairs.

A distinct theoretical disadvantage to using epithelium formed into vesicles like this is that the basal surface, through which an epithelium normally interacts with its vascular supply, is sealed off from the medium. Practically, however, this does not seem to be critical for survival of the epithelium. There apparently is active pumping across the cells vectorially in a reverse direction in culture, resulting in the continued swelling of some vesicles. We have maintained vesicles in hanging drops for up to a week, during which time vesicles that were punctured with a needle had resealed and were reinflated by the following day. A theoretically more acceptable culture system would be one in which the basal surface had direct access to the culture medium, and some of the newer advances in epithelial cell culture may make such a system possible. For example, uterine epithelia have been grown on floating collagen gels (Sengupta et al., 1986) or over filters on reconstructed basement membrane (Glasser et al., 1987). Under such conditions the cells appear to be structurally very similar to those in vivo, and they transport in a basal to apical direction. However, it has yet to be shown whether blastocysts will attach and interact normally with the cells in preference to the culture surface.

These observations support the notion that implantation is blocked prior to estradiol induction by a negatively charged coat on the uterine epithelium that repels the coat on the blastocyst. Most evidence indicates that the coat on the uterine surface (Hewitt et al., 1979; Enders et al., 1980; Morris and Potter, 1984), and possibly on the blastocyst surface (Jenkinson and Searly, 1977; Nilsson et al., 1973; Nilsson and Hjerten, 1982), somehow must either be lost or internalized for adhesion to occur. The fact that estradiol induces a turnover of heparan sulfate proteoglycan on the uterine epithelial basolateral surface (Morris et al., 1988a, b) is consistent with the idea that the entire plasma membrane may, in fact, be turning over more rapidly at the time of implantation. One possible result of this induced turnover may be the reduction in surface charge, which permits the embryo and uterine epithelium to approach close enough to permit molecular interaction either by specific adhesion molecules on the surface or by nonspecific van der Waals interactions.

SUMMARY

Trophectodermal cells of developing blastocysts tend to be very adhesive for a variety of culture surfaces, including glass and plastic. Therefore, if the goal of a culture system is to examine the initial adhesive interactions between the trophectoderm and uterine epithelial apical surface, it is essential to have no competing surfaces present. We have avoided this problem by investigating adhesion between single mouse blastocysts paired with single vesicles of uterine epithelium within 10-15 µl hanging drops of medium. (1) Within one to two days in hanging drop culture, adhesive interactions have been observed, starting initially with an intertwining of microvilli on the two surfaces and leading to close membrane-to-membrane apposition and adhesion over restricted regions. The frequency of adhesion is widely variable, owing most likely to difficulties of maintaining a constant degree of contact between the embryo and epithelium. (2) Attempts to modify the frequency of adhesion in hanging drops by various additives to the medium gave results that were too variable to permit definitive conclusions. The impression was that enzyme or inhibitor treatments that appeared to enhance adhesion also damaged the epithelium, and may have resulted in enhanced phagocytotic uptake by trophectoderm of epithelially-derived debris. Estradiol with progesterone may have had a slight stimulatory effect on adhesion frequency, but this needs further study, especially with respect to the possibility that contaminating stromal cells may have been responsible for the instances of increased adhesion. (3) Experiments with positively charged beads (DEAE-Sephadex) reveal a strong negative charge on the surface of epithelial vesicles prepared from 3.5-day pregnant mice and a strong decline in charge on vesicles prepared from mice a day later. The charge is predominantly due to sialic acid residues on oligosaccharide chains that probably are associated with both glycoproteins and glycolipids. (4) Other work from this laboratory (Morris et al., 1988a, b) indicates that estradiol stimulates a turnover and degradation of cell-surface heparin sulfate proteoglycan in mouse uterine epithelial cells. The results taken together support the conclusion that a blastocyst normally is prevented from attaching prematurely to the apical epithelial surface by a negatively charged glycocalyx and that the charge is reduced at the time of hormonally induced attachment.

ACKNOWLEDGEMENTS

The authors wish to thank Georgeen Gaza-Bulseco and Lin Chai, who have helped with the preparation of the manuscript. This work was supported by a grant (HD-19530) from the National Institutes of Health, U.S. Public Health Service.

REFERENCES

Böving, B.G. (1971) Biomechanics of implantation. In: *The Biology Of The Blastocyst*, (ed.), R.J. Blandau, Chicago: University Chicago Press, pp. 423-442.

Chávez, D.J. and Anderson, T.L. (1985) The glycocalyx of the mouse uterine luminal epithelium during estrus, early pregnancy, the peri-implantation period, and delayed implantation. I. Acquisition of Ricinus communis I binding sites during pregnancy. *Biol. Reprod.* 32, 1135-1142.

Chávez, D.J. and Van Blerkom, J. (1981) In vitro attachment and outgrowth of mouse trophectoderm. In: *Cellular and Molecular Aspects of Implantation*, (eds.), S.R. Glasser and D.W. Bullock, New York: Plenum Publ. Corp., pp. 457-460.

Cunha, G.R., Bigsby, R.M., Cooke, P.S., and Sugimura, Y. (1985) Stromal-epithelial interactions in adult organs. *Cell Differentiation* 17, 137-148.

Dziadek, M., Fujiwara, S., Paulsson, M., and Timpl, R. (1985) Immunological characterization of basement membrane types of heparan sulfate proteoglycan. *EMBO J.* 4, 905-912.

Enders, A.C., Chávez, D.J., and Schlafke, S. (1981) Comparison of implantation in utero and in vitro. In: *Cellular and Molecular Aspects of Implantation.* (ed.) S.R. Glasser and D.W. Bullock, New York: Plenum Publ.Corp., pp. 365-382.

Enders, A.C. and Schlafke, S. (1974) Surface coats of the mouse blastocyst and uterus during the preimplantation period. *Anat. Rec.* 180, 31-45.

Enders, A.C. and Schlafke, S. (1979) Comparative aspects of blastocyst-endometrial interactions at implantation. In: *Maternal Recognition of Pregnancy*, Ciba Fdn. Ser. 64 (new series), pp. 3-32.

Enders, A.C., Schlafke, S., and Welsh, A.O. (1980) Trophoblastic and uterine luminal epithelial surfaces at the time of blastocyst adhesion in the rat. *Am. J. Anat.* 159, 59-72.

Farach, M.C., Tang, J.P., Decker, G.L., and Carson, D.D. (1987) Heparin/heparan sulfate is involved in attachment and spreading of mouse embryos in vitro. *Dev. Biol.* 123, 401-410.

Giaever, I. and Keese, C.R. (1983) Behavior of cells at fluid interfaces. *Proc. Natl. Acad. Sci. USA* 80, 219-222.

Glass, R.H., Aggeler, J., Spindle, A., Pedersen, R.A., and Werb, Z. (1983) Degradation of extracellular matrix by mouse trophoblast outgrowths: A model for implantation. *J. Cell Biol.* 96, 1108-1116.

Glass, R.H., Spindle, A.I., and Pedersen, R.A. (1979) Mouse embryo attachment to substratum and interaction of trophoblast with cultured cells. *J. Exp. Zool.* 208, 327-335.

Glass, R.H., Spindle, A.I., and Pedersen, R.A. (1980) The free surface of mouse trophoblast in culture is non-adhesive for other cells. *J. Reprod. Fertil.* 59, 403-407.

Glasser, S.R., Lampelo, S., Munir, M.I., and Julian, J.A. (1987) Expression of desmin, laminin and fibronectin during in situ differentiation (decidualization) of rat uterine stromal cells. *Differentiation* 35, 132-142.

Glenister, T.W. (1961) Organ culture as a new method for studying the implantation of mammalian blastocysts. *Proc. Roy. Soc. Lond. (Biol.)* 154, 428-431.

Glenister, T.W. (1966) Nidation processes in organ culture. *Int. J. Fertil.* 11, 412-423.

Glenister, T.W. (1967) Organ culture and its combination with electron microscopy of the nidation process. In: *Fertility and Sterility*, (ed.) B. Westsin and N. Wiqvist, Ser. 133, Amsterdam: Exerpta Medica Fdn., pp. 385-394.

Glenister, T.W. (1971) Methods for studying ovoimplantation and early embryo-placental development in vitro. In: *Methods in Mammalian Embryology*, (ed.) J.C. Daniel Jr., San Francisco: W.H. Freeman and Co., pp. 320-333.

Grant, P.S. (1973a) The effect of progesterone and oestradiol on blastocysts cultured within the lumina of immature mouse uteri. *J. Embryol. Expt. Morphol.* 29, 617-638.

Grant, P.S. (1973b) The effect of progesterone and oestradiol on immature mouse uteri maintained as organ cultures. *J. Endocrinol.* 57, 171-174.

Grant, P.S., Ljungkvist, I., and Nilsson, O. (1975) The hormonal control and morphology of blastocyst invasion in the mouse uterus in vitro. *J. Embryol. Exp. Morph.* 34, 299-310.

Heifetz, A., Lennarz, W.J., Libbus, B., and Hsu, Y.-C. (1980) Synthesis of glycoconjugates during the development of mouse embryos in vitro. *Dev. Biol.* 80, 398-408.

Hewitt, K., Beer, A.E., and Grinnell, F. (1979) Disappearance of anionic sites from the surface of the rat endometrial epithelium at the time of blastocyst implantation. *Biol. Reprod.* 21, 691-707.

Hohn, H.-P., Donner, A., and Denker, H.-W. (1987) Specific receptivity of rabbit endometrium for attachment of trophoblast but not of tumor cells in vitro. *J. Cell Biol.* 105, 306a.

Hsu, Y.-C. (1979) In vitro development of individually cultured whole mouse embryos from blastocyst to early somite stage. *Dev. Biol.* 68, 453-461.

Jenkinson, E.J. and Searle, R.F. (1977) Cell surface changes on the mouse blastocyst at implantation. *Exp. Cell Res.* 106, 386-390.

Karst, W. and Merker, H.-J. (1988) The differentiation behaviour of MDCK cells grown on matrix components and in collagen gels. *Cell Differentiation* 22, 211-224.

McCormack, S.A. and Glasser, S.R. (1980) Differential response of individual uterine cell types from immature rats treated with estradiol. *Endocrinol.* 163, 1634-1649.

Morris, J.E. and Potter, S.W. (1984) A comparison of developmental changes in surface charge in mouse blastocysts and uterine epithelium using DEAE beads and dextran sulfate in vitro. *Dev. Biol.* 103, 190-199.

Morris, J.E., Potter, S.W., and Buckley, P.M. (1982) Mouse embryos and uterine epithelia show adhesive interactions in culture. *J. Exp. Zool.* 222, 195-198.

Morris, J.E., Potter, S.W., and Gaza-Bulseco, G. (1988a) Estradiol induces an accumulation of free heparan sulfate glycosaminoglycan chains in uterine epithelium. *Endocrinol.* 122, 242-253.

Morris, J.E., Potter, S.W., and Gaza-Bulseco, G. (1988b) Estradiol-stimulated turnover of heparan sulfate proteoglycan in mouse uterine epithelium. *J. Biol. Chem.* 263, 4712-4718.

Morris, J.E., Potter, S.W., Rynd, L.S., and Buckley, P.M. (1983) Adhesion of mouse blastocysts to uterine epithelium in culture: A requirement for mutual surface interactions. *J. Exp. Zool.* 225, 467-479.

Nilsson, B.O. and Hjerten, S. (1982) Electrophoretic quantification of the changes in the average net negative surface charge density of mouse blastocysts implanting in vivo and in vitro. *Biol. Reprod.* 27, 485-493.

Nilsson, O., Lindqvist, I., and Ronquist, G. (1973) Decreased surface charge of mouse blastocysts at implantation. *Exp. Cell Res.* 83, 421-423.

Paul, J. (1975) *Cell and Tissue Culture*, 5th ed., Edinburgh: Churchill Livingstone, pp. 1-2.

Potter, S.W. and Morris, J.E. (1985) Development of mouse embryos in hanging drop culture. *Anat. Rec.* 211, 48-56.

Salamonsen, L.A., Doughton, B.W., and Findlay, J.K. (1986) The effects of the preimplantation blastocyst in vivo and in vitro on protein synthesis and secretion by cultured epithelial cells from sheep endometrium. *Endocrinol.* 119, 622-628.

Salamonsen, L.A., Healy, D.L., and Findlay, J.K. (1987) Progesterone in vitro stimulates secretion of a specific protein by ovine epithelial endometrial cells. *J. Steroid. Biochem.* 28, 285-288.

Salomon, D.S. and Sherman, M.I. (1975) Implantation and invasiveness of mouse blastocysts on uterine monolayers. *Exp. Cell Res.* 90, 261-268.

Schlafke, S. and Enders, A.C. (1975) Cellular basis of interaction between trophoblast and uterus at implantation. *Biol. Reprod.* 12, 41-65.

Sengupta, J., Given, R.L., Carey, J.B., and Weitlauf, H.M. (1986) Primary culture of mouse endometrium on floating collagen gels: A potential in vitro model for implantation. *Ann. N.Y. Acad. Sci.* 476, 75-94.

Sherman, M.I. (1978) Implantation of mouse blastocyst in vitro. In: *Methods in Mammalian Reproduction*, (ed.) J.C. Daniel, Jr. New York: Academic Press, pp. 247-257.

Sherman, M.I. and Wudl, L.R. (1976) The implanting mouse blastocyst. In: *The Cell Surface in Animal Embryogenesis and Development*, (ed.) G. Poste and G.L. Nicolson, Amsterdam: North-Holland Publ., pp. 81-125.

Spindle, A. and Pedersen, R.A. (1973) Hatching, attachment, and outgrowth of mouse blastocysts in vitro: Fixed nitrogen requirements. *J. Exp. Zool.* 186, 305-318.

Tang, J.-P., Julian, J., Glasser, S.R., and Carson, D.D. (1987) Heparan sulfate proteoglycan synthesis and metabolism by mouse uterine epithelial cells cultured in vitro. *J. Biol. Chem.* 262, 12832-12842.

Wiley, L.M. and Pedersen, R.A. (1977) Morphology of mouse egg cylinder development in vitro: A light and electron microscopy study. *J. Exp. Zool.* 200, 389-402.

Wimsatt, W.A. (1975) Some comparative aspects of implantation. *Biol. Reprod.* 12, 1-40.

A THREE-DIMENSIONAL ORGAN CULTURE MODEL FOR THE STUDY OF IMPLANTATION OF RABBIT BLASTOCYST IN VITRO

Hans-Peter Hohn[1] and Hans-Werner Denker[1]

Institut für Anatomie der RWTH Aachen
Melatener Strasse 211
D-5100 Aachen, Federal Republic of Germany

INTRODUCTION

Implantation of the mammalian embryo in the uterine wall involves the attachment of the trophoblast to the uterine epithelial cells followed, in most species, by invasion of the trophoblast. The present state of knowledge about implantation results from morphological and histochemical examinations of embryos in situ and from limited experimental studies performed predominantly in vivo (c.f. Denker, 1977, 1983; Schlafke and Enders, 1975). Implantation is thought to start when the invasive phase of the trophoblast coincides with a "receptive state" of the endometrium (Psychoyos and Casimiri, 1980). While there is good evidence that endometrial receptivity is controlled by estrogens and progesterone (Psychoyos, 1976), the molecular interactions of the trophoblast and the endometrium as well as the changes in their cell physiological state that are required, remain largely unknown.

The analysis of the molecular basis of implantation would be facilitated by an in vitro model which would allow better experimental control while eliminating many of the ill-defined influences of maternal homeostasis. A number of model systems have been described in the literature and, as in vitro systems in general, have their limitations (see Discussion). These systems include: 1) the "blastocyst (trophoblast) outgrowth" model and its variants as very simplified systems which allow to study blastocyst attachment to matrix materials or to monolayers of uterine or other cells; 2) vesicles derived from uterine epithelium offering a similar potential for experimental studies; and 3) three-dimensional models comprising both the uterine epithelium and the stroma, where attachment as well as invasion may be investigated. The current investigation is developing a new model of the latter type with organotypic culture of uterine tissue. It is based on former studies establishing culture conditions under which the progestational differentiation of endometrial explants continues to a certain extent (Hohn et al., 1984, 1898). When blastocysts are cocultured with these endometrial explants their trophoblast shows a degree of differentiation that is morphologically comparable to the situation in vivo (Hohn et al., submitted). An important element of the endometrial organ culture model is that the explants are subject to a preculture phase resulting in the regeneration of a complete epithelial lining all around the fragments. During subsequent confrontation with blastocysts this prevents atypical direct contact of the trophoblast with uterine stroma exposed during explantation. Endometrial tissue and blastocysts are taken from rabbits because this species has been used

[1]Present Address: Institut für Anatomie, Universitätsklinikum, Hufelandstrasse 55, D-4300 Essen 1, FR Germany

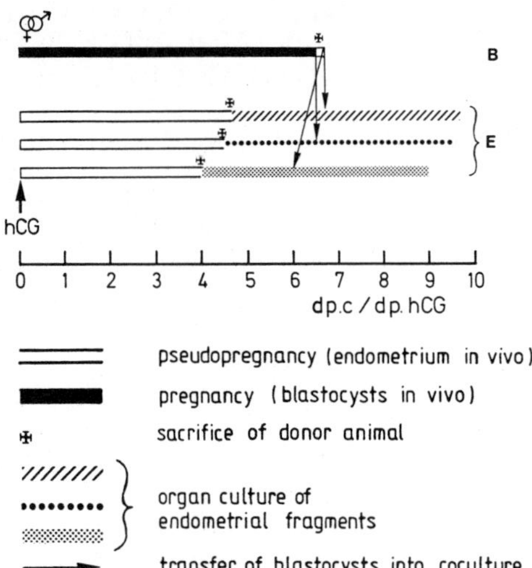

Figure 1. Experimental design: time schedule. The three different lines for organ culture represent three different time schedules used. d = days, p.c. = post coitum, p. hCG = after injection of hCG; B = donor animals for blastocysts, E = donor animals for endometrium.

extensively as a model for morphological and biochemical investigations of implantation (Denker, 1970, 1977, 1983).

MATERIALS AND METHODS

Preparation Of Endometrium And Blastocysts

Sexually mature rabbits were kept in single cages in air conditioned quarters under a light/dark cycle of 12/12 hours with a standard pellet feed and water ad libitum (Altromin, Lage, FRG). In order to obtain uteri and blastocysts, animals were killed by stunning and exsanguination.

On day 0 of an experiment (for experimental design see Figure 1), one group of females was treated with a single i.v. injection of 75 IU human chorionic gonadotropin (hCG; Prolan®, Leverkusen, FRG) to induce pseudopregnancy. In a second group, females were mated with two males each and ovulation was stimulated by the i.v. injection of 75 IU hCG (pregnancy). After 4 days and 16 hours, endometrial fragments of about 1 mm in diameter were explanted from the mesometrial and the antimesometrial portion of the uterus and transferred to an organ culture system as described previously (Hohn et al., 1984; Hohn et al., 1989). The cultivation was performed in 25 ml Erlenmeyer flasks containing 10 ml medium (cMEM, see below) on a gyratory shaker at 100 rpm at 37°C. The flasks were gassed continuously with 95% air, 5% CO_2. The endometrial fragments were precultured in this way until the coculture with blastocysts was initiated. During this phase, debris was eliminated from the surface, and the epithelium regenerated all around the fragment, which thus prevented direct attachment of the trophoblast to endometrial stroma.

Blastocysts were obtained from pregnant animals at 6 days and 12 hours post coitum (p.c.) or at 6 days and 16 hours p.c.. For this purpose the segments containing blastocysts were separated from uteri after explantation and blastocysts were recovered by carefully tearing apart the antimesometrial part of the uterine wall with fine surgical forceps. The blastocysts were collected in culture medium and transferred to and maintained in different types of co-culture with precultured endometrial fragments (Figure 2) for up to 3 days in 12 ml cMEM in 25 ml Erlenmeyer flasks. The cultures were incubated generally on a gyratory shaker at 37°C in a humidified atmosphere with 95% CO_2. For a limited series of experiments, the blastocyst coverings were removed at the start of the experiment, with fine surgical forceps.

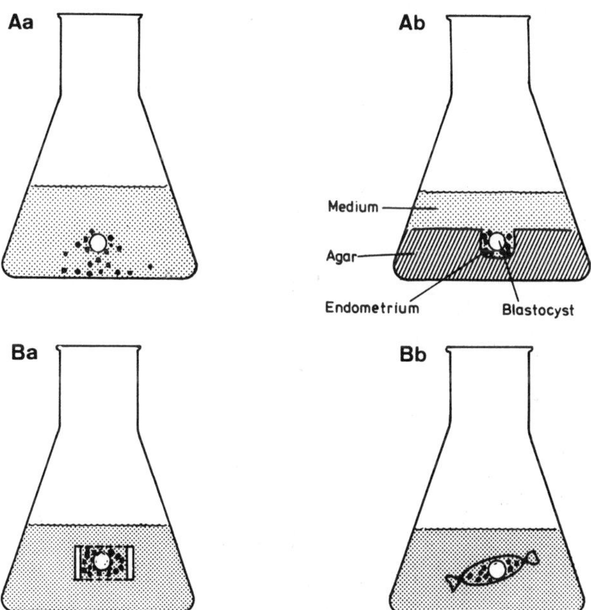

Figure 2. Culture systems used. Endometrial fragments were precultured in Erlenmeyer flasks on a gyratory shaker. Cocultures were performed in the same device with blastocysts (large open circle) and endometrial fragments (small filled circles) either floating free in the culture medium (Aa) or after placing both of them in a well of 8 mm in diameter that was formed in a layer of agar, 1 cm thick (Ab). In order to keep them in permanent close contact they were enclosed either in short plastic tubes (8 x 8 mm) lined by dialysis membrane on both sides (Ba) or in dialysis tubings (Bb).

Table 1

	Total Number of Blastocysts in Co-Culture	Blastocysts Collapsed after Co-Culture (%)		Blastocysts Expanded after Co-Culture (%)		Expanded Blastocysts with Adhering Endom. Fragments (% of expanded Blastocysts)*	
Open System Aa[1]	23	14	(60.9)	9	(39.1)	0	(0)
Open System Ab[1]	18	11	(61.2)	7	(38.8)	0	(0)
Chamber System Ba[1]	34	24	(70.6)	10	(29.4)	8	(80.0)
Chamber System Bb[1]	40	29	(72.5)	11	(27.5)	9	(81.8)
Chamber System Bb with reduced chamber pressure[2]	8	4	(50)	4	(50)	1	(25)
Confrontations in the Presence and Absence of FBS[3]							
Medium with FBS	21	15	(71.4)	6	(28.6)	5	(83.3)
Serum-Free Medium	19	13	(68.4)	6	(31.6)	6	(100)
Confrontations with Endometrial Fragments of Different Origin[4]							
Mesometrial Endom.	19	14	(73.76)	5	(26.3)	4	(80.0)
Antimesom. Endom.	21	15	(71.4)	6	(28.6)	5	(83.3)
Confrontations with Endometrium Precultured for Different Periods[5]							
6 d	13	9	(69.2)	4	(30.8)	3	(75)
6 d 12 h	21	15	(71.4)	6	(28.6)	5	(83.3)
6 d 16 h	26	18	(69.2)	8	(30.8)	7	(87.5)
Confrontations of Blastocysts with or without Blastocyst Coverings[6]							
Coverings Removed	17	15	(82.4)	2	(17.6)	2	(100)
Coverings Intact	21	15	(71.4)	6	(28.6)	5	(83.3)
Influence of Stage of Blastocysts at Start of Co-Culture[7]							
6 d 12 h	21	15	(71.4)	6	(28.6)	5	(83.3)
6 d 16 h	26	18	(69.2)	8	(30.8)	7	(87.5)

Co-Culture Systems

In order to achieve implantation in vitro blastocysts and endometrium were combined in two different types of co-culture (Figure 2), a system with "random" collision of both partners on one hand and a system where they were kept in a close contact throughout the experiment on the other hand. For each type, two different experimental designs were tested with the following arrangements:

A. Open Systems

a) The endometrial fragments and the blastocyst remained free floating in the medium. The agitation was 0 or 100 rpm (Figure 2Aa).
b) Both were placed in a 7 mm diameter depression in a layer of 9-10mm thick agar (2% in MEM) (Figure 2Ab) on the bottom of an Erlenmeyer flask. The shaker was set at 30 rpm.

Table 1 Legend

1) The different systems of confrontation were compared using mesometrial and antimesometrial endometrial fragments in cMEM, blastocysts with intact coverings, and synchronous co-culture starting at 6 days and 12 hours.

2) Co-culture as in 1) starting at 6 days and 12 hours while some space was left free around blastocysts and uterine fragments.

3) Co-culture in chamber system Bb, antimesometrial explants only, other details as in 1).

4) Conditions for confrontation: system Bb, medium, etc. as in 1).

5) Uterine fragments explanted at different stages were precultured for two days until 6 days, 6 days and 12 hours, or 6 days and 16 hours and were then confronted with blastocysts (6 days and 12 hours) with intact coverings in dialysis tubings (system Bb) in cMEM.

6) Synchronous confrontation culture starting 6 days and 12 hours in system Bb using cMEM.

7) Synchronous confrontation as in 6) with intact blastocyst coverings starting either at 6 days and 12 hours or 6 days and 16 hours.

*) These values represent all blastocysts with adhering endometrial fragments disregarding the state of preservation of the latter.

B. Chamber Systems

a) Endometrial fragments and blastocysts were kept together in short plastic tubes of 8 x 8 mm which were filled with medium and closed with dialysis membranes (Visking, 0.1 mm, cutoff 16,000 dalton) on both sides (Figure 2Ba). Tubes were incubated in 12 ml of medium under the conditions described above at 60 rpm.
b) Blastocysts and endometrial fragments were enclosed in dialysis tubing (Visking, 6 mm in diameter, 0.2 mm, cutoff 16,000 dalton) filled with culture medium (Figure 2Bb). The tubing was sealed with thread at each side. The incubation was performed as described in Ba).

The main purpose of the present study was to establish the system so that numerous variations had to be tried within these groups resulting in numbers of blastocysts and ratios of implantation being too limited for statistical calculations for each group. In addition to the four different confrontation systems described above, variations included the use of: 1a) serum-containing or 1b) serum-free medium; co-culture with 2a) intact or 2b) removed blastocyst coverings; confrontation of blastocysts explanted on 6 days and 12 hours p.c. with endometrial fragments precultured for two days to the state of 3a) 6 days or 3b) 6 days and 12 hours p.i. hCG. For actual numbers, see Table 1.

Culture Media

Eagle's minimal essential medium supplemented with 100 IU/ml penicillin, 100 μg/ml streptomycin and non-essential amino acids (MEM-REGA-1; Gibco, Karlsruhe, FRG) was completed immediately before use with 10% fetal bovine serum (FBS-309, Gibco), 3.4 mmole/l L-glutamine (Serva, Heidelberg, FRG) and 3.2 x 10^{-8} M progesterone (Merck, Darmstadt, FRG). This will be referred to as cMEM (completed MEM). The serum-free medium used was Wissler BM 86 (Boehringer, Mannheim) supplemented with glutamine, antibiotics, and progesterone at the same concentrations as cMEM, and in addition with 3.7 x 10^{-11}M estradiol-17-β, and 1% SerXtend (NEN, DuPont) as a serum substitute. The media were changed daily.

Histology

After culture, endometrial fragments, blastocysts, and blastocysts with adhering endometrial fragments were transferred to 2.5% (v/v) glutaraldehyde in 0.1 M cacodylate buffer (pH 7.4) for fixation overnight. During fixation, microphotographs were taken of blastocysts and blastocysts with adhering endometrial fragments. In order to reduce the shear forces during processing for histology, the areas of blastocysts, where endometrial fragments were attached to the trophoblast, were cut off the remaining parts of the blastocyst with microsurgical scissors while the specimens were still in the fixation solution. These fragments were embedded in 1.5% agar (w/v, in phosphate buffered saline) after the agar solution had cooled to 40°C. The fixation was continued overnight. Then the tissue was postfixed in 2% (w/v) osmium tetroxide, dehydrated in a graded series of ethanol and embedded in araldite. Semithin sections (0.75 μm) were stained with 0.5% toluidine blue and 0.5% (w/v) pyronin in 0.5% (w/v) sodium borate. Semithin sections were examined and photographed in a Zeiss Photomicroscope II. Thin sections (silver-gold) contrasted with lead citrate and uranyl acetate were examined in a Zeiss EM 10 electron microscope.

RESULTS

Influences Of The Time Of Blastocyst Explantation

A problem with this system was caused by the fact that rabbit blastocysts expand considerably during the periimplantation phase and become very fragile. As a result, some are lost already in their loss during explantation. The explantation of intact blastocysts was more difficult when the dissolution of the blastocyst coverings was advanced, i.e., at 6 days and 16 hours p.c. as compared to 6 days and 12 hours p.c. The rate of implantation was not influenced noticeably by the time of explantation (Table 1).

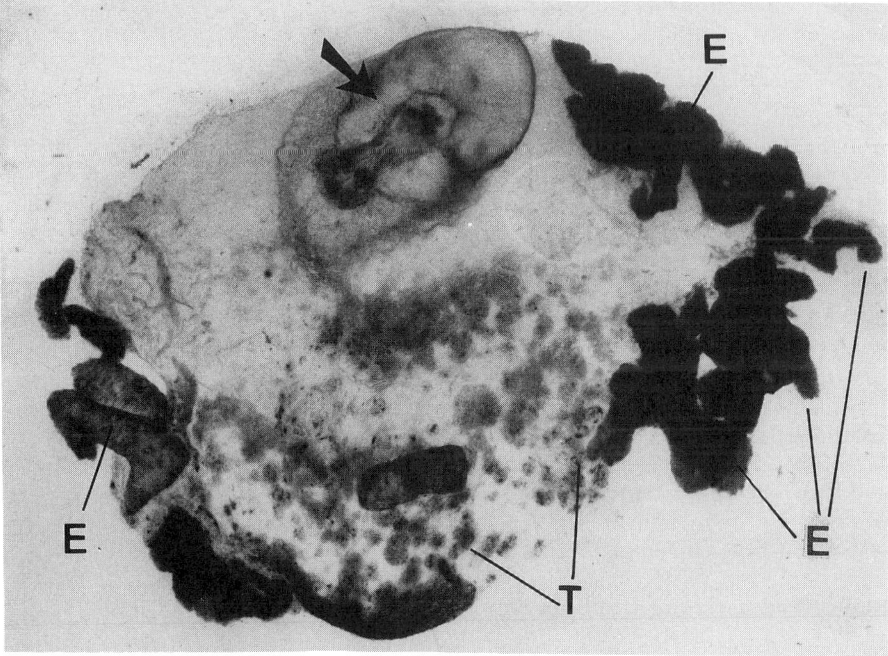

Figure 3. Low magnification micrograph of a blastocyst with numerous endometrial fragments (E) attached to the trophoblast in the abembryonic region where trophoblastic knobs (T) are well developed. The embryonic anlage (arrow) shows an advanced stage of differentiation. Coculture was performed in system Bb (see Figure 2) from 6.5 days to 9.5 days p.c. X24.

Comparison Of The Different Co-Culture Systems

During cultivation, the majority of blastocysts were lost due to collapse and degeneration with ratios differing in the two general systems. In the open systems (Figure 2Aa, b) about 40% of the blastocysts were still expanded after 3 days while only 30% remained expanded when blastocysts and endometrial fragments were kept in close contact in plastic tubes (System Ba, Figure 2) or in dialysis tubings (System Bb, Figure 2). Expansion of blastocysts, however, appeared to be a major prerequisite for implantation in vitro since attachment of trophoblast to endometrial fragments was never observed when blastocysts were found to be collapsed and degenerated. The larger proportion of expanded blastocysts as seen in open systems, however, were of no advantage since attachment to endometrial fragments was not obtained in this case (see below).

In the uterus, blastocysts and uterine epithelium are in permanent close contact after apposition has been achieved in the "implantation chamber". In our experiments, this was mimicked by enclosing both partners in the microchambers described above. Such confrontations yielded endometrial fragments attached to blastocysts (Figure 3) if the latter had remained expanded. Attachment was found in about 85% of the expanded blastocysts. When contact in the chamber was decreased (by leaving some space around blastocysts and endometrial fragments) the attachment rates were significantly lower with endometrium attached only to about 25% of expanded blastocysts although more blastocysts were expanding. Attachment rates were in the same range a) when serum-free medium was used throughout an experiment, b) with endometrium explanted from the mesometrial or the antimesometrial part of the uterus, c) with endometrial fragments precultured for two days until 6 days and 16 hours, 6 days and 12 hours, or 6 days p.c., and d) with blastocysts obtained at 6 days and 12 hours or 6 days and 16 hours. When blastocyst coverings were removed before confrontation the percentage of collapsed blastocysts was increased while the proportion of attachment was not clearly altered (cf. Table 1), although the low numbers of blastocysts do not allow statistical evaluation.

In open systems only random collision of both partners could occur while they were floating free in the culture fluid. This was also the case for attempts at co-culture in agar wells; the blastocysts started to float during agitation, and the contact between endometrium and blastocysts was not sufficient in this system even when the incubation was performed without any shaking. Trophoblast attachment to endometrial epithelium was never observed under these conditions.

State Of Endometrium After Confrontation

The preservation of endometrial fragments was heterogeneous in the different culture systems. In simple co-culture (Figure 2 Aa) the morphology was found to be the same as described by Hohn et al. (1984; 1989) for the culture of endometrial fragments alone. In chamber systems, however, degeneration of groups of cells was seen more often (40-60%, varying in single experiments) although other fragments were well preserved.

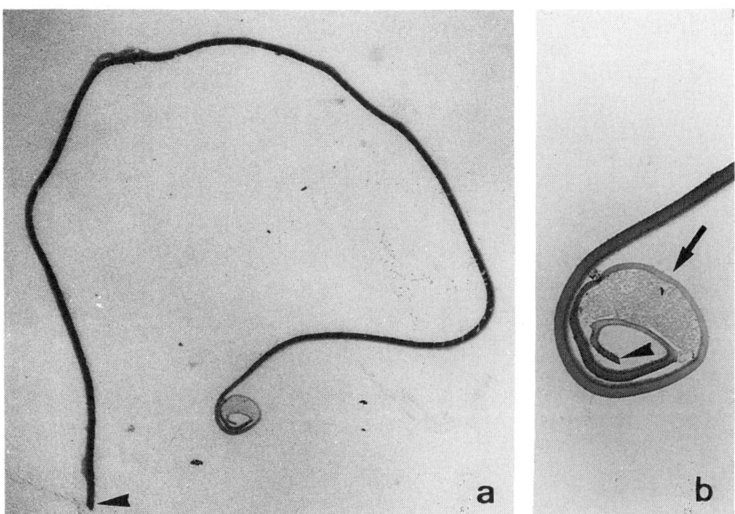

Figure 4. Disintegration of the blastocyst coverings is a prerequisite for trophoblast attachment to the endometrium. In vitro this process appears to involve enzymatic dissolution like in vivo as indicated by uneven thickness of the different layers and smooth thinned edges with disappearance of the layered structure (b, arrow) as well as mechanical rupture forming sharp edges (arrowheads). a) X110, b) X350.

State Of Blastocysts After Confrontation

In open systems as well as in chamber systems, expanded blastocysts developed remarkably well and reached advanced stages of differentiation (Figure 3) as described by Hohn et al. (unpublished), in the embryonic as well as in the abembryonic regions. Of particular relevance for the confrontation experiments is that the potentially invasive elements of the trophoblast, i.e., the trophoblastic knobs in the abembryonic region and large syncytia in the vicinity of the embryonic disc were well differentiated.

Fate Of Blastocyst Coverings

When blastocyst coverings were not removed before the confrontation of blastocysts with endometrium in vitro, they did not interfere with attachment of the trophoblast to the uterine epithelium, i.e., a process of shedding of the coverings took place in vitro. Remnants of blastocyst coverings were examined after culture for morphological signs of enzymatic dissolution or mechanical rupture. Although mechanical rupture could not be excluded, morphological evidence suggested an enzymatic process as some fragments of blastocyst coverings had smooth and tapered rather than broken edges (Figure 4). Broken edges could also have been generated during processing for histology.

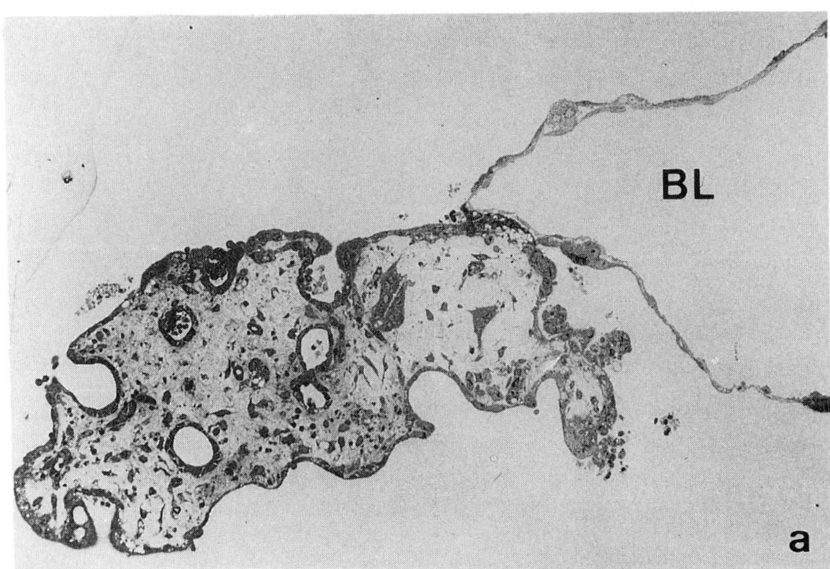
BL
a

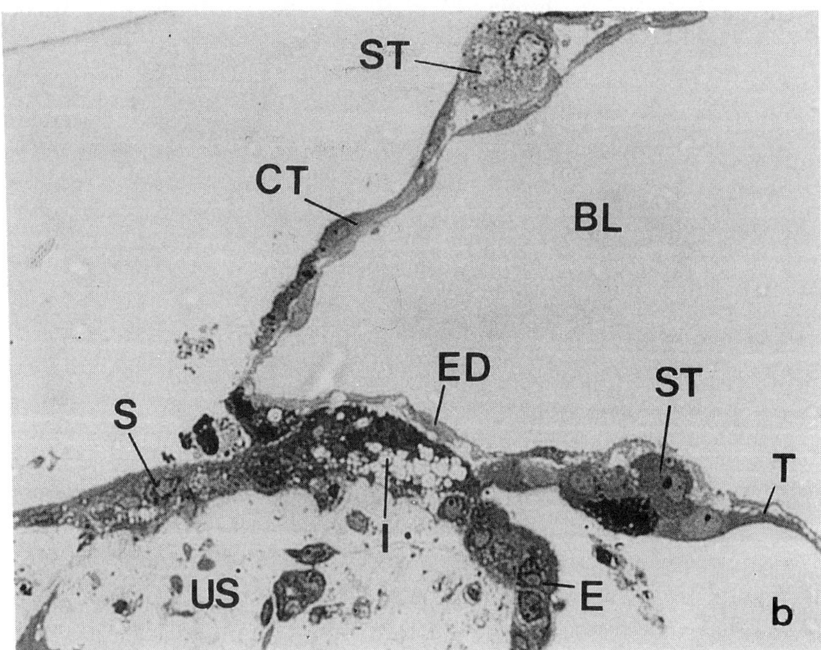
ST
CT
BL
ED
ST
S
T
I
US
E
b

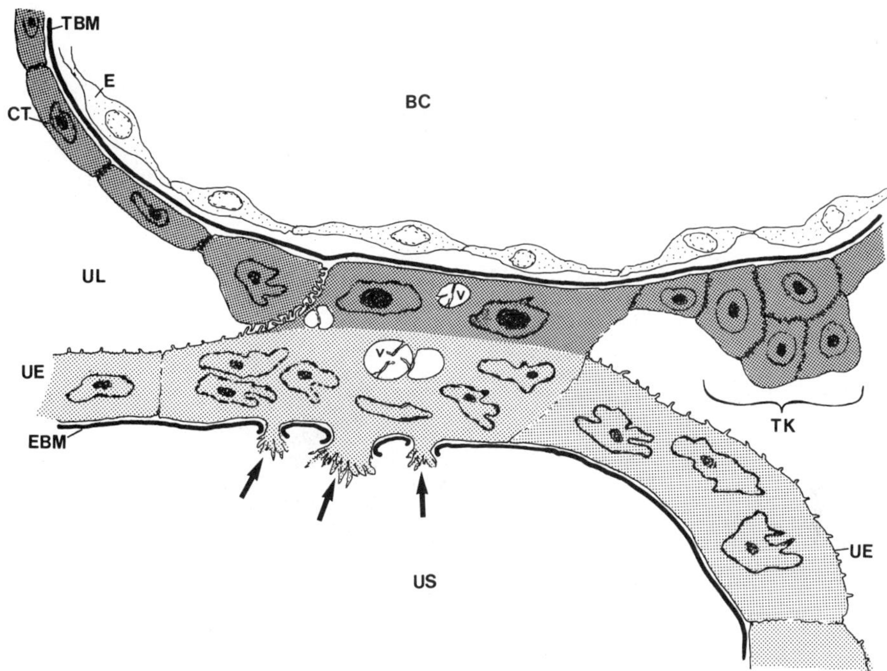

Figure 6. Schematic summary of light and electron microscopic observations of implantation sites obtained in vitro. The border line between trophoblastic and endometrial areas in the trophoblastic-endometrial syncytium was defined according to electron microscopical findings. BC = blastocyst, CT = cytotrophoblast, E = endoderm, EBM = endometrial basement membrane being penetrated in the area of the implantation site (arrows), TBM = trophoblastic basement membrane, TK = unattached trophoblastic knob, UE = uterine epithelium with symplasms and individual cells, US = uterine stroma, V = vacuoles.

Figure 5. Light microscopical histology of in vitro "implantation" site. The trophoblast (T) consisting of cytotrophoblast (CT) and syncytiotrophoblast (ST) has attached to the uterine epithelium (E) composed of individual cells (E) and symplasms (S). At the site of attachment the fusion of uterine epithelial cells and trophoblast cells has resulted in the heterogeneous symplasm of the "implantation" site (I). ED = endoderm, BL = blastocyst lumen, US = uterine stroma. a) X140, b) X450.

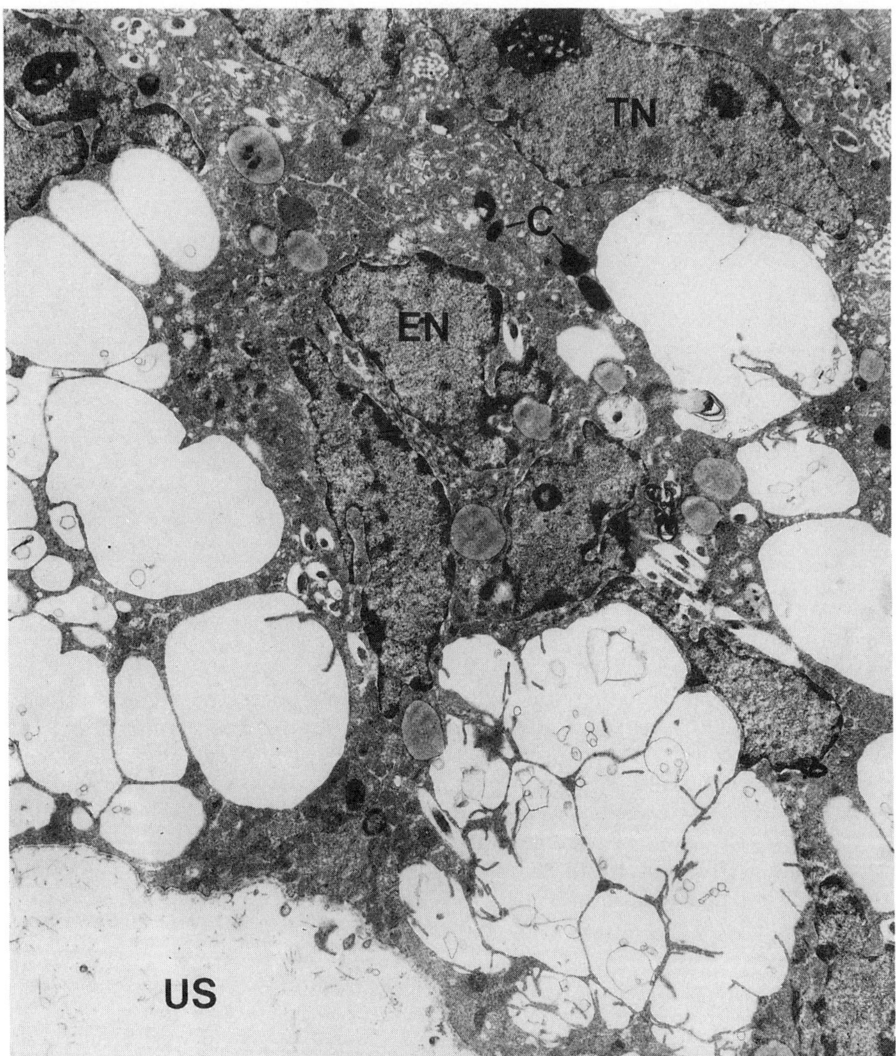

Figure 7. After fusion of trophoblast with uterine epithelium nuclei of the trophoblastic (TN) and the endometrial (EN) type can be observed in close relation without any separating cell membrane in between. Trophoblastic nuclei have a more or less elliptical shape, little chromatin, but large and strongly staining nucleoli, while endometrial nuclei are more irregular and lobulated with dense marginal heterochromatin. The dark ovoid structures are the crystalloids (C) which are typical for trophoblast cytoplasm. US = uterine stroma. X6700.

Morphology Of Trophoblast Attachment And Implantation In Vitro

Antimesometrial as well as mesometrial endometrial fragments were found attached exclusively to trophoblastic knobs of the abembryonic hemisphere but never to the syncytiotrophoblast of the embryonic pole (Figure 3) although the confrontation culture was maintained one day beyond the time of initiation of implantation at the embryonic pole in vivo.

Attachment of blastocysts was found both to endometrial fragments with intact morphology, i.e., with little signs of cell degeneration, as well as to fragments that showed more extensive cell necrosis (ratio of normal morphology to degeneration of about 40% to 60%). Interestingly, even in the latter cases trophoblast attachment did not seem to be totally non-specific/artificial since attachment was always by abembryonic trophoblastic knobs but not by any part of the interposed cytotrophoblast. The trophoblast never showed degeneration at those sites. In the contact area the endometrial epithelium representing several circles of cells around the implantation site appeared to be preserved thus suggesting that the endometrium was still in a normal state when attachment occured.

In endometrial fragments that showed normal morphology, the attachment site appeared to result from fusion of endometrial and trophoblastic elements (Figure 5). The typical electron microscopic findings are summarized in Figure 6 (for comparison with in vivo morphology see Enders and Schlafke, 1971, and Denker, 1977, 1983). At the attachment point, the trophoblast and uterine epithelium merged and here a cytoplasmic mass was located that was common to both of them. In this common cytoplasm, two different types of nuclei were located in close relationship: Homogeneously stained, ellipsoid nuclei of the trophoblastic type with relatively smooth, undulating membranes and little heterochromatin but usually two large strongly stained nucleoli, and on the other hand, more bizarre shaped nuclei of the endometrial type with marginal chromatin, a deeply indented nuclear membrane and small nucleoli.

Nuclei of the two types were still grouped together in different zones as were differing types of mitochondria, although no plasma membrane was found between them. Trophoblastic mitochondria appeared ovoid and swollen with only a few cristae while endometrial mitochondria had a lesser diameter but were more elongated with numerous cristae. Intervening cell membranes could only be detected adjacent to the endometrial and to the trophoblastic basement membrane but not in the intermediate zone where the common symplasm was located. Here a) trophoblast-type nuclei were not separated by a membrane from endometrial-type nuclei (Figure 7); b) in some areas endometrial-type nuclei were observed associated with mitochondria of the trophoblastic type (Figure 8), c) rod-like (endometrial-type) mitochondria were sometimes associated with trophoblastic-type nuclei. Large, irregular vacuoles were a common feature in these symplasms, some of them lined by a smooth membrane, others by membranes with microvillous-like projections. Some of the vacuoles may have resulted from enclosure of extracellular space where the surfaces of the trophoblast and of endometrial epithelium had come in contact but the fusion was still incomplete. However, the location of the vacuoles in the common symplasm was quite irregular, only few of them marking the borderline between trophoblast and endometrial type cytoplasm/organelles. A feature that is quite reminiscent of the in vivo process of implantation was the penetration of the endometrial basement membrane by processes of the trophoblastic-endometrial syncytium (Figure 9).

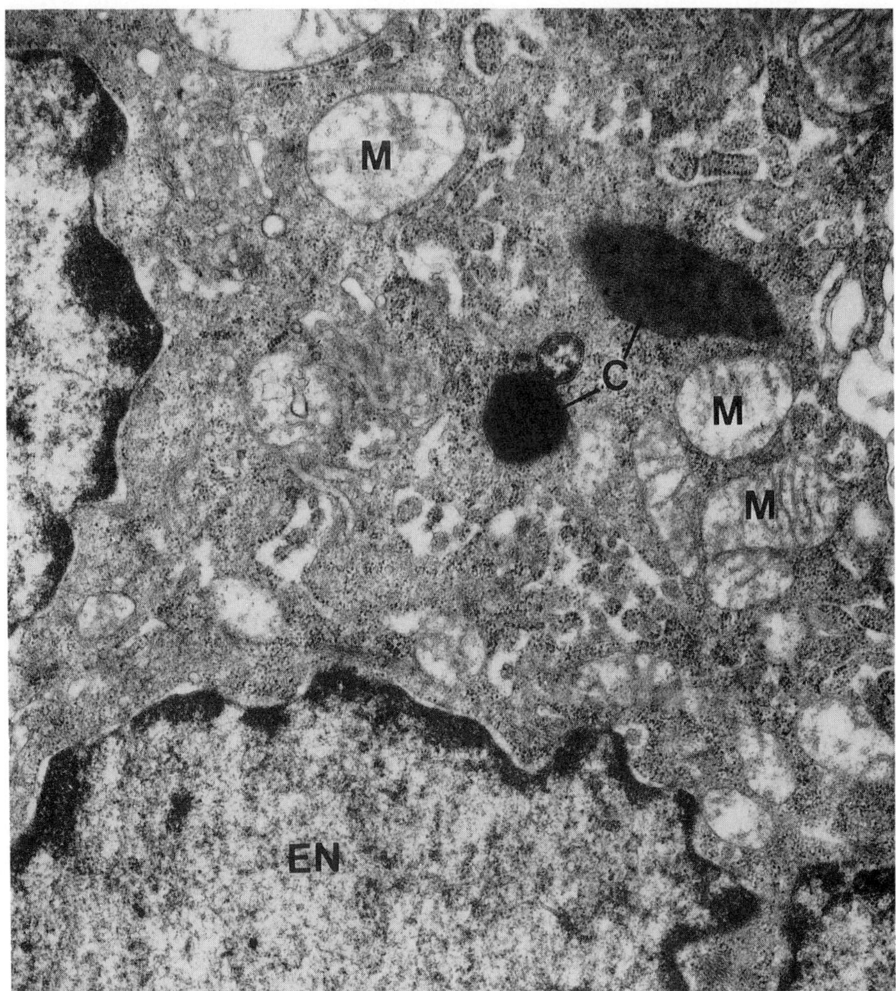

Figure 8. At the fusion site, nuclei with features typical for the uterine epithelium (irregular lobular shape, small nucleoli, large islands of marginal chromatin; EN) are often found associated with mitochondria (M) of a character that is generally observed for trophoblastic mitochondria as well as with crystalloids (C) also typical for trophoblast. X26800.

DISCUSSION

Studies of embryo implantation in vitro described in the literature have been performed in two-dimensional and three-dimensional model systems. The majority of experiments has used two-dimensional systems which are generally referred to as the "blastocyst (trophoblast) outgrowth" model. The results that have been obtained with this model have been reviewed extensively by Jenkinson (1977), Enders et al. (1981), and to some degree by Glasser (1985) (See also Morris et al., 1990). In general blastocysts are allowed to attach to plastic surfaces that are either uncoated or coated with different proteins such as constituents of the extracellular

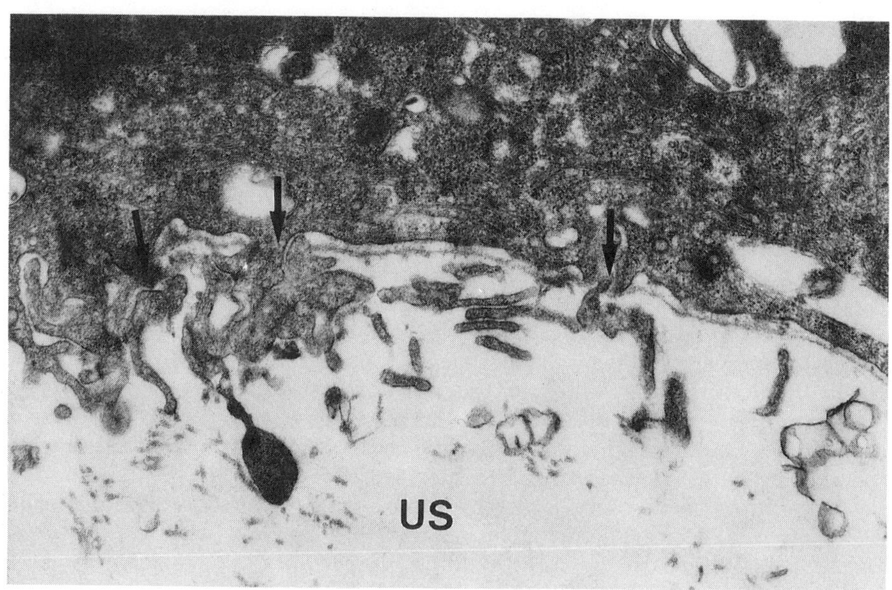

Figure 9. Penetration of the endometrial basement membrane during implantation in vitro. The basement membrane (BM) underneath the implantation site is penetrated at various spots by processes (arrows) of the symplasm resulting from fusion of trophoblast and uterine epithelium. US = uterine stroma. X24200.

matrix (Armant et al., 1986); alternatively they are placed onto cell monolayers derived from the endometrial stroma and/or epithelium or even other tissues (mouse: Jenkinson and Wilson, 1973; Glass et al., 1976, 1979, 1980; Sherman, 1978; Salomon and Sherman, 1975; Sherman and Wudl, 1976; Sherman and Mathaei, 1980; Sherman et al., 1979, 1981; Chávez, 1984; Chávez and Van Blerkom, 1981; human uterine epithelial cells and human blastocysts: Lindenberg et al., 1986; for detailed reviews see Jenkinson, 1977; Enders et al., 1981; Glasser, 1985). These models are very simplified; they can be expected to be of value for the study of cell biological elements of the attachment phase but not for the invasion phase when cell-cell and cell-matrix interactions are certainly more complex. Moreover, monolayer cultures of uterine epithelial and stromal cells fail to show typical responses to physiological doses of steroid hormones. This may be due to the absence of normal cell-cell or cell-matrix interactions. In the rabbit, large scale fusion of uterine epithelial cells with each other, a typical in vivo phenomenon, has not been observed in monolayer cultures (Mulholland et al., 1988). Likewise, synthesis and release of uteroglobin, a progesterone-dependent uterine protein, is obtained at constant rates only during the first 24 hours after explantation and declines in longer term cultures (Rajkumar et al., 1983a, b; Ricketts et al., 1983). The importance of stroma-epithelial interactions for epithelial differentiation and for the mediation of steroid hormone effects has been demonstrated in other systems by Kratochwil et al. (1972, 1976, 1979) and Cunha (1985). In the mouse uterus, biochemical evidence for the existence of such a relationship was found in studies of estradiol binding which was higher to stromal cells than to epithelial cells (Korach and Lamb, 1981). In the rabbit endometrium, the stromal cells have significantly higher levels of progesterone receptors than the epithelial cells (Perrot-Applanat et al., 1985, 1986). Bigsby and Cunha (1986) demonstrated that in the mouse uterus the proliferative response of the epithelium to estrogens is mediated by the stromal cells. These investigations indicate that the normal relationship of endometrial stroma and epithelium should be a prerequisite for an in vitro model which aims at simulating the cell biology of blastocyst attachment to the uterine epithelium and the subsequent process of invasion into the stroma.

An interesting model intermediate between two-dimensional and three-dimensional systems using the mouse was introduced by Morris et al. (1982, 1983). Vesicles of uterine epithelium are confronted with blastocysts in hanging drop cultures. It could be argued that the cell biological state of the epithelium in these vesicles is not very different from that in monolayer cultures. However, the behavior of the epithelium in the two types of culture does appear to be different since adhesion of blastocysts to uterine epithelial cells was observed with these vesicles but not with monolayers where trophoblastic processes quickly displaced the uterine epithelial cells and attached to the culture dish (Morris et al., 1983). The latter phenomenon had been observed also by other authors (Chávez and Van Blerkom, 1981; Chávez, 1984; reviewed by Jenkinson, 1977, by Glass et al., 1979, and by Enders et al., 1981). On the other hand, the importance of adhesion between the trophoblast and the uterine epithelium in the displacement type of implantation as seen in the mouse as well as in the rat (Schlafke and Enders, 1975) is not clear; not only do the uterine epithelial cells tend to retract from or become displaced by the trophoblast in vivo but also the epithelium detaches very easily from its basal lamina during the periimplantation phase, in these species, perhaps aided by decidual cell processes (Schlafke et al., 1985). This does not need the presence of a blastocyst but also occurres around "inert" bodies (Blandau, 1949). Consistent with this behavior in vivo is that in vitro no basement membrane was observed in the uterine epithelial vesicles, and blastocysts adhered preferably to degenerating cells (Morris et al., 1983).

A three-dimensional organ culture model was proposed by Glenister (1960, 1961a, b, 1963, 1965, 1967, 1970) who investigated rabbit embryo implantation using large fragments of endometrium freshly explanted on artificial supports at the air-culture medium interface. However, the following problems encountered with the system raise doubts as to its usefulness for studying the mechanisms of trophoblast attachment and invasion: a) Signs of degeneration were found not only at the site of "implantation" but also in other parts of the endometrial explants. b) Blastocysts tended to attach to clots of serum positioned between themselves and the endometrium, to exposed stroma, and to areas of cell degeneration as well as to healthy epithelium. c) Only cellular trophoblast invaded into stroma while syncytiotrophoblast followed necrotic tracks. d) No evidence was shown for the fusion of trophoblast and uterine epithelium which precedes the invasion into stroma in the rabbit in vivo (Larsen, 1961; Enders and Schlafke, 1971). e) "Implantation" occurred in nonreceptive endometrium from virgin or from estrous rabbits without steroid hormone treatment. f) "Implantation" in vitro did not depend on the addition of female sex steroid hormones to the culture media. g) "Implanted" blastocysts showed grossly abnormal morphology and signs of degeneration. Major reasons for these deviations from the normal might have been the following: 1) damage of endometrial tissue occurring during explantation was not corrected by precultivation allowing regeneration before confrontation with blastocysts; 2) nutrition of the uterine explants and of the blastocysts may have been insufficient due to limited diffusion and through unfavorable ratios of medium to tissue volumes. The resulting cell degeneration probably caused stroma to become exposed locally due to defects in the epithelium. It appears reasonable to assume that this has to atypical attachment and invasion since Cowell (1969, see also below) had shown that attachment did occur atypically and independent of steroid hormones if no intact uterine epithelium was interposed.

The same reservations apply to the experiments of Grant (1973) who kept whole uterine horns of immature mice in organ culture for studies of embryo implantation. In this system problems of diffusion must have been particularly severe for the uterine epithelium as well as for blastocysts since they were located in the center of the explanted organ. Indeed, "implantation" was observed in uteri of immature mice and was not influenced clearly by estrogens or progesterone. This probably points to an impaired barrier function of the uterine epithelium in the explanted uteri, since Cowell (1969) found that, in vivo, implantation did occur in immature mice after mechanical irritation of the endometrium resulting in degeneration of the epithelium. This must be considered a serious objection to Grant's experiments. Another indication for the lack of typical endometrial reactions in the latter model is that decidualization in vitro of the stroma was not induced by any hormonal regimen. In addition, the blastocysts undergoing implantation were generally described to show various degrees of degeneration and irregular development. This model was later on modified (Grant et al., 1975). Ovariectomized immature and adult animals with or without hormone treatment were used as donors for complete uterine horns. The conditions were adjusted to maintain more normal uterine morphology in organ culture, using chemically defined media and by stretching the uteri in order to reduce their diameter to provide shorter diffusion distances. Although cell degeneration was observed in uterine tissue and disturbed differentiation of the blastocysts was described, trophoblast invasion was now found to depend on the presence of progesterone in vitro. The morphology of the uteri, however, was not influenced by the addition of hormones to the culture media except for the closure reaction, and decidualization occurred only to a limited degree and only if it had already been triggered on in vivo before explantation.

Grant's model appears to be suitable for studying selected aspects of the hormonal control of implantation in the mouse, in particular epithelial penetration of the trophoblast. It may be less helpful for investigating the hormonal control of trophoblast invasion into the stroma. In mice as well as in rats, however, the uterine epithelium becomes very fragile in the periimplantation phase and it readily retracts and degenerates upon slight stimulation (as discussed above). Therefore, investigations using this model may be of limited value for embryo implantation in other species that have a more stable uterine epithelium and where firm attachment of the trophoblast to it may be more critical than in the mouse and the rat.

The present communication describes a model for embryo implantation in the rabbit designed to take these points into consideration. The uterine epithelium is much more resistant to mechanical disruption in the rabbit than in the mouse and rat, as shown by IUD (intrauterine device) experiments (Denker, 1976 and unpublished observations). It was possible to regenerate a complete epithelial lining around endometrial fragments during pre-culturing on a gyratory shaker, thus making sure that the trophoblast had to interact with healthy-appearing intact epithelium without contacting stroma directly.

Trophoblast attachment to the epithelium of uterine fragments and beginnning implantation were obtained in this model when blastocysts and endometrial fragments were kept permanently in close contact as they are in the uterus. Simple co-cultures where only random collision of both partners was possible were not useful. This is consistent with Enders' (1976) suggestion that continuous pressure of the endometrium against the blastocyst may be important for adhesion and also with the observations of Mitchell et al. (1987) that in the rabbit the contact between the trophoblast and the uterine epithelium is enforced by pressure exerted by the uterine wall and counteracted by the intraconceptus pressure that is even slightly higher than the pressure within the adjacent uterine lumen.

In the in vitro system described here, close contact between blastocyst and endometrium was accomplished of by enclosing them in chambers (dialysis bags). This method, although successful, cannot be considered completely satisfactory since: 1) handling is difficult; and 2) cell degeneration is seen more frequently than in free-floating endometrial cultures, probably due to limited diffusion of nutrients and metabolites or increased pressure in the chambers. Consequently, the patterns of trophoblast attachment were heterogeneous dependent on the preservation of endometrial fragments. When the fragments were well preserved (little cell degeneration), morphology of the implantation site was quite close to the in vivo situation (c.f. Larsen, 1961; Enders and Schlafke, 1971; Schlafke and Enders, 1975). When, however, degeneration was seen in endometrial fragments where trophoblast attachment had occurred it remained unclear whether attachment had originally been normal before the degeneration of the endometrial tissue began. In these cases, however, the trophoblast always appeared to be well organized. In many cases, the epithelium was preserved in the area of the attachment site indicating that adhesion had taken place initially in the normal way to an intact epithelium.

The failure of attachment of the syncytiotrophoblast at the embryonic pole of the blastocyst which normally starts at about 8 days p.c. in vivo (cf. Larsen, 1961; Enders and Schlafke, 1971; Schlafke and Enders, 1975) may be caused by inappropriate development of the embryonic syncytiotrophoblast or asynchronous development of mesometrial endometrial fragments and the embryonic syncytiotrophoblast in vitro. A retardation of embryonic development including the formation of the embryonic syncytiotrophoblast was generally observed in vitro (Hohn et al., unpublished). Even though this retardation was decreased in the presence of endometrial fragments it might have been sufficient to cause asynchronous development of both partners thus disabling their interaction.

In additional investigations using this system and in vivo experiments in the rabbit, tumor cells of varying origin including the rabbit, did not attach to the uterine epithelium under any conditions but were able to invade exposed stroma (Hohn et al., 1985; Donner et al., unpublished). This is in agreement with observations made with tumor cells in other systems. Tumor cells cannot attach to the apical surface of epithelia (except for the mesothelium and endothelia) which in general is non-adhesive for other cells (de Ridder et al., 1975). The trophoblast in contrast can attach to the uterine epithelium, during the "receptive phase" of the endometrium. According to this terminology of Psychoyos and Casimiri (1980), the uterine epithelium of the rabbit seems to develop "specific receptivity" for trophoblast attachment during the periimplantation phase. In rats and mice, however, the uterine epithelium does not seem to be as selective since tumor cells were found to invade into the uterine stroma from the uterine lumen via the epithelium during the implantation phase (Short and Yoshinaga, 1967; Wilson and Potts, 1970).

In conclusion, this model appears to be of value for the in vitro investigation of elements of the process of embryo implantation including trophoblast attachment to the uterine epithelium as well as incipient invasion into the endometrial stroma. The system would benefit from further improvement, in particular the method to maintain the continuous close contact between blastocysts and endometrial fragments has to be modified in a way that normal endometrial morphology is preserved in the endometrial fragments throughout an experiment. It will be interesting to see to what extent the model may facilitate studies on molecular mechanisms involved in embryo implantation.

SUMMARY

Embryo implantation in the uterus involves attachment of the trophoblast to uterine epithelial cells, followed by penetration through the epithelium and its basal membrane, and by the invasion into the stroma. Implantation starts when the invasive phase of the trophoblast coincides with a "receptive state" of the endometrium, the latter being controlled by estrogens and progesterone. The molecular basis for these processes is largely unknown so far. To study these mechanisms in vitro, a novel model for embryo implantation has been developed. The system is based on an organ culture model for rabbit endometrium where tissue fragments continue transformation as typical for pregnancy under the influence of progesterone (Hohn et al., 1989). Under the same conditions rabbit blastocysts show slightly retarded but otherwise fairly normal development in the presence of endometrial fragments.

Prior to confrontation with blastocysts, endometrial fragments are precultured in order to eliminate debris on the surface. In addition, the epithelium regenerates during this preculture phase and thus prevents direct attachment of the trophoblast to endometrial stroma. Subsequently late preimplantation stage blastocysts (6 1/2 - 6 2/3 days p.c.) are added for confrontation culture. Attachment of blastocysts to endometrium in vitro appears to be possible only if both are kept constantly in as close contact as they are in the uterus. Random collision as during floatation in simple culture systems with and without gyration is not sufficient. Better contact is provided by co-culture in small plastic tubes closed with dialysis membranes, or in dialysis tubings. In these systems, however, the increased pressure or the decreased exchange of metabolites and nutrients caused increased rates of cell degeneration in the uterine tissue. Adhesion of endometrial fragments occured only when blastocysts remained expanded. It was generally seen in the abembryonic region of the blastocyst and included typical trophoblastic knobs fusing with uterine epithelial cells. When the endometrial fragments showed normal morphology the attachment of the trophoblast to and its invasion into the endometrium were in many respects similar to the cytological details of implantation in utero.

ACKNOWLEDGEMENTS

The authors are very grateful to Ms. Gerda Helm and Ms. Elisabeth Hölscher for skillful technical assistance, Ms. Gisela Mathieu for secretarial help, Ms. Gabriele Bock for photographic work, Mr. Wolfgang Graulich for the drawing, and PD Dr. Elke Winterhager for numerous helpful comments on the TEM images. The investigations were supported by Deutsche Forschungsgemeinschaft grants De 181/9-5 + 6 and Ho 1059/1-7.

REFERENCES

Armant, D.R., Kaplan, H.A., and Lennarz, W.J. (1986) Fibronectin and laminin promote in vitro attachment and outgrowth of mouse blastocysts. *Dev. Biol.* 116, 519-523.

Bigsby, R.M. and Cunha, G.R. (1986) Estrogen stimulation of deoxyribonucleic acid synthesis in uterine epithelial cells which lack estrogen receptors. *Endocrinol.* 119, 390-396.

Blandau, R.J. (1949) Embryo-endometrial interrelationship in the rat and guinea pig. *Anat. Rec.* 104, 331-359.

Chávez, D.J. (1984) Embryology of the mouse from ovulation through periimplantation stages in vitro. *Scan. Electr. Microsc.* II, 729-735.

Chávez, D.J. and Van Blerkom, J. (1981) In vitro attachment and outgrowth of mouse trophectoderm. In: *Cellular And Molecular Aspects Of Implantation*, (eds.), S.R. Glasser and D.W. Bullock, D.W., New York: Plenum Press, pp. 457-460.

Cowell, T.P. (1969) Implantation and development of mouse eggs transferred to the uteri of non-progestational mice. *J. Reprod. Fert.* 19, 239-245.

Cunha, G.R. (1985) Mesenchymal-epithelial interactions during androgen-induced development of the prostate. In: *Developmental Mechanisms: Normal And Abnormal*, (eds.), J.W. Lash and L. Saxén, New York: Alan R. Liss Inc, pp. 15-24.

Denker, H.-W. (1970) Topochemie hochmolekularer Kohlenhydratsubstanzen in Frühentwicklung und Implantation des Kaninchens. I. Allgemeine Lokalisierung und Charakterisierung hochmolekularer Kohlenhydratsubstanzen in frühen Embryonalstadien. *Zool. Jahrb. Physiol.* 75, 141-245.

Denker, H.-W. (1976) Copper IUD-induced loss of endometrial arylamidase activity in the rabbit. *Biol. Reprod.* 15, 519-522.

Denker, H.-W. (1977) Implantation: The role of proteinases, and blockage of implantation by proteinase inhibitors. Berlin, Heidelberg, New York: Springer-Verlag (*Adv. Anat. Embryol. Cell Biol.* 53, Part 5).

Denker, H.-W. (1980) Embryo implantation and trophoblast invasion. In: *Cell Movement And Neoplasia*, (eds.), M. De Brabander, M. Mareel, and L. De Ridder, Oxford, New York: Pergamon Press, pp. 151-162.

Denker, H.-W. (1983) Basic aspects of ovoimplantation. In: *Obstetrics And Gynecology Annual*, Vol. 12, (ed.) R.M. Wynn, Norwalk, Connecticut: Appleton Century Crofts, pp. 15-42.

de Ridder, L., Mareel, M., and Vakaet, L. (1975) Adhesion of malignant and nonmalignant cells to cultured embryonic substrates. *Cancer Res.* 35, 3164-

Enders, A.C. (1976) Anatomical aspects of implantation. *J. Reprod. Fert. Suppl.* 25, 1-15.

Enders, A.C., Chávez, D.J., and Schlafke, S. (1981) Comparison of implantation in utero and in vitro. In: *Cellular And Molecular Aspects Of Implantation*, (eds.), S.R. Glasser and Bullock, D.W., New York: Plenum Publishing Corporation, pp. 365-382.

Enders, A.C. and Schlafke, S. (1971) Penetration of the uterine epithelium during implantation in the rabbit. *Am. J.Anat.* 132, 219-240.

Glass, R.H., Spindle, A.I., and Pedersen, R.A. (1976) Differential inhibition of trophoblast outgrowth and inner cell mass growth by actinomycin D in cultured mouse embryos. *J. Reprod. Fert.* 48, 443-445.

Glass, R.H., Spindle, A.I., and Pedersen, R.A. (1979) Mouse embryo attachment to substratum and interaction of trophoblast with cultured cells. *J. Exp. Zool.* 208, 327-336.

Glass, R.H., Spindle, A.I., Maglio, M., and Pedersen, R.A. (1980) The free surface of mouse trophoblast in culture is non-adhesive for other cells. *J. Reprod. Fert.* 59, 403-407.

Glasser, S.R. (1985) Laboratory models of implantation. In: *Reproductive Toxicology, Target Organ Toxicol. Ser.*, (ed.), R.L. Dixon, New York: Raven Press, pp. 219-238.

Glenister, T.W. (1960) Experimental nidation of blastocysts in organ culture. *Bull. Soc. Roy Belge Gynec. Obstet.* 30, 635-640.

Glenister, T.W. (1961a) Organ culture as a new method for studying the implantation of mammalian blastocysts. *Proc. R. Soc. B.* 154, 428-431.

Glenister, T.W. (1961b) Observations on the behaviour in organ culture of rabbit trophoblast from implanting blastocysts and early placentae. *J. Anat.* 95, 474-485.

Glenister, T.W. (1963) Observations on mammalian blastocysts implanting in organ culture. In: *Delayed Implantation*, (ed.), A.C. Enders, Rice University, Semicentennial Publications, pp. 171-183.

Glenister, T.W. (1965) The behaviour of trophoblast when blastocysts effect nidation in organ culture. In: *The Early Conceptus, Normal And Abnormal*. Papers and Discussions Presented at a Symposium Held at Queen's College, Dundee, 1964. University of St. Andrew, 1965, (ed.) W.W. Park, Edinbourgh and London: E. and S. Livingstone Ltd, pp. 24-26.

Glenister, T.W. (1967) Organ culture and its combination with electron microscopy in the study of nidation processes. In: *Fertility And Sterility*, Proceedings of the 5th World Congress, 1966 Stockholm, (eds.), B. Westin and N. Wiqvist, International Congress Series No. 133, Excerpta Medica Foundation, pp. 385-394.

Glenister, T.W. (1970) Ovo-implantation in vitro and its relation to normal implantation. In: *Ovo-Implantation, Human Gonadotropin And Prolactin*, (eds.), P.O. Hubinont, F. Leroy, C. Robyn, and P. Leleux, Basel, Munich, New York: S. Karger, pp. 73-85.

Grant, P.S. (1973) The effect of progesterone and oestradiol on blastocysts cultured within the lumina of immature mouse uteri. *J. Embryol. Exp. Morphol.* 29,

Grant, P.S., Ljunqvist, I., and Nilsson, O. (1975) The hormonal control and morphology of blastocyst invasion in the mouse uterus in vitro. *J. Embryol. Exp. Morphol.* 34, 299-310.

Hohn, H.-P., Busch, L.C., and Denker, H.-W. (1984) Endometrium in organ culture: A model host tissue for studies of trophoblast invasion. *Eur. J. Cell Biol.* (Suppl. 5) 33, 17.

Hohn, H.-P., Donner, A., and Denker, H.-W. (1985) Evaluation of an in vitro model for embryo implantation: Selective receptivity of rabbit endometrium for trophoblast attachment. *Eur. J. Cell Biol. (Suppl. 12)* 39, 18.

Hohn, H.-P., Winterhager, E., Busch, L.C., Mareel, M.M., and Denker, H.-W. (1989) Rabbit endometrium in organ culture: Morphological evidence for progestational differentiation in vitro. *Cell Tissue Res.* 257, 505-518.

Jenkinson, E.J. (1977) The in vitro blastocyst outgrowth system as a model for the analysis of peri-implantation development. In: *Development In Mammals*, Vol. 2, (ed.), M.H. Johnson, Amsterdam, New York, Oxford: North Holland Publishing Company, pp. 151-172.

Jenkinson, E.J. and Wilson, I.B. (1973) In vitro studies on the control of trophoblast outgrowth in the mouse. *J. Embryol. Exp. Morphol.* 30, 21-30.

Korach, K.S., Lamb, J.C. (1981) Estrogen action in the mouse uterus: Differential nuclear localization of estradiol in uterine cell types. *Endocrinol.* 108, 1989-1991.

Kratochwil, K. (1972) Tissue interaction during embryonic development: General properties. In: *Tissue Interactions In Carcinogenesis*, (ed.), D. Tavin, London, New York: Academic Press, pp. 1-47.

Kratochwil, K., Durnberger, H., Heuberger, B., and Wasner, G. (1979) Hormone induced cell and tissue interaction in the embryonic mammary gland. *Cold Spring Harbour Conference on Cell Proliferation* 6, 717-726.

Kratochwil, K. and Schwartz, P. (1976) Tissue interaction in androgen response of embryonic mammary rudiment of mouse: Identification of target tissue of testosterone. *Proc. Natl. Acad. Sci. USA*, 73, 4041-4044.

Larsen, J.F. (1961) Electron microscopy of the implantation site in the rabbit. *Am. J. Anat.* 109, 319-334.

Lindenberg, S., Hyttel, P., Lenz, S., and Holmes, P.V. (1986) Ultrastructure of the early human implantation in vitro. *Human Reprod.* 1, 533-538.

Mitchell, J.A., Driessen, G., Scheidt, H., and Denker, H.-W. (1987) Changes in intraconceptus pressure in the rabbit during days 7-10 of pregnancy. *Biol. Reprod.* 36, 669-675.

Morris, J.E. and Potter, S.W. (1990) An in vitro model for studying interactions between mouse trophoblast and uterine epithelial cells. A brief review of in vitro systems and observations on cell-surface changes during blastocyst attachment. *Trophoblast Res.* 4, 51-69.

Morris, J.E., Potter, S.W., and Buckley, P.M. (1982) Mouse embryos and uterine epithelia show adhesive interactions in culture. *J. Exp. Zool.* 222, 195-198.

Morris, J.E., Potter, S.W., Rynd, L.S., and Buckley, P.M. (1983) Adhesion of mouse blastocysts to uterine epithelium in culture: A requirement for mutual surface interactions. *J. Exp. Zool.* 225, 467-479.

Mulholland, J., Winterhager, E., and Beier, H.M. (1988) Changes in proteins synthesized by rabbit endometrial epithelial cells following primary culture. *Cell Tissue Res.* 252, 123-132.

Perrot-Applanat, M., Groyer-Picard, M.-T., Logeat, F., and Milgrom, E. (1986) Ultrastructural localization of the progesterone receptor by an immunogold method: effect of hormone administration. *J. Cell Biol.* 102, 1191-1199.

Perrot-Applanat, M., Logeat, F., Groyer-Picard, M.-T., and Milgrom, E. (1985) Immunocytochemical study of mammalian progesterone receptor using monoclonal antibodies. *Endocrinol.* 116, 1473-1484.

Psychoyos, A. (1976) Hormonal control of uterine receptivity for nidation. *J. Reprod. Fert. Suppl.* 25, 17-28.

Psychoyos, A. and Casimiri, V. (1980) Factors involved in uterine receptivity and refractoriness. In: *Progress In Reproductive Biology*, Vol. 7, (ed.), P.O Hubinont, Basel: Karger, pp. 143-157

Rajkumar, K., Bigsby, R., Lieberman, R., and Gerschenson, L.E. (1983a) Uteroglobin production by cultured rabbit uterine epithelial cells. *Endocrinol.* 112, 1490-1498.

Rajkumar, K., Bigsby, R., Lieberman, R., and Gerschenson, L.E. (1983b) Effect of progesterone and 17β-estradiol on the production of uteroglobin by cultured rabbit uterine epithelial cells. *Endocrinol.* 112, 1499-1505.

Ricketts, A.P., Hagensee, M., and Bullock, D.W. (1983) Characterization in primary monolayer culture of separated cell types from rabbit endometrium. *J. Reprod. Fert.* 67, 151-160.

Salomon, D.S. and Sherman, M.I. (1975) Implantation and invasiveness of mouse blastocysts on uterine monolayers. *Exp. Cell Res.* 90, 261-268.

Schlafke, S. and Enders, A.C. (1975) Cellular basis of interaction between trophoblast and uterus at implantation. *Biol. Reprod.* 12, 41-65.

Schlafke, S., Welsh, A.O., and Enders, A.C., (1985) Penetration of the basal lamina of the uterine luminal epithelium during implantation in the rat. *Anat. Rec.* 212, 47-56.

Sherman, M.I. (1978) Implantation of mouse blastocysts in vitro. In: *Methods In Mammalian Reproduction*, (ed.), Daniel, J.C., Jr., New York, San Francisco, London: Academic Press, pp. 247-257.

Sherman, M.I. and Mattaei, K.I. (1980) Factors involved in implantation-related events in vitro. *Prog. Reprod. Biol.* 7, 43-53.

Sherman, M.I., Sellens, M.H., Atienza-Samols, S.B., Pai, A.C., and Schindler, J. (1980) Relationship between the programs for implantation and trophoblast differentiation. In: *Cellular And Molecular Aspects Of Implantation*, (eds.), S.R. Glasser and D.W. Bullock, New York, Plenum Publishing Corporation, pp 75-89.

Sherman, M.I., Shalgi, R., Rizzino, A., Sellens, M.H., Gay, S., and Gay, R. (1979) Changes in the surface of the mouse blastocyst at implantation. *Maternal Recognition Of Pregnancy, Ciba Foundation Series* 6, 33-52.

Sherman, M.I. and Wudl, L.R. (1976) The implanting mouse blastocyst. In: *The Cell Surface In Animal Embryogenesis And Development*, (eds.), G. Poste and G.L. Nicolson, Amsterdam: Elsevier/North Holland Biomedical Press, pp. 81-125.

Short, R.V. and Yoshinaga, K. (1967) Hormonal influences on tumor growth in the uterus of the rat. *J. Reprod. Fert.* 14, 287-293.

Wilson, J.B. and Potts, D.M. (1970) Melanoma invasion in the mouse uterus. *J. Reprod. Fert.* 22, 429-434.

CHORIOCARCINOMA CELL SPHEROIDS: AN IN VITRO MODEL FOR THE HUMAN TROPHOBLAST

Ruth Grümmer[1], H.-P. Hohn[1], and H.-W. Denker[1]

Institut für Anatomie, RWTH Aachen
Melatener Strasse 211
D-5100 Aachen, Federal Republic Germany

INTRODUCTION

Cell differentiation and cell behavior are known to be strongly influenced by cell-cell and cell-matrix interactions (Grover et al., 1983; Landry and Freyer, 1984). Accordingly, among in vitro models proposed for studies of invasion, multicellular three-dimensional systems have been found to give more reliable data than two-dimensional (monolayer) systems (Mareel et al., 1979). Spheroids offer many tissue characteristics that cannot be obtained in monolayer or suspension culture (Sutherland and Durand, 1976; Yuhas et al., 1977). They represent tissue-like structures with various types of cell-to-cell contacts and can form an organized extracellular matrix (Nederman et al., 1984).

In order to study the interactions of human trophoblast with host tissue during invasion, a human choriocarcinoma cell line (BeWo) growing as multicellular spheroids in vitro is being used. The BeWo human trophoblastic tumor cell line was established in culture in 1966 (Pattillo and Gey, 1968) from postpartal choriocarcinoma which had been serially transplanted in the hamster cheek pouch by Hertz (1959). In continuous culture, this strain has maintained several characteristics of the normal trophoblast. The cells produce the placental hormones human chorionic gonadotropin (hCG), human placental lactogen (hPL), progesterone (P), and estrogens (E) (Pattillo et al., 1968, 1971).

In a previous communication this laboratory gave some preliminary data on a BeWo cell spheroid system (Denker et al., 1987). Here data are presented on the morphology, growth characteristics, and hormone production of these spheroids.

MATERIALS AND METHODS

Cell Culture

Choriocarcinoma cells (BeWo, CCL 98, American Type Culture Collection) were routinely cultured as monolayers through serial passaging in Ham's F-12 medium (without glutamine) (Flow Laboratories) supplemented with 15% FCS (Gibco) and 0.17% glutamine (Gibco) at 37°C in 95% air, 5% CO_2 in 50 ml plastic cell culture flasks (Greiner). Cells were subcultured by dispersion with 0.05% trypsin/0.02% EDTA in phosphate-buffered saline (PBS) (Biochrom). Culture media were filter-sterilized prior to use. Medium was changed three times a week.

[1]Present address: Institute für Anatomie, Universitätsklinikum, Hufelandstr. 55, D-Essen 1, Federal Republic of Germany

For initiating spheroid culture, exponentially growing cells were harvested by incubating monolayers for 15 minutes with a solution of 0.05% trypsin/0.02% EDTA at 37°C in 95% air/5% CO_2. Cells were diluted in Ham's F-12 medium containing 15% FCS, then the single cell suspension was aliquoted into 25 ml Erlenmeyer flasks at a concentration of 6 x 10^5 cells/6 ml and incubated on a gyratory shaker at 37°C in 95% air/5% CO_2. The rotor speed was set at 70 rpm. The medium was changed on the third day and subsequently every other day.

Morphology

Representative multicellular spheroids were fixed in 2.5% glutaraldehyde in 0.1 M cacodylate buffer. Tissue was postfixed in 2% osmium tetroxide, dehydrated in ethanol and embedded in Araldite. Semithin sections were stained with toluidine blue, thin sections were contrasted with uranyl acetate and lead citrate and studied with a Zeiss EM 10 electron microscope.

For SEM, glutaraldehyde-fixed spheroids were dehydrated in acetone and critical-point dried, sputter-coated with gold, and examined in a Leitz AMR 1000 scanning electron microscope.

Proliferation Studies

For the detection of proliferating cells, multicellular spheroids of defined size were incubated in medium containing 5'-bromo-2'-deoxyuridine at a final concentration of 10 μM for 30 minutes at 37°C in 95% air/5% CO_2. They were then fixed in ethanol: methanol (1:1) for 30 minutes, dehydrated in ethanol and embedded in paraffin. Serial 5 μm sections were obtained from the spheroids and histochemistry was performed on those sections nearest to the center of the spheroid. Sections were deparaffinized, washed in PBS (pH 7.2), and endogenous peroxidase activity was quenched with 0.5% hydrogen peroxide in methanol. Sections were incubated with anti-BrdU (Becton Dickinson) at a dilution of 1/50 for 30 minutes, followed by peroxidase-conjugated rabbit anti-mouse IgG (Dakopatts) as a secondary antibody at a dilution of 1/75 for 30 minutes. All incubations were carried out at room temperature. Peroxidase was detected with 3, 3'-diaminobenzidine (Sigma). Sections were then dehydrated and mounted with DePeX (Serva). Parallel sections were stained with hematoxylin-eosin.

Growth And Hormone Production Of Spheroids

For the study of growth and hormone production, one series of experiments was run in which multicellular spheroids of a mean diameter of 226 ± 39 μm (x̄ ± SD) were selected three days after initiation of culture, measured individually with an eyepiece micrometer and transferred into new flasks. Each culture vessel received 200 spheroids in 6 ml medium. At intervals of 24 hours, the diameter of each spheroid in one culture vessel was measured. Medium was replaced daily and used medium was collected and stored at -20°C until assayed.

HCG values were determined by the hCG MAIA clone kit (Serono) which recognizes the hCG molecule and its β-subunit. Progesterone values were quantitated by a Coat-A-Count progesterone RIA (Diagnostic Products Corporation), estradiol values by a ^{125}I-estradiol-RIA with a double antibody test in uncoated tubes (Baxter Merz and Dade AG). Coefficients of correlation were computed after Pearson and mean curves were evaluated from both regression lines.

Cell Number

For determining the cell number, multicellular spheroids were removed periodically from the stirring vessels, measured by an eyepiece micrometer and disaggregated by incubating with 0.05% trypsin/0.02% EDTA for 30 minutes at 37°C in 5% CO_2 in air. Then numbers of cells in these monodispersed suspensions were counted with a hemocytometer.

RESULTS

During shaker culture, BeWo choriocarcinoma cells consistently produced a high proportion of uniformly shaped solid spheroids (Figures 1 and 4a). More than 300 multicell spheroids were always produced within the first three days of culture with an inoculum of 6 x 10^5 cells/6 ml. A growth curve for the development of spheroids is shown in Figure 2. During the observed period of culture the diameter of spheroids increased linearly from a mean of 226 ± 49 µm on day 1 to 524 ± 117 µm on day 9. The number of cells per spheroid (Figure 3) increased correspondingly from 813 cells/spheroid at a diameter of 226 µm (volume 0.006 mm^3) to 8913 cells/spheroid at a diameter of 524 µm (volume 0.075 mm^3).

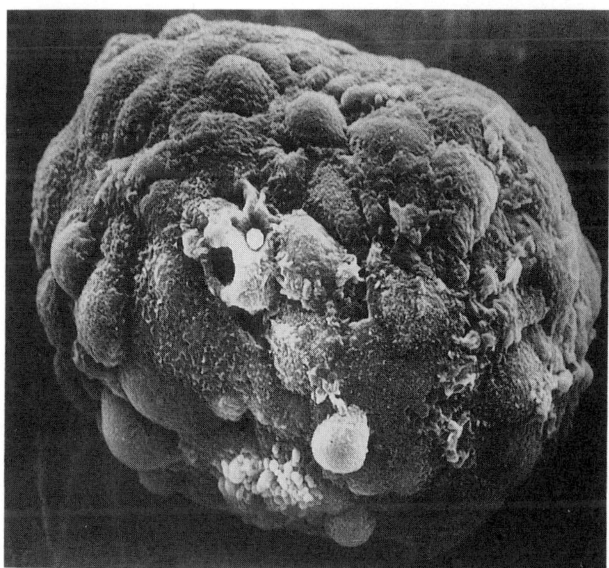

Figure 1. Scanning electron micrograph of a BeWo choriocarcinoma multicell spheroid after 4 days in shaker culture. Cells of the superficial layer appear only slightly flattened and of different size. Their external membrane is covered with microvilli and microridges. Remnants of some degenerating cells can also be seen (blebbing, protrusions).

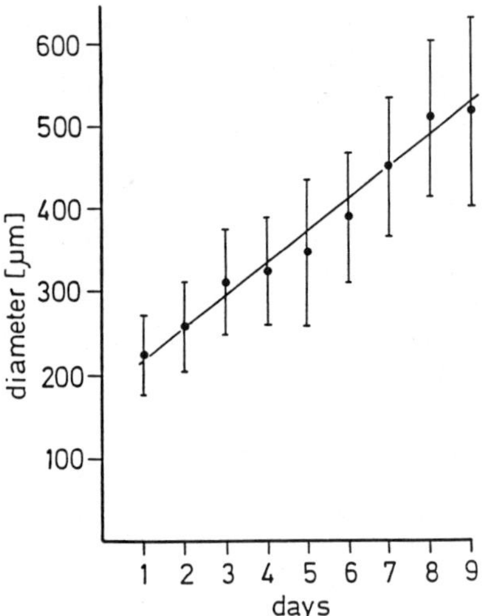

Figure 2. Correlation between diameter of spheroids of BeWo choriocarcinoma cells ($\bar{x} \pm$ SD) and days in shaker culture (r = 0.99)

Morphology

Figure 1 shows a scanning electron micrograph of a BeWo spheroid after 3 days of culture. The surface (outer contour) was relatively smooth and covered with rounded cells which seemed to be of different size; their external membrane was covered with abundant microvilli.

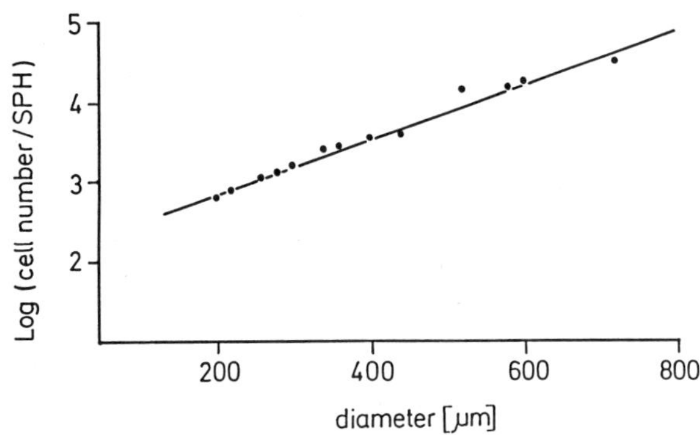

Figure 3. Correlation between logarithm of cell number per spheroid and diameter of BeWo cell spheroids (r = 0.99).

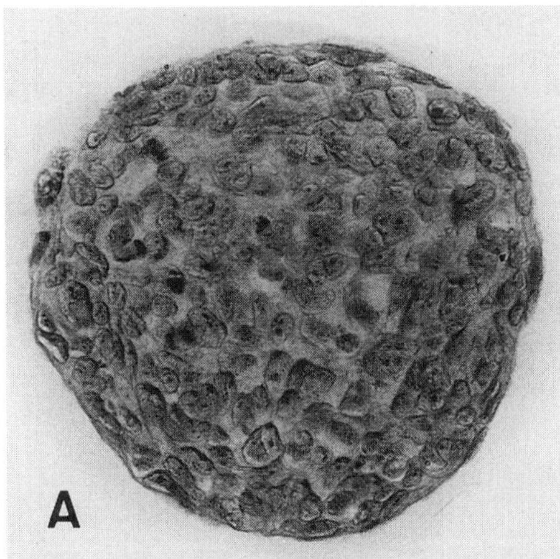
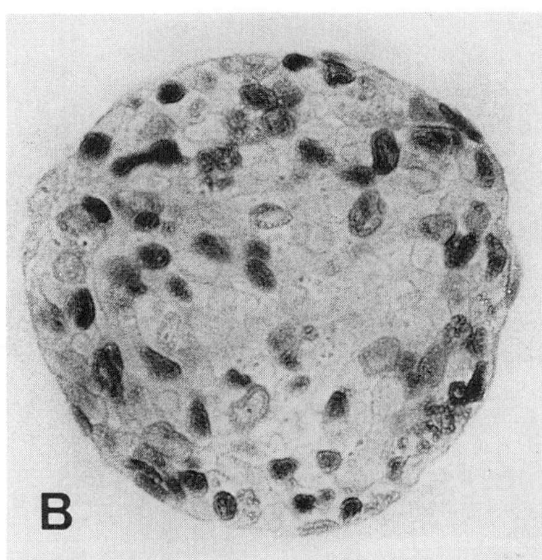

Figure 4. Cell proliferation demonstrated in a BeWo spheroid using the BrdU method. BeWo cell spheroids (280 µm in diameter) were taken after 3 days in shaker culture and were incubated for 30 minutes. Paraffin sections, X200.
A. Morphology, hematoxylin-eosin. Note compact cell mass; no sign of central necrosis can be seen. B. Parallel section stained with anti-BrdU antibody showing proliferating cells distributed evenly all over the spheroid.

Light microscopic examination of sections of BeWo spheroids revealed that the interior structure was that of a compact cell mass (Figure 4A). Single necrotic cells were uniformly distributed over the spheroid, but there was no central zone of necrosis in spheroids up to 580 µm in diameter after 9 days of culture (end of observation period). The cells were very heterogeneous in size and shape and stained to different degrees with hematoxylin-eosin. Though in a few cases, enlarged binucleated cells could be seen, no multinucleated giant cells could be identified clearly. Also there was no obvious regularity in the distribution of differently sized cells.

Transmission electron microscopic examination (Figures 5 and 6) revealed that the majority of cells in the BeWo spheroids showed morphological characteristics of relatively undifferentiated, mononuclear cells. Their nuclei tended to be round with few invaginations; they were large in relationship to the cytoplasmic volume. They had a homogeneous nucleoplasm with only sparse marginal heterochromatin but a prominent nucleolus. The cytoplasm of these cells contained abundant free ribosomes with few profiles of endoplasmic reticulum, and the Golgi apparatus was relatively small, features of rapidly dividing undifferentiated cells. In addition, another type of cell was present in the BeWo spheroids which showed ultrastructural features of more differentiated cells. Their cytoplasm was more dense and contained many dilated cisternae of rough endoplasmic reticulum. The nucleus was irregularly shaped and showed more heterochromatin. Cells with these characteristics that were located at the surface of the spheroid had microvilli at the outer plasma membrane, whereas other cells with a similar cytoplasmic and nuclear morphology which were scattered in the interior of the spheroid rarely showed patches of microvilli.

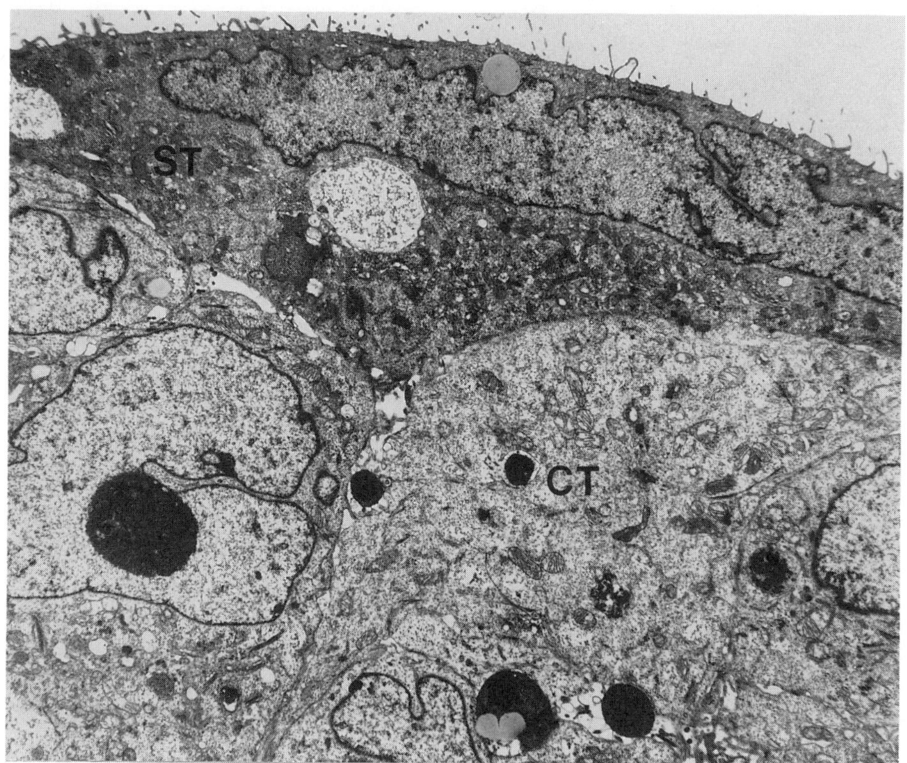

Figure 5. TEM showing a sector of the outer surface of a BeWo multicell spheroid after 4 days in shaker culture. Two types of cells can be distinguished in these spheroids. Cytotrophoblast-like cells (CT) have a round shape and contain electron-lucent ground cytoplasm, a nucleus with few invaginations, and a prominent nucleolus. At the outer surface a flattened cell can be seen which shows a denser cytoplasm and an irregularly shaped nucleus which is richer in heterochromatin (syncytiotrophoblast-like cell, ST). Its external surface is covered by numerous microvilli (X4900).

In semithin sections many mitotically active cells could be seen uniformly distributed throughout the spheroids. For more exact localization of proliferating cells, spheroids were incubated with medium containing 5'-bromo-2'-deoxyuridine, a thymidine analogue which is incorporated into replicating DNA. Anti-BrdU was then used for histochemical identification of those cells that had undergone DNA synthesis. In spheroids up to 280 µm in diameter proliferating cells were distributed all over the spheroid (Figure 4B). With increasing size of spheroids stained cells were located predominantly in the outer layer but proliferation could still be observed in the central region of the spheroids in all stages and size classes up to 520 µm in diameter.

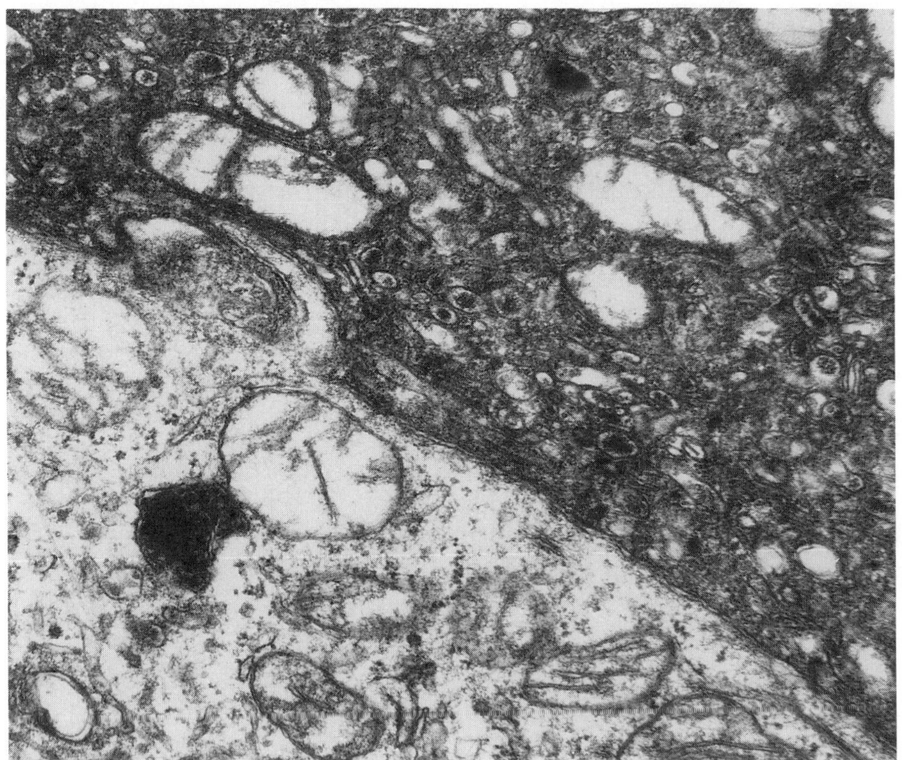

Figure 6. Cytoplasmic structure of different types of BeWo cells in a multicell spheroid. CT-like cells (below) show an electron-lucent, loosely structured ground cytoplasm, scattered polyribosomes, sparse endoplasmic reticulum, and round, swollen mitochondria. The ground cytoplasm of ST-like cells (above) is electron-dense and contains numerous cisternae of smooth and rough endoplasmic reticulum partly filled with flocculent material. (X30,000).

Hormone Production

BeWo choriocarcinoma cells grown as multicellular spheroids in vitro were capable of producing the placental hormones human chorionic gonadotropin (hCG), progesterone (P) and 17-β-estradiol (E_2) and of releasing them into the culture medium. Hormone production per spheroid increased corresponding to the increase in spheroid size (Figure 7). Spheroids of a mean volume of 0.006 mm^3 produced 11.15 mU β-hCG, 0.97 pmol P, and 8.4 x 10^{-15} mol E_2/SPH/24 hours, while spheroids of a mean volume of 0.075 mm^3 produced 21.24 mU β-hCG, 3.35 pmol P, and 10.8 x 10^{-15} mol E_2/SPH/24 hours. Whereas progesterone production was directly proportional to spheroid volume (r = 0.97), correlation coefficients for estradiol and β-hCG production and volume were 0.71 and 0.75, respectively (Figure 7).

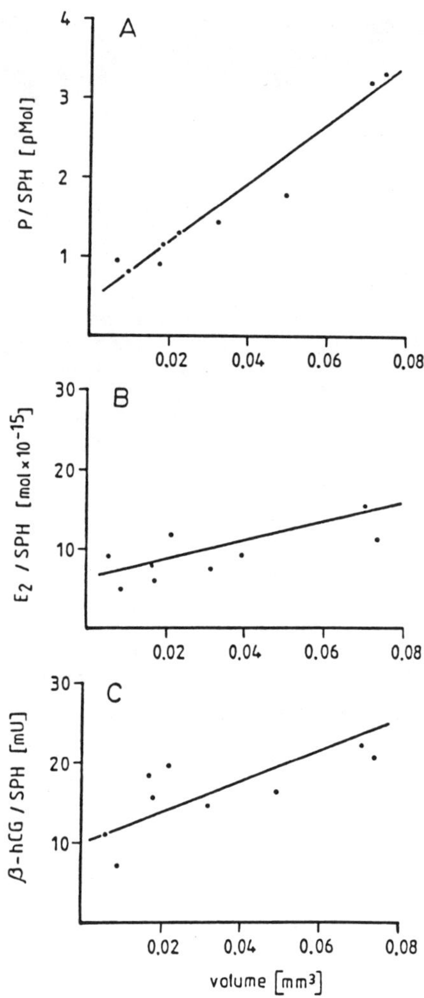

Figure 7. Correlation between production of progesterone (A, r = 0.97), 17-β-estradiol (B, r = 0.71), and β-hCG (C, r = 0.73) per spheroid and volume of BeWo spheroids.

On a per cell basis, hormone production values declined with increasing spheroid size and time of culturing (Figure 8). After 1 day (mean volume 0.006 mm³) levels of hormone production were 13.7 μU β-hCG, 1.2 x 10⁻¹⁵ mol P, and 10.3 x 10⁻¹⁸ mol E₂/cell/24 hours; after 9 days of culture (mean volume 0.075 mm³) values were 2.4 μU β-hCG, 0.4 x 10⁻¹⁵ mol P, and 1.2 x 10⁻¹⁸ mol E₂/cell/24 hours.

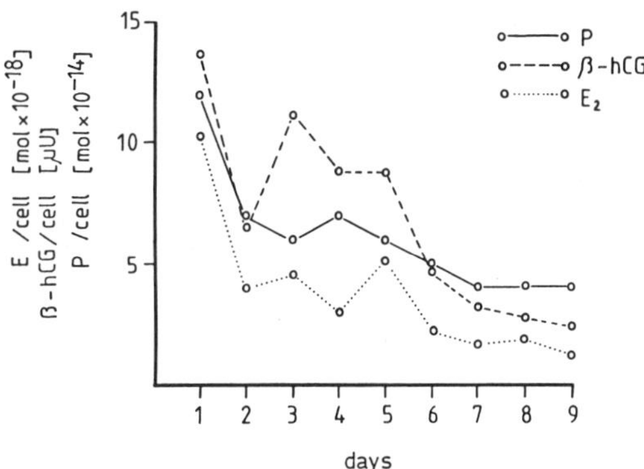

Figure 8. Hormone production (progesterone P, 17-β-estradiol E, and β-hCG) per cell in BeWo choriocarcinoma spheroids dependent on days in shaker culture.

DISCUSSION

The BeWo choriocarcinoma cell line has previously been studied in monolayer culture with respect to morphological signs of differentiation (Skehan and Friedman, 1978; Burres and Cass, 1986) and hormone production (Pattillo et al., 1971; Hussa, 1977; Simpson et al., 1978; Ritvos et al., 1987). This laboratory has now established BeWo multicellular spheroids which maintain morphological characteristics and hormonal functions of malignant trophoblast.

Under the conditions described above, BeWo cells are easily and reproducibly grown as spheroids. During the observed period of nine days (spheroid diameter 226 - 524 μm) spheroid growth was linear. Sutherland and Durand (1976) and Yuhas and Li (1978) have observed with other cell types that spheroids enlarge exponentially for a few days and then continue on a linear growth curve before reaching a critical diameter of about 1600 μm (Landry et al., 1982) or even 4000 μm (Folkman and Hochberg, 1973) beyond which there is no further expansion. The development of central necrosis varies considerably in spheroids of different cell lines. Sutherland et al. (1971) observed a necrotic core in V79 spheroids 'of more than 250 μm in diameter, while spheroids of other cell lines could reach sizes of about 500 μm (Erlichman and Tannock, 1986) or even 700 μm (Haji-Karim and Carlsson, 1978) in diameter before any visible sign of central necrosis occurred. Though single uniformly distributed necrotic cells could be seen in BeWo spheroids under the described culture conditions no central necrosis occurred in spheroids with diameters of up to 524 μm. Obviously diffusion of nutrients and metabolites is still sufficient at this size. This corresponds to a distribution of proliferating cells all over the spheroid. BeWo cell spheroids up to 524 μm in diameter have not yet reached a dormant phase in which there is no further expansion; here spheroids would be expected to establish an outer viable rim distinguishable from an inner necrotic zone (Yuhas and Li, 1978; Franko and Sutherland, 1979).

BeWo choriocarcinoma cells have previously been shown to be able to differentiate in vitro in monolayer cultures (Friedman and Skehan, 1979; Burres and Cass, 1986). Two BeWo phenotypes coexist in these cultures. The predominant form consists of cytotrophoblast-like (CTL) cells that are proliferative, mononuclear, and moderate in size. The remaining cells are syncytiotrophoblast-like (STL); they are nonproliferative, multinuclear, and giant. This heterogeneity in cell morphology could also be observed in our BeWo multicell spheroids although cells with more than 2 nuclei were rare. Many cell types are able to differentiate spontaneously in spheroid culture (Seeds, 1973; Honegger and Richelson, 1977; Trapp et al., 1979). The extent of cell-cell-contact developing in this three-dimensional culture system may influence the spontaneous differentiation of BeWo cells (Friedman et al., 1984). Furthermore, the production of an extracellular matrix has been demonstrated within spheroids (Nederman et al., 1984) and its direct role in differentiation and morphogenesis in these structures has been documented (Grover et al., 1983). On the other hand, it has been reported by Friedman and Skehan (1979) that formation of STL cells appears to be reduced at high densities in monolayer cultures.

One important characteristic of differentiating trophoblastic cells is the production of placental hormones (Pierce and Midgley, 1963; Enders, 1965; Hoshina et al., 1985; Kliman et al., 1986) which are secreted by the syncytiotrophoblast (Horne et al., 1976; Heyderman et al., 1981; Hamasaki et al., 1987). In addition to preservation of cell morphology of early trophoblast the BeWo choriocarcinoma cell line has also maintained functional hormone synthesis in vitro (Pattillo and Gey, 1968). In this study, BeWo cells growing as spheroids clearly preserve the ability to secrete hCG, progesterone, and estradiol into the culture medium. At a spheroid diameter of 226 - 524 µm, values of progesterone and estradiol production correspond approximately to those found by Saltzman et al. (1987) for multicellular spheroids of the JAr choriocarcinoma cell line. Values of hCG production measured are higher in BeWo spheroids, but these should be compared with caution because information about the specifity of the antibody used by Saltzman et al. was not provided.

The production of hCG per cell by spheroids was about 4-fold higher than by cells grown in monolayer cultures (Grümmer, unpublished results) and about 7-fold higher than values reported by Takazama and Sekija (1984) in BeWo monolayer cultures.

Production of hCG is commonly considered a useful marker for differentiation of trophoblastic cells and is typically found in syncytiotrophoblast cells (Morrish et al., 1987). Whether the observed higher production of hCG per cell in spheroids than in monolayer culture is due to an increased formation of STL-cells in spheroids or to a higher hCG-production per STL-cell will have to be examined in further experiments. It has been reported that cells representing an intermediate stage of differentiation can also produce hCG (Hoshina et al., 1984; Kliman et al., 1987).

With regard to steroid hormone production there was more progesterone than estradiol produced in BeWo spheroids. This is in accordance with observations of Bahn et al. (1981) that progesterone is the major steroid hormone synthesized by the BeWo cell line and is produced in higher quantities than estradiol in BeWo monolayer cultures.

On a per cell basis, hormone production declined with increasing number of cells per spheroid. This could, on one hand, be explained by a reduced proportion of STL cells (c.f. Friedman and Skehan 1979, see above). Alternatively, conditions for diffusion of precursors for hormone synthesis, especially to centrally located cells, should be better in smaller spheroids.

BeWo choriocarcinoma cells grown as spheroids, like monolayer cultures (Pattillo and Gey, 1968; Bahn et al., 1981), have maintained several characteristics of malignant trophoblast. Compared with monolayers, however, they have the unique advantage of preserving a three-dimensional arrangement of cells which mimicks the in vivo situation (Sutherland et al., 1971). They allow in vitro investigations of differentiation, of interactions between normal and malignant cells, and of invasive potential in confrontation with normal tissue (Schleich, 1973; Schleich et al., 1976; Mareel et al., 1979). The advantage of using multicellular tumor spheroids versus single cell suspensions for observing invasion has been demonstrated in the studies cited here. Choriocarcinoma cell spheroids of the type described in the present communication may be useful for studies of trophoblast attachment and invasion when confronted with human endometrium grown in an organ culture model (Grümmer et al., 1989).

SUMMARY

In an attempt to establish a model that could be used in studies in vitro of the cell biology of human trophoblast-endometrial interactions, the formation and properties of multicellular spheroids produced from a choriocarcinoma cell line (BeWo) was investigated. Growth characteristics and morphological and hormonal differentiation of cells in these spheroids were observed for a period of 9 days. Ultrastructural investigations revealed that BeWo spheroids were composed of different cell types. The majority of cells showed morphological characteristics of relatively undifferentiated, mononuclear cytotrophoblast-like cells, while other cells showed ultrastructural features of more differentiated syncytiotrophoblast-like cells, although cell fusion remained very limited. Studies on BrdU incorporation revealed that cell proliferation continued throughout the spheroids until the end of the observation period (9 days), but was higher at the periphery when spheroids reached a diameter of 300 µm or more. Spheroid size increased linearly from a mean diameter of 226 ± 49 µm ($\bar{x} \pm$ SD) on day 1 to a mean diameter of 524 ± 117 µm on day 9. The number of cells per spheroid increased correspondingly. The ability to secrete placental hormones into the medium was well maintained in these spheroids: values determined by RIA for P, E_2, and β-hCG increased in a linear manner per SPH/24 hours during the observation period of 9 days. On a per cell basis, however, hormone production declined for all three hormones.

These morphological and functional investigations show that BeWo choriocarcinoma spheroids may be a useful 3-dimensional model for studies of trophoblast differentiation and could be of value for investigations on trophoblast attachment and invasion.

ACKNOWLEDGEMENTS

The authors thank Ms. M. von Bentheim for excellent technical assistance throughout this work, PD. Dr. E. Winterhager and Dr. U. Mootz for cooperating in TEM and SEM, respectively, and Dr. T. Stein and the staff of the Institut für Klinische Chemie and Pathobiochemie der RWTH Aachen for performing RIA determinations. Technical help by Ms. G. Bock and Mr. W. Graulich is likewise gratefully acknowledged. This work was supported by a grant from the Minister für Wissenschaft und Forschung, NRW (Proj. Nr. 500 019 88).

REFERENCES

Bahn, R.S., Worsham, A., Speeg, K.V., Ascoli, M., and Rabin, D. (1981) Characterization of steroid production in cultured human choriocarcinoma cells. *J. Clin. Endocrinol. Metab.* 52, 447-450.

Burres, N.S. and Cass, C.E. (1986) Density-dependent inhibition of expression of syncytiotrophoblast markers by cultured human choriocarcinoma (BeWo) cells. *J. Cell. Physiol.* 128, 375-382.

Denker, H.-W., Hohn, H.-P., and Winterhager, E. (1987) Dreidimensionale Aggregate von Chorionkarzinomzellen als ein In-vitro-Modell für den invasiven Trophoblasten. *Verh. Anat. Ges. 81, Anat. Anz. Suppl.* 162, 499-

Enders, A.C. (1965) Formation of syncytium from cytotrophoblast in the human placenta. *Obstet. Gynecol.* 25, 378-386.

Erlichman, C. and Tannock, I.F. (1986) Growth and characterization of multicellular tumor spheroids of human bladder carcinoma origin. *In Vitro Cell Dev. Biol.* 22, 449-456.

Folkman, J. and Hochberg, M. (1973) Self regulation of growth in three dimensions. *J. Exp. Med.* 138, 745-753.

Franko, A.J. and Sutherland, R.M. (1979) Oxygen diffusion distance and development of necrosis in multicell spheroids. *Radiat. Res.* 79, 439-453.

Friedman, S.J. and Skehan, D. (1979) Morphological differentiation of human choriocarcinoma cells induced by methotrexate. *Canc. Res.* 39, 1960-1967.

Friedman, S.J., Galuszka, D., Gedeon, I., Dewar, C.L., Skehan, D., and Heckman, C.A. (1984) Changes in cell-substratum adhesion and nuclear-cytoskeletal anchorage during cytodifferentiation of BeWo choriocarcinoma cells. *Exp. Cell Res.* 154, 386-393.

Grover, A., Oshima, R.G., and Adamson, E.D. (1983) Epithelial layer formation in differentiating aggregates of F9 embryonal carcinoma cells. *J. Cell Biol.* 96, 1690-1696.

Grümmer, R., Hohn, H.-P., and Denker, H-W. (1989) Investigations on the invasion of choriocarcinoma cells grown as spheroids. *Placenta* 10, 511-512.

Haji-Karim, M. and Carlsson, J. (1978) Proliferation and viability in cellular spheroids of human origin. *Canc. Res.* 38, 1457-1464.

Hamasaki, K., Okamura, Y., Ueda, H., Kagawa, H., and Fujimoto, S. (1987) Immunocytochemistry of human chorionic gonadotropin in human chorionic villi. *Am. J. Obstet. Gynecol.* 156, 479-483.

Hertz, R. (1959) Choriocarcinoma of women maintained in serial passage in hamster and rat. *Proc. Soc. Expl. Biol. Med.* 102, 77-80.

Heyderman, E., Gibbons, A.R., and Rosen, S.W. (1981) Immunoperoxidase localization of human placental lactogen: A marker for the placental origin of the giant cells in 'syncytial endometritis' of pregnancy. *J. Clin. Pathol.* 34, 303-307.

Honegger, P. and Richelson, E. (1977) Biochemical differentiation in aggregating cell cultures of different fetal brain regions. *Brain Res.* 133, 329-339.

Horne, C.H.W., Towler, C.M., Pugh-Humphreys, R.G.P., Thomson, A.W., and Bohn, H. (1976) Pregnancy specific β_1-glycoprotein - a product of the syncytiotrophoblast. *Experientia* 32, 1197-1199.

Hoshina, M., Hussa, R., Pattillo, R., Camel, M., and Boime, I. (1984) The role of trophoblast differentiation in the control of hCG and hPL genes, In: *Human Trophoblast Neoplasms*, (eds.), R.A. Pattillo and R.O. Hussa, Plenum Press, New York/London, pp 299-312.

Hoshina, M., Boothby, M., Hussa, R., Pattillo, R., Camel, H.M., and Boime, I. (1985) Linkage of human chorionic gonadotropin and placental lactogen biosynthesis to trophoblast differentiation and tumorigenesis. *Placenta* 6, 163-172.

Hussa, R.O. (1977) Immunologic and physical characterization of human chorionic gonadotropin and its subunits in cultures of human malignant trophoblast. *J. Clin. Endocrinol. Metab.* 44, 1154-1162.

Kliman, H.J., Nestler, J.E., Sermasi, E., Sanger, J.M., and Strauss, J.F. (1986) Purification, characterization and in vitro differentiation of cytotrophoblasts from human term placenta. *Endocrinol.* 118, 1567-1582.

Kliman, H. J., Feinman, M. A., and Strauss, J. F., III (1987) Differentiation of human cytotrophoblast into syncytiotrophoblast in culture. *Tropho. Res.* 2,

Landry, J. and Freyer, J.P. (1984) Regulatory mechanisms in spheroidal aggregates of normal and cancerous cells. In: *Spheroids in Cancer Research* (eds.), H. Acker, J. Carlsson, R. Durand, and R. M. Sutherland, Springer-Verlag, pp. 50-60.

Landry, J., Freyer, J.P., and Sutherland, R.M. (1982) A model for the growth of multicellular spheroids. *Cell Tissue Kinet.* 15, 585-594.

Mareel, M., Kint, J., and Meyvisch, C. (1979) Methods of study of the invasion of malignant C3H-mouse fibroblasts into embryonic chick heart in vitro. *Virchows Arch. B Cell Pathol.* 30, 95-111.

Morrish, D.W., Manickavel, V., Jewell, Z.D., and Siy, O. (1987) Immunolocalization of alpha and beta chains of human chorionic gonadotropin, placental lactogen and pregnancy-specific beta-glycoprotein in term placenta by a touch preparation method. *Histochem.* 88, 57-60.

Nederman, R., Norling, B., Glimelius, B., Carlsson, J., and Bruuk, U. (1984) Demonstration of an extracellular matrix in multicellular tumor spheroids. *Canc. Res.* 44, 3090-3097.

Pattillo, R.A. and Gey, G.O. (1968) The establishment of a cell line of human hormone-synthesizing trophoblastic cells in vitro. *Canc. Res.* 28, 1231-1236.

Pattillo, R.A., Gey, G.O., Delfs, E., Huang, W.Y., Hause, L., Garancis, J., Knoth, M., Amatruda, J., Bertino, J., Friesen, H..G., and Mattingly, R.F. (1971) The hormone-synthesizing trophoblastic cell in vitro: A model for cancer research and placental hormone synthesis. *Ann. NY Acad. Sci.* 172, 288-

Pierce, G.B. and Midgley, A.R. (1963) The origin and function of human syncytiotrophoblastic giant cells. *Am. J. Pathol.* 43, 153-173.

Ritvos, O., Jalkanen, J., Huhtaniemi, I., Stenmark, O.-H., Alfthan, H., and Ranta, T. (1987) 12-O-tetradecanoyl phorbol-13-acetate potentiates adenosine 3', 5'-monophosphate-mediated chorionic gonadotropin secretion by cultured human choriocarcinoma cells. *Endocrinol.* 120, 1521-1526.

Saltzman, R.A., White, T.E., Miller, R.K., Sutherland, R.M., and Keng, P.C. (1987) Human choriocarcinoma (JAr) spheroid culture: A model system for studying the trophoblast. *Teratol.* 35, 10, 36A abstract.

Schleich, A. (1973) The confrontation of normal and malignant cells in vitro. An experimental system in tumor invasion studies. In: *Chemotherapy Of Cancer Dissemination And Metastasis*, (eds.), S. Garathini and G. Franchi, New York: Raven Press, pp. 51-58.

Schleich, A.B., Frick, M., and Mayer, A. (1976) Patterns of invasive growth in vitro. Human decidua graviditatis confronted with established human cell lines and primary human explants. *J. Natl. Cancer Inst.* 56, 221-237.

Seeds, N.W. (1973) Differentiation of aggregating brain cell cultures. In: *Tissue Culture Of The Nervous System*, (ed.), G. Sato, New York: Plenum Press, pp. 35-53.

Simpson, E.R., Porter, J.C., Milewich, L., Bilheimer, D.W., and MacDonald, P.C. (1978) Regulation by plasma lipoproteins of progesterone biosynthesis and 3-hydroxy-3-methyl glutaryl coenzyme A reductase activity in cultured human choriocarcinoma cells. *J. Clin. Endocrin. Metabol.* 47, 1099-1105.

Skehan, P. and Friedman, S.J. (1978) Differentiation of human choriocarcinoma cells by methotrexate. *In Vitro* 14, 340.

Sutherland, P.M. and Durand, R.E. (1976) Radiation effect on mammalian cells grown as an in vitro tumor model. *Curr. Top. Radiat. Res.* 11, 87-139.

Sutherland, R.M., McCredie, J.A., and Inch, W.R. (1971) Growth of multicell spheroids in tissue culture as a model of nodular carcinomas. *J. Natl. Cancer Inst.* 46, 113-117.

Takamizawa, H. and Sekiya, S. (1984) Cell biology of choriocarcinoma. *Asia-Oceania J. Obstet. Gynaecol.* 10, 245-256.

Trapp, B.D., Honegger, P., Richelson, E., and de Webster, H.F. (1979) Morphological differentiation of mechanically dissociated fetal brain in aggregating cell cultures. *Brain Res.* 150, 117-130.

Yuhas, J.M. and Li, A.P. (1978) Growth fraction as the major determinant of multicellular tumor spheroid growth rates. *Canc. Res.* 38, 1528-1532.

Yuhas, J.M., Li, A.P., Martinez, A.O., and Ladman, A.J. (1977) A simplified method for production and growth of multicellular tumor spheroids. *Canc. Res.* 37, 3639-3643.

CELL BIOLOGY AND
IMMUNOLOGY OF THE
INVASIVE TROPHOBLAST

ADHESIVE INTERACTIONS OF MURINE AND HUMAN TROPHOBLAST CELLS

Susan J. Fisher[1,3,4,6], Ann Sutherland[3], Lenny Moss[3],
Lynn Hartman[1], Eileen Crowley[2], Merton Bernfield[5],
Patricia Calarco[3], and Caroline Damsky[2,3]

[1]Divisions of Oral Biology and [2]Periodontology, School of Dentistry
[3]Department of Anatomy, School of Medicine
[4]Department of Pharmaceutical Chemistry, School of Pharmacy
University of California, San Francisco, California 94143 USA
[5]Joint Program in Neonatology, Harvard Medical School
Boston, Massachusetts 02115 USA

INTRODUCTION

Successful implantation of the embryo requires a complex and ordered series of adhesive interactions occurring between the trophoblast cells of the blastocyst and the uterus. First, the blastocyst hatches from the acellular zona pellucida and attaches to the uterine epithelium. These initial adhesive interactions are probably transient since in mice and in humans the cells of the trophoblast outgrowth are also invasive. As a result, these cells rapidly penetrate the uterine epithelium and its associated basement membrane, then invade the endometrium where they contact decidual cells, each of which is surrounded by a specialized basement membrane collar (Wewer et al., 1985, 1986). Invasion stops once the trophoblast cells have penetrated the uterine arterioles and tapped the maternal blood supply (Ramsey et al., 1976). The result is formation of the hemochorial placenta, in which blood from the maternal circulation constantly bathes the fetal chorionic villi. Thus, both cell-cell and cell-matrix interactions are important in the adhesive interactions that occur during placental development.

Since the dynamic aspects of many of the processes involved in implantation and placental development are difficult to study in vivo, several types of in vitro models have been used for this purpose. With regard to studying murine trophoblast behavior in vitro, mouse blastocysts have been cultured on plastic, glass, or collagen (Mintz, 1964; Cole and Paul, 1965; Gwatkin, 1966; Sherman and Barlow,1972; Spindle and Pedersen, 1973; Nilsson, 1974; Sherman, 1975a and b). Trophoblast outgrowth on more complex substrates such as extracellular matrices (ECMs, Glass et al., 1983) and cell monolayers (Cole and Paul, 1965; Salomon and Sherman, 1975; Sherman and Salomon, 1975; Sherman, 1975a, b; Glass et al., 1979) has also been studied. Investigators using such systems have consistently found that the outgrowing cells can invade and migrate through extracellular matrices (Glass et al., 1983), uterine stromal cells (Salomon and Sherman, 1975; Sherman and Salomon, 1975; Sherman, 1975a, b; Glass et al., 1979), or various other cell types (Cole and Paul, 1965; Salomon and Sherman, 1975; Sherman and Salomon, 1975; Glass et al., 1979). Recently, several studies have shown that the adhesive

[6]To Whom Correspondence Should Be Addressed: HSW 604, Box 0512, University of California-San Francisco, San Francisco, California 94143-0512 USA

molecules fibronectin (Fn), laminin (Ln), collagen type IV (Col IV) and vitronectin will also support trophoblast spreading in serum-free medium (Armant et al., 1986a, b; Farach et al., 1987; Sutherland et al., 1988).

The ability of blastocysts to attach and spread on these defined ligands suggests the presence of specific cell surface receptors to mediate these adhesive interactions. The most extensively studied class of extracellular matrix receptors is the integrin superfamily (Tamkun et al., 1986; Hynes, 1987). This family consists of a large number of heterodimeric receptor complexes that interact with their respective ligands via cell recognition sites containing the tripeptide Arg-Gly-Asp (Pierschbacher et al., 1984b). Among these are the fibronectin, vitronectin, collagen, and fibrinogen receptors (Pytela et al., 1985a, b; see reviews by Ruoslahti and Pierschbacher, 1986, 1987; Hynes, 1987). Recent studies (Sutherland et al., 1988) using anti-ECMr, a polyclonal antiserum against adhesion-related glycoproteins (Knudsen et al., 1981), suggest that members of the integrin family mediate blastocyst attachment and outgrowth on several defined extracellular matrix molecules in vitro and are likely to play a major role in embryo implantation in vivo. Expression of heparin/heparan sulfate on the surface of mouse blastocysts has also been implicated in their attachment and spreading (Farach et al., 1987).

With regard to studying human trophoblast behavior, hatched blastocysts have been cultured on monolayers of human endometrial epithelium (Lindenberg et al., 1986). Work from this laboratory has shown that chorionic villi (Fisher et al., 1985) and cytotrophoblast cells isolated from first trimester human placentae (Fisher et al., submitted for publication) can adhere to and rapidly degrade the extracellular matrices on which they are grown. The timetable of invasive behavior observed in vitro paralleled that which occurs in vivo. Villi and cells isolated from second trimester placentae could adhere to, but not invade, the same matrices. This model has been extremely useful in studies designed to determine stage-specific adhesive and invasive mechanisms relevant to the developmentally regulated behavior of trophoblast cells during placentation.

The current studies using mouse blastocysts focused on defining mechanisms that function during the peri-implantation period: the initial stages of trophoblast-ECM interactions. Human cytotrophoblast cells, isolated from first trimester placentae, were used to study adhesive interactions that are probably more relevant to the events subsequent to implantation, when trophoblast cells interact with components of the endometrium and myometrium. Thus, by studying both murine and human trophoblast cells we were able to examine factors that could play an important role in the cell-matrix adhesive interactions of trophoblast cells during several stages of early placental development.

MATERIAL AND METHODS

Antisera, Proteins, and Peptides

Antibodies used included: rabbit anti-fibronectin (anti-Fn; gift of Dr. Richard Hynes, Massachusetts Institute of Technology), rabbit anti-laminin (anti-Ln; gift of Dr. Deborah Hall, University of California, San Francisco), rabbit anti-type IV collagen (anti-Col IV, purchased from Dr. Heinz Furthmayr, Yale

University), rabbit anti-talin (gift of Dr. Mary Beckerle, University of Utah), rabbit anti-vinculin (anti-Vnc, Miles Scientific, Naperville IL) and goat anti-ECMr, (formerly called anti-GP 140; Knudsen et al., 1981; Tomaselli et al., 1987, 1988; Sutherland et al., 1988), made against hamster fibroblast adhesion-related glycoproteins of 120-160 kDa. Monoclonal antibodies to syndecan, a cell surface proteoglycan, were prepared in rats (Jalkanen et al., 1985). Polyclonal antibodies to the purified ectodomain of syndecan were raised in a rabbit (Saunders and Bernfield, 1988). Anti-human fibronectin receptor (anti-FnR) antiserum was a gift of Drs. Erkki Ruoslahti and Michael Pierschbacher (La Jolla Cancer Research Foundation). Monoclonal antibodies against the human FnR (anti-FnR Mab) and against integrins containing β_1 chains (anti-β_1) were produced in rats using the JAr choriocarcinoma cell line as the immunogen (Hall et al., 1989). All of the antibodies were used as the purified IgG fraction except anti-ECMr which was used as an antiserum in some experiments.

Purified Ln was a gift of Dr. Deborah Hall. Purified Fn and Col IV were purchased from Collaborative Research, Lexington, MA. Two synthetic hexapeptides were used: Gly-Arg-Gly-Asp-Ser-Pro (GRGDSP) containing the tripeptide cell binding region of fibronectin (Pierschbacher and Ruoslahti, 1984a; Yamada and Kennedy, 1984; Ruoslahti and Pierschbacher, 1986) and Gly-Arg-Gly-Glu-Ser-Pro (GRGESP) which does not prevent the adhesion of cultured cells to fibronectin (Pierschbacher and Ruoslahti, 1984b). The peptides were the kind gift of Drs. Michael Pierschbacher and Erkki Ruoslahti (La Jolla Cancer Research Foundation).

Embryo Culture And Outgrowth On Substrates Of ECM Molecules

Mouse embryos were cultured as described by Sutherland et al. (1988). The extent of spreading on ECM substrates was determined on prints of embryos using a Numonics digitizer as described by Sutherland et al. (1988). Embryo attachment in the absence of spreading was quantified by gently pipetting a small amount of medium on each embryo using a glass pipette pulled to a very fine bore. Those that did not move were considered to be attached.

Immunofluorescence And Immunoprecipitation Of Embryos

Immunofluorescence was performed on intact embryo outgrowths as described by Sutherland et al. (1988). In order to localize syndecan on sections of embryos, embryos were embedded in PEG-1000, sectioned at 1 μm on a Reichert microtome, and incubated with antibodies as described previously for whole mounts.

Surface ^{125}I-labeling of hatched blastocysts and 72 hour outgrowths and their subsequent immunoprecipitation and analysis by SDS-PAGE was performed according to Sutherland et al. (1988).

Cell Culture

Human cytotrophoblast cells were isolated from first and second trimester placentae obtained immediately after vacuum aspiration. Third trimester placentae, either preterm (27-37 weeks gestation) or full term, were obtained immediately after delivery. The fetal tissue was dissected free of adherent decidua and rinsed in several volumes of PBS. The villi were transferred to Dulbecco's modified Eagle's medium (DME) H-21 containing 10% fetal calf serum (FCS) and 50 μg/ml gentamicin. After centrifugation (180 g, 5 minutes) the medium was aspirated and the wet weight of the tissue determined. The villi were washed in two more changes of medium and resuspended (5/1, v/wt) in enzymatic dissociation solution I, consisting of PBS containing 500 U/ml collagenase (Sigma, Type IV), 200 U/ml hyaluronidase (Sigma, Type 1-S), 0.2 mg/ml DNase (Sigma, Type IV) and 1 mg/ml bovine serum albumin (BSA). The tissue was gently shaken in a 37°C water bath. The incubation time necessary to remove the layer of syncytiotrophoblast cells covering the chorionic villi was assessed histologically for placentae of different ages and for every lot of collagenase. In general, first trimester tissue was incubated for 20 minutes, and tissue from the second trimester onward for times ranging between 20 and 30 minutes. The first dissociation solution, containing syncytium, was removed and the villi were resuspended in dissociation solution II, consisting of PBS containing 0.25% trypsin (Sigma, Type XIII), 2 mM EDTA and 0.2 mg/ml DNase. After incubation in a shaking water bath at 37°C for 10 minutes, the mixture of villus cores and dissociated cells was diluted with an equal volume of medium containing 10% FCS. The villi and medium were poured through gauze to remove large tissue fragments, and the supernatant was centrifuged as described previously. Again, the incubation time necessary to remove the cytotrophoblast layer was assessed histologically for placentae of different ages. A single cycle removed the cytotrophoblast cell layer from first trimester chorionic villi, but additional cells could be isolated if this procedure was repeated once for second trimester placentae and twice for placentae from 27 weeks gestation onward. Cytotrophoblast cells were purified from the cell pellets using a Percoll gradient according to the method of Kliman et al. (1986). The resulting cells were resuspended in MEM D-valine medium (Gilbert and Migeon, 1975) containing 20% dialyzed FCS, 1% glutamine, and 50 μg/ml gentamycin, counted in a hemocytometer and adjusted to a final concentration of 5 x 10^5/ml. One ml of the cell suspension was added to each tissue culture well (15 mm diameter). In all cases either the wells, or coverslips (12 mm diameter) placed within the wells, were coated with a metabolically labeled ECM produced by PF HR9 cells as previously described (Fisher et al., 1985).

Human fibroblasts were isolated by further trypsin treatment (dissociation solution II, 30 minutes at 37°C) of the chorionic villus connective tissue cores retained during the gauze filtration step of the cytotrophoblast isolation procedure. The resulting cells were resuspended in DME H16 containing 10% FCS, 1% glutamine and 50 μg/ml gentamicin and plated onto 75 mm^2 tissue culture flasks (Costar, Cambridge, MA). The cells were passaged at least six times before they were used. The JAr human choriocarcinoma cell line (Patillo et al., 1971) was grown in DME H16 containing 10% FCS, 1% glutamine, and 50 μg/ml gentamicin. The BeWo human choriocarcinoma cell line (Patillo and Gey, 1968) was grown in Ham's F12 medium containing 10% FCS, 1% glutamine and 50 μg/ml gentamicin. Mouse mammary tumor epithelial cells (MMTE) were grown as monolayer cultures in DME H-21 medium supplemented with insulin, gentamycin, penicillin-streptomycin, 10% FCS, and 1% glutamine.

Adhesion Of Human Placental Cells To Nitrocellulose Replicas

The method used was modified from that of Hayman et al. (1982). Freshly collected human plasma was diluted (1:5, v/v) in electrophoresis sample buffer and separated under reducing conditions on 5% SDS-PAGE gels (Laemmli, 1970). The proteins were transferred to nitrocellulose (Towbin et al.,1979) and the replicas incubated overnight in 5% nonfat dry milk containing 1% thimerosal and 1% antifoam A to reduce non-specific adherence. Fibroblasts used in the overlay experiments were removed from the tissue culture flasks using 0.1 mg/ml trypsin and the cell pellets were washed in PBS containing 1% soybean trypsin inhibitor. Cytotrophoblasts were used immediately after isolation. All the cells were resuspended (1 x 10^6/ml) in RPMI medium containing 2% nutridoma, 1% glutamine, and 1% Na pyruvate, then allowed to attach to the nitrocellulose replicas for 2 hours. The replicas were washed three times in PBS and fixed in freshly made 3% paraformaldehyde-PBS. Bands to which the cells adhered were visualized by staining with amido black (Schaffner and Weissman, 1973). The position of fibronectin was localized by incubating the replica of an adjacent plasma-containing lane with antifibronectin antibodies and peroxidase-conjugated secondary antibodies.

Immunoprecipitation From Cultured Cells

Cytotrophoblasts and placental fibroblasts were metabolically labeled by incubation with [^{3}H]glucosamine (50 µCi/ml medium, American Radiolabeled Chemicals, 25 Ci/mmol) for 48 hours. The cultures were washed once with PBS, dissolved in lysis buffer (25 mM Tris, pH 7.5, containing 0.5% NP-40, 1 mM EDTA, and 150 mM NaCl) and incubated 30 minutes on ice, during which they were vortexed 5 times. The lysate was centrifuged (1200 g, 20 minutes; 10,000 g, 10 minutes) to remove insoluble material. The supernatant was precleared with unconjugated Sepharose 4B and immunoprecipitated adding 10% (v/v) culture supernatant containing anti-FnR or anti-β_1. Following a 1.5 hour incubation at 4°C with constant mixing, immune conjugates were bound by goat anti-rat antibody conjugated to agarose (Sigma). The beads were then washed in a series of buffers as described by Tomaselli et al. (1987). Finally the bound proteins were solubilized in Laemmli electrophoresis sample buffer containing 2% SDS but no reducing agents.

SDS-Polyacrylamide Gel Electrophoresis

Immunoprecipitates were analyzed under non-reducing conditions using 7.5% polyacrylamide slab gels (Laemmli, 1970). Prestained molecular weight markers (BRL) were used, and consisted of myosin (200 kDa), phosphorylase B (97.4 kDa), and BSA (68 kDa). The gels were fixed, infiltrated with Enhance (New England Nuclear), dried, and exposed to Kodak XAR-5 X-ray film.

Chromatography Of [^{3}H]Labeled Oligosaccharides

[^{3}H]Glucosamine-labeled oligosaccharides were isolated from pronase digests (Fisher and Laine, 1979) of the immunoprecipitates. The solubilized

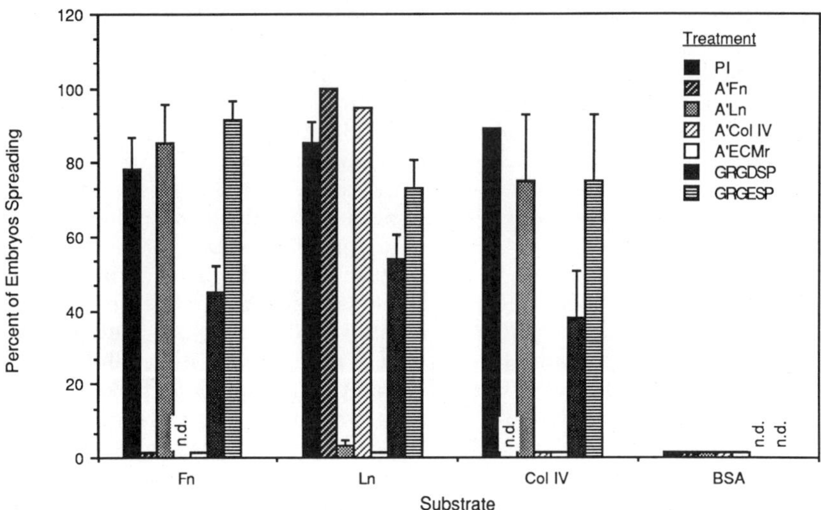

Figure 1. Graph of embryo outgrowth on Fn (25 µg/ml), Ln (25 µg/ml, Col IV (25 µg/ml) or BSA (25 µg/ml). Embryos were cultured, without prior incubation, in the continuous presence of either substrate-specific antibodies, anti-ECMr, or hexapeptides. PI, preimmune IgG (160 µg/ml); A'Fn, antifibronectin (30 µg/ml); A'Col IV, anti-collagen type IV (25 µg/ml); A'ECMr, anti-extracellular matrix receptor (30 µg/ml).

oligosaccharides were applied to a 2.5 x 60 cm column of Sephades G-50 (fine) and 1 ml fractions were collected. Radioactivity was determined by counting in a Beckman Scintillation Counter (Model 1701).

RESULTS

Extracellular Matrix Proteins In Implantation

Blastocyst outgrowth on substrates coated with Ln, Fn, or Col IV, and the effects of adding specific anti-ligand and anti-receptor antibodies, as well as RGD-containing peptides, on this process have been studied. The results of these experiments are graphed in Figure 1. Photographs of results for embryos cultured on Fn are also included (Figure 2). Immediately after hatching from the zona pellucida, blastocysts were not adhesive. In most cases embryos cultured in serum-free medium attached and initiated spreading on Ln and Fn by 24-36 hours, but this process was slightly delayed on Col IV. Maximal spreading on all the substrates occurred by 48-60 hours. Embryos cultured on BSA substrates neither attached nor spread. The interactions of embryos with Fn, Ln, and Col IV were ligand-specific as demonstrated by the addition of the relevant anti-ligand sera (Figure 1). In particular, embryo outgrowth on Ln was also unaffected by the presence of antibodies against Col IV and against entactin (data not shown), two common contaminants of Ln preparations.

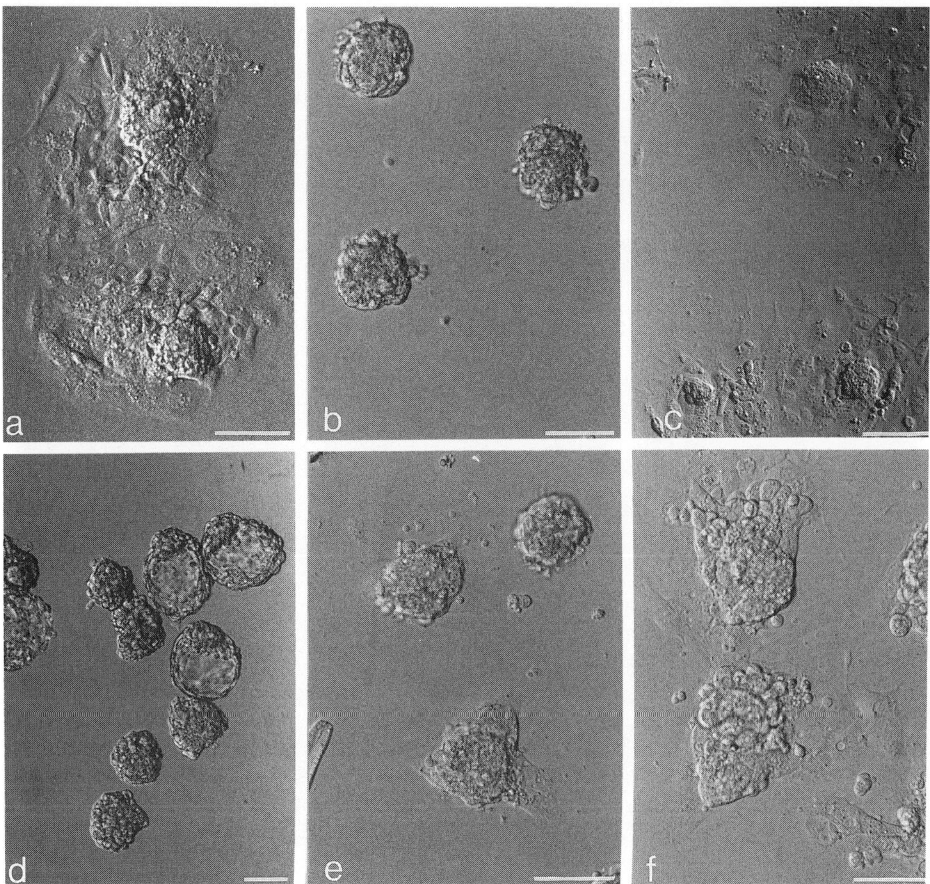

Figure 2. Embryo outgrowth on Fn (25 µg/ml) in serum-free medium under various treatments. (a) Control; no antibody. (b) Anti-Fn. (c) Anti-Ln. (d) Anti-ECMr. (e) GRGDSP hexapeptide. (f) GRGESP hexapeptide. Bars, 100 µm .

Effects Of Anti-ECMr Antibodies And Synthetic Peptides On Blastocyst-Ligand Interactions

Anti-ECMr has been shown to recognize the β_1 family of integrin adhesion receptors (Tomaselli et al., 1987, 1988), and to inhibit attachment of a wide variety of rodent cells to Fn, Ln, and collagens. This antibody completely inhibited the attachment of embryos on Fn (Figures 1 and 2), Ln (Figure 1) and Col IV (Figure 1). The effects of this antibody were reversible. When embryos cultured for 48 hours in anti-ECMr-containing medium were rinsed and replated in control medium, spreading was initiated within 4 hours and outgrowths were evident within 12 hours. Peptides from two sources (see Materials and Methods) were used to examine the role of the Arg-Gly-Asp-containing cell-recognition site in embryo adhesion and outgrowth formation. Neither the active (GRGDSP), nor the control (GRGESP) peptide prevented attachment of embryos to Fn or Ln substrates at the concentrations tested (250 and 500 µg/ml). However, GRGDSP had a pronounced

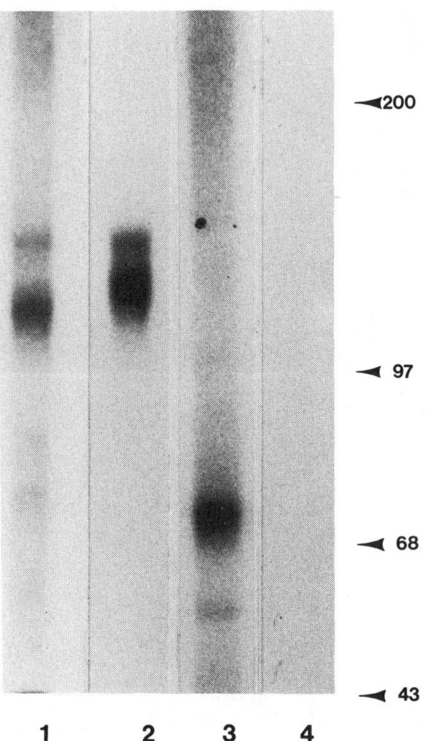

Figure 3. Immunoprecipitation of anti-ECMr antigens from embryos and cells. ^{125}I-labeled cell surface proteins were precipitated from 74 hour mouse embryo outgrowth (Lane 1), MMTE (Lane 2), and mouse hatched blastocysts (Lane 3). (Lane 4) Immunoprecipitation by normal goat serum of ^{125}I-labeled cell surface proteins from MMTE cells. Cells and embryos were surface-labeled with ^{125}I, and then incubated in medium containing either anti-ECMr or preimmune antibodies in order to bind integrin molecules exposed on the external surface of the embryo or cells prior to lysis (see Materials and Methods).

effect on trophoblast outgrowth, greatly reducing the incidence and extent of outgrowth on both Ln and Fn (Figure 1). The control peptide, GRGESP, had no effect on outgrowth on Fn substrates, but inhibited to a small extent embryo outgrowth on Ln.

Synthesis And Expression Of Adhesion Receptors By Embryo Outgrowth

Next, this laboratory investigated whether mouse trophoblasts express integrin-related ECM receptors recognized by anti-ECMr at a time that is consistent with their proposed function in outgrowth formation. Embryos were ^{125}I-labeled at either the hatched blastocyst stage or after culturing for 72 hours on Fn. Immunoprecipitates of cell-surface ^{125}I-labeled proteins obtained using anti-ECMr (Figure 3, Lane 1) showed two bands: a 144 kDa component and a diffuse band from

120 - 125 kDa. A similar pattern was recognized on cultured murine mammary epithelial cells (MMTE; Figure 3, Lane 2). Immunoprecipitates from embryos harvested shortly after hatching, at a time when they are nonadhesive (Figure 3; Lane 3), contained no bands in the 120 - 160 kDa region of the gel above background, although an 80 kDa band was detected.

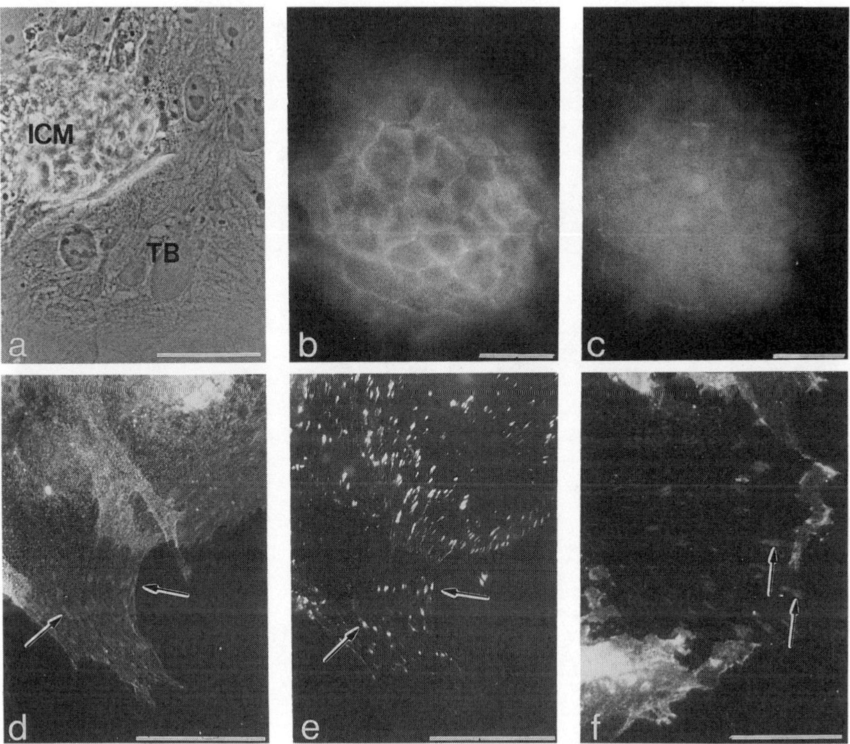

Figure 4. Immunofluorescent staining of antigens recognized by anti-ECMr, anti-FnR, and anti-Vnc on blastocyst outgrowths. (a) Phase photograph of an outgrowth, demonstrating the regions shown in the other photographs. ICM, inner cell mass. TB, trophoblast cells. (b and c) Double staining for anti-FnR and anti-Vnc in the ICM region of a 72 hour outgrowth. Anti-FnR is localized at the surface of cells in the top of the ICM, while anti-Vnc staining is diffuse and only slightly above background. (d and e) Double staining for anti-FnR and anti-Vnc in the TB region of a 72 hour outgrowth. In the highly spread trophoblast cells at the periphery of the outgrowth, both anti-FnR and anti-Vnc staining are organized into a series of discrete cell-matrix contacts (arrows). (f) In areas where the trophoblast has been torn away from the substrate, anti-ECMr staining is seen on the substrate in bright strips and dots (arrows). Bars, 25 μm.

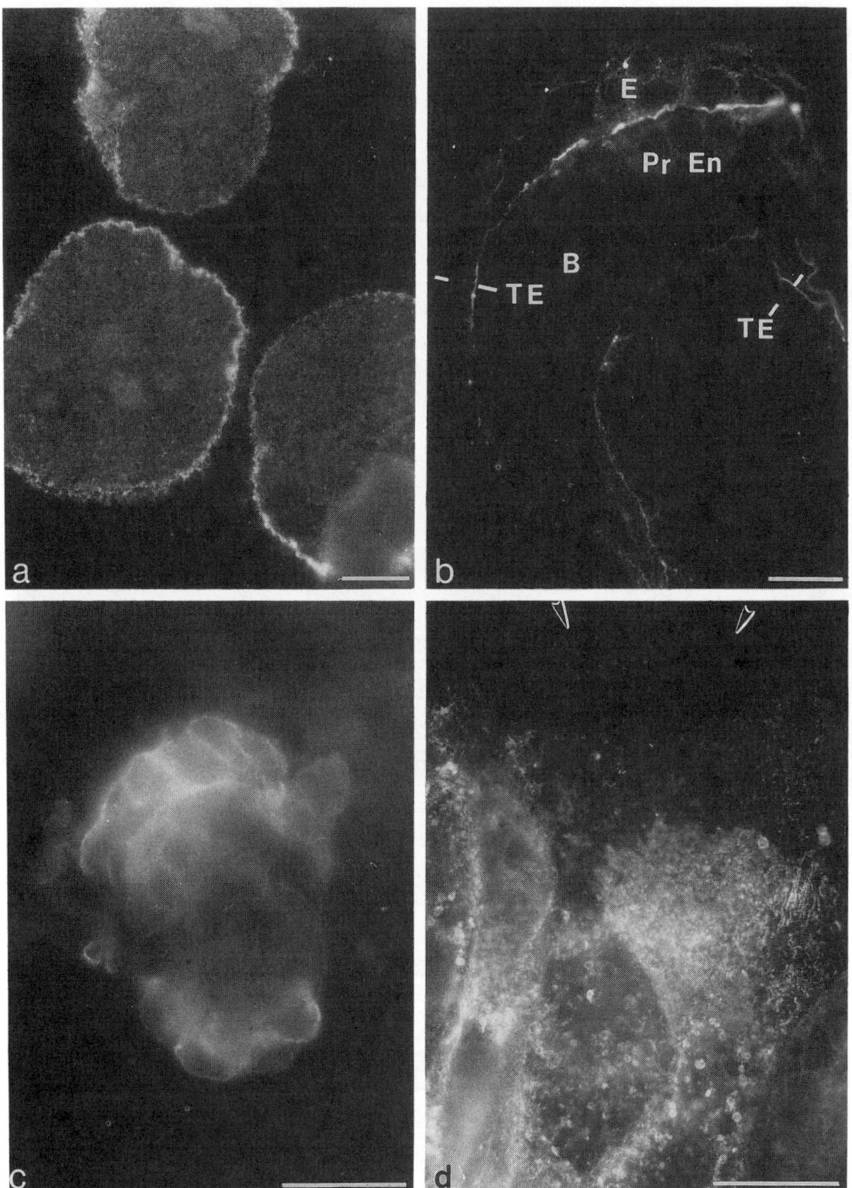
a
b
E
Pr En
B
TE
TE
c
d

Next, it was determined whether the localization of ECMr antigens was consistent with their proposed function in outgrowth formation. Embryos, hatched blastocysts and embryo outgrowths were stained with anti-ECMr and a polyclonal antiserum against the human Fn receptor (anti-FnR). Both antibodies recognize the β_1 family of integrins. Indirect immunofluorescence was performed on hatched blastocysts and blastocysts cultured 48 and 72 hours post-hatching on Fn substrates. In hatched blastocysts stained with either anti-ECMr or anti-FnR, no fluorescent signal above the control level was seen (data not shown). The 48 hour and 72 hour embryo outgrowths were highly three-dimensional with a rounded inner cell mass (ICM) and a surrounding flattened area of trophoblast outgrowth (Figure 4a). Results from single-staining experiments using anti-ECMr alone gave staining patterns similar to those with anti-FnR. At both early (48 hours post-hatching) and later (72 hours post-hatching) periods, anti-ECMr and anti-FnR staining were strongly enriched at the surface of cells in the top part of the ICM (Figure 4b). In contrast, staining for both Vnc (Figure 4c) and talin (not shown) in this region was diffuse and only slightly above the preimmune control. At the earliest time points that could be assayed during the outgrowth period, the pattern of cell-matrix contact sites was highly organized in the lamellopodial protrusions of the spreading trophoblast cells, as detected by staining with anti-FnR (Figure 4d), anti-Vnc (Figure 4e), and anti-ECMr (Figure 4f).

Figure 5. Immunofluorescent staining of the cell surface proteoglycan syndecan recognized by polyclonal antibodies on sections of PEG-embedded 8-cell embryos and hatched blastocysts, or on whole-mount embryo outgrowths. (a) Anti-syndecan staining is prominent on the external surface of 8-cell embryos, but is not detectable in areas of cell-cell contact. (b) In the hatched blastocyst, anti-syndecan staining is lost from the external surface of the embryo, but is prominent on the internal surfaces of both polar and mural trophectoderm (TE, white bars demark the internal and external surfaces of the trophectoderm), as well as at the interface of primitive endoderm (PrEn) and primitive ectoderm (E). B, blastocoele. (c) In the ICM region of embryo outgrowths, anti-syndecan staining can be seen outlining many, but not all of the cells of the ICM. (d) In the TB region anti-syndecan staining is punctate, and absent from the most peripheral cells. Arrowheads indicate edge of outgrowth. Bars; a and b, 10 μm; c and d, 25 μm.

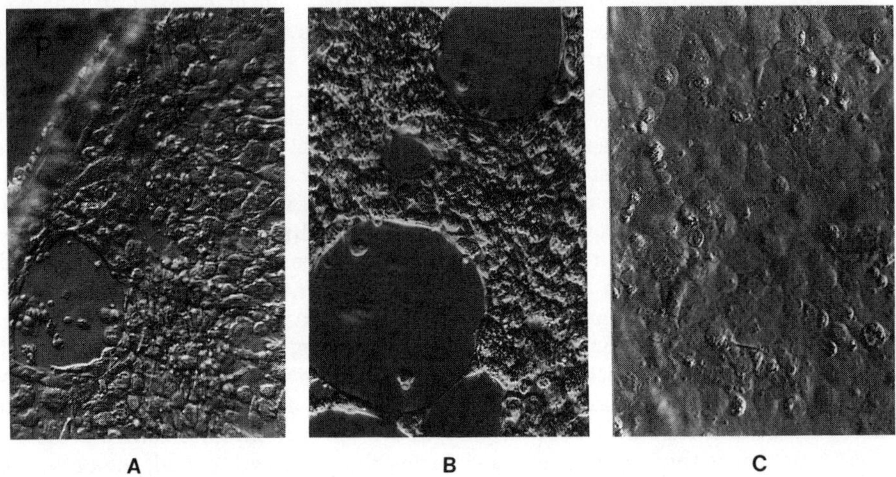

Figure 6. Isolated cytotrophoblasts cultured on PF HR9 ECM. (A) First trimester cells attached and spread for 24 hours on the matrix-coated coverslip, but failed to adhere to the uncoated tissue culture plastic (P). A small hole, cleared of matrix, is visible at the left edge of the coverslip. (B) The cleared areas were rapidly expanded during the next 48 hours in culture. Second trimester human cytotrophoblast cells (C) did not invade the matrix substrates on which they were cultured and remained as confluent monolayer even after 72 hours in culture.

Localization Of Syndecan, A Cell Surface Proteoglycan

An integral-membrane cell-surface proteoglycan, syndecan, has been shown to interact with the heparan binding domain of Fn, as well as with interstitial collagens. To determine whether syndecan might also play a role in blastocyst-ECM interactions, polyclonal antibodies against this component were used to determine the onset of its expression and its distribution in preimplantation embryos. Examination of PEG-embedded sections stained with anti-syndecan showed that it appeared at cell surfaces as early as the four-cell stage when it surrounded the external surface of the blastomeres. It was not found in areas of cell-cell contact. This pattern continued during cleavage (Figure 5a shows a section of an eight-cell embryo). In the morula, syndecan was also expressed at sites of cell-cell contact. Just prior to hatching, syndecan was on all cell surfaces: it was prominent at the external surface of the embryo and outlined all the cells of the inner cell mass and the trophectoderm (data not shown). After hatching and with the differentiation of primitive endoderm, syndecan became polarized. It was lost from the external surface of the embryo and from the surface of the primitive endoderm facing the blastocoele cavity. It remained on the internal surface of both the mural and polar trophectoderm and was especially prominent at the interface between primitive ectoderm and primitive endoderm (Figure 5b). In stained whole mounts of 72 hour blastocyst outgrowths, syndecan was present on the surface of the cells in the rounded inner cell mass (Figure 5c). In the outgrowth area it was not present on the trophoblast cells at the periphery of the outgrowth, but was present on the endoderm cells that migrated on top of these trophoblast cells (Figure 5d). In contrast to integrin, no syndecan staining was detected on the substrate in areas where embryos were torn away from the dish (data not shown). These results suggest that syndecan may not play a role in initial mouse embryo-ECM interactions.

Human Cytotrophoblasts Adhere To And Invade Extracellular Matrices In Vitro

When first trimester human cytotrophoblasts were plated onto 15 mm tissue culture wells containing 12 mm glass coverslips coated with the ECM synthesized by the PF HR9 teratocarcinoma cell line, the cells preferentially attached to the matrix-coated coverslips. Few, if any, adhered to the uncoated tissue culture plastic, which was exposed at the perimeter of the well (Figure 6a). In addition, the behavior of cytotrophoblasts cultured from first trimester human placentae was markedly different from that of cytotrophoblasts cultured from placentae of the second trimester onward. When first trimester cytotrophoblast cells were plated as a confluent monolayer, small, usually circular holes in both the monolayer and the matrix substrate (as determined by scanning electron microscopy, data not shown) appeared after the first 24 hours in culture (Figure 6a). Such gaps formed only when the cells were plated as a confluent monolayer and were never observed in more sparse cultures. In addition, only a subpopulation of the cells appeared to be involved in the initial process of rapid clearing since islands of cells that retained the morphological characteristics of an intact monolayer were always present in these cultures. During the next 24 hours the "holes" rapidly expanded and coalesced until these cleared areas occupied a major portion of the tissue culture well (Figure 6b). After 72 hours significant expansion of the existing holes usually stopped, and no new holes formed in the discrete islands that remained of the original cell monolayer. Since cytotrophoblasts adhered poorly in the absence of matrix, they accumulated at the periphery of the cleared areas. In contrast, when cytotrophoblasts isolated from second trimester (Figure 6c) onward were plated on matrix, no holes in the monolayer were seen, even after 1 week in culture. Together, these results suggest that human cytotrophoblast cells retain in vitro the stage-specific invasive behavior they exhibit in vivo.

Degradation Of ^{3}H-Labeled PF HR9 Matrix By Cytotrophoblasts, Placental Fibroblasts And Human Choriocarcinoma Cells

Cytotrophoblasts were plated on ^{3}H-labeled matrices to confirm that formation of holes in confluent monolayer cultures of first trimester cells was indicative of active matrix degradation and not solely the result of mechanical penetration. The experiment was repeated 6 times using matrices labeled with $[^3H]$ glucosamine, $[^3H]$proline and $[^3H]$leucine. In each case at least half the total radioactivity incorporated into the matrix was solubilized by the first trimester cytotrophoblast cells during the first 3 days in culture. However, the absolute counts released varied among experiments due to differences in ^{3}H-label incorporation into the matrix. Figure 7 summarizes the results from one such experiment and shows the pattern of release into the medium of $[^3H]$glucosamine-(panel A), $[^3H]$proline-(panel B), and $[^3H]$leucine-(panel C) labeled matrix components as a function of time in culture. During the first 3 days, first trimester cytotrophoblast cells rapidly solubilized both protein- and carbohydrate-containing matrix components.

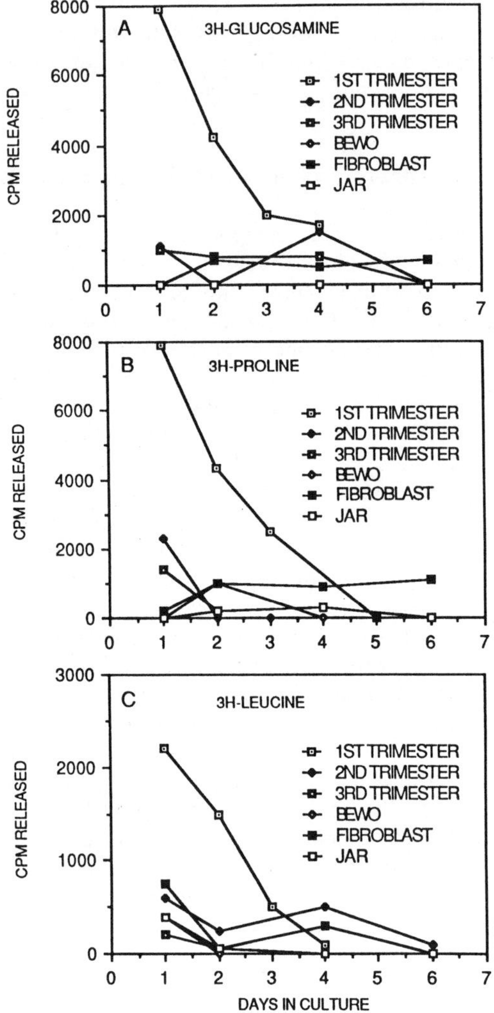

Figure 7. Degradation of [3H]-labeled PF HR9 matrix by cytotrophoblasts, placental fibroblasts and the human choriocarcinoma cell lines BeWo and JAr. Counts released daily from (A) [3H]glucosamine-, (B) [3H]proline-, and (C) [3H]leucine-labeled matrices are shown. Each data point is the mean of the [3H]-matrix components released into the medium from three wells. The bars represent standard error of the mean. Each data point was corrected for the spontaneous release of labeled matrix components by subtracting the mean of counts released into the medium from three wells that contained [3H]-labeled matrices but no cells.

Second and third trimester cytotrophoblast cells showed a greatly reduced ability to liberate glucosamine-containing matrix material. The cumulative release of [3H]glucosamine components by cells isolated from a 15 week human placenta was approximately 30% of that released by first trimester cells (Figure

7A). Similar results were obtained with [³H]proline- (Figure 7B) and [³H]leucine-labeled (Figure 7C) matrix components. Cells isolated from later second trimester (data not shown) and third trimester placentas showed degradation levels similar to that of the early second trimester cytotrophoblasts. Thus, cytotrophoblast-associated degradative activity was reduced substantially early in the second trimester.

   To determine whether the ability to degrade matrix substrates was a unique characteristic of first trimester human cytotrophoblast cells, first trimester placental fibroblasts and the human choriocarcinoma cell lines BeWo and JAr were also plated on radiolabeled PF HR9 extracellular matrices (Figure 7). For all radiolabeled matrices tested the release of radioactivity by all three cell types was either comparable to or less than that of the second trimester cytotrophoblast cells. In addition, daily observation of the cultures using phase contrast microscopy showed that no gaps formed in any of the fibroblast or choriocarcinoma monolayer cultures.

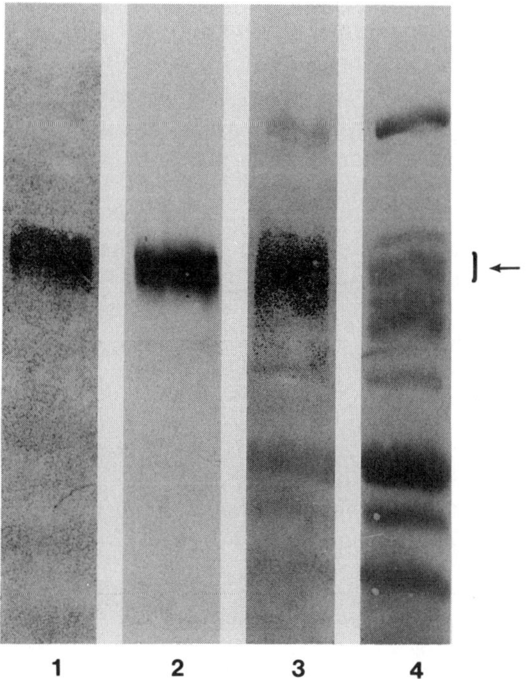

Figure 8. Attachment of cytotrophoblasts to nitrocellulose transfers of plasma glycoproteins. Second trimester human cytotrophoblasts (Lane 1) adhered to a band which migrated in the 200 kDa region of the gel and which reacted with anti-Fn serum (Lane 2). Fibroblasts isolated from first trimester human placentae (Lane 3) also adhered to fibronectin. The transfer in Lane 4 was stained with amido black. The fibronectin bands are bracketed and indicated by an arrow.

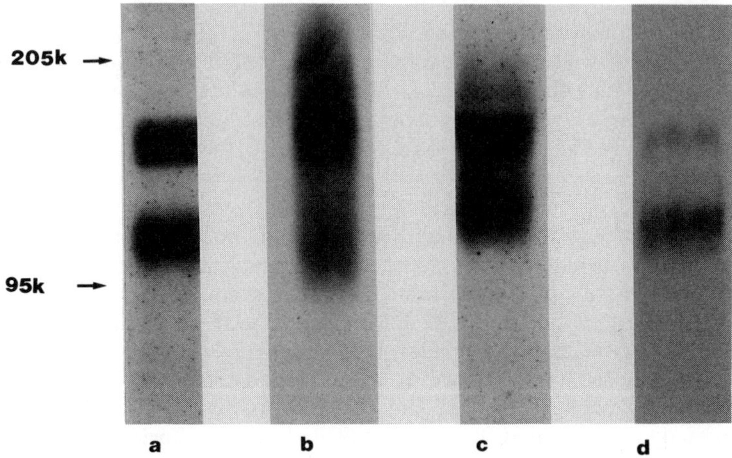

Figure 9. Immunoprecipitation of anti-FnR and anti-β_1-integrin antigens from placental cells. Fibroblasts contained clearly distinguishable bands of 120 and 150 kDa (Lane a) typical of the FnR in many cell types. In contrast, the corresponding bands present in immunoprecipitates from first and second trimester cytotrophoblast cells (Lanes b and c) were extremely broad and overlapped one another. However, the immunoprecipitates from term cytotrophoblasts (Lane d) again contained two distinct bands.

Adherence Of Human Cytotrophoblasts To Electrophoretically Separated Plasma Glycoproteins

The adhesive properties of human cytotrophoblast cells were examined in several ways. The relative affinity of cytotrophoblast cells for adhesive glycoproteins in plasma was determined by overlaying nitrocellulose transfers with freshly isolated cells (Figure 8). Both first and second trimester human cytotrophoblasts (Lane 1) preferentially adhered to a band which migrated in the 200 kDa region of the gel and which reacted with anti-Fn serum (Lane 2). Fibroblasts isolated from first trimester human placentae adhered to a 70 kDa protein, probably vitronectin (data not shown on this 5% gel, Hayman et al., 1982), and fibronectin (Lane 3). However, the area of attachment of the fibroblasts was always broader than that of the trophoblast cells. When fibronectin was removed from the plasma by passage through a column of gelatin-Sepharose, no adherence of either the fibroblasts or the trophoblasts to the 200 kDa region of the transfer was observed (data not shown).

Human Cytotrophoblast Integrins

Cytotrophoblast cells isolated from first trimester and from term human placentae attached to both Fn and Ln (data not shown). In order to examine the integrin-related adhesion receptors present in these cells, two monoclonal antibodies, one recognizing specifically the FnR (anti-FnR Mab) and the other the β_1 integrin family (anti-β_1), were used in immunoprecipitation experiments (Hall et al., 1989). Anti-FnR Mab was used to immunoprecipitate this antigen from ^{125}I-labeled first, second and third (term) trimester human cytotrophoblasts

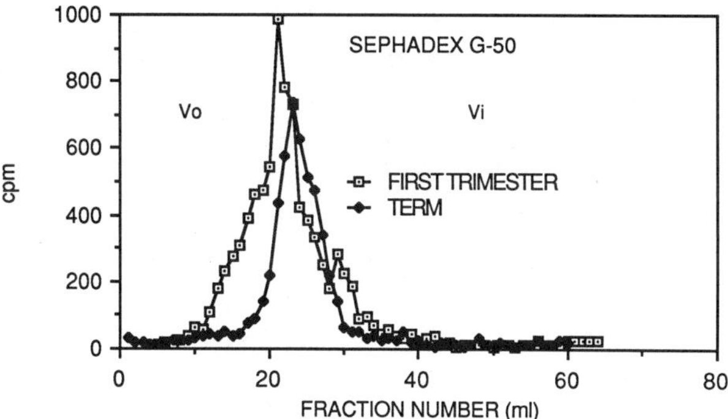

Figure 10. Sephadex G-50 chromatography of integrin ^{3}H-glycopeptides. The ^{3}H-glycopeptides isolated from first trimester cytotrophoblast cells chromatographed near the void volume of the column. By comparison, the ^{3}H-glycopeptides isolated from term cytotrophoblasts were less heterogeneous and had a lower average molecular weight. V_o, void volume; V_i, inclusion volume.

and from ^{125}I-labeled first trimester human placental fibroblasts. Electrophoretic separation of the immunoprecipitates from all the cells (Figure 9) contained two bands in the 100 to 200 kDa region of the gel. Interesting differences in the appearance of these bands were observed among the various placental cells examined. Fibroblasts contained clearly distinguishable bands of 120 and 150 kDa (Lane a) typical of the FnR in many cell types. In contrast, the corresponding bands present in immunoprecipitates from first and second trimester cytotrophoblast cells (Lanes b and c) were extremely broad and overlapped one another. However, the immunoprecipitates from term cytotrophoblasts (Lane d) again contained two distinct bands. Thus, the electrophoretic characteristics of the FnR from term cytotrophoblasts resembled that of placental fibroblasts rather than those of cytotrophoblasts isolated from first and second trimester placentae.

The diffuse appearance of the bands from early cytotrophoblast cells suggested that both chains of the FnR were more heavily glycosylated in first and second trimester human cytotrophoblast cells than in either fibroblasts or term cytotrophoblasts. To determine whether this was the case the anti-β_1 monoclonal antibody (which recognizes the FnR as well as other β_1 heterodimers) was used to immunoprecipitate these receptor molecules from cytotrophoblasts and from fibroblasts. The immunoprecipitates were subjected to extensive pronase digestion to isolate the carbohydrate portion of the molecules. The resulting glycopeptides were chromatographed on a column of Sephadex G-50 (Figure 10). A heterogeneous population of ^{3}H-labeled glycopeptides, spanning a similar molecular weight range, was isolated from both first (Figure 10) and second trimester (data not shown) human cytotrophoblast cells. These glycopeptides included high molecular mass components (8-10 kDa) that eluted near the void volume of the column. Interestingly, the chromatographic properties of ^{3}H-glycopeptides isolated from term cytotrophoblasts (Figure 10) and first trimester human placental fibroblasts (data not shown) were nearly identical. In both cases a relatively more homogeneous peak that lacked the higher molecular weight population of oligosaccharides was observed. Thus, the distinctive electrophoretic migration of the FnR from early placental cytotrophoblasts is due, at least in part, to increased glycosylation of this molecule.

DISCUSSION

Our experiments focused on the interactions of trophoblast cells with ECM components. These interactions are an important part of the cascade of events crucial for the successful implantation of the embryo. Data obtained using the implantation stage mouse embryo demonstrate that blastocysts interact in vitro with several defined ECM ligands (e.g., Fn, Col IV, and Ln) that they are likely to encounter in vivo. All of these interactions are inhibited by antibodies that recognize the β_1 family of cell surface integrin adhesion receptors (Hynes, 1987; Sutherland et al., 1988). These integrins are present on the surface of the blastocyst at the time they become adhesive, but not before. Following attachment, these receptors can be localized to adhesion plaques at the periphery of the trophoblast outgrowth, as well as on other cells of the implanting embryo. Together, the antibody inhibition, immunoprecipitation, and localization experiments (Figures 1-4) suggest that integrins are likely to play a significant role in the initial interactions of the embryo with components of the uterine stroma and basement membranes.

Adhesion receptors other than those belonging to the integrin superfamily also play a role in embryo-ECM interactions. Recently, heparin/heparan sulfate proteoglycan was localized to the surface of periimplantation stage blastocysts. Furthermore, soluble heparin inhibited the rate at which embryos attach and grow out on both ECM components and monolayers of uterine epithelial cells (Farach et al., 1987). A cell surface proteoglycan that could influence blastocyst interactions with ECM is syndecan, an integral membrane component containing both heparan sulfate and chondroitin sulfate moieties (Rapraeger et al., 1985). It is proposed to be a receptor for the matrix components Fn, thrombospondin, and interstitial collagens (Saunders and Bernfield, 1988). Syndecan is found mainly on epithelial cell surfaces of mature tissue and its expression is developmentally regulated during epithelial-mesenchymal interactions. The current results (Figure 5) show that syndecan is expressed on all cells of the blastocyst prior to hatching, but is absent from the external surface of the embryo at the time of implantation. Rather, at this time it is polarized to sites of earliest matrix accumulation within the embryo where it is in a position to play a role in primitive endoderm differentiation and/or parietal endoderm migration (Sutherland et al., 1989). Thus, syndecan is not likely to play a role in the initial interactions between the embryo and the uterine ECM. This does not rule out a role for this molecule in later stages of implantation, nor does it imply that other heparan sulfate-bearing molecules do not play a role in initial embryo-ECM interactions.

The current experiments using human cytotrophoblasts suggest that integrin-ECM interactions are likely to play an interesting role in the stages subsequent to implantation during which the placenta becomes established. Cell migration, as well as ECM degradation, are required in order for the trophoblast to contact the maternal circulation. These processes were studied in an in vitro model that uses cytotrophoblasts isolated from placentae of different gestational ages, and a complex basal lamina-like matrix laid down by PF HR9 teratocarcinoma endodermal cells (Fisher et al., 1985; Fisher et al., submitted for publication). The results shown in Figures 6 and 7 demonstrate that the invasive activity of human cytotrophoblasts is developmentally regulated in vitro, as well as in vivo. This suggests a concomitant developmental regulation of the proteases that play a role in trophoblast invasion. Preliminary evidence from our laboratory indicates that a group of metalloproteases is present in extracts and conditioned medium from first trimester cytotrophoblasts, but is absent in later periods of gestation (Fisher et al., 1989).

Migration of the trophoblast cells through the endometrium and/or myometrium requires highly specialized and developmentally regulated interaction with a variety of components present in basement membranes and interstitial matrices. Analysis to date of the adhesion mechanisms of cytotrophoblasts shows that they adhere to Fn and Ln and express at least three sets of heterodimers of the β_1 integrin family of adhesion receptors (Moss et al., in preparation). It is not yet clear whether there are differences in the amount of these integrin receptors expressed during gestation. However, Figures 9 and 10 show that glycosylation of the integrins changes dramatically during gestation. The two components of the fibronectin receptor heterodimer isolated from first and second trimester cytotrophoblasts and recognized by the B1E5 fibronectin receptor monoclonal antibody appear much more diffuse by SDS-PAGE than those isolated from term cytotrophoblast or first trimester fibroblasts. Analysis of glycopeptides isolated from the integrins indicates that those from first trimester cells are significantly larger and more heterogeneous than those from later stages (Moss et al., 1989). One well documented role of the oligosaccharide moieties of glycoproteins is their ability to protect the polypeptide backbone from proteolysis, prolonging the half-life of the glycoprotein (Olden et al., 1978). In the case of first trimester cytotrophoblasts, specialized oligosaccharides may protect the integrin adhesion receptors from degradation in the protease-rich environment present during the early stages of placentation. Interestingly, fibronectin isolated from early embryonal carcinoma cells (Cossu and Warren, 1983) and from human placenta (Zhu et al., 1984) carries branched polylactosamine chains. The oligosaccharides of placental fibronectin increase in chain length and complexity during pregnancy (Zhu and Laine, 1987). This post-translational modification makes placental fibronectin more protease resistant than plasma fibronectin, whose carbohydrate moieties are in the form of less complex, biantennary chains (Zhu et al., 1984). However, increased glycosylation of the integrins could also have a more direct effect on trophoblast interactions with ECM. In the case of placental fibronectin, for example, the presence of complex polylactosamine chains reduced its affinity for collagen (Zhu and Laine, 1985). Thus, in future experiments we will address the issue of whether alterations in the glycosylation of the integrins during gestation can modulate the adhesive interactions between these receptors and their ECM ligands.

SUMMARY

Evidence is presented suggesting that the trophoblast-ECM interactions involved in establishing the placenta, including the crucial steps of blastocyst adhesion and trophoblast invasion, are likely to be extremely complex. This is consistent with the fact that these processes require the trophoblast cells to interact with a variety of maternal cells and their associated basement membranes, as well as components of the interstitial matrix. For the trophoblast cells to reach and ultimately penetrate the uterine vessels, both directed migration and penetration must be facilitated. At the same time these processes must be carefully controlled so as to limit invasion physically to the endometrium and/or superficial segments of the myometrium. These specialized requirements suggest that interesting regulatory processes are involved. This laboratory's approach for understanding these mechanisms has been to correlate the developmentally regulated changes in the adhesive and invasive properties of the trophoblast cells with changes in the expression or regulation of the molecules that mediate these processes. These studies thus far show that the β_1 integrin adhesion receptors play a significant role in the initial interactions of the blastocyst with the ECM. The fact that glycosylation of human cytotrophoblast integrins is developmentally regulated during gestation also suggests that integrins play an important role in later stages of placentation. Similarly, evidence for the developmental regulation of metalloproteases during gestation has been developed. Whether these mechanisms are unique to trophoblast cells or are common to other invasive cells remains to be determined. For example, tumor cells may use the mechanisms, which are transiently employed by the trophoblast cells to gain access to the maternal circulation, to enter the blood stream during metastasis.

ACKNOWLEDGEMENTS

The authors are grateful to Ms. Ann Burlage and Ms. Evelyn Nakagawa for excellent technical assistance and to Mr. Albert Tai for photographic preparation. This work was supported by grants (HD22210 and HD22518) from the National Institutes of Health.

REFERENCES

Armant, D.R., Kaplan, H.A., and Lennarz, W.I. (1986a) Fibronectin and laminin promote in vitro attachment and outgrowth of mouse blastocysts. *Dev. Biol.* 116, 519-523.

Armant, D.R., Kaplan, H.A., and Lennarz, W.I. (1986b) The effect of hexapeptides on attachment and outgrowth of mouse blastocysts cultured in vitro: evidence for the involvement of the cell recognition tripeptide Arg-Gly-Asp. *Proc. Natl. Acad. Sci.* 83, 6751-6755.

Cole, R.J. and Paul, J. (1965) Properties of cultured preimplantation mouse and rabbit embryos, and cell strains derived from them. In: *Properties of Preimplantation Stages of Pregnancy*, (eds), G.E.W. Wolstenholme and M. O'Conner, New York: Academic Press, pp. 82-122.

Cossu, G. and Warren, L. (1983) Lactosaminoglycans and heparan sulfate are covalently bound to fibronectins synthesized by mouse stem teratocarcinoma cells. *J. Biol. Chem.* 258, 5603-5607.

Farach, M.C., Tang, J.P., Decker, G.L., and Carson, D.D. (1987) Heparin/heparan sulfate is involved in attachment and spreading of mouse embryos in vitro. *Dev. Biol.* 123, 401-410.

Fisher, S.J. and Laine, R.A. (1979) Carbohydrate structure of the major glycopeptide from human cold-insoluble globulin. *J. Supramol. Struct.* 11, 391-399.

Fisher, S.J., Leitch, M.S., Kantor, M.S., Basbaum, C.B., and Kramer, R.H. (1985) Degradation of extracellular matrix by trophoblastic cells of first trimester human placentas. *J. Cell Biochem.* 27, 31-41.

Fisher, S.J., Cui, T.-Y., Zhang, L., Grahl, K., Hartman, L., Zhang, G.-Y., Tarpey, J., and Damsky, C.H. (1989) Adhesive and invasive properties of human placental cytotrophoblast cells in vitro. *J. Cell Biol.* 109. 891-892.

Fisher, S.J., Cui, T.-Y., Zhang, L., Hartman, L., Esteves, R., Tarpey, J., and Damsky, C. (1989) Developmental regulation of human cytotrophoblast extracellular matrix-degrading proteases. *Abstract. American Society for Cell Biology.*

Gilbert, S.F. and Migeon, B.R. (1975) D-valine as a selective agent for normal human and rodent epithelial cells in culture. *Cell* 5, 11-17.

Glass, R.H., Aggeler, J., Spindle, A.I., Pedersen, R.A., and Werb, Z. (1983) Degradation of extracellular matrix by mouse trophoblast outgrowths: A model for implantation. *J. Cell Biol.* 96, 1108-1116.

Glass, R.H., Spindle, A.I., and Pedersen, R.A. (1979) Mouse embryo attachment to substratum and interaction of trophoblast with cultured cells. *J. Exp. Zool.* 208, 327-336.

Gwatkin, R.B.L. (1966) Amino acid requirements for the attachment and outgrowth of the mouse blastocyst in vitro. *J. Cell Physiol.* 68, 335-344.

Hall, D.E., Crowley, E., Reichardt, L., and Damsky, C. (1989) β_1 integrin adhesion receptors mediate cell attachment to two major attachment promoting domains of laminin. *Abstract. American Society for Cell Biology.*

Hayman, E.G., Engvall, E., A'Hearn, E., Barnes, D., Pierschbacher, M., and Ruoslahti, E. (1982) Cell attachment on replicas of SDS polyacrylamide gels reveals two adhesive plasma proteins. *J.Cell Biol.* 95, 20-23.

Hynes, R.O. (1987) Integrins: A family of cell surface receptors. *Cell* 48, 549-554.

Jalkanen, M., Nguyen, H., Rapraeger, A., Kern, N., and Bernfield, M. (1985) Heparan sulfate proteoglycans from mouse mammary epithelial cells: Localization on the cell surface with a monoclonal antibody. *J. Cell Biol.* 101, 976-984.

Kliman, H.J., Nestler, J.E., Sermasi, E., Sanger, J.M., and Strauss, J.F. (1986) Purification, characterization and in vitro differentiation of cytotrophoblast from human term placentae. *Endocrinol.* 118, 1567-1582.

Knudsen, K.A., Horowitz, A.F., and Buck, C.A. (1981) A monoclonal antibody identifies a glycoprotein complex involved in cell-substratum adhesion. *Exp. Cell Res.* 157, 218-226.

Laemmli, U.K. (1970) Cleavage of structural proteins during assembly of the head of bacteriophage T4. *Nature (Lond.)* 227, 680-685.

Lindenberg, S., Hyttel, P., Lenz, S., and Holmes, P. (1986) Ultrastructure of the early human embryo. *Human Reprod.* 1, 533-538.

Moss, L., Fisher, S., and Damsky, C.H. (1989) Stage and tissue specificity in the glycosylation of human trophoblast integrins. *Abstract. American Society for Cell Biology*.

Mintz, B. (1964) Formation of genetically mosaic mouse embryos and early development of 'lethal (t^{12}/t^{12})-normal' mosaics. *J. Exp. Zool.* 157, 273-292.

Nilsson, O. (1974) The morphology of blastocyst implantation. *J. Reprod. Fertil.* 39, 187-194.

Olden, K., Pratt, R.M., and Yamada, K. (1978) Role of carbohydrate in protein secretion and turnover: Effects of tunicamycin on the major cell surface glycoprotein of chick embryo fibroblasts. *Cell* 13, 461-473.

Pattillo, R.A. and Gey, G.O. (1968) The establishment of a cell line of human hormone-synthesizing trophoblastic cells in vitro. *Cancer Res.* 28, 1231-1236.

Pattillo, R.A., Ruckert, A.C., Hussa, R.O., Bernstein, R., and Delfs, E. (1971) The JAR cell line: Continuous human multihormone production and controls. *In Vitro* (Rockville) 6, 398-405.

Pierschbacher, M.D. and Ruoslahti, E. (1984a) Cell attachment activity of fibronectin can be duplicated by a small synthetic fragment of the molecule. *Nature (Lond.)* 309, 30-33.

Pierschbacher, M.D. and Ruoslahti, E. (1984b) Variants of the cell attachment site of fibronectin that retain attachment-promoting activity. *Proc. Natl. Acad. Sci. USA* 81, 5985-5988.

Pytela, R., Pierschbacher, M.D., and Ruoslahti, E. (1985a) Identification and isolation of a 140 kd glycoprotein with properties expected of a fibronectin receptor. *Cell* 40, 191-198.

Pytela, R., Pierschbacher, M.D., and Ruoslahti, E. (1985b) A 125/115 kDa cell surface receptor specific for vitronectin interacts with the arginine-glycine-aspartic acid adhesion sequence derived from fibronectin. *Proc. Natl. Acad. Sci. USA* 82, 5766-5770.

Ramsey, E.M., Houston, M.L., and Harris, J.W.S. (1986) Interactions of the trophoblast and maternal tissues in three closely related primate species. *Am. J. Obstet. Gynecol.* 124, 647-652.

Rapraeger, A., Jalkanen, M.M., Endo, E., Koda, J., and Bernfield, M. (1985) Cell surface proteoglycan from mouse mammary epithelial cells bears chondroitin and heparan sulfate glycosaminoglycans. *J. Biol. Chem.* 260, 11046-11052.

Ruoslahti, E. and Pierschbacher, M.D. (1986) Arg-Gly-Asp: A versatile cell recognition signal. *Cell* 44, 517-518.

Salomon, D.S. and Sherman, M.I. (1975) Implantation and invasiveness of mouse blastocysts on uterine monolayers. *Exp. Cell Res.* 90, 261-268.

Saunders, S. and Bernfield, M. (1988) Cell surface proteoglycan binds mouse mammary epithelial cells to Fn and behaves as a receptor for interstitial matrix. *J. Cell Biol.* 106, 423-430.

Schaffner, W. and Weissmann, C. (1973) A rapid, sensitive, and specific method for the determination of proteins in dilute solution. *Anal. Biochem.* 56, 502-514.

Sherman, M.I. (1975a) Long term culture of cells derived from mouse blastocysts. *Differentiation* 3, 51-67.

Sherman, M.I. (1975b) The role of cell-cell interactions during early mouse embryogenesis. In: *Early Development of Mammals*, (eds.), M. Balls and A.E. Wild, London: Cambridge University Press, pp. 145-165.

Sherman, M.I. and Barlow, P.W. (1972) Deoxyribonucleic acid content in delayed mouse blastocysts. *J. Reprod. Fertil.* 29, 123-126.

Sherman, M.I. and Salomon, D.S. (1975) The relationships between the early mouse embryo and its environment. In: *The Developmental Biology Of Reproduction*, (eds.), C.L. Market and J. Papaconstantinov, New York: Academic Press, pp. 277-309.

Spindle, A.I. and Pedersen, R.A. (1973) Hatching, attachment and outgrowth of mouse blastocysts in vitro: Fixed nitrogen requirements. *J. Exp. Zool.* 186, 305-318.

Sutherland, A.E., Calarco, P.G., and Damsky, C.H. (1988) Expression and function of cell surface extracellular matrix receptors in mouse blastocyst attachment and outgrowth. *J. Cell Biol.* 106, 1331-1348.

Sutherland, A.E., Calarco, P.G., Bernfield, M., and Damsky, C.H. (1989) Syndecan, a cell surface proteoglycan, is developmentally regulated during preimplantation mouse development. *Abstract. American Society for Cell Biology.*

Tamkun, J.W., DeSimmone, D.W., Fonda, D, Patel, R.S., Buck, C., Horwitz, A.F., and Hynes, R.O. (1986) Structure of integrin, glycoprotein involved in the transmembrane linkage between fibronectin and actin. *Cell* 46, 271-282.

Tomaselli, K., Damsky, C.H., and Reichardt, L. (1987) Interactions of neuronal cell (PC12) with the extracellular matrix proteins laminin, collagen IV, and fibronectin: Identification of cell surface glycoproteins involved in attachment and outgrowth. *J. Cell Biol.* 105, 2347-2358.

Tomaselli, K., Damsky, C.H., and Reichardt, L. (1988) Purification and characterization of mammalian antigens expressed by a rat neuronal cell line (PC12): Evidence that they function as receptors for laminin on type IV collagen. *J. Cell Biol.* 107, 1241-1252.

Towbin, H., Staehelin, T., and Gordon, J. (1979) Electrophoretic transfer of proteins from polyacrylamide gels to nitrocellulose sheets: procedure and some applications. *Proc. Natl. Acad. Sci. USA* 76, 4350-4354.

Wewer, U., Damjanov, A., Weiss, J., Liotta, L., and Damjanov, I. (1986) Mouse endothelial stromal cells produce basement membrane components. *Differentiation* 32, 49-58.

Wewer, U., Faber, M., Liotta, L., and Albrechtsen, R. (1985) Immunochemical and ultrastructural assessment of the nature of the pericellular basement membrane of human decidual cells. *Lab. Invest.* 53, 624-633.

Yamada, K. and Kennedy, D. (1984) Dualistic nature of adhesive protein function: Fibronectin and its biologically active peptide fragments can autoinhibit fibronectin function. *J. Cell Biol.* 99, 29-36.

Zhu, B. C.-R., Fisher, S., Pande, J., Calaycay, J., Shively, J., and Laine, R. (1984) Human placental (fetal) fibronectin: Increased glycosylation and higher protease resistance than plasma fibronectin. *J. Biol. Chem.* 259, 3962-3970.

Zhu, B. C.-R. and Laine, R. (1985) Polylactosamine glycosylation on human fetal placental fibronectin weakens the binding affinity of fibronectin to gelatin. *J. Biol. Chem.* 260, 4041-4045.

Zhu, B. C.-R. and Laine, R. (1987) Developmental study of human fetal placental fibronectin: alterations in carbohydrates of tissue fibronectin during gestation. *Arch. Biochem. Biophys.* 252, 1-6.

THE ROLE OF MATRIX MACROMOLECULES IN THE INVASION OF DECIDUA BY TROPHOBLAST: MODEL STUDIES USING BeWo CELLS

John D. Aplin[1] and Anna K. Charlton

Departments of Obstetrics and Gynecology,
Biochemistry and Molecular Biology
University of Manchester
Manchester M13 0JH, United Kingdom

INTRODUCTION

Light and electron microscopic studies of the human placental bed suggest that the invasion of trophoblast into decidua depends in some measure on interaction with decidual extracellular matrix (Boyd and Hamilton, 1970; Pijnenborg et al., 1980). This applies both to the initial stages of interstitial implantation and to the invasion of decidual stroma and arteries that occurs during the late first and second trimesters of pregnancy (Pijnenborg et al., 1980; Sheppard and Bonnar, 1974). One important aspect of this phenomenon is the inherent invasiveness of trophoblast. However, it is clear that many - even most - cell types, given a suitable environment, have a capacity to migrate across surfaces (Abercrombie, 1980) and some cells also invade and migrate through three-dimensional matrices (Mareel, 1979; Schor, 1980; Schor et al., 1981; Jones et al., 1981). The expression of these properties depends on the relationship of each cell to its neighbors, that is, on cell-cell contacts (Aplin and Foden, 1985; Schor et al., 1985), and also on the surrounding extracellular matrix (Glass et al., 1983; Jones et al., 1981; Schor et al., 1981). The matrix provides a structural framework for the tissue, but many of its components are also used by cells as anchors against which to exert tractional forces during movement (Harris et al., 1981). These cell-matrix interactions are mediated by specific plasma membrane receptor molecules (Hynes, 1987; Argraves et al., 1987; Wayner and Carter, 1987) that interact with matrix components outside the cell and signal the reorganization of the cytoskeleton required for change of shape. Shape change is, in turn, a prerequisite of motility (Abercrombie, 1980).

Thus alterations in extracellular matrix could give rise to cellular environments that either promote or inhibit invasiveness. A permissive environment might be a loose-knit three-dimensional structure containing cavities and channels sufficiently large to accept cells. At the same time, anchorage molecules should be present to promote the formation of receptor-mediated contacts with the cell surface. An inhibitory environment, by contrast, might be densely packed and/or deficient in anchorage molecules. It is also important to note that cellular invasiveness may be modulated by altered expression of cell surface receptors (Dedhar et al., 1987). This in turn may be responsive to humoral factors.

---

[1]To Whom Correspondence Should be Addressed: Research Floor, St.Mary's Hospital, Whitworth Park, Manchester M13 0JH, U.K.

Many aspects of the invasive process cannot be studied by the examination of tissue biopsies. Cell culture models allow direct observation of the time-dependant processes of cellular motility, and their experimental manipulation. It is unfortunate that recovery of normal extravillous cytotrophoblast in quantities sufficient for experimental study poses technical problems. In the present study, some adhesive and migratory properties of a choriocarcinoma cell line, BeWo, are examined on substrata containing extracellular matrix components found in decidual tissue. Tentative conclusions about mechanisms involved in trophoblastic invasion of decidua are also drawn.

MATERIALS AND METHODS

Cells

BeWo cells were obtained from the American Type Culture Collection and grown in an equal mixture of Ham's F12 and Dulbecco's Minimal Eagle's Medium. To this was added 10% fetal calf serum, glutamine (2 mM), Hepes buffer (15 mM, pH 7.2), penicillin, and streptomycin. Endothelial cells from the umbilical vein were isolated as previously described (Gimbrone et al., 1974) and grown in the same medium with added endothelial cell growth factor. FL amnion cells were cultured as described previously (Aplin et al., 1983).

Substrata

Fibronectin was purified from human plasma by gelatin affinity chromatography as previously described (Pena et al., 1980). Laminin from the mouse Englebreth-Holm-Swarm (EHS) tumor was purchased from Bethesda Research Laboratories (Cambridge, U.K.). Fibrinogen was purchased from Kabi Vitrum (Stockholm, Sweden). Fibronectin was removed from the fibrinogen, in which it was present as a minor contaminant, by passage through gelatin-Sepharose. Collagens for coating experiments were purified from pepsinized human placenta using standard techniques (Abedin et al., 1982). Rat tail collagen isolated by standard procedures (Elsdale and Bard, 1974) was used for forming gels. It was a kind gift from Dr.S.Campbell, Manchester. It consisted of 95% type I, 5% type III collagen. Gels were formed on the bottom of culture flasks as previously described (Elsdale and Bard, 1974). Collagens, fibrinogen, fibronectin and laminin were used to coat plastic culture dishes at 50 µg/ml, 60 minutes. A crude extract of placental extracellular matrix was prepared and used to coat culture dishes as described previously (Aplin and Foden, 1982). Fibrinoid material was dissected from term placenta under sterile conditions and smeared as a thin layer over discrete areas of tissue culture dishes.

Subcellular matrices from confluent cell cultures were prepared by treatment at ambient temperature with 2% sodium deoxycholate, 0.02% w/v disodium EDTA as previously described (Aplin et al., 1984). The cell layer dissolved immediately and the solution was aspirated. The dishes were then washed extensively (at least 10 changes) with phosphate-saline, pH 7.2.

Adhesion and Migration Assays

For short term assay of adhesion to extracellular matrices, cells were harvested in trypsin/EDTA, washed in warm medium without serum and then seeded onto prepared substrata in the absence of serum as described previously (Aplin et al., 1983). The assays were conducted over periods of up to 3 hours during which cell viability was maintained. Assays of cell migration were performed on prepared substrata in the presence of serum. Fields were located and the flask marked to allow their relocation. For coculture assays (BeWo-endothelial cell or cell-fibrinoid), concentrated cell suspensions were seeded as a drop (approx. 1E6 per 100 µl) on an area of substratum adjacent to the confrontation zone. Preparations were incubated for 2 hours to allow cells to attach, then the container was flooded and washed thoroughly with medium (3 changes) to remove poorly attached cells.

Immunohistochemistry

Wax sections (5 µm) of first trimester decidua were passed through xylene, ethanol, a graded series of aqueous ethanols, water, methanol containing H_2O_2 (1\100 v\v), distilled water, 50 mM Tris HCl pH 7.4, 0.15 M NaCl (Tris-saline) and then normal rabbit serum, 1\10 v\v in tris-saline (30 minutes, 20°C). They were incubated (1 hour, 20°C) in sheep antibody to human plasma fibronectin (Bethesda Research Laboratories; 1:500), rinsed thoroughly in 3 changes of Tris-saline and then incubated (1 hour, 20°C) in peroxidase-conjugated antibody to sheep immunoglobulin (Dako, Copenhagen; 1:100). After further rinsing bound antibody was revealed using N,N'-diaminobenzidine (50 mg in 100 ml Tris-saline containing 15 µl H_2O_2). The results were comparable with fibronectin distributions observed using immunofluorescence of cryosectioned decidua (Aplin et al., 1988).

RESULTS

BeWo Cell Adhesion to Matrix Components

Cells were seeded in the absence of serum onto plastic dishes coated with matrix components, and attachment and change of shape monitored for up to 3 hours. Various matrix components were used including fibronectin, laminin, fibrinogen, and collagens I, III, IV, and V. The results are summarized in Figures 1 and 2. Many cells (50-75%) attached to each of the matrix components in the course of the assays, though attachment to fibronectin was the most efficient, being essentially complete after 15 minutes. Cells also attached to plastic (Figure 2d) and, to a lesser extent, to surfaces coated with albumin (not shown); thus the data do not allow an assessment to be made concerning the specificity of cell attachment under serum-free conditions to laminin, fibrinogen or the collagens under serum-free conditions.

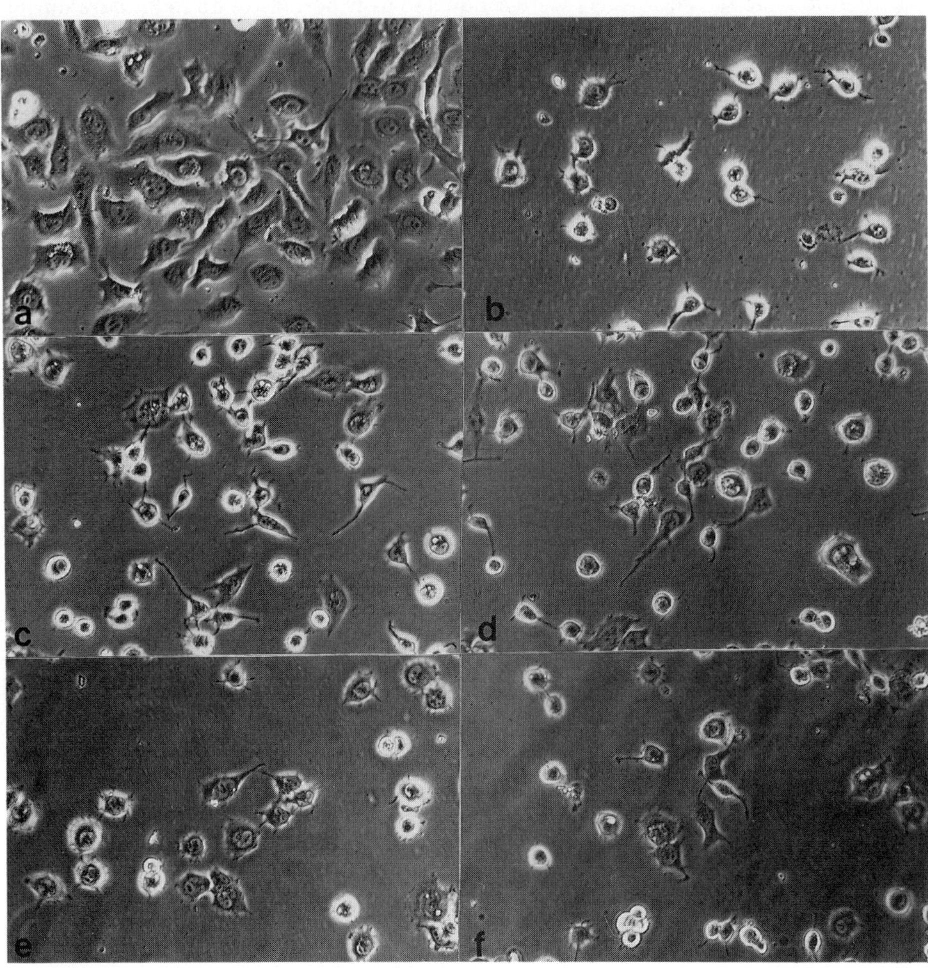

Figure 1. Phase contrast micrographs of BeWo cells adhering to dishes coated with: (a) plasma fibronectin; (b) laminin; (c) collagen type I; (d) collagen type III; (e) collagen type IV; (f) collagen type V. Photographs were taken 2 hours after adding the cells. Fibronectin provides more effective support for attachment and spreading than any other substrate. (X310)

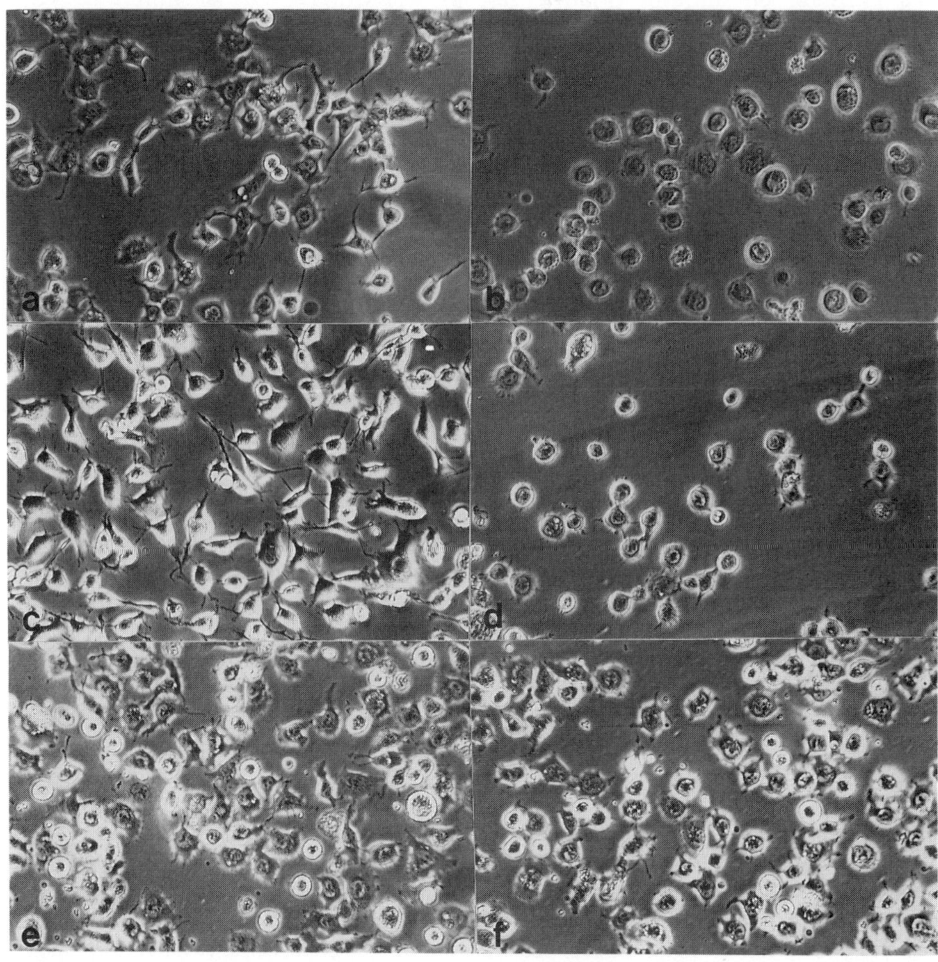

Figure 2. BeWo cells adhering to dishes coated with (a) fibrinogen-fibronectin; (b) fibrinogen; (c) placental matrix extract; (d) control, bare plastic; (e) subcellular matrix (SCM) produced by BeWo cells in a previous confluent culture removed by treatment with detergent; (f) SCM from confluent FL cells. Photographs were taken at 90 minutes or 30 minutes (b,f). The dendritic morphology characteristic of cells on fibrinogen-fibronectin is also observed on the placental matrix extract. It is no longer present after removing fibronectin. (X310).

Some flattening or elongation of cells was observed on all of the surfaces examined, but the most extensive and rapid flattening was observed on fibronectin (Figure 1a). Spreading on fibronectin gave rise to polygonal cells within 45 minutes. On the collagens, spreading was less extensive, but more polarized cells appeared, sometimes extending long processes across the substratum. Spreading was minimal on laminin (Figure 1b), plastic (Figure 2d) or albumin. On fibrinogen (Figure 2b) cells were more symmetrically flattened than on the collagens or fibronectin, though spreading was again very limited with many cells remaining refractile after 90 minutes. On fibrinogen-fibronectin substrata (approx. 50:1 molar ratio; Figure 2a) shape change was as rapid as on fibronectin. However, the cells took on more elogated shapes, with long thin dendritic extensions.

Adhesion assays were also conducted in which BeWo cells were seeded onto surfaces containing more complex, but perhaps more physiologically appropriate matrix preparations. A crude preparation made by extracting term placental extracellular matrix with urea, previously shown to contain fibronectin- and fibrinogen-derived antigens (Aplin and Foden, 1982) as well as type I collagen (Aplin and Charlton, unpublished), produced rapid and efficient cell attachment followed by elongation to a highly dendritic morphology (Figure 2c). This resembles the behavior of other cell types on the same surface (Aplin and Foden, 1982; Campbell et al., 1984).

Matrix preparations were also generated from cells in monolayer culture (subcellular matrix, SCM). Two cell lines were utilized: BeWo cells and the transformed amnion epithelioid cell line FL. Cells at confluence were detached using detergent to leave a substratum onto which freshly trypsinized BeWo cells could be seeded. SCMs left by both cell lines promoted efficient attachment of BeWo cells followed by significant spreading (Figures 2e,f). Cell spreading was clearly less than on fibronectin, and immunological and biochemical investigations confirmed that neither cell type deposited fibronectin into endogenous SCM (unpublished observations).

When adhesion assays were conducted in the presence of serum, many of the observed differences in attachment and morphology were lost. Cells attached to all the surfaces (including bare plastic) and spread to polygonal shapes within 60 minutes. More elongated cell forms were visible on laminin and, especially, on placental matrix extract and fibrinogen-fibronectin (not shown). However, after 1 day in culture these slight differences were no longer observable (Figure 3).

Island Formation by BeWo Cells in Culture

Cultured BeWo cells grew in islands. These were formed by a combination of cell division and joining up of adjacent cells into groups, the contribution of the latter process depending on the initial plating density. After several days central parts of the islands started to form cellular bilayers as a result of cell division and expulsion from the basal layer (Figures 4, 5). Although some binucleate and multinucleate cells were always present, the majority (over 95%) of cells was always mononucleate.

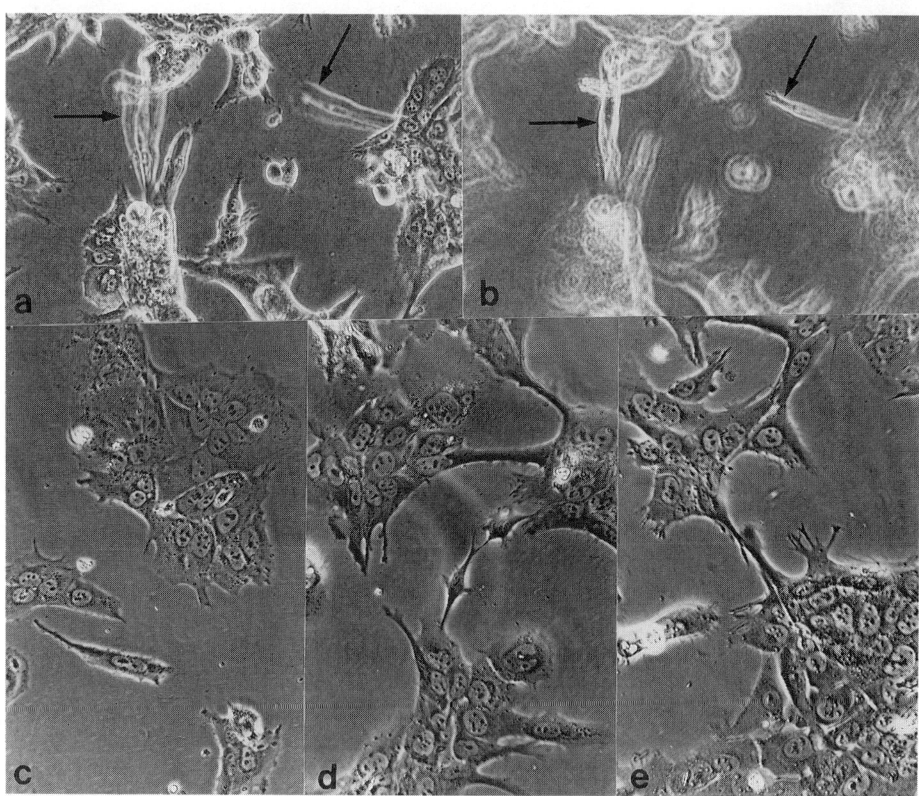

Figure 3. BeWo cultures at 29 hours. Cells were seeded onto substrata of: (a, b) collagen gel; (c) laminin; (d) fibronectin; (e) plastic. Culture proceeded in medium containing serum. Ragged-edged islands comprising some polygonal and some elongated cells are evident in each culture. On the collagen gel, a few bipolar cells are present attached by one end to the underside of islands, projecting downward into the gel matrix; (a) and (b) show the same field focused respectively in the plane of the gel surface or below it. Two burrowing cells are visible, (arrows) but these occur in only 5% of islands. The cells never escape fully into the gel. (X310)

The formation and morphology of cell islands during several days of culture in serum-containing medium occurred in similar fashion on substrata of fibronectin, laminin, or collagen (Figure 3). In the early stages of island development before multilayering began, the edges of islands were generally ragged, comprising elongated (often bipolar) cells (Figure 3). After 3-4 days, islands were smooth-edged with occasional elongated cells at the periphery (Figures 4, 5). Cells cultured on collagen gels underwent the same progression (Figure 3).

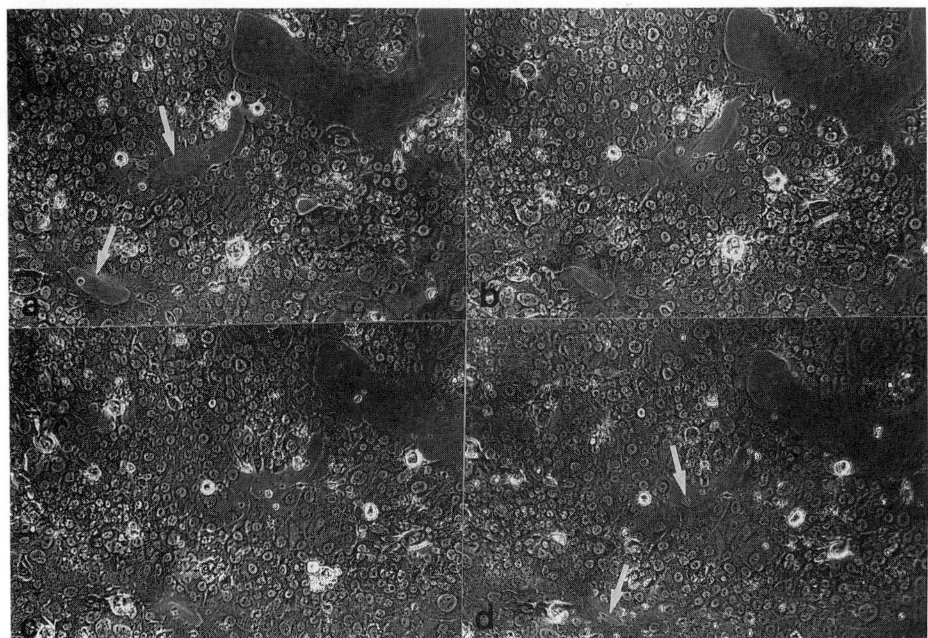

Figure 4. Infilling of vacant substratum by BeWo cell islands. Four time points are shown from a culture on plastic: (a) 1 hour; (b) 2 hours; (c) 3 hours; (d) 4 hours. Two areas of substratum, indicated by arrows in (a) and (d), are progressively occupied by the spreading cell sheet. Note that no significant infilling of the vacant area at top right occurred during the experiment. (X150).

No migration of cells from the surface to the interior of the gels occurred at any stage of culture on gels of type I/III (95/5) collagen. However, careful examination revealed that in a few (about 5%) islands, elongated cells were oriented downward from the basal surface of the island into the gel (Figure 3a, b). These cells never fully detached to escape into the gel.

Migratory Behavior of BeWo Cells In Vitro

By far the dominant migratory process in BeWo cultures was the spreading of islands. This occurred by an epithelial-like mechanism, cells at the edges becoming more well-spread without detaching from neighboring cells. In this way infilling of unoccupied zones of substratum occurred in dense cultures (Figure 6). Eventually, by cell division, a newly populated region of substratum regained the cell density characteristic of mature islands.

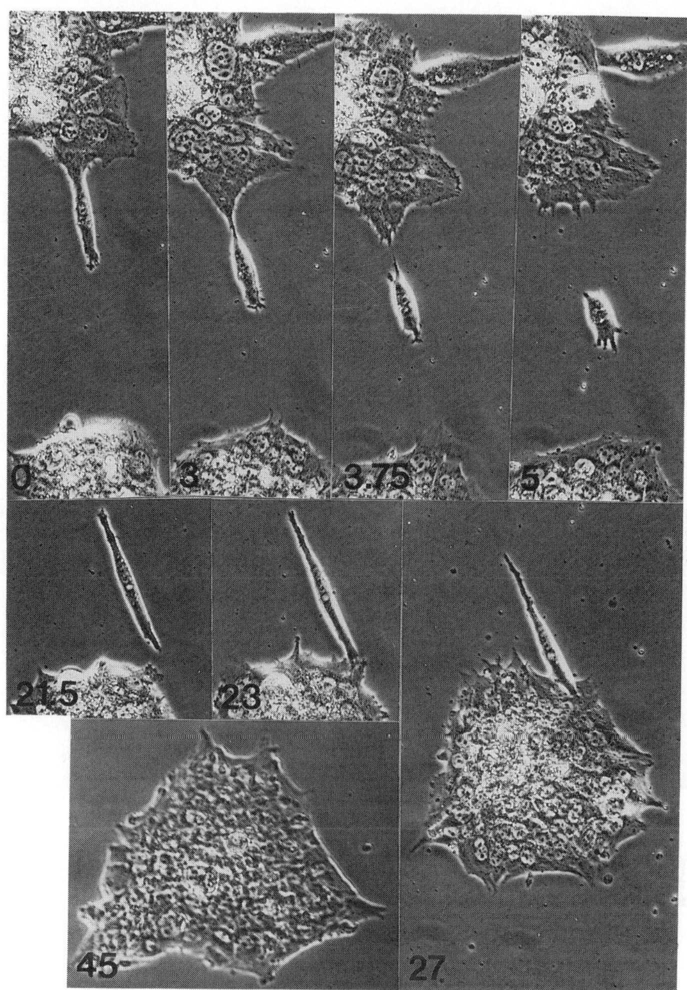

Figure 5. Time lapse series showing the escape of a single cell from an island in a BeWo cell culture. Times are given in hours. After its escape the cell retains its polarity and direction of migration, although a period of about 15 hours elapses during which no translocation occurs. Subsequently it coalesces with an adjacent island, and 18 hours later no trace of its entry remains. Note also the onset of cell bilayer formation in the upper island. (X310).

However, careful observation revealed that at all stages of culture, single migratory cells could be found. An example is shown in Figure 5. In cultures comprising mainly medium-sized or large cell islands, these migratory cells arose by elongation and detachment from the periphery of colonies. After a period of migration, one of two things happened: either the cell divided, giving rise to a nascent cell island, or it coalesced with a neighboring island (Figure 5).

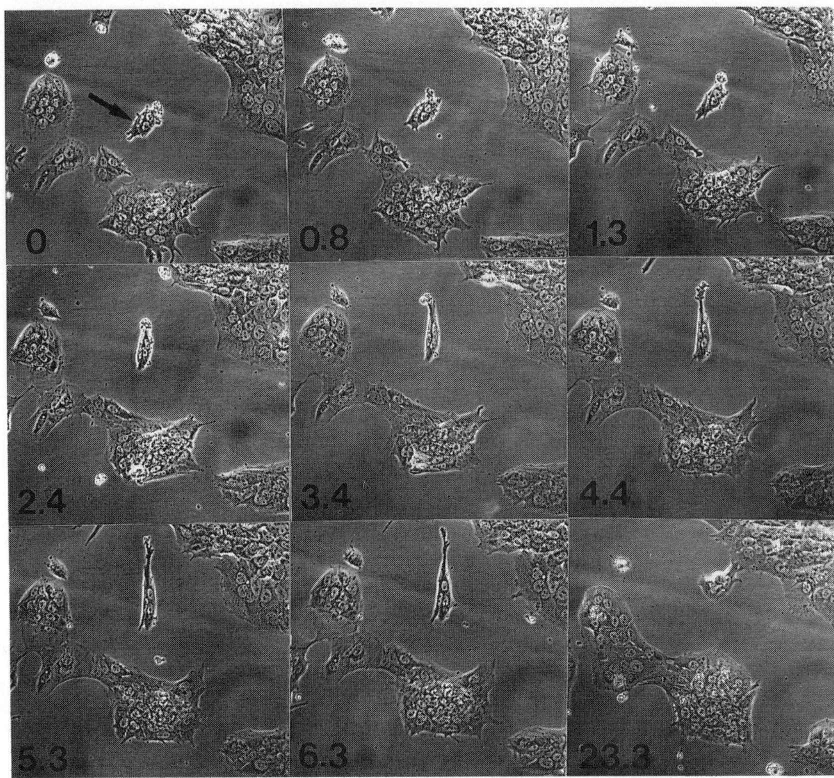

Figure 6. Motile behavior of a small island of BeWo cells in vitro. Time points are given in hours. The frame shifts slightly and so a useful reference marker is provided by a diagonal line on the substrate in the upper part of each picture. An island of 3 cells, arrowed at time zero, undergoes significant change of shape during 1 day. Some of the shapes are reminiscent of polarized, migrating single cells; for example, at 6.3 hours the island exhibits a trailing edge. However, little net displacement occurs. The eventual coalescence of the small colony with the larger island at top right occurs mainly as a result of the spreading of the latter. (X150).

Attempts were made to discover whether small groups of cells could migrate in a coordinate fashion as described for certain types of epithelial cells (Albrecht-Buehler, 1979; Kolega, 1981). Islands of 2-16 cells were highly motile, often adopting polygonal or bipolar shapes some of the features of which were consistent with directional migration. For example, the island shown in Figure 6 (6.3 hour point) has an obvious retraction fibril which resembles the trailing edge of a migratory fibroblast (Heath, 1982). However, careful examination of small groups of cells failed to reveal any significant translocation (e.g., Figure 6).

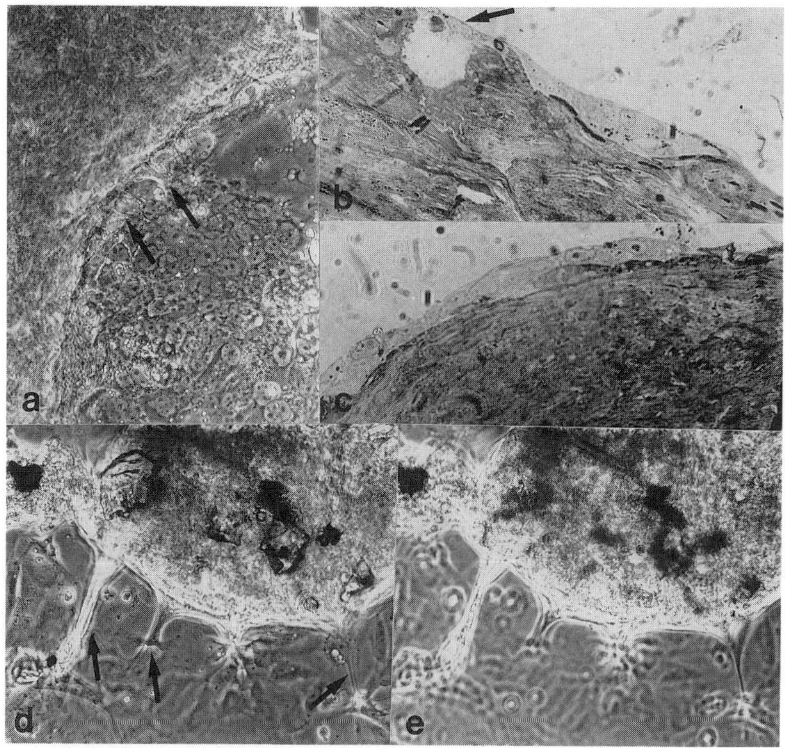

Figure 7. Migration of BeWo and vascular endothelial cells onto aggregates of fibrinoid material obtained from term placenta. (a) a dense culture of BeWo cells at bottom right in contact with a fibrinoid aggregate (top left). Cells are visible (arrows) bridging upwards to mount the aggregate (X310). (b, c) thin sections from the same experiment stained with toluidine blue show BeWo cell monolayers advancing across the surface of the fibrinoid. A bridging cell is seen in b (arrow) (X1800). The cells do not invade into the fibrinoid; (d, e) Vascular endothelial cells are capable of similar behavior. In d the plane of focus is just above the substratum, while in e it is in a higher plane of the same field where cells are making contact with the fibrinoid aggregate (top). Several elongated bridging cells are present (arrows) stretching from the substratum diagonally upwards to contact the fibrinoid. (X310)

Cell Migration onto Fibrinoid Matrix

In order to model the interaction of cytotrophoblast with fibrinoid matrix at the tips of anchoring villi (cf. Figure 9) and the walls of transformed blood vessels, dense cultures of BeWo cells were allowed to contact substrate-attached fibrinoid material obtained by dissection from term placenta. The results, shown in Figure 7, indicate that BeWo cells are capable of spreading on, and migrating across the surface of fibrinoid.   At the edges of fibrinoid deposits, where contact was established with the advancing cell colony, bridging cells formed and passed from the plastic substratum onto the surface of the fibrinoid aggregate.   Cells then crawled further onto the aggregate.   In this way streams of cells were transferred to the fibrinoid, and its surface became colonized.   No migration into the fibrinoid mass was observed.   It is notable that a similar process of colonization of fibrinoid could be observed in migrating vascular endothelial cells (Figure 7).

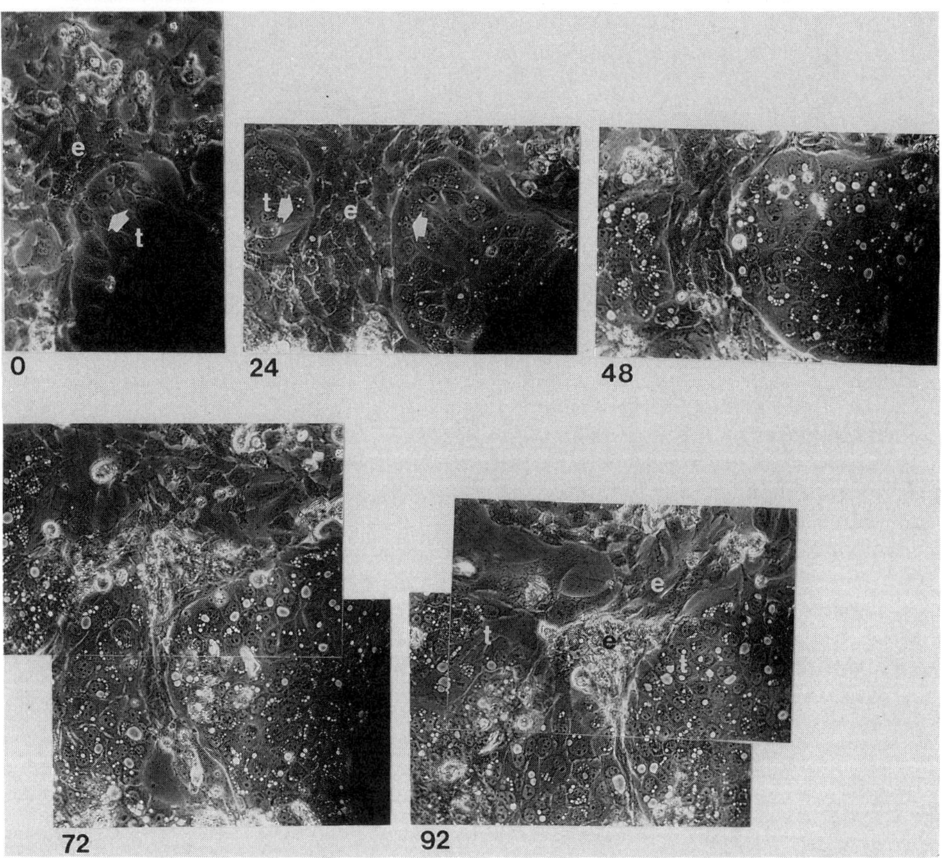

Figure 8.   Monolayer coculture of vascular endothelial cells (e) and BeWo (t). Times are given in hours.   The dark area at bottom right in each frame is a reference mark out of the plane of focus.   Two tongues of advancing BeWo cells progressively displace the interposed endothelial cells from the substratum by a 'pincer' movement.   At 92 hours, the two BeWo colonies have nearly merged, and a large aggregate of loosely adherent, displaced endothelial cells is present.   Arrows indicate the direction of movement of the BeWo cells. (X310)

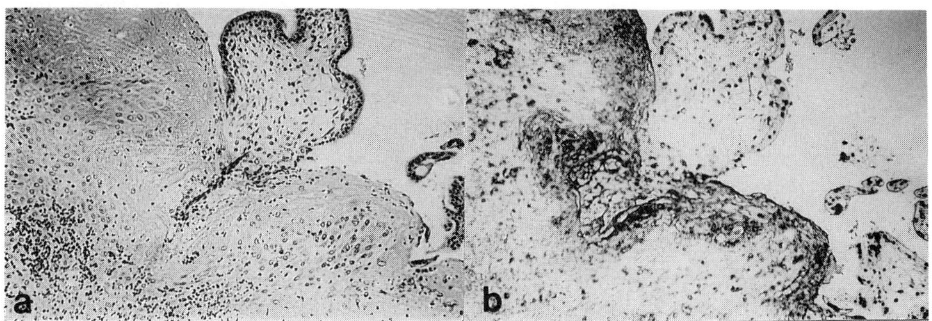

Figure 9. Human maternofetal interface at the 10th week of gestation. Serial sections stained with (a) hematoxylin and eosin or (b) anti-fibronectin antibody using the immunoperoxidase technique, and hematoxylin. Cytotrophoblast is shown escaping from an anchoring villus (top right) into a zone of fibrinoid (Rohr's stria) rich in fibronectin. Deeper areas of decidua also contain fibronectin. (X70)

BeWo-Endothelial Cell Coculture

In order to model processes occurring during the invasion of cytotrophoblast into maternal spiral arteries, coculture assays were set up in which advancing BeWo cell colonies challenged monolayers of umbilical vein endothelial cells. The results are illustrated in Figure 9, and demonstrated that BeWo cells displaced endothelial cells from the growth substratum. Electron microscopic examination of thin sections through confrontation zones (not shown) showed that this occurred by underlapping and prising away of the endothelial cells as observed in similar experiments with other cell types (Ireland and Stern, 1982). Eventually aggregates of endothelial cells detached and were lost.

Fibronectin In The Placental Bed

In previous immunofluorescence studies the presence of fibronectin was demonstrated throughout decidual interstitium as well as in maternal arteries (Aplin et al., 1988). It is also present in maternal fibrinoid and collagenous matrices at the tips of anchoring villi from where cytotrophoblast invasion of decidual tissue occurs (Figure 9).

DISCUSSION

Of the several physiological substrata tested, fibronectin (alone or in complexes with fibrinogen) produced the most efficient attachment and rapid shape change in newly adherent BeWo cells in the absence of serum. In the presence of serum, equally good adherence was achieved on all the tested matrices; this may be a result of the presence of fibronectin in serum, although other adhesion-producing components of serum such as vitronectin (Suzuki et al., 1985) may also play a role. Adhesion in the presence of serum probably represents a closer approach to physiological conditions at the maternofetal interface. Studies of decidual extracellular matrix have demonstrated the presence of fibronectin in all of the tissue areas to which trophoblast migration occurs (Figure 9 and Aplin et al., 1988). The present results therefore support the concept that fibronectin may play a role in the establishment of points of anchorage between migrating trophoblast and surrounding extracellular matrix either in arteries or the interstitium. Previous studies (Aplin and Foden, 1982; Campbell et al., 1984) showed that the elongated or dendritic morphology assumed by various other cell types on fibronectin-fibrinogen or placental matrix extract are associated with enhanced motility and this now requires investigation in BeWo cell cultures. Flat, elongated cytotrophoblast can frequently be seen in thin sections of the maternofetal interface on the surface of fibrinoid material (Boyd and Hamilton, 1970). The results suggest that complexation of fibronectin with fibrinogen may alter its adhesion-promoting potential, and this is now under investigation.

The growth of BeWo cells in islands and their associated motile behavior has been examined. A naive parallel can be drawn between cells at the periphery of these islands and at the tips of anchoring villi from where cells migrate into either blood vessels or the decidual stroma (Figure 9; Pijnenborg et al., 1980). The origin of BeWo as a choriocarcinoma might give rise to the expectation that its invasive behavior would be more pronounced than that of normal cytotrophoblasts; despite this, no invasion into the interstices of collagen gels or fibrinoid aggregates occurred in vitro. Similar findings have been reported for other carcinoma cell lines (Schor, 1980). For the present, therefore, no adequate in vitro model has been constructed to enable the interstitial migration of cytotrophoblast to be investigated. It may be that the influence of humoral factors - derived from pregnancy plasma, maternal decidua or the embryo - is required both to initiate and, later, to limit the interstitial invasion of trophoblast. Some support for this notion arises from the observation that phorbol ester (in the presence of protease inhibitor) stimulates endothelial cells to invade fibrin matrices; in its absence the cells form a surface monolayer (Montesano et al., 1987).

Our model system is more useful in analyzing intravascular invasion. This process should be seen as an active migration across a surface, the subendothelial matrix. Adhesion and motility are clearly required to move the stream of cells against the direction of blood flow in the vessel, and so no notion of passive colonization can be entertained. Vessels are frequently observed in the basal plate in which trophoblast appears to have replaced completely the endothelium (Pijnenborg et al., 1980). Cocultures of BeWo and endothelial cells suggest a mechanism for this process in which the latter are progressively displaced from the subendothelium by the advancing edge of a BeWo colony.

The observation that the predominant mode of trophoblast migration in BeWo cell cultures is the cooperative sheet sliding or spreading reminiscent of most epithelial cell types (DiPasquale, 1975; Heath, 1982) gives rise to clear predictions concerning the expected distribution of trophoblast within maternal spiral arteries. If this behavior holds true in normal spiral arteries, colonization might be expected to occur as a cell continuum from the cytotrophoblast shell along the arterial walls. Serial sections of maternal arteries have been reported to demonstrate discontinuous trophoblastic colonization (Pijnenborg et al., 1980); this is inconsistent with the above prediction and requires an explanation.

Occasionally single cells are observed to escape from BeWo cell islands (Figure 6). It should be emphasized that this has been observed in pure BeWo cell cultures and not in cocultures at the BeWo-endothelial cell interface. However, the appearance of single `wandering' cells raises the possibility that, within the arteries, single trophoblast cells could, after escaping from the main mass, migrate to deeper locations and, by division, establish new, isolated colonies. These would be separated from the main area of trophoblast infiltration by regions of unmodified vessel. However, such colonies could equally well arise from the penetration of interstitially invading trophoblast into the vessel from within its walls.

The processes by which the transformation of vessel walls follows their colonization by trophoblast have yet to be elucidated. However, it is clear that the smooth muscle cell layer is frequently fully replaced by a fibrinoid matrix that allows freer passage of blood along the vascular lumen. In this context, it is interesting to note that endothelial cells can colonize placental fibrinoid material in vitro (Figure 8). This may suggest that, after vascular transformation has been affected, endothelial cells may recolonize the surface of the vessel.

Decidual extracellular matrix consists of a sparsely fibrillar network containing collagen types I, III, V and fibronectin (Wienke et al., 1968; More et al., 1974; Wynn, 1974; 1977; Dallenbach-Hellweg, 1981; Cornillie et al., 1985; Kisalus et al., 1987; Aplin et al., 1988). In immunofluorescence these components appear to occupy the pericellular spaces surrounding enlarged, differentiated decidual cells (Kisalus et al., 1987; Aplin et al., 1988). Type VI collagen, which is abundant in pre-implantation endometrial stroma, is present in the walls of blood vessels but absent from most other locations (Aplin et al., 1988). Type VI collagen microfibrils have been suggested to play a role in cross-linking banded fibrils of the major collagens (Bruns et al., 1986) and Aplin et al. (1988) speculated that its selective loss might allow local expansion of intercellular spaces while retaining the tissue stability afforded by the major collagens. This might in turn give rise to a matrix environment conducive to interstitial migration of trophoblast. TEM (transmission electron microscopy) examination demonstrates that banded collagen fibrils are more sparsely distributed in decidua than in preimplantation endometrium (Aplin and Jones, 1989), indicating that some breakdown of major collagens must also occur. Enhanced production of glycosaminoglycans (GAGS) may be required to provide hydrated ground substance; little information is yet available on GAG distribution in human endometrium, but it is noteworthy that in the mouse, production of hyaluronate is increased 5-fold on the day of implantation (Carson et al., 1987).

At the placental site the endometrially-derived interstitial matrix is extensively modified by the deposition of fibrinoid material to form the striae of Rohr and Nitabuch (Boyd and Hamilton, 1970). It seems likely that some components of the fibrinoid originate in maternal plasma. The fibrinoid matrix provides part of the extracellular environment into which cytotrophoblast migration occurs from the tips of anchoring villi in late first and second trimester. In addition to fibrin-related components, the fibrinoid material is rich in fibronectin (Figure 9; Aplin and Foden, 1982; Sutcliffe et al., 1982) and supports the adhesion and migration of BeWo cells.

Decidual cells in man and rodents also synthesize and deposit into the pericellular space basement membrane components (Wewer et al., 1985; 1986; Faber et al., 1986; Kisalus et al., 1987; Aplin et al., 1988; Aplin, 1989). These include type IV collagen, laminin, and heparin sulphate proteoglycan. They become organized into a characteristic bilaminar basal lamina (`aura') that surrounds the mature decidual cell. The lamina is not continuous, but contains holes through which protrude club-shaped cellular processes (Wynn, 1977; Kisalus et al., 1987) containing secretory bodies. Some areas of the decidua also contain elongated, undifferentiated stromal cells that lack the pericellular lamina. Two functions have been proposed for this capsular lamina: it may act to organize or stabilize the surrounding interstitial matrix; fibrils are frequently to be found in close association with the stromal aspect of the lamina densa and sometimes appear to act as 'stitching' between adjacent cells (Aplin and Jones, 1989). Secondly, the presence of a capsular lamina is likely to render immotile the mature decidual cell. This idea is supported by the lack of cell-free laminal material in the interstitium. There is also a third possibility that the capsular lamina plays a role in trophoblast anchorage and migration, but no evidence to this effect has been obtained in the present study. However, we believe that decidual cells function to provide a matrix environment that fulfills the requirements of penetrability and adhesiveness required for migration of trophoblast.

SUMMARY

The BeWo choriocarcinoma cell line was used to investigate aspects of cell adhesion and migration thought to be relevant to the invasion of maternal decidua by normal cytotrophoblast. BeWo cells attach and flatten or elongate rapidly on model matrices containing fibronectin. In contrast, adhesion is less efficient on fibrinogen, collagens I, III, IV, V, and laminin. Fibronectin is present in decidua and may play a role in promoting trophoblast invasion. The migratory phenotype of BeWo cells is complex; cells grow in islands which colonize adjacent unoccupied surfaces by an epithelial-like process of coordinated sheet spreading. At the same time, single cells occasionally become detached from their neighbors and migrate away. No invasion into three-dimensional collagen or fibrinoid matrices occurs, but cells do migrate in the plane of the surface. The processes involved in migration across surfaces are suggested to provide a model of events during intravascular invasion of cytotrophoblast in the first and second trimesters of pregnancy. This model is extended by investigating the interaction of BeWo cells with vascular endothelial cells in vitro. The vascular cells are displaced from the growth substratum by advancing sheets of BeWo cells. The observations lead to clear predictions concerning the distribution of cytotrophoblast in vivo. Further analysis of this tissue distribution should enable the utility of the model to be judged.

ACKNOWLEDGEMENTS

This work was supported by a grant from the Wellcome Trust.

REFERENCES

Abercrombie, M.(1980) The crawling movement of metazoan cells. *Proc. Royal Soc.* B 207, 129-147.

Abedin, M.A., Ayad, S., and Weiss, J. (1982) Isolation and characterisation of cysteine-rich collagens from bovine placental tissues and uterus and their relationship to types IV and V collagen. *Biosci. Rep.* 2, 493-502.

Albrecht-Buehler, G. (1979) Group locomotion of PtK1 cells. *Exp. Cell Res.* 122, 402-407.

Aplin, J.D. (1989) Cellular biochemistry of the endometrium. In: *Biology of the Uterus*, (ed.), R.M.Wynn, Plenum:New York.

Aplin, J.D., Bardsley, W.G., and Niven, V.M. (1983) Kinetic analysis of cell spreading. *J. Cell. Sci.* 61, 375-388.

Aplin, J.D., Campbell, S., and Foden, L.J. (1984) Adhesion of human amnion cells to extracellular matrix. *Exp. Cell Res.* 153, 425-438.

Aplin, J.D. and Foden, L.J. (1982) A cell spreading factor,abundant in human placenta, contains fibronectin and fibrinogen. *J.Cell Sci.* 58, 287-302.

Aplin, J.D. and Foden, L.J. (1985) Defective adhesion to extracellular matrix leads to altered social behaviour in cultured fibroblasts. *J. Cell Sci.* 76, 199-211.

Aplin, J.D., Charlton, A.K., and Ayad, S. (1988) An immunohistochemical study of human endometrial extracellular matrix during the menstrual cycle and first trimester of pregnancy. *Cell Tiss. Res.* 253, 231-240.

Aplin, J.D. and Jones, C.J.P. (1989) Extracellular matrix in endometrium and decidua. In: *Placental- And Decidual-Specific Protein Synthesis And Secretion: Regulation, Role And Interactions*, (eds.), O. Genbacev and A. Klopper, Plenum: New York, in press.

Argraves, W.S., Suzuki, S., Arai, H., Thompson, K., Pierschbacher, M.D., and Ruoslahti, E. (1987) The amino acid sequence of the fibronectin receptor. *J.Cell Biol.* 105, 1183-1190.

Boyd, J.D. and Hamilton, W.J. (1970) *The Human Placenta*. Heffer and Sons, Cambridge, U.K.

Bruns, R.R., Press, W., Engvall, E., Timpl, R., and Gross, J. (1986) Type VI collagen in extracellular, 100 nm periodic fibrils and filaments: Identification by immunoelectron microscopy. *J. Cell Biol.* 103, 393-404.

Campbell, S., Allen, T.D., and Aplin, J.D. (1984) Novel substratum-dependent morphology and motility in epithelial cells. *Eur. J. Cell Biol.* 34, 275-280.

Carson, D.D., Dutt, A., and Tang, J.P. (1987) Glycoconjugate synthesis during early pregnancy: hyaluronate synthesis and function. *Dev. Biol.* 120, 228-235.

Cornillie, F.J., Lauweryns, J.M., and Brosens, I.A. (1985) Normal human endometrium. An ultrastructural survey. *Gynecol. Obstet. Invest.* 20, 113-129.

Dallenbach-Hellweg, G. (1981) *Histopathology Of The Endometrium.* Springer-Verlag, Berlin.

Dedhar, S., Argraves, W.S., Suzuki, S., and Ruoslahti, E. (1987) Human osteosarcoma cells resistant to detachment by an RGD-containing peptide overproduce the fibronectin receptor. *J.Cell Biol.* 105, 1175-1182.

DiPasquale, A. (1975) Locomotory activity of epithelial cells in culture. *Exp. Cell Res.* 94, 191-215.

Elsdale, T. and Bard, J.B.L. (1972) Collagen substrata for studies on cell behavior. *J.Cell Biol.* 54, 626-637.

Faber, K., Wewer, U.M., Berthelson, J.G., Liotta, L.A., and Albrechtsen, R. (1986) Laminin production by human endometrial stromal cells relates to the cyclic and pathological state of the endometrium. *Am. J. Pathol.* 124, 384-391.

Gimbrone, M.A., Cotran, R.S., and Folkman, J. (1974) Human vascular endothelial cells in culture. *J. Cell Biol.* 60, 673-684.

Glass, R.H., Aggeler, J., Spindle, A., Pederson, R.A., and Werb, Z. (1983) Degradation of extracellular matrix by mouse trophoblast outgrowths: A model for implantation. *J. Cell Biol.* 96, 1108-1116.

Harris, A.K., Stopak, D., and Wild, P. (1981) Fibroblast traction as a mechanism for collagen morphogenesis. *Nature* 290, 249-251.

Heath, J.P. (1982) Adhesions to substratum and locomotory behaviour of fibroblastic and epithelial cells in culture. In: *Cell Behaviour,* (eds.), R.Bellairs, A.Curtis, and G. Dunn, Cambridge University Press, pp. 77-108.

Hynes, R.O. (1987) Integrins: A family of cell surface receptors. *Cell* 48, 549-554.

Ireland, G.W. and Stern, C.D. (1982) Cell-substrate contacts in cultured chick embryonic cells: an interference reflection study. *J. Cell Sci.* 58, 165-183.

Jones, P.A., Neustein, H.B., Gonzales, F., and Bogenmann, E. (1981) Invasion of an artificial blood vessel wall by human fibrosarcoma cells. *Cancer Research* 41, 4613-4620.

Kisalus, L.L., Herr, J.C., and Little, C.D. (1987) Immunolocalisation of extracellular matrix proteins and collagen synthesis in first trimester human decidua. *Anat. Rec.* 218, 402-415.

Kolega, J. (1981) The movement of cell clusters in vitro: morphology and directionality. *J.Cell Sci.* 49, 15-32.

Mareel, M.M.K. (1979) Is invasiveness in vitro characteristic of malignant cells? *Cell Biol. Intl. Rep.* 3, 627-640.

Montesano, R., Pepper, M.S., Vassalli, J.D., and Orci, L. (1987) Phorbol ester induces cultured endothelial cells to invade a fibrin matrix in the presence of fibrinolytic inhibitors. *J. Cell. Physiol.* 132, 509-516.

More, I.A.R., Armstrong, E.M., Carty, M., and McSeveney, D. (1974) Cyclical changes in the ultrastructure of the normal endometrial stromal cell. *Br. J. Obstet. Gynaecol.* 81, 337-347.

Pena, S.D.J., Mills, G., Hughes, R.C., and Aplin, J.D. (1980) Polypeptide heterogeneity of hamster and calf fibronectins. *Biochem. J.* 189, 337-347.

Pijnenborg, R., Dixon, G., Robertson, W.B., and Brosens, I. (1980) Trophoblastic invasion of human decidua from 8 to 18 weeks of pregnancy. *Placenta* 1, 3-19.

Schor, S.L. (1980) Cell proliferation and migration on collagen substrata in vitro. *J. Cell Sci.* 41, 159-175.

Schor, S.L., Schor, A.M., and Bazill, G.W. (1981) The effects of fibronectin on the migration of human foreskin fibroblasts and Syrian hamster melanoma cells into three-dimensional gels of native collagen fibres. *J. Cell Sci.* 48, 301-314.

Schor, S.L., Schor, A.M., Rushton, G., and Smith, L. (1985) Adult, fetal and transformed fibroblasts different migratory phenotypes on collagen gels: evidence for an isoformic transition during fetal development. *J. Cell Sci.* 73, 221-234.

Sheppard, B.L. and Bonnar, J. (1974) The ultrastructure of the arterial supply of the human placenta in early and late pregnancy. *J. Obstet. Gynaecol. Br. Cmwlth.* 81, 487-511.

Sutcliffe, R.G., Davies, M., Hunter, J.D., Waters, J.J., and Parry, J.E.(1982) The protein composition of the fibrinoid material at the human uteroplacental interface. *Placenta* 3, 297-308.

Suzuki, S., Oldberg, A., Hayman, E.G., Pierschbacher, M.D., and Ruoslahti, E. (1985) Complete amino acid sequence of vitronectin deduced from cDNA. Similarity of cell attachment sites in vitronectin and fibronectin. *EMBO J* 4, 2519-2524.

Wayner, E.A. and Carter, W.G. (1987) Identification of multiple cell adhesion receptors for collagen and fibronectin in human fibrosarcoma cells possessing unique and common subunits. *J. Cell Biol.* 105, 1873-1884.

Wewer, U.M., Faber, M., Liotta, L.A., and Albrechtsen, R. (1985) Immunochemical and ultrastructural assessment of the nature of the pericellular basement membrane of human decidual cells. *Lab. Invest.* 53, 624-633.

Wewer, U.M., Damjanov, A., Weiss, J., Liotta, L.A., and Damjanov, I. (1986) Mouse endometrial cells produce basement membrane components. *Differentiation* 32, 49-58.

Wienke, E.C., Filiberto Calvazos, B.S., Hall, D.S., and Lucas, L.V. (1968) Ultrastructure of the endometrial stromal cell during the menstrual cycle. *Am. J. Obstet. Gynecol.* 102, 65-77.

Wynn, R.M. (1974) Ultrastructural development of the human decidua. *Am. J. Obstet. Gynecol.* 118, 652-670.

Wynn, R.M. (1977) Histology and ultrastructure of the human endometrium. In: *Biology of the Uterus*, (ed.), R.M.Wynn, Plenum:New York, pp. 341-376.

INTERACTIONS BETWEEN THE HUMAN TROPHOBLAST CELLS AND THE EXTRACELLULAR MATRIX OF THE ENDOMETRIUM. SPECIFIC EXPRESSION OF α-GALACTOSE RESIDUES BY INVASIVE HUMAN TROPHOBLASTIC CELLS

J.M. Foidart[1,3], Y. Christiane[1], and H. Emonard[2]

[1]Laboratory of Biology, University of Liege
Tower of Pathology (B35), Sart Tilman
B-4020 Liege, Belgium

[2]Laboratoire de Pathologie Cellulaire
CNRS UA 602, Institut Pasteur
F-69365 Lyon, France

INTRODUCTION

Mammalian embryo implantation takes place through a complex series of events resulting in apposition and attachment of the blastocyst to the uterine epithelium and subsequent invasion of the maternal endometrium. During invasion, the embryo penetrates the epithelium and basal lamina and invades the underlying stroma. Endometrial stromal cells secrete and deposit complex extracellular matrices (ECM) which play important roles in various physiologic and pathologic processes. In recent years, knowledge about the structure and function of ECM and their components has increased considerably. ECM are supramolecular complexes which constitute an extracellular environment influencing the differentiation, proliferation, organization, and attachment of cells. They play a key role in organogenesis, embryogenesis, and post-traumatic healing as well as in tumorigenesis, tumoral invasion and homing of metastatic tumor cells (Liotta, 1986; Bruijn et al., 1988).

In this article, a concise review of the most recent advances in ECM-trophoblast interactions is given. An attempt is made to demonstrate how the study of the ECM in human reproductive biology should lead to a better understanding of the molecular events occurring during human embryo implantation and of the key role of biochemically defined ECM components upon implantation, nidation, and trophoblast invasion.

MATERIALS AND METHODS

Human Tissues And Cells

Human Tissues

Human tissues were obtained from 10 uncomplicated term pregnancies immediately after delivery and from 10 therapeutic abortions for psychological reasons (blood groups: 6 type O, 4 type A patients). Pathologic specimens were

---

[3]To Whom Correspondence Should Be Addressed.

obtained from 6 spontaneous or 13 therapeutic abortions for a well-known pathology (3 XO monosomias, 5 molar placentae, 4 tubal pregnancies, and 1 choriocarcinoma). Placental bed biopsies of first trimester pregnancies were obtained by echoguided chorionic villi sampling forceps.

Human Choriocarcinoma Cells

Human choriocarcinoma cells (BeWo) (American type culture collection - Rockville, MD, USA) were cultured on glass coverslips in HAM F10 95% - FCS 5% medium and briefly fixed in ethanol (4°C, 10 minutes) before processing for immunofluorescence.

Human Trophoblast Cells

Trophoblast cells were isolated by mild trypsin treatment of first trimester placental villi according to Thiede (1960). Cells were applied to a Ficoll-Hypaque gradient (Pharmacia, Uppsala, Sweden), centrifuged at 1200 rpm for 45 minutes and pelleted cells discarded. Cells remaining at the interphase were plated in Linbro culture dishes (Flow Lab., USA) with TC199 medium containing 10% fetal calf serum. Trophoblast identity of these isolated cells was established by the following criteria: a) their morphology by light and electron microscopy; b) their ability to produce placental polypeptide hormones (4.5 ± 1 µg human placental lactogen, HPL/10^6cells/day and 2 ± 0.3 IU human chorionic gonadotropin, HCG/10^6cells/day); c) their immunostaining after 4 days in culture of over 90% of the cells by an anti-HPL antibody using immunofluorescence and/or immunoperoxidase technique.

Immunohistochemical Studies

Laminin and fibronectin were localized in human decidual tissue (7 weeks of pregnancy) and in biopsies of human postovulatory endometrium by an indirect peroxidase-antiperoxidase technique using previously described procedures and reagents (Foidart et al., 1980a, 1980b).

Adhesion Studies

Adhesion studies were performed according to previously described techniques (Farach et al., 1987). Freshly isolated trophoblast cells or freshly trypsinized BeWo cells were plated on plastic or on types I, III, or IV collagen (1 ml/well of 10 µg/ml in 0.1 M acetic acid) prepared either from the EHS tumor or from human placenta. Acetic acid was removed by evaporation for 48 hours under ultraviolet irradiation. Laminin (10 µg/well in phosphate buffered saline (PBS)) or fibronectin (10 µg/well in PBS) were also added. After overnight incubation at 4°C, unbound material was removed and non-specific binding sites were saturated with bovine serum albumin (500 µg/well in PBS). After overnight incubation, all wells were extensively washed with PBS. For the cell attachment assay, 2×10^5 trophoblast cells were added to the culture medium (0.9 ml of TC199 medium without serum). After various periods of time, unbound cells were removed by washing with fresh culture medium. Attached cells were trypsinized and counted in a Thoma cell and an electronic cell counter (Coulter Counter). At least six assays were performed for each experimental condition. Results are expressed as the mean ± standard deviation of all assays.

Purification Of A Natural Anti-α-Galactosyl IgG Antibody

The natural anti-α-galactosyl antibody was isolated from AB blood group patients by affinity chromatography on a melibiose-Sepharose column as described previously (Galili et al., 1984b). The reactivity of the antibody was assessed by its interaction with α-galactosyl residues on rabbit erythrocytes. The antibody was coupled with peroxidase (Tijssen and Kurstak, 1984) or labeled with fluorescein isothiocyanate as described previously (Unanue and Dixon, 1967). Griffonia Simplicifolia isolectin B4 (GS IB4) is a lectin commonly used to demonstrate certain types of α-linked galactose residues (Hayes and Goldstein, 1974; Wood et al., 1979). Peroxidase-labeled GS IB4 and peroxidase-labeled rabbit anti-human hCG antibody were purchased from Sigma (Saint-Louis, MO, USA).

Localization Of α-Galactose Residues

Fresh-frozen tissue sections (10 μm thick) were cut in a cryostat and reacted with fluorescein isothiocyanate-conjugated human anti-α-galactosyl IgG antibody (25 μg/ml in PBS) for 30 minutes at room temperature in a moist chamber. After extensive washing (5 times 5 minutes each in PBS), sections were mounted and examined in a Leitz Ortholux epifluorescence microscope. In some studies, peroxidase-labeled GS IB4 lectin (25 μg/ml in PBS) and peroxidase-conjugated anti-α-galactosyl IgG (10 μg/ml in PBS) were used to localize this carbohydrate on paraformaldehyde-fixed tissues according to previously described techniques (Sternberger et al., 1970). Control sections were treated with a solution of the specific antibody or of the GS IB4 lectin saturated with galactose.

Double Labeling Studies

Peroxidase-conjugated rabbit anti-human chorionic gonadotropin antibody (50 μg/ml in PBS) was applied to paraformaldehyde-fixed sections for 1 hour at room temperature in a moist chamber. After extensive washing (5 times 5 minutes with PBS) the sections were incubated in the presence of 0.01% diaminobenzidine (0.01% in Tris-saline (50 mM Tris HCl pH 7.6, 150 mM NaCl) with 0.1% H_2O_2 for 5 minutes. They were then incubated for 1 hour at room temperature in methanol (80%) to block endogenous peroxidase activity. After 5 washes in PBS, tissues were incubated for 30 minutes at room temperature in a moist chamber with either peroxidase-conjugated human anti-α-galactosyl antibody (10 μg/ml in PBS) or peroxidase-conjugated GS IB4 (25 μg/ml in PBS). After washing, the sections were incubated for 10 minutes in the dark in Tris-saline containing amino-ethyl-carbazole (0.01%). In this buffer, no staining reaction occurred when the human antibody was omitted. This control demonstrates inactivation of the tissue-bound peroxidase conjugated to the rabbit anti-HCG antibody.

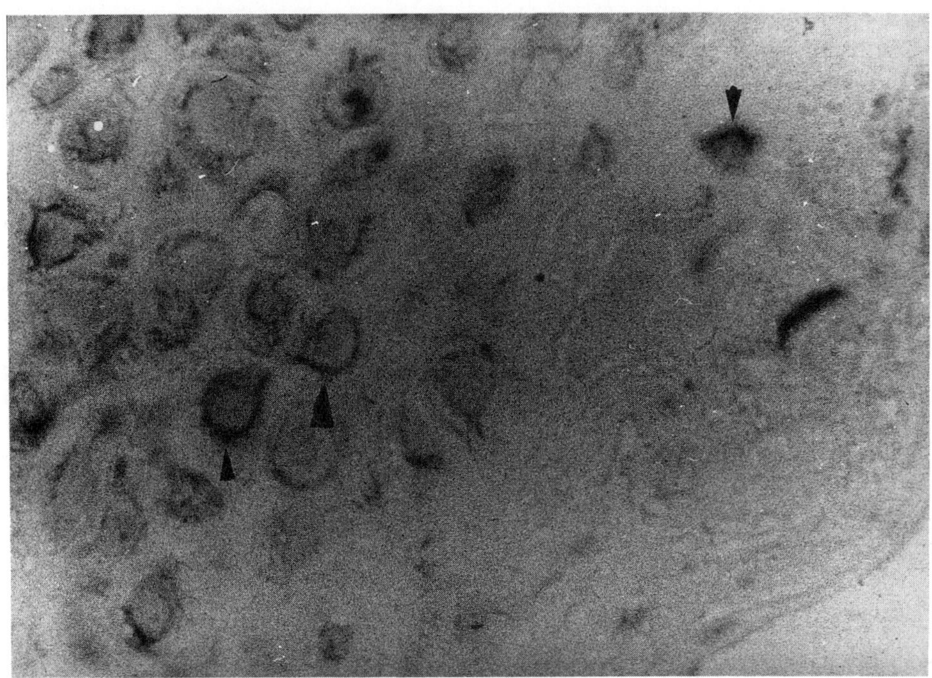

Figure 1. Distribution of laminin in decidual tissue at 7 weeks of pregnancy in an area with intermediate-sized decidual cells. The immunoperoxidase staining discloses focal immunoreactivity along the cell surfaces (arrowheads) (X1260).

RESULTS

Immunohistochemical Characterization Of The Pericellular Matrix of Human Decidual Cells

Sequential hormonal events following conception initiate a series of changes in the endometrium involving the uterine glands and stroma (Wynn, 1977). Stromal cells undergo a division and differentiation sequence, resulting in a large number of end cells which are thought to play a role in endocrine secretion, interaction with trophoblast and protection of the embryo from rejection by a maternal immune response (Frame et al., 1979; Dodd et al., 1980; Pijnenborg et al., 1980; Golander et al., 1981; Kearns and Lala, 1983). Decidual cells arise from stromal cells under the influence of hormones, primarily progesterone. The mature decidual cells are separated from one another and from trophoblast cells by an extracellular matrix composed of granular and fibrillar proteic material, fibrin, and collagen fibers. That much of the fibrinoid material is probably polymerized fibrin is suggested by the characteristic ultrastructural features of that protein (Robertson and Warner, 1974).

The mature decidual cells are surrounded by a dense granular material that by electron microscope (EM) examination resembles basement membrane (Lawn et al., 1971; Charpin et al., 1985). Immunohistochemical studies, at the light and EM level demonstrate the presence of laminin, entactin, type IV collagen, heparan sulfate proteoglycan, and fibronectin in this material (Wan et al., 1984; Charpin et al., 1985; Wewer et al., 1985, 1986; Kisalus et al., 1987) (Figure 1). Biochemical studies further confirm that human decidual cells synthesize basement membrane proteins (Wewer et al., 1986). It appears therefore that endometrial stromal cells upon decidual transformation synthesize and secrete all the collagenous and noncollagenous matrix proteins previously described in authentic basement membranes of epithelial, endothelial tissues or that limit neoplastic epithelia (Foidart et al., 1980b).

Decidual cells originate from stromal mesenchymal cells which, in the proliferative endometrium, do not produce basement membrane but rather interstitial matrix macromolecules, including types I and III collagen plus fibronectin (Kisalus et al., 1987) (Figure 2). When suitably stimulated, the cells transform into active basement membrane producers, presumably by selective gene activation and repression elicited by endocrine changes or pathologic conditions of the endometrium (Faber et al., 1986).

The coordinated increase in basement membrane protein expression as stromal cells decidualize may be considered preparatory to trophoblast invasion (Glasser et al., 1987). The ECM probably plays a key role in trophoblast attachment and regulation of its subsequent invasion into maternal endometrium (Faber et al., 1986; Wewer et al., 1986). Breakdown of type VI (stromal) collagen at the time of implantation may be important in trophoblast invasion (Aplin et al., 1988). Loke and Burland (1988) recently demonstrated that human trophoblast cells preferentially adhere to a complex extracellular matrix and differentiate into extravillous rather than villous trophoblast. It is therefore plausible that the ECM deposited by the decidual cells may anchor the trophoblast cells and even exert morphogenetic effects on them. It has also been suggested that this basement membrane matrix may function as a barrier to trophoblast invasion just as authentic basement membranes are an obstacle to invasion by malignant cells (Bell, 1985; Foidart et al., 1985). Finally, the ECM may also act as a permeability barrier, restricting access of macromolecules to the trophoblast cells (Wewer et al., 1986).

Role Of Laminin And Fibronectin On Human Trophoblast Cell Attachment

Laminin

Laminin is a large glycoprotein of about 900 KDa, comprised of three disulfide-linked polypeptide chains termed A (405 KDa), B1 (225 KDa), and B2 (205 KDa) chains. The complete sequence of the human and mouse B1 chains have recently been determined (Tryggvason et al., 1987). Laminin is considered important for the attachment of cells to the basement membrane (BM) and it promotes attachment of a variety of cell lines in vitro (Foidart et al., 1982). Laminin has a high affinity for type IV (BM) collagen, BM proteoglycan, and entactin. It has been suggested that laminin may be utilized by highly invasive and metastatic tumor cells to promote their attachment to the BM (type IV) collagen. Tumor cells with high capability to use laminin for attachment are more metastatic than the control tumor cells (Terranova et al., 1982).

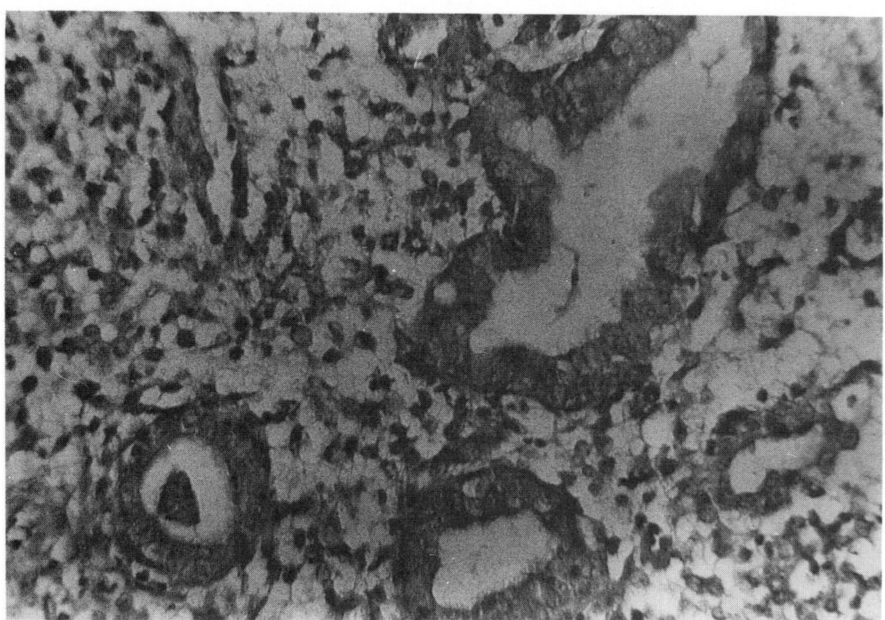

Figure 2. Immunohistochemical localization of fibronectin in human postovulatory endometrium (day 24). Fibronectin is extremely abundant throughout the stromal matrix (X320).

A better understanding of laminin-mediated cell adhesion came with the discovery of a specific laminin-binding protein at the cell surface of human breast carcinoma cells (Terranova et al., 1983). In vivo, many BM-associated cells use this laminin-binding protein to attach to the matrix (Lesot et al., 1983; Malinoff and Wicha, 1983; Wicha and Huard, 1983; Hand et al., 1985). This laminin receptor was recently identified in human endometrium during the secretory phase where it was localized to the surface of the predecidual cells (Faber et al., 1986). It is therefore possible that normal human trophoblast cells which also synthesize and secrete laminin (Aplin and Campbell, 1985; Dziadek and Timpl, 1985) utilize this protein to attach to the ECM of the endometrium.

Addition of exogenous laminin stimulates attachment in vitro of human trophoblast cells but not fibroblasts to type IV collagen coated dishes (Table 1). Addition of antibodies to laminin inhibits the adhesion of human trophoblast cells to type IV collagen and also abolishes the stimulating effect of this protein on the attachment of trophoblast cells. Normal IgG does not influence the binding of these cells to type IV collagen. Furthermore, human dermal fibroblast attachment to type IV collagen is not modified by either laminin or anti-laminin antibodies (Table 1) (Foidart et al., 1986). Previously it was shown that purified rabbit antibody to laminin administered intravenously to pregnant mice induces proteinuria as well as a high incidence of abortion, retroplacental hematoma, and fetal deaths (Foidart et al., 1983). It has also been demonstrated that laminin promotes in vitro attachment and outgrowth of mouse blastocysts plated on culture dishes coated with laminin (Armant et al., 1986). It appears therefore that laminin is probably utilized by young trophoblast cells during implantation to anchor to the pericellular matrix of the decidual cells. It could also promote subsequent invasion of the trophoblast cells in the maternal endometrium in a complex sequence of events comparable to those observed during malignant tumor cell infiltration in which laminin is known to play a key role (Liotta, 1986).

Fibronectins

Fibronectins form a class of multifunctional glycoproteins present at the cell surface, in the pericellular and extracellular matrix, BMs, plasma, and many other body fluids (Bruijn et al., 1988). These proteins also mediate cell adhesion, bind to extracellular matrix components such as collagen or proteoglycans and to cell surface receptors termed integrins (Ruoslahti et al., 1981). Fibronectins are present in the human endometrium in the interstitial stroma and in association with the glandular BM (Kisalus et al., 1987). It is therefore possible that this glycoprotein also contributes to the implantation of the blastocyst.

Carson et al. (1988) recently demonstrated that mouse embryos acquire the ability to attach to laminin, fibronectin, hyaluronate, and collagen types II and VI relatively early in their developmental program. The blastocysts acquire the ability to outgrow on other collagen types at a later time in culture. Binding to collagen type II and denatured collagen type IV was hampered by prior incubation of the blastocysts with a synthetic tetrapeptide Arg-Gly-Asp-Ser. In contrast, binding of the embryo to collagen types I, V, and VI was not inhibited by this peptide. It appears thus that embryos use multiple mechanisms to attach to collagens. Among these are adhesion systems that have a peptide recognition specificity similar to those of fibronectin receptors. This peptide does not interfere with laminin-mediated trophoblast outgrowth suggesting that trophoblast expresses different cell surface receptors for fibronectin and laminin. These findings suggest that differentiation of cells of the trophoectoderm into trophoblast cells with an adherent and invasive phenotype may involve the production of cell surface receptors for fibronectin, laminin, and possibly other matrix proteins (Armant et al., 1986).

The capacity of laminin and fibronectin to promote human normal or malignant trophoblast attachment was compared in culture (Figures 3A and 3B). In these assays, culture dishes were coated with purified laminin or fibronectin or various collagens (types I-IV). After overnight incubation, the dishes were extensively rinsed with PBS. Non-specific binding sites on the culture dishes were then saturated with bovine serum albumin (500 pg/ml) in order to prevent non-specific adhesion to the plastic. Freshly harvested human trophoblast cells or choriocarcinoma cells (2×10^5 BeWo cells/dish) were then plated in the presence of TC199 medium without serum. Plating efficiency was monitored by counting the number of cells attached after various periods of time as previously described in detail (Foidart et al., 1980; 1986).

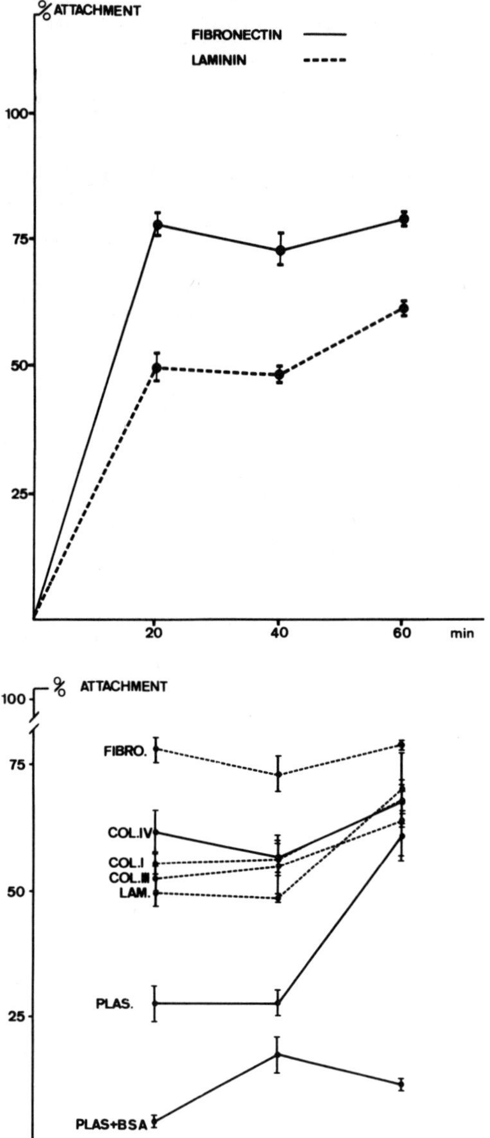
% ATTACHMENT
FIBRONECTIN
LAMININ
100
75
50
25
20
40
60
min
% ATTACHMENT
100
75
FIBRO.
COL.IV
COL.I
COL.III
LAM.
50
PLAS.
25
PLAS+BSA
20
40
60
min

When normal or malignant trophoblast cells were plated on plastic dishes coated with bovine serum albumin less than 20% of the cells attached within the first hour. Fibronectin, laminin, and various collagens considerably stimulated attachment of the human trophoblast cells within the first 20 minutes after plating. After one hour, 75% of the normal or malignant trophoblast cells adhered to fibronectin-coated dishes while about 60% attached to laminin-coated dishes (Figures 3A and 3B). It appears therefore that normal or malignant human trophoblast cells, like murine blastocysts, have the capacity to utilize at least two molecules to attach to the endometrium.

Expresssion Of α-Galactosyl Residues By Human Trophoblastic Cells

During embryo implantation, trophoblast cells of the blastocyst invade the maternal tissues and interact with the extracellular matrix of the endometrium in a process that somewhat resembles tumor invasion. It is logical to speculate that cellular functions normally expressed by the trophoblast cells during early steps of development correspond to some biological properties of cancerous cells with invasive potential.

The carbohydrate moieties of cell surface glycoconjugates are involved in cell-cell interactions during fertilization, placentation, morphogenesis, development, tumorigenesis, and metastasis (Wassarman, 1987; Grabel, 1984; Milos and Zalik, 1981; Gabius et al., 1986a, 1986b; Monsigny et al., 1983; Raz et al., 1984; 1986). Polylactosamine groups on human fetal placental fibronectin alter its biological properties (Zhu and Laine, 1985; Zhu et. al., 1984). The carbohydrate sequence in fibronectin and various glycoconjugates is modified during malignant transformation (Hakomori, 1985; Cossu and Warren, 1983; Nichols et al., 1986; Feizi, 1985). α-Galactosyl residues were recently described in murine laminin and at the cell surface of murine tumor cells (McCoy et al., 1985; Arumugham et al., 1986). In human, this carbohydrate is normally only expressed on surface of group B erythrocytes (Watkins, 1966), and on senescent, thalassemic, and sickle red cells (Kabat, 1976; Galili et al., 1983; 1984a; 1984b). In 1984, Galili et al. isolated a natural IgG antibody with a distinct anti-α-galactosyl reactivity (Galili et al., 1984b). This is present in high titer in the serum of every normal individual irrespective of blood group. It appears to contribute to extravascular lysis of normal and some pathological red cells since it interacts with normal senescent red cells and more extensively with pathological red cells of thalassemia and sickle cell anemia patients. The same antibody reduces lung colonization by murine malignant cells (Castronovo et al., 1987).

Figure 3. Adhesion of human normal (A) or malignant (B) trophoblast cells. The number of attached cells after plating on various substrates was estimated after 20, 40, or 60 minutes. At least six assays were performed for each condition.

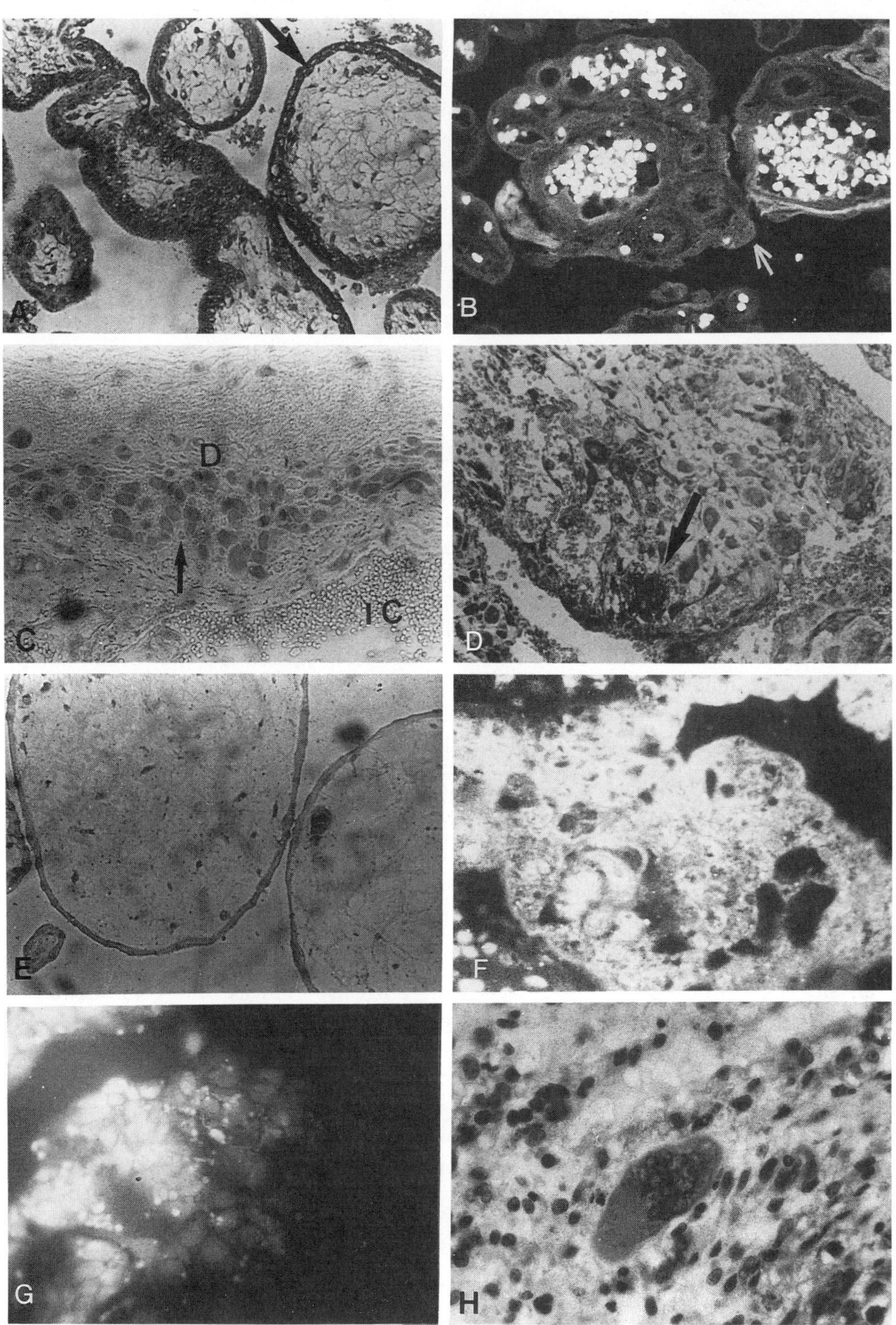

The possibility that α-galactosyl residues could be expressed by trophoblast cells at different stages of placentation and in various diseases was examined as it is related to trophoblast invasion. The α-Gal was clearly demonstrated on the trophoblastic layer of first trimester chorionic villi up to 10 weeks of gestation by using the GS IB4 or the anti-α-galactosyl IgG (Figure 4A). After 10 weeks of pregnancy, the staining became progressively weaker and disappeared almost completely in 13-14 week-old specimens. In term placental villi, staining was completely absent from villous trophoblast (Figure 4B). In the decidua from first trimester pregnancy, α-Gal was demonstrated at the surface of trophoblast cells lining the maternal tissues (Figure 4C). Interstitial trophoblast and vascular trophoblast which colonized the maternal spiral arteries were also strongly stained (Figure 4D). At term, the parietal, interstitial, and endovascular trophoblasts cells also displayed cell surface α-Gal residues. Using the human anti α-Gal IgG and anti-hCG rabbit IgG in double labeling studies, the α-Gal-positive cells were identified in maternal endometrium as trophoblast secreting hCG. No difference could be observed between the staining patterns with peroxidase-labeled GS IB4 lectin and the natural peroxidase- or FITC-conjugated anti-α-Gal antibody.

Trophoblast from spontaneous abortion and XO monosomia (Figure 4E) regularly displayed weak to negative staining. In tubal pregnancy, the trophoblast cells appeared also strongly stained in contrast to maternal cells which were negative. Trophoblast cells from molar proliferation (Figure 4F) and choriocarcinoma reacted strongly indicating a high degree of expression of alpha Gal residues. In culture, human choriocarcinoma cell surface also bound the human anti-α Gal antibody (Figures 4G and 4H) (Table II). Control sections reacted with GS IB4 lectin or human anti-α-Gal antibodies preincubated with galactose displayed no staining.

Villous interstitial and vascular trophoblast cells express α-Gal residues at their surface during early pregnancy (up to 10 weeks). Thereafter, the staining disappears progressively from villous trophoblast and becomes completely negative at 13-14 weeks amenorrhea. On the contrary, endovascular trophoblast and trophoblast cells that penetrate the endometrium and the myometrium display cell surface α-Gal up to term.

Figure 4. A. Staining with GS IB4-peroxidase on placental villi at 8 weeks of gestation. α-Gal is clearly demonstrated on the trophoblastic layer (arrow) (X120). B. Immunofluorescence staining with human anti-α-Gal IgG on term placental villi. The α-Gal is completely absent from villous trophoblast (arrow). Note the bright unspecific autofluorescence of the fetal erythrocytes (X120). C. Peroxidase staining with GS IB4 on maternal decidua. α-Gal is demonstrated on trophoblast cells lining maternal tissues (arrow) (X300).(D: decidua - IC: intervillous chamber). D. Peroxidase-lectin staining with GS IB4 on a maternal artery. Endovascular trophoblast exhibits α-Gal residues (arrow) (X300). E. Peroxidase-lectin staining with GS IB4 on chorionic villi from XO monosomia. In this pathology, α-Gal is absent or very weak on the trophoblastic layer (X120). F. Immunofluorescence staining with anti-α-Gal IgG on molar placenta. Trophoblastic cells display α-Gal residues in a proliferative area (arrow) (X300). G. Immunofluorescence staining with anti-α-Gal IgG on human choriocarcinoma cells in culture (X300). H. Peroxidase staining with GS IB4 on a tumoral trophoblast cell from a human choriocarcinoma (X500).

Table I

In Vitro Effect Of Human Anti-Laminin Antibodies On Human Trophoblast Cell Plating Efficiency

Medium	Percentage of cells attached to BM (type IV) collagen	
	Trophoblast cells	Fibroblasts
Alone	49 ± 5	79 ± 10
Containing soluble laminin (5 µg/ml)	71 ± 7*	85 ± 9
Containing antibody to laminin (5 µg/ml)	35 ± 8**	78 ± 12
Containing non-specific human IgG (5 µg/ml)	52 ± 7	75 ± 7
Plus laminin (5 µg/ml) + antibody to laminin (5 µg/ml)	44 ± 8	83 ± 8

* Significantly different from medium alone, $P < 0.01$, Student's "t" test.
** $P < 0.05$

Table II

Diseases	n	Expression of α-Gal residues by trophoblastic cells
Spontaneous abortion	6	0
Tubal pregnancy	4	+
XO monosomia	3	0
Molar proliferation	5	++
Choriocarcinoma	1	++
Choriocarcinoma cells in culture		++

n = number of cases

The density of α-Gal at the surface of malignant animal cells is closely correlated with an increased metastatic capacity (McCoy et al., 1985; Yogeeswaran, 1983; Schirrmacher, 1985). Trophoblastic cells, like malignant cells, invade the maternal endo- and myometrium. In addition, the density of staining is increased in highly invasive trophoblast (molar proliferation, choriocarcinoma, choriocarcinoma cells in culture). It is thus tempting to speculate that this carbohydrate could play a role in trophoblast invasion and placentation.

The natural anti α-Gal antibody, present in high titer in the serum of normal individuals (Galili et al., 1984) persists throughout pregnancy. It could bind to the trophoblast cells and block the staining reaction of the villous trophoblast. This antibody could ultimately contribute to limit trophoblast invasion of the maternal tissues and thereby participate in the immunological tolerance of pregnancy.

SUMMARY

During implantation human trophoblast cells come into contact with the stromal cells and extracellular matrix of the endometrium. Matrix proteins contribute to regulate cell attachment, differentiation, and invasion. Endocrine changes following conception initiate decidualization of the endometrium, resulting in a coordinated increase in basement membrane proteins at the periphery of stromal decidual cells. This newly deposited matrix may play a significant role in anchoring the trophoblast cells and in regulating their invasion of the endometrium. Fibronectin, laminin, and collagens promote trophoblast cell attachment in vitro by several mechanisms. Trophoblast cells exhibit at their surface α-Gal residues which are able to bind a naturally occurring human serum antibody. This antigen is absent from other normal human cells except invasive malignant cells. Its expression is prominent in invasive molar tissue and choriocarcinoma. It is speculated that α-Gal plays a role in trophoblast invasion.

ACKNOWLEDGEMENTS

This study has been supported by grants of the CGER in Belgium, of the FRSM in Belgium (n° 1.5067.87F and n° 3.4514.88), of the "Association contre le Cancer" (Belgium), of the "Fonds de la Recherche Facultaire" (a grant of the Faculty of Medicine) and of the "Centre Anti-Cancéreux".

REFERENCES

Aplin, J.D. and Campbell, S. (1985) An immunofluorescence study of the extracellular matrix associated with cytotrophoblast of the chorion laeve. *Placenta* 6, 469-479.

Aplin, J.D., Charlton, A.K., and Ayad, S. (1988) An immunohistochemical study of human endometrial extracellular matrix during the menstrual cycle and first trimester of pregnancy. *Cell Tissue Res.* 253, 231-240.

Armant, D.R., Kaplan, H.A., and Lennarz, W.J. (1986) Fibronectin and laminin promote in vitro attachment and outgrowth of mouse blastocysts. *Dev. Biol.* 116, 519-523.

Arumugham, R.G., Hsieh, T.C.Y., Tanzer, M.L., and Laine, R.A. (1986) Structures of the asparagine-linked sugar chains of laminin. *Biochim. Biophys. Acta* 883, 112-126.

Bell, S.C. (1985) Comparative aspects of decidualization in rodents and human: Cell types, secreted products and associated function. In: *Implantation Of The Human Embryo,* (eds.), R.G. Edwards, J.M. Purdy, and P.C. Steptoe, Academic Press, London, pp. 71-122.

Bruijn, J.A., Hogendoorn, P.C.W., Hoedemaeker, Ph.J., and Fleuren, G.J. (1988) The extracellular matrix in pathology. A review. *J. Lab. Clin. Med.* 111, 140-149.

Carson, D.D., Tang, J.P., and Gay, S. (1988) Collagens support embryo attachment and outgrowth in vitro: Effects of the Arg-Gly-Asp sequence. *Dev. Biol.* 127, 368-375.

Castronovo, V., Foidart, J.M., Li Vecchi, M., Foidart, J.B., Bracke, M., Mareel, M., and Mahieu, P. (1987) Human anti-alpha-galactosyl IgG reduces the lung colonization by murine MO4 cells. *Invasion Metast.* 7, 325-345.

Charpin, C., Kopp, F., Pourreau-Schneider, N., Lissitzky, J.C., Lavaut, M.N., Martin, P.M., and Toga, M. (1985) Laminin distribution in human decidua and immature placenta. *Am. J. Obstet. Gynecol.* 151, 822-826.

Cossu, G. and Warren, L. (1983) Lactosaminoglycans and heparan sulfate are covalently bound to fibronectins synthesized by mouse stem teratocarcinoma cells. *J. Biol. Chem.* 258, 5603-5607.

Dodd, M., Andrew, T.A., and Coles, J.S. (1980) Functional behavior of skin allografts transplanted to rabbit deciduomata. *J. Anat.* 130, 381-386.

Dziadek, M. and Timpl, R. (1985) Expression of nidogen and laminin in basement membranes during mouse embryogenesis and in teratocarcinoma cells. *Dev. Biol.* 111, 372-382.

Faber, M., Wewer, U.M., Berthelsen, J.G., Liotta, L.A., and Albrechtsen, R. (1986) Laminin production by human endometrial stromal cells relates to the cyclic and pathologic state of the endometrium. *Am. J. Pathol.* 124, 384-391.

Farach, M.C., Tang, J.P., Decker, G.L., and Carson, D.D. (1987) Heparin/heparan sulfate is involved in attachment and spreading of mouse embryos in vitro. *Dev. Biol.* 123, 401-410.

Feizi, T. (1985) Demonstration by monoclonal antibodies that carbohydrates structures of glycoproteins and glycolipids are onco-developmental antigens. *Nature* 314, 53.

Foidart, J.M., Bere, E.W., Yaar, M., Rennard, S.I., Gullino, M., Martin, G.R., and Katz, S.I. (1980a) Distribution and immunoelectron microscopic localization of laminin, a noncollagenous basement membrane glycoprotein. *Lab Invest.* 42, 336-342.

Foidart, J.M., Berman, J.J., Paglia, L., Rennard, S., Abe, S., Perantoni, A., and Martin, G.R. (1980b) Synthesis of fibronectin, laminin, and several collagens by a liver-derived epithelial line. *Lab. Invest.* 42, 525-532.

Foidart, J.M., Hunt, J., Lapière, C.M., Nusgens, B., De Rycker, C., Bruwier, M., Lambotte, R., Bernard, A., and Mahieu, Ph. (1986) Antibodies to laminin in preeclampsia. *Kidney Int.* 29, 1050-1057.

Foidart, J.M., Timpl, R., Furthmayr, H., and Martin, G.R. (1982) Laminin, a glycoprotein from basement membranes. In: *Immunochemistry Of The Extracellular Matrix - I. Methods*, (ed.), H. Furthmayr, CRC Press, Inc., Boca Raton, Florida, Vol. 1, pp. 125-134.

Foidart, J.M., Verly, M., Castronovo, V., Van Cauwenberge, J.R., Colin, C., Mahieu, P., Kinet, J.P., Timpl, R., Lapière, C., Nusgens, B., Pierard, D., and Birembaut, Ph. (1985) Interaction of human breast adenocarcinoma with the extracellular matrix: A new concept to study cancer invasion. In: *Evaluation du Risque de Cancer Mammaire. Chimiotherapie Première?*, (eds.), C. Colin and W. Gordenne, Pierre Mardaga, Liege, pp. 239-247.

Foidart, J.M., Yaar, M., Figueroa, A., Wilk, A., Brown, K.S., and Liotta, L.A. (1983) Abortion in mice induced by intravenous injections of antibodies to type IV collagen or laminin. *Am. J. Pathol.* 110, 346-357.

Frame, L.T., Wiley, L., and Rogol, A.D. (1979) Indirect immunofluorescent localization of prolactin to the cytoplasm of decidua and trophoblast cells in human placental membranes at term. *J. Clin. Endocrinol. Metab.* 49, 435-437.

Gabius, H.J., Brehler, R., Schauer, A., and Cramer, F. (1986a) Localization of endogenous lectins in normal human breast, benign breast lesions and mammary carcinomas. *Virchow's Arch. (Cell. Pathol.)* 52, 107.

Gabius, H.J., Engelhardt, R., and Cramer, F. (1986b) Endogenous tumor lectins: Overview and perspectives. *Anticancer Res.* 6, 573.

Galili, V., Clark, M., Mohandas, N., Rachmilewitz, E.A., and Shohet, S.B. (1984a) The natural anti-alpha-galactosyl IgG on red cells in sickle cell disease. *Blood 64 (Suppl.)*, 102.

Galili, V., Korhesh, A., Kahane, I., and Rachmilewitz, E.A. (1983) Demonstration of a natural antigalactosyl IgG antibody on thalassemic red blood cells. *Blood* 61, 1258.

Galili, V., Rachmilewitz, E.A., Peleg, A., and Flechner, I. (1984b) A unique natural human IgG antibody with anti-alpha-galactosyl specificity. *J. Exp. Med.* 160, 1519.

Glasser, S.R., Lampelo, S., Munir, M.I., and Julian, J.A. (1987) Expression of desmin, laminin and fibronectin during in situ differentiation (decidualization) of rat uterine stromal cells. *Differentiation* 35, 132-142.

Golander, A., Zakuth, V., Schecter, Y., and Spirez, Z. (1981) Suppression of lymphocyte reactivity in vitro by a soluble factor secreted by explants of human decidua. *Eur. J. Immunol.* 11, 849-855.

Grabel, L.B. (1984) Isolation of a putative cell adhesion mediating lectin from teratocarcinoma stem cells and its possible role in differentiation. *Cell Differ.* 15, 121.

Hakomori, S.I. (1985) Aberrant glycosylation in cancer cell membranes as focused on glycolipids. Overview and perspectives. *Cancer Res.* 45, 2405.

Hand, P.H., Thor, A., Schlom, J., Rao, C.N., and Liotta, L. (1985) Expression of laminin receptor in normal and carcinomatous human tissues as defined by a monoclonal antibody. *Cancer Res.* 45, 2713-2719.

Hayes, C.E. and Goldstein, I.J. (1974) An alpha-D-galactosyl binding lectin from Bandeiraea Simplicifolia seeds: Isolation by affinity chromatography and characterization. *J. Biol. Chem.* 249, 1904.

Kabat, E. (1976) *Structural Concepts In Immunochemistry* (2nd edition). Holt, Rinchart and Winston, New York, pp. 174.

Kearns, M. and Lala, P.K. (1983) Life history of decidual cells: A review. *Am. J. Reprod. Immunol.* 3, 78-82.

Kisalus, L.L., Herr, J.C., and Little, C.D. (1987) Immunolocalization of extracellular matrix proteins and collagen synthesis in first trimester human decidua. *Anat. Rec.* 218, 402-415.

Lawn, A.M., Wilson, E.W., and Finn, C.A. (1971) The ultrastructure of human decidual and predecidual cells. *J. Reprod. Fertil.* 26, 85-93.

Lesot, H., Kuhl, U., and von der Mark, K. (1983) Isolation of a laminin binding protein from muscle cell membranes. *EMBO J.* 2, 861-865.

Liotta, L.A. (1986) Tumor invasion and metastasis. Role of the extracellular matrix. Rhoads Memorial Award Lecture. *Cancer Res.* 46, 1-6.

Loke, Y.W. and Burland, K. (1988) Human trophoblast cells cultured in modified medium and supported by extracellular matrix. *Placenta* 9, 173-182.

McCoy, J., Goldstein, I.J., and Varani, J. (1985) A review of studies in our laboratory regarding Ella methodology for the study of cell surfaces carbohydrates from tumors of varying metastatic potential. *Tumor Biol.* 6, 99.

Malinoff, H.L. and Wicha, M.S. (1983) Isolation of a cell surface receptor protein for laminin from murine fibrosarcoma cells. *J. Cell. Biol.* 96, 1475-1479.

Milos, N. and Zalik, S.E. (1981) Effect of the alpha-galactose binding lectins on cell to substratum and cell to cell adhesion of cells from the extraembryonic endoderm of the early chick blastoderm. *Roux's Arch. Dev. Biol.* 190, 259.

Monsigny, M., Kieda, C., and Roche, A.C. (1983) Membrane glycoproteins, glycolipids and membrane lectins as recognition signals in normal and malignant cells. *Biol. Cell* 47, 95.

Nichols, E.J., Fenderson, B.A., Carter, W.G., and Hakomori, S.I. (1986) Domain-specific distribution of carbohydrates in human fibronectins and the transformation-dependent translocation of branched type 2 chain defined by monoclonal antibody C6. *J. Biol. Chem.* 261, 11295-11301.

Pijnenborg, R., Dixon, G., Robertson, W.B., and Brosens, I. (1980) Trophoblastic invasion of human decidua from 8 to 18 weeks of pregnancy. *Placenta* 1, 3-19.

Raz, A., Meromsky, L., Carmi, P., Karkash, R., Lotan, P., and Lotan, R. (1984) Monoclonal antibodies to endogenous galactose specific tumor cell lectins. *EMBO J.* 3, 2979.

Raz, A., Meromsky, L., and Lotan, R. (1986) Differential expression of endogenous lectins on the surface of non tumorigenic, tumorigenic and metastatic cells. *Cancer Res.* 46, 3667.

Robertson, W.B. and Warner, B. (1974) The ultrastructure of the human placental bed. *J. Pathol.* 112, 203-211.

Ruoslahti, E., Engvall, E., and Hayman, E.G. (1981) Fibronectin: Current concepts of its structure and functions. *Coll. Res.* 1, 95-128.

Schirrmacher, V. (1985) Cancer metastasis. Experimental approaches, theoretical concepts and impacts for treatment strategies. *Adv. Cancer Res.* 43, 1.

Sternberger, L.A., Hardy, P.A. Jr., Cuculis, J.J., and Meyer, H.G. (1970) The unlabeled antibody enzyme method of immunohistochemistry: Preparation and properties of the soluble antigen-antibody complex (horseradish peroxidase - antihorseradish peroxidase) and its use in identification of Spirochetes. *J. Histochem. Cytochem.* 18, 315.

Terranova, V.P., Liotta, L.A., Russo, R.G., and Martin, G.R. (1982) Role of laminin in the attachment and metastasis of murine tumor cells. *Cancer Res.* 42, 2265-2269.

Terranova, V.P., Rao, C.N., Kalebic, T., Margulies, I.M., and Liotta, L.A. (1983) Laminin receptor on human breast carcinoma cells. *Proc. Natl. Acad. Sci. (USA)* 80, 444-448.

Thiede, H.A. (1960) Studies of the human trophoblast in tissue culture. *Am. J. Obstet. Gynecol.* 79, 636-647.

Tijssen, P. and Kurstak, E. (1984) Highly efficient and simple methods for the preparation of peroxidase and active peroxidase-antibody conjugates for enzyme immunoassays. *Anal. Biochem.* 136, 451.

Tryggvason, K., Hoyhtya, M., and Salo, T. (1987) Proteolytic degradation of extracellular matrix in tumor invasion. *Biochim. Biophys. Acta* 907, 191-217.

Unanue, E. and Dixon, F. (1967) Experimental glomerulonephritis: Immunological events and pathogenic mechanisms. *Adv. Immunol.* 6, 1.

Wan, Y.J., Wu, T.C., Chung, A.E., and Damjanov, I. (1984) Monoclonal antibodies to laminin reveal the heterogeneity of basement membranes in the developing and adult mouse tissues. *J. Cell. Biol.* 98, 971-979.

Wassarman, P.M. (1987) The biology and chemistry of fertilization. *Science* 235, 553.

Watkins, W. (1966) Blood group substances. *Science* 152, 172.

Wewer, U.M., Damjanov, A., Weiss, J., Liotta, L.A., and Damjanov, I. (1986) Mouse endometrial stromal cells produce basement membrane components. *Differentiation* 32, 49-58.

Wewer, U.M., Faber, M., Liotta, L.A., and Albrechtsen, R. (1985) Immunochemical and ultrastructural assessment of the nature of the pericellular basement membrane of human decidual cells. *Lab. Invest.* 53, 624-633.

Wicha, M.S. and Huard, T.K. (1983) Macrophages express cell surface laminin. *Exp. Cell Res.* 143, 475-479.

Wood, C., Kabat, E.A., Murphy, L.A., and Goldstein, I.J. (1979) Immunochemical studies of the combining sites of the two lectins A4 and B4 isolated from Bandeiraea Simplicifolia. *Arch. Biochem. Biophys.* 198, 1.

Wynn, R.M. (1977) *Biology Of The Uterus*. Plenum Press, New York.

Yogeeswaran, G. (1983) Cell surface glycolipids and glycoproteins in malignant transformation. *Adv. Cancer Res.* 38, 289.

Zhu, B.C.R., Fisher, S.F., Pande, H., Calaycay, J., Shively, J.E., and Laine, R.A. (1984) Human placental (fetal) fibronectin: Increased glycosylation and higher protease resistance than plasma fibronectin. *J. Biol. Chem.* 259, 3962-3970.

Zhu, B.C.R. and Laine, R.A. (1985) Polylactosamine glycosylation on human fetal placental fibronectin weakens the binding affinity of fibronectin to gelatin. *J. Biol. Chem.* 260, 4041-4045.

APPEARANCE, SHEDDING, AND ENDOCYTOSIS OF A BLASTOCYST SURFACE GALACTOSE-GALACTOSAMINE DERIVATIVE DETECTED WITH A MONOCLONAL ANTIBODY

Mats Hjortberg and B. Ove Nilsson

Department of Human Anatomy
Biomedical Center, Box 571
Uppsala University
S-751 23 Uppsala, Sweden

INTRODUCTION

Mouse monoclonal antibodies against mouse blastocysts have been produced by intrasplenic immunization (Nilsson et al., 1986; Nilsson et al., 1987). One antibody, N.63, detects a derivative involving terminal D-galactose on the surface of the mouse trophectoderm (Svalander et al., 1989). The epitope is stage-specific both for normal blastocysts and those from experimentally delayed implantation. In delayed implantation, staining by the ABC technique is similar for delayed, inactive blastocysts and estrogen-activated, implanting blastocysts. However, it is not excluded that the membrane-bound molecule has different properties in these two different states of the trophectoderm.

Pilot experiments showed that the antigen N.63 is shedded by implanting blastocysts. The present paper reports on the appearance and shedding of the N,63 antigen and on the endocytosis of immune complexes during the development of blastocysts from preimplantation to implantation.

MATERIALS AND METHODS

Mouse Embryonic Stages

Outbred NMRI strain mouse embryos of defined developmental stage were purchased from Alab, Sweden. The animals were housed under standard conditions at the Animal-Care Department of the Biomedical Center in Uppsala, Sweden. To induce pregnancy, females aged about two months were caged with males overnight. The following morning was designated as day 1 of pregnancy if a vaginal plug was found. To obtain delayed and adhesive blastocysts lacking zona pellucida (zona-free), the method of facultative delay of implantation was used (Mayer, 1963; Bergström, 1978).

Two groups of blastocysts were used: delayed and adhesive. Delayed inactive blastocysts were collected 8-10 days after induction of delay by ovariectomy on day 3 of pregnancy and treatment every fifth day with 0.1 mg progesterone (Depo-Provera; Upjohn, Sweden). Adhesive blastocysts were collected 18 hours after the reactivation from delay by an injection of 0.1 µg estrogen (Estradiol- 17β; Sigma).

Culture

Modified Brinster's medium for ovum culture (Brinster, 1963) with the addition of 1 mg/ml of glucose (Sigma), 1 % fetal calf serum (FCS) (SVA, Sweden), dialyzed against isotonic NaCl for 3 days, and amino acids (Sigma), supplemented according to the amount described for Eagle's Basal Medium (Eagle, 1955), were used as a blastocyst culturing medium. For delayed blastocysts, glucose, arginine and leucine were omitted (diapause medium) to prevent the blastocysts from outgrowing (Naeslund, 1979).

Appearance Of Antigen On The Blastocyst Surface

The appearance of antigen N.63 on the surface of blastocysts was estimated by an indirect immunofluorescence (IIF) method using antigen-neutralized blastocysts recovered after different times in culture. The following groups were used: (1) delayed and (2) adhesive blastocysts cultured between 0, 1, 2, 3, and 4 hours after antigen-neutralization, making a total of 10 groups. The number of blastocysts used for each experiment varied between 3 and 4 for each group, and each experiment was repeated three times.

The immunohistochemical protocol used to detect the surface antigen was as follows. (1) Flush and rinse 3 x 10 minutes with Dulbecco's phosphate buffered saline (DPBS) supplemented with 10 % FCS at room temperature. (2) Incubate with undiluted monoclonal antibody supernatant N.63 for 60 minutes at room temperature (antibody neutralization). (3) Rinse 2 x 10 minutes with DPBS + 10 % FCS. (4) Rinse 2 x 10 minutes in culture medium (see above). (5) Culture blastocysts for 0 to 4 hours in a plastic tissue culture chamber measuring 35 x 10 mm (Nunc., Denmark) under humidified conditions with 5 % CO_2 in air (see Table 1). (6) Rinse 3 x 10 minutes in DPBS + 1 % bovine serum albumin RIA-grade (BSA) (Sigma) + 0.01 % NaN_3 (BDH) (A-buffer). (7) Incubate for 90 minutes in monoclonal antibody supernatant N.63, diluted 1:2 in A-buffer. (8) Rinse 3 x 10 minutes in A-buffer. (9) Incubate for 30 minutes in TRITC-conjugated rabbit anti-mouse Ig antibodies (Dakopatts A/S, Denmark), diluted 1:40 in A-buffer. (10) Rinse 3 x 10 minutes in A-buffer. The blastocysts were examined with a Nikon Diaphot microscope equipped with an epi-fluorescence system for TRITC application.

Control experiments were performed by replacing mAb N.63 at step 2, step 7 or both steps with supernatants from the mouse plasmacytoma (Sp 2/0) cells used for the antibody production. The results of the control experiments were negative.

Antigen Shedding

The shedding of antigen N.63 from cultured blastocysts was estimated by ABC-staining of dot-blots of culture medium recovered after different periods of culture. The groups studied were medium samples taken from delayed and adhesive blastocysts at various time of culture (see Table 2). Two dot-blots from each group were used. The experiments were repeated three times for a total of 6 dot-blots for each group.

Table 1

Appearance Of Antigen N.63 On The Blastocyst Surface As Detected By IIF

Duration of Culture	State of Blastocysts	
(in hours)	Delayed	Adhesive
0	- (9)	- (10)
1	- (11)	+ (9)
2	- (10)	+ (10)
3	± (12)	+ (9)
4	± (12)	+ (9)

Indicates + strong, ± weak, and - negative immunolabeling;
number of examined specimens in parenthesis.

Table 2

Shedding Of Antigen N.63 By Blastocysts As Observed By
ABC-Staining Of Dot Blots Of Spent Medium

Duration of Culture	State of Blastocysts	
(in hours)	Delayed[1]	Adhesive[2]
1	- (6)	+ (6)
2	- (6)	+ (6)
3	- (6)	+ (6)
4	- (6)	+ (6)

[1]Refers to delayed blastocyst in diapause medium.
[2]Refers to adhesive blastocysts cultured in fully supplemented Brinster's medium.
[3]Refers to number of microdrops.

The blastocysts were carefully rinsed 3 x 10 minutes in culture medium at room temperature. For the inactive blastocysts diapause medium was used. Incubation was performed for 1 to 4 hours in microdrops covered with a layer of paraffin oil at humidified conditions with 5 % CO_2. One µl of medium was used for each blastocyst and about 10 blastocysts were prepared for each microdrop. After various periods of culture, the culture media were removed and dot-blotted onto pieces of NC-paper (BA- 85, 0.45 µm; Schleicher and Schuell). The papers were air-dried at 37° for about 60 minutes and stored at room temperature for two days until used. The avidin-biotin-peroxidase complex (ABC) method was applied for the detection of antigen N.63 on NC-papers. The staining was performed by passing the dot-blotted NC-papers with microtweezers through reagents or washing media in series of 1 ml wells (Disposo-Tray; Flow).

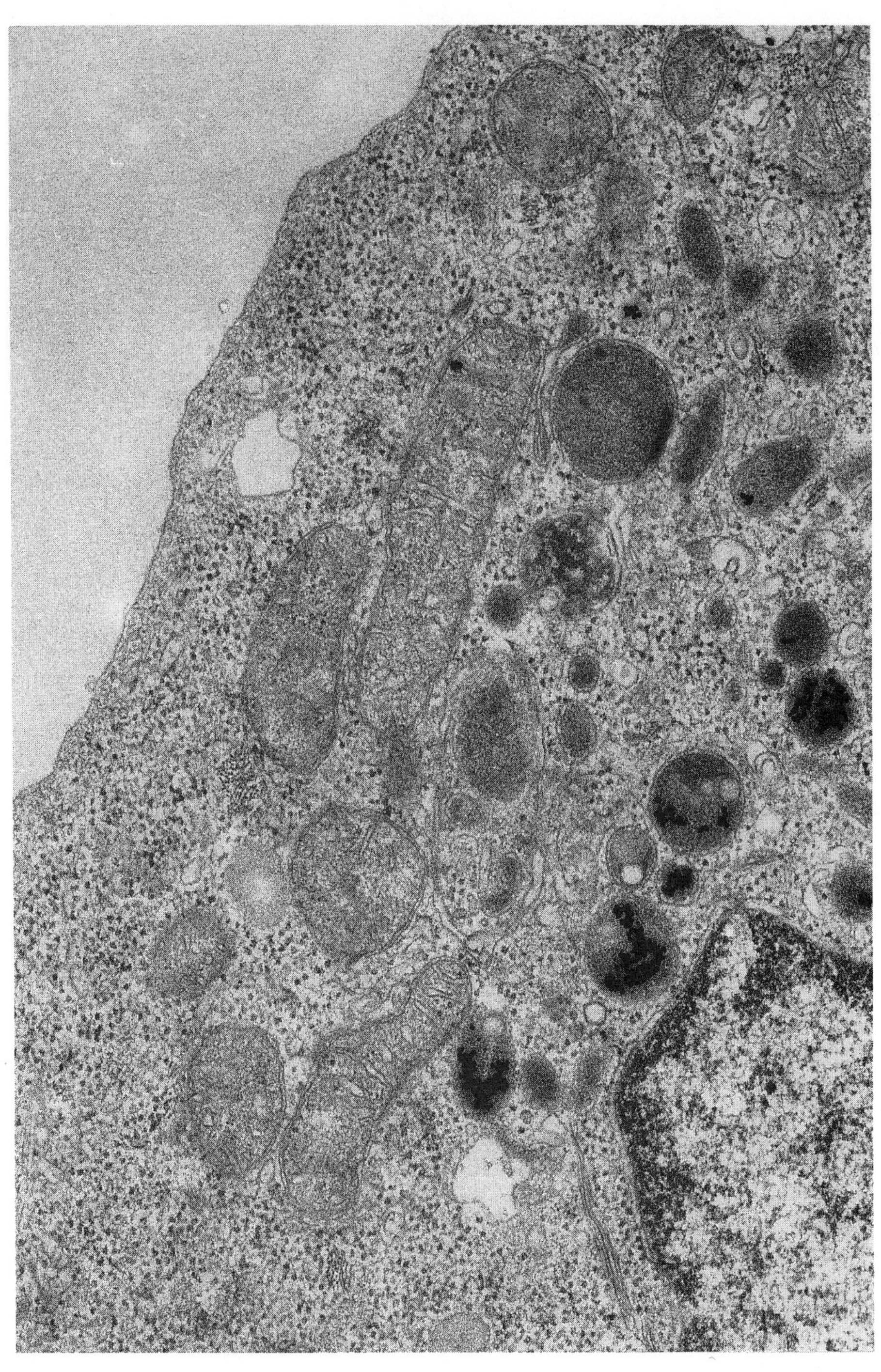

The protocol for dot-blotted NC paper pieces used to detect the shedded surface antigen in the culture medium was as follows. (1) Block the NC binding capacity by preincubation in DPBS with 3% BSA and 0.01% NaN_3 at 4°C overnight. (2) Rinse for 3 x 10 minutes in DPBS and 5 minutes in distilled water. (3) Quench for endogenous peroxidase 5 minutes in 3% H_2O_2 (Merck). (4) Wash for 10 minutes in distilled water and 2 x 10 minutes in DPBS. (5) Block unspecific secondary antibody binding for 30 minutes with 5% normal goat serum (SVA) in DPBS. (6) Incubate in mAb N.63 supernatant, diluted 1:10 in DPBS with 0.1% BSA, for 30 minutes. (7) Wash for 3 x 10 minutes in DPBS. (8) Incubate for 30 minutes in biotinylated goat-anti-mouse IgM secondary antibody (Vector), diluted 1:200 in DPBS with 1% normal goat serum added. (9) Wash for 3 x 10 minutes in DPBS. (10) Incubate in ABC reagent (Vectastain; Vector), diluted 1:200 in DPBS with 1% normal goat serum, for 30 minutes. (11) Wash for 3 x 10 minutes in DPBS. (12) Incubate for 15 minutes in peroxidase substrate complex, prepared immediately before use by mixing 10 mg of 3-amino-9-ethylcarbazole (Sigma) dissolved in 6 ml of DMSO (Merck) with 50 ml of 20 mM sodium acetate buffer, pH 5, and adding 4 µl of 30% H_2O_2. (13) Wash for 3 x 10 minutes in distilled water. A brownish red color on the blotted dot indicated a positive immunolabeling reaction.

In order to test the specificity of the immunocytochemical method, the following control experiments were conducted: (a) omission of mAb N.63, (b) omission of the biotinylated goat-anti-mouse antibody, (c) dot-blots from culture medium with no blastocysts present during incubation, (d) incubation in supernatant from Sp2/0 cell cultures instead of mAb N.63 supernatant. The control experiments were negative.

Endocytosis Of Immune Complexes

Preembedding staining with immunogold marker for TEM was used for estimating endocytotic activity in delayed and adhesive blastocysts. The blastocysts were carefully rinsed 3 x 10 minutes in DPBS + 10% FCS and incubated in undiluted primary antibody supernatant for 60 minutes at room temperature. After rinsing 3 x 10 minutes, incubation in secondary goat-anti-mouse IgM coupled to 15 nm colloidal gold particles (Janssen), diluted 1:20 was performed in DPBS + 10% FCS for 120 minutes at room temperature. The blastocysts were then rinsed in culture medium for 3 x 10 minutes and further incubated in a culture chamber, 37°C, 5% CO_2 in air for 0, 1, 2, 3, and 4 hours. Afterwards the embryos were rinsed in DPBS supplemented with 2.5% saccarose (BDH) and fixed in 2.5% glutaraldehyde (Merck), 2.5% saccharose overnight. The blastocysts were postfixed in 1% osmium tetroxide (Jonsson Matthey) for 1 hour, dehydrated in ethanol and embedded in Epon 812 (Agar Aids). Four to five embryos were embedded in each of ten groups. Thin sections (LKB ultrotome) were cut at four to six levels from each specimen, stained with uranyl acetate and lead citrate and observed with a Philips 301 transmission electron microscope. Control experiments were performed where no antibody or an irrelevant primary antibody was used instead of antibody N.63.

Figure 1. Delayed blastocyst labeled with antibody N.63 gold complex and cultured 1h in diapause medium. The antibodies are located in the lysosomes of the trophoblast cells. X65,000

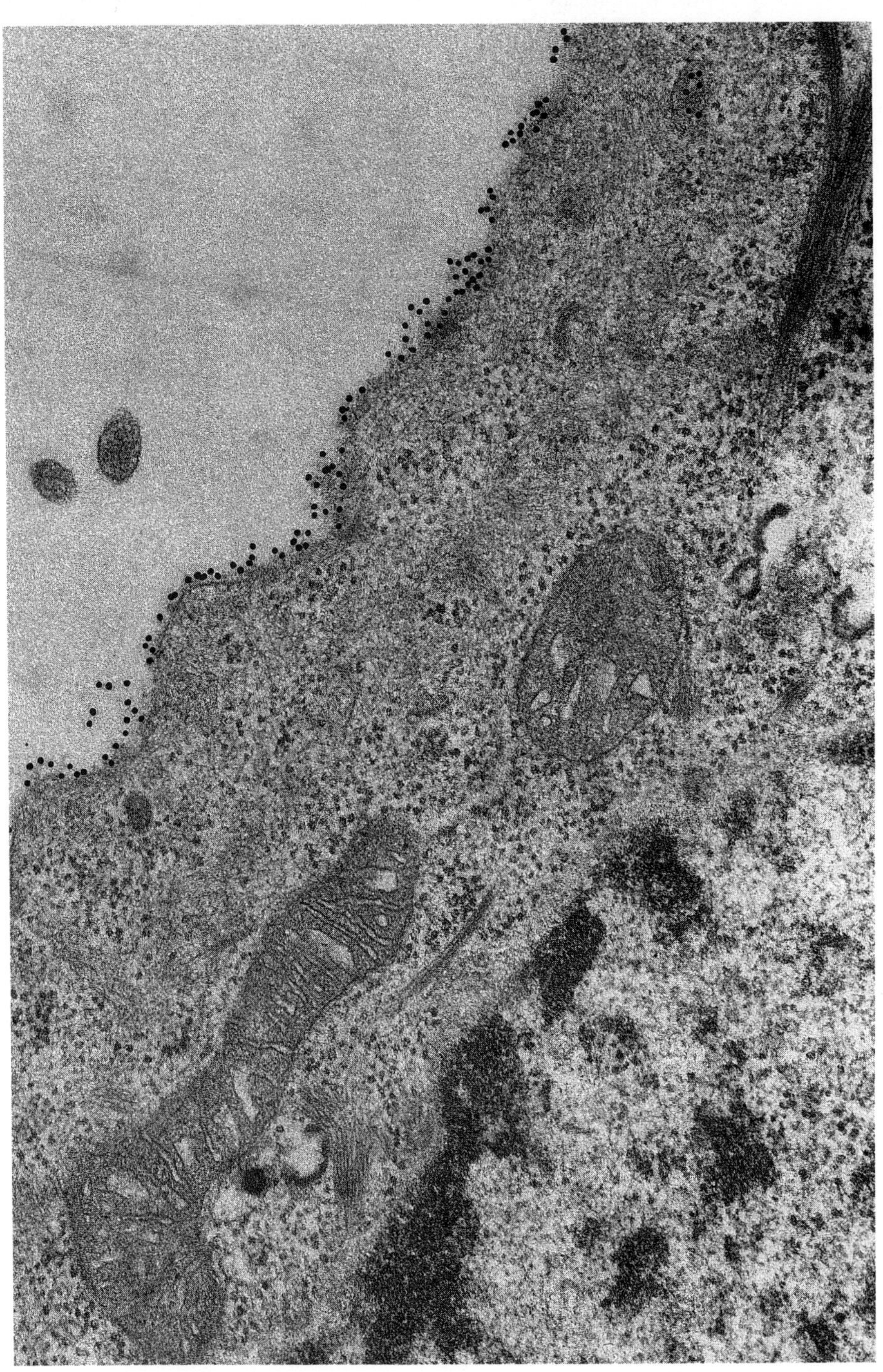

RESULTS

Appearance Of Antigen On The Blastocyst Surface

The results are summarized in Table 1. Blastocysts at diapause showed low activity in synthesizing new antigen N.63 at the trophoblast surface and not until a period of 3 hours in culture, weak fluorescence was detectable.

For adhesive blastocysts new antigen N.63 was detected as soon as 1 hour in culture. The fluorescence intensity was similar to that of adhesive control blastocysts cultured 0 hour.

Antigen Shedding

Delayed blastocysts showed no shedding activity at all. After 1 hour of culturing, the medium which contained the adhesive blastocysts was positive for antigen N.63. The results are summarized in Table 2.

Endocytosis Of Immune Complexes

Delayed blastocysts internalized the antigen-antibody complex N.63 rapidly. After 1 hour of culture most of the complex was visible in the lysosomes of the trophoblast cells (Figure 1). Adhesive blastocysts on the other hand showed a low endocytotic activity (Figures 2 and 3).

The endocytotic pattern was changed when an unspecific primary antibody was used. No binding and internalization by delayed blastocysts was detected. However in adhesive blastocysts the gold antigen-antibody complex was present in the lysosomes of the trophoblast already after 1 hour.

DISCUSSION

A mouse blastocyst which has reached the uterine lumen remains for about a day before it attaches to the uterine surface and begins to implant. During the period in contact with uterine secretions, the blastocyst has lost its zona pellucida and has prepared for implantation by increasing adhesiveness, initiating systems for signalling its presence to the endometrium, and finding ways to escape an immunological attack.

A blastocyst experimentally delayed from implantation also loses its zona pellucida early during the delayed state and, after being activated for implantation by an injection of estrogen, prepares for implantation in a manner similar to that of a non-manipulated blastocyst. Since the zona pellucida hinders immunological recognition at the surface of the trophectoderm, experimentally delayed implantation is used in order to obtain zona-free blastocysts of stages varying from delay to implantation.

Figure 2. Adhesive blastocyst labeled with antibody N.63 gold complex and cultured 0 hour in fully supplemented Brinster's medium. The antibodies are located at the trophoblast surface. X62,000

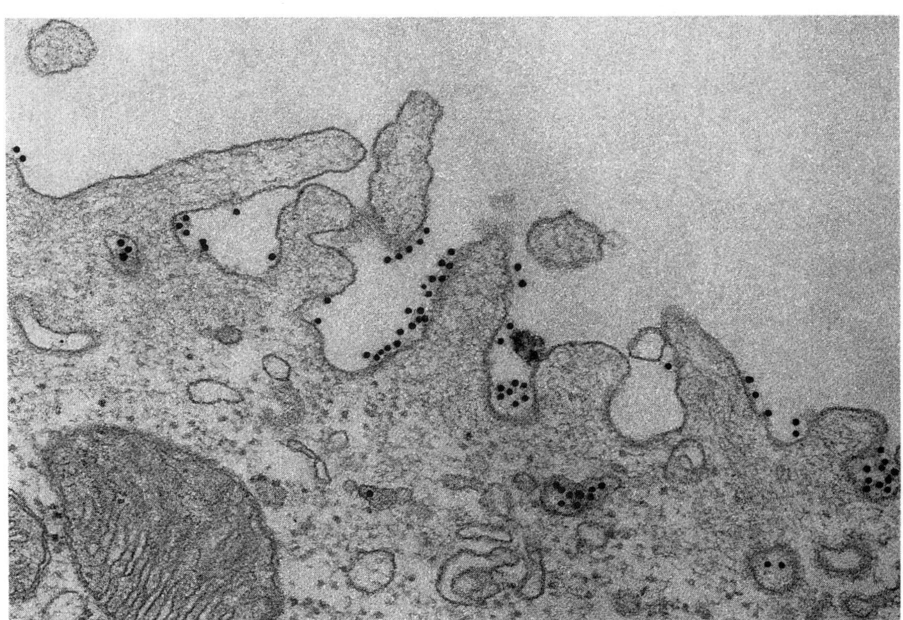

Figure 3. Adhesive blastocyst labeled with antibody N.63 gold complex and cultured 1h in fully supplemented Brinster's medium. The antibodies are located mainly at the trophoblast surface. The labeling indicates a low endocytotic activity. X60,000

A monoclonal antibody, denoted N.63, which was raised against implanting blastocysts, detects an epitope that is expressed at the trophectoderm surface of both delayed and implanting blastocysts (Svalander et al., 1989). Since the pilot experiments demonstrated that the rate of turnover of the antigen N.63 differed between delayed and implanting blastocysts, its appearance on the blastocyst surface, and its shedding, endocytosis were studied to determine its possible function at implantation.

The appearance of antigen N.63 at the blastocyst surface was found to be much more rapid in implanting blastocysts then in delayed ones. One reason for this could be that implanting blastocysts generally have a high metabolic activity when measured by their uptake of oxygen (Nilsson et al., 1982) and consumption of glucose (Nilsson et al., 1980), and have a well developed rough endoplasmatic reticulum (Naeslund et al., 1980). Another reason could be that antigen N.63 plays a part in the process of implantation, for instance, by facilitating blastocyst attachment or by inducing responses in the endometrium. This possibility raised the question whether N.63 was restricted to the cell membrane or intended for export. It did not remain localized at the cell surface but that it was shed into the culture medium.

The shedding of N.63 by implanting blastocysts was markedly greater than that of delayed blastocysts. Shedding of surface materials by cells is a common phenomenon (for review, see Black, 1980). One shedded molecule, the plasminogen activator, has been identified in blastocysts (Sherman et al., 1976; Strickland et al., 1976). Experiments by Nieder et al. (1987) indicate shedding of proteins which seem to be involved in maternal recognition of pregnancy. Thus, since the antigen N.63 is actively shedded at implantation, that molecule may also have the capacity to act as a blastocyst signal substance to initiate implantation-facilitating responses in the endometrium.

The epitope detected by the antibody N.63 is a galactose-galactosamine derivative which is carried on a glycolipid when located in the cell membrane (Svalander et al., 1989). However, it is not yet known what type of shed molecule contains the epitope N.63.

Immune complexes are formed at the blastocyst surface when the antibody N.63 is added. These complexes may be shed, but alternatively they may be internalized by the trophoblast. To localize the immune complexes, we used an anti-mouse IgM antibody labeled with gold and examined blastocysts with an electron microscope. The delayed blastocysts engulfed most of the immune complexes, whereas the implanting blastocysts shedded most of the complexes. This finding points out an important difference between delayed and implanting blastocysts in processing of immune complexes that might affect the ability of the trophoblast to escape immunological rejection. Thus, endocytosis of immune complexes by the delayed blastocyst might be used to eliminate antigens for which antibodies are present in the uterine lumen, since endocytosis occurs when the trophoblast is surrounded by uterine secretion and is not yet in contact with the maternal tissue. Shedding by the implanting blastocyst, however, occurs when the blastocyst is about to attach and implant into the endometrium, that is, when it soon will be exposed to antibodies within the maternal tissue. Thus, one possible explanation for shedding immune complexes at this stage is that it is the most rapid way to get rid of potentially damaging immune complexes.

Alternatively, the handling of non-immune complexes (second Ig-Au complex with or without an irrelevant first antibody) by delayed and implanting blastocysts showed a difference in that under these conditions, the delayed blastocysts were weakly endocytotic, while the implanting ones were strongly endocytotic. Thus, the endocytotic activity of the trophectoderm differs depending both upon the state of the blastocyst (in delay or implanting), and upon the character of the protein (immune complex or not). It seems that immune complexes are taken care of in a special way by the blastocysts.

SUMMARY

A monoclonal (IgM) antibody N.63 detects a galactose derivative in the surface of the trophectoderm of mouse blastocysts. Both inactive, delayed blastocysts and adhesive, implanting ones were positive. When comparing the two types of blastocysts in terms of antigen appearance, antigen shedding, and endocytosis of immune complexes, several differences were observed.

The appearance and shedding of the antigen were more rapid in implanting blastocysts than delayed ones. Thus, the antigen could have a specific function during implantation, for instance, acting as a signal substance.

The endocytosis of immune complexes was more pronounced among delayed than among implanting blastocysts, while the reverse was true for the endocytosis of irrelevant antibody-gold complexes. Further, the shedding of immune complexes was highly active in implanting blastocysts compared to delayed ones. Thus, the handling of specific antibodies by the trophoblast differs depending upon at what stage of implantation the trophoblast encounters the antibody.

ACKNOWLEDGEMENTS

We are indebted to Ms. Barbro Einarsson and Ms. Marianne Ljungqvist for providing excellent technical assistance. This investigation was supported financially by the Swedish Medical Research Council (project no. 00070), and the "Expressen" Prenatal Research Foundation. Depo-Provera was kindly donated by Upjohn AB, Sweden.

REFERENCES

Bergström, S. (1978) Experimentally delayed implantation. In: *Methods In Mammalian Reproduction*, (ed.), Daniel , J.C., Jr., Academic Press, New York, U.S.A., pp. 419-435.

Black, P.H. (1980) Shedding from the cell surface of normal and cancer cells. *Adv. Cancer Res.* 32. 75-199.

Brinster, R.L. (1963) A method for in vitro cultivation of mouse ova from two-cell to blastocyst. *Exp. Cell Res.* 32, 205-208.

Eagle, H. (1955) Nutrition needs of mammalian cells in tissue culture. *Science* 122, 501-504.

Mayer, G. (1963) The experimental control of ovum implantation. In: *Techniques In Endocrine Research*, (eds.), P. Eichstein and F. Knowles, Academic Press, New York, U.S.A., pp. 245-261.

Naeslund, G. (1979) The effect of glucose-, arginine- and leucine- deprivation on mouse blastocyst outgrowth in vitro. *Upsala J. Med.Sci.* 84, 9-20.

Naeslund, G. Lundkvist, O., and Nilsson B.O. (1980). Transmission electron microscopy of mouse blastocysts activated and growth-arrested vivo and in vitro. *Anat. Embryol.* 159, 33-48.

Nieder, G.L. Weitlauf, H., and Suda-Hartman, H. (1987). Synthesis and secretion of stage-specific proteins by pre-implantation mouse embryos. *Biol. Reprod.* 36, 687-699.

Nilsson, B.O., Magnusson, C., Widéhn, S. and Hillersjö, T. (1982) Correlation between blastocyst oxygen consumption and trophoblast cytochrome oxidase reaction at implantation of delayed mouse blastocysts. *J. Embryol. Exp. Morphol.* 71, 75-82.

Nilsson, B.O., Ostensson, C.G., Eick, S., and Hellerström, C. (1980) Utilization of glucose by the implanting mouse blastocyst activated by oestrogen. *Endocrinologie* 76, 82-93.

Nilsson, .B.O., Svalander, P.C., Andersson, J., Grönvik, K.-O., and Larsson A. (1986) Methods for production and detection of monoclonal antibodies against surface components of adhesive implanting mouse blastocysts. *Upsala J. Med.Sci.* 91, 317-322.

Nilsson, B.O., Svalander, P.C., and Larsson, A. (1987) Immunization of mice and rabbits by intrasplenic deposition of nanogram quantities of protein attached to sepharose beads or nitrocellulose paper strips. *J. Immunol. Methods* 99, 67-75.

Sherman, M.I., Strickland, S., and Reich, E. (1976) Differentiation of early mouse embryonic and teratocarcinoma cells in vitro: Plasminogen activator production. *Cancer Res.* 36, 4208-4216.

Strickland, S., Reich, E., and Sherman, M.I. (1976) Plasminogen activator in early embryogenesis: Enzyme production by trophoblast and parietal endoderm. *Cell* 9, 231-240.

Svalander, P.C., Hjortberg, M., Grönvik, K.-O., and Nilsson, B.O. (1989) Mouse blastocyst surface expression of galactose-containing epitopes coinciding with trophoblast differentiation. *Cell Different. Develop.* 26, 191-200.

ANTIGENIC EXPRESSION BY MIGRATING TROPHOBLAST AND ITS RELEVANCE TO IMPLANTATION

- A Review -

Y.W. Loke, Ashley King, and Anna Grabowska

Division of Cellular and Genetic Pathology
Department of Pathology
University of Cambridge
Cambridge, England

INTRODUCTION

In the formation of the hemochorial placenta, an essential feature is the elaboration of waves of migratory trophoblast which invade through the basal decidua to reach as far as the myometrium (Robertson, 1987). The functional significance of this migration is not known nor is it understood which factors influence this pattern of behavior. Control is likely to be multifactorial, and the different facets are addressed by other authors in this volume. The concern of this chapter is the role played by trophoblast antigens in the interaction between invasive trophoblast and uterine tissues. The hypothesis of the placenta being antigenically neutral has, in recent years, been modified by observations that the extravillous migratory population of trophoblast does in fact express antigens, although the exact nature of these molecules is unclear. The MHC-related and blood group-related antigens expressed by invasive trophoblast populations will be reviewed, and their role in trophoblast growth and survival within the uterine environment will be examined.

Expression Of MHC Antigens By Extravillous Trophoblast

Placental trophoblast is not a homogeneous population but comprises different subsets, each with its own phenotypic and functional characteristics (Loke and Butterworth, 1987). While it is generally accepted that the trophoblast layers lining the human chorionic villi (both cytotrophoblast and syncytiotrophoblast) are devoid of MHC Class I and Class II antigens, this is not the case among the extravillous population. This trophoblast subpopulation migrates into the uterus. It is these cells with which this chapter is chiefly concerned. Sunderland and his colleagues (1981) made the initial observation that extravillous trophoblast, unlike the villous variety, was stained by the monoclonal antibody (Mab) W6/32 which is directed at a monomorphic determinant of the MHC Class I heavy chain molecule, and this observation was subsequently confirmed by other investigators (Faulk and McIntyre, 1983; Wells et al., 1984). From a study of sections of early placental bed biopsies, the transition from Class I-negative to Class I-positive trophoblast occurs at the cytotrophoblast columns where the cells fan out towards the decidua (Butterworth et al., 1985). Even the placental bed giant cells are observed to stain with Mab W6/32. This phenotype distinguishes the cells from the Class I-negative and highly secretory villous syncytiotrophoblast in spite of their common multinuclear characteristics. Thus, trophoblast would seem capable of developing along one of two directions. It may either differentiate into syncytiotrophoblast enveloping the chorionic villi, or it may proliferate outwards from the villus into uterine tissues and ultimately undergo transformation to placental bed giant cells. It remains to be determined whether this divergence in developmental pathways between villous and extravillous trophoblast is an inherent property of the two populations or is due to maternal influence. In this context, it is interesting to note that the intravascular trophoblast which invades and replaces the endothelial lining of many of the uterine spiral arteries, is also reactive with Mab W6/32 (Loke and Butterworth, 1987). Thus, at the hemochorial interface of the human placenta, maternal blood comes into contact with extravillous cytotrophoblast located in the artery which expresses Class I antigens, and with villous syncytiotrophoblast which does not. What this immunological relationship means for the trophoblast-maternal interaction is unclear.

Observations concerning hydatidiform moles are identical to those for the normal placenta, with villous trophoblast being unreactive with W6/32 (Fisher and Lawler, 1984) but the extravillous population staining with this Mab (Sunderland et al., 1985a). The data on choriocarcinomas are conflicting. While Yamashita et al. (1984) did not find any W6/32 positive cells from a primary tumor and a metastatic deposit of choriocarcinoma, Sunderland et al. (1985b) observed 40-70% of cells from a uterine choriocarcinoma to express Class I antigens. Unpublished findings in this laboratory from a case of testicular teratocarcinoma are in accord with those of Sunderland and his colleagues (1985b). Choriocarcinomatous elements observed among the metastatic deposits in the lung and regional lymph nodes all stained with W6/32 (Loke et al., unpublished observations). Interestingly, the primary seminoma in the testes did not. These disparate findings are reflected in cultured cell lines derived from choriocarcinomas which appear to be equally heterogeneous. The JAr cell line is usually found to be Class I negative (Jones and Bodmer, 1980), but the BeWo cell line can express a certain amount of these antigens (Trowsdale et al., 1980). Studies on Class I mRNA content in Northern blot hybridization measurements confirm that the JAr line has no detectable HLA transcriptional activity, while the BeWo cells have about 13% Class I mRNA relative to lymphocytes. At present, there are insufficient data to compare the behavioral pattern of choriocarcinomas that express Class I antigens with those that do not. When available, this should give us some insight into the contribution of these antigens towards trophoblast proliferation and invasiveness.

Essentially similar patterns of MHC expression by murine trophoblast are observed with Class I antigens being present on the invading spongiotrophoblast but absent on the labyrinthine population which lines the venous channels (Singh et al., 1983). It seems, therefore, that in both human and murine trophoblast, down-regulation of MHC Class I expression is specifically directed at that population which is predominantly involved in fetal-maternal exchange. The only conflicting finding is in the baboon where it is reported that the villous syncytiotrophoblast is the only cellular component of the placenta in this species which reacts with Mab W6/32 (Stern et al., 1987). It may be best to defer a conclusion until it can be certain that this is not due to cross-reactivity between Mab W6/32 and some non-MHC baboon antigens.

Nature Of Extravillous Trophoblast HLA Molecule

There is currently much discussion concerning the nature of the Class I antigen expressed by extravillous trophoblast (Head et al., 1987). This was initiated by the observation of Redman and his colleagues (1984) that while these extravillous trophoblast cells react readily with an anti-monomorphic Class I Mab like W6/32, they fail to bind to antisera directed at HLA-A and HLA-B polymorphic determinants. Similarly, Wells et al. (1984) have reported that extravillous trophoblast is unreactive with another Mab 61D2 which is also directed at HLA common determinants, but possibly at an epitope which is situated nearer the distal end of the allotypic region of the Class I heavy chain molecule than that recognized by W6/32. This differential staining pattern with W6/32 and 61D2 is also seen in some of the cytotrophoblast populations in the amniochorion (Hsi et al., 1984) and in ectopic implantation (Earl et al., 1985). These immunocytochemical observations seem to indicate that the Class I molecule expressed by extravillous trophoblast is non-polymorphic. In vitro studies in this laboratory on trophoblast susceptibility to complement-mediated cytolysis are in accord with this conclusion. This trophoblast is not lysed by human multispecific anti-HLA antisera while cultured fetal skin cells are readily killed (Grabowska et al., unpublished observations).

The situations in the mouse and in the rat are somewhat different. Although murine trophoblast cells appear to be resistant to lysis by allospecific cytotoxic T cells, they are, nevertheless, killed by anti-paternal alloantibody and complement (Zuckermann and Head, 1987). In the rat, Billington and Burrows (1986) demonstrated the expression of paternal Class I MHC antigens on spongiotrophoblast indicating that polymorphic determinants are present on the rat equivalent of human extravillous trophoblast. There is the additional complication that besides these classical Class I determinants, extravillous trophoblast in the rat is also said to express 'pregnancy-associated' or Pa antigen which exhibits a broadly shared polymorphism (Macpherson et al., 1986). The authors have postulated that only the Pa antigen is exposed to maternal recognition while the Class I RTI.A^a antigen is hidden on the fetal side of the trophoblast cell membrane (Ho et al., 1987).

Biochemical data have recently been provided by Ellis and her colleagues (1986) who reported that immunoprecipitation of extracts from human amniochorion trophoblast and from the BeWo choriocarcinoma cell line using Mabs to monomorphic determinants of HLA Class I gave an electrophoretically non-polymorphic glycoprotein associated with β2 microglobulin. The glycoprotein is, at 40,000 molecular weight, smaller than the classical Class I heavy chain. These data, however, differ from those of Anderson and Berkowitz (1985) who observed that W6/32 immunoprecipitated from the BeWo cell line a Class I antigen of 44 kD associated with β2 microglobulin. The Pa antigen in the rat, also, is of conventional size (Ghani et al., 1984). Strangely, it has not been possible to immunoprecipitate the W6/32 reactive molecule from our cultured trophoblast although this Mab readily identified bands of 45,000 and 12,000 molecular weights from surface labeled proteins extracted from other cell types cultured in a similar way (Loke et al., unpublished). Whether this is due to a quantitative difference in antigen expression or antigen-antibody affinity has not yet been established. Thus, with the present available evidence, it is still uncertain as to whether Class I antigens are merely sparsely distributed on the cell surface of extravillous trophoblast or qualitatively different from those expressed by other somatic cells.

The suggestion has been made that the unusual Class I-like antigen expressed by extravillous trophoblast may represent a non-classical HLA Class I molecule encoded within the MHC region (Ellis et al., 1986). This may be analogous to the murine Qa/Tla system that directs the synthesis of similar sized Class I molecules with little or no polymorphism and restricted tissue distribution. A feature of the murine Qa region is the production of secreted, soluble Class I molecules (Robinson, 1987). Many of these antigens have a molecular weight of around 40,000 in association with a 12,000 β2 microglobulin, and do not exhibit detectable alloreactivity. The gene is consistently expressed only in certain tissues like the liver. Molecular analyses reveal that, while the nucleotide sequences that encode these secreted molecules are highly homologous to those of classical Class I antigens, there are nevertheless several regions of disparity. These include the sequence encoding the transmembrane domain where hydrophobic amino acid residues are replaced by polar and charged ones that prevent the molecule from being inserted in the membrane, and also the sequence for the cytoplasmic domain which exhibits a deletion of about 39-66 bases (Barra et al., 1985). Similar non-classical Class I soluble antigens have also been observed in rat liver and serum. Spencer and Fabre (1987) made the interesting observation that these molecules did not react with any of the well characterized antibodies directed against polymorphic or monomorphic rat Class I antigens except one, designated as MRCOX18, which is specific for monomorphic determinants. The similarity between this pattern of reactivity in the rat and the findings that human extravillous trophoblast antigens are not recognizable by any anti-Class I antibodies except Mab W6/32 is immediately apparent. This raises the question of what epitopes are delineated by MRCOX18 and W6/32 and consequently makes interpretation of data relating to these two antibodies rather difficult. Secreted Class I molecules have also been demonstrated in a variety of human tumors. Initial studies of a mutant B cell line identified a secreted water soluble form of HLA-A2 resulting from an alternative RNA splicing pathway that excluded exon 5, which encodes the transmembrane region. Instead, exon 4 is joined directly to exon 6 leading to a transcript that encodes a polypeptide lacking 39 amino acids including those which would normally anchor the molecule to the cell membrane (Krangel, 1986).

Molecular analyses such as those just described have not been possible on human trophoblast because it has been extremely difficult to obtain these cells with a sufficient degree of homogeneity. Recently, a culture method based on the use of a modified medium and a substrate of extracellular matrix has been developed which consistently yields trophoblast of high purity (Loke and Burland, 1988). It is hoped that these cells will provide the necessary starting point from which to explore the nature of their Class I-like molecule.

Besides the expression of interesting Class I molecules, Starkey (1987) has recently reported that cytotrophoblast cells of the term amniochorion and also those of the cell columns of first and second trimester chorionic villi are reactive with a Mab B7/21 specific for the MHC Class II antigen HLA-DP, although these cells do not stain with other Mabs directed against HLA-DR or HLA-DQ. Whether this illustrates another unusual MHC antigen expressed by human extravillous trophoblast awaits clarification.

Regulation Of MHC Expression By Trophoblast

There is now mounting evidence to indicate that the tissue distribution of Class I antigens is not as ubiquitous as it was once thought to be. Thus, the situation in trophoblast is not unique. Very little is known about the factors that regulate cellular MHC expression. Indeed, the question of whether this is constitutive or inductive has not yet been resolved (Halloran et al., 1986). However, it seems likely from available data that a variety of physiological and pathological stimuli can at least modulate MHC expression. One such stimulus which may have an important regulatory effect on MHC expression during embryogenesis and trophoblast development is interferon (IFN), a powerful inducer of both Class I and Class II antigens on bone marrow as well as other cell types (Fabre et al., 1987; Moller, 1987). Although IFN is usually only produced in significant quantities during the acute inflammatory response (particularly in viral infections), it has been proposed that low levels of IFN may be present continuously even in health (Bocci, 1985). In their study of the ontogeny of Class I MHC expression by the murine embryo, Ozato et al. (1985) observed that these antigens are demonstrable at low levels by day 10 of gestation with the appropriate mRNA detectable by dot hybridization on day 9. Murine α, β, and γ-IFN will induce Class I mRNA and antigen expression on embryonic cells from day 8, but not if administered at day 6 which suggests that induction is stage-specific and will not occur in early development. Since the study shows the onset of responsiveness to IFN precedes Class I gene expression, the possibility arises that the latter could be regulated by the endogenous production of IFN. Similar effects of IFN on mouse trophoblast have been reported (Drake et al., 1987; King et al., 1987); purified recombinant IFN was observed to induce the appearance of Class I antigen on the surface of secondary trophoblast giant cells of 7.5 day ectoplacental cones, but had no effect on primary trophoblast giant cells from early blastocysts. This again confirms that, at the time of implantation, there is a period during which MHC antigens are neither expressed constitutively nor inducible by IFN.

Trophoblast inducibility by IFN, however, may not be merely a reflection of the stage of development, but could be a characteristic inherent to different populations. For instance, γ-IFN will enhance Class I expression in the JEG-3 and BeWo choriocarcinoma cell lines but not in the JAr line (Anderson and Berkowitz, 1985; Hunt et al., 1987). Interestingly, the BeWo cell line normally expresses low levels of Class I antigens, while JAr cell lines are completely devoid of these antigens. These observations indicate that Class I antigens are inducible by IFN only in those trophoblast subsets which are capable of expressing them in the first place, albeit at low levels. This could explain the disparity between the findings of Zuckermann and Head (1986) who observed IFN induction of Class I antigens in trophoblast cells isolated by enzymic disaggregation from murine placentae, and those of Hunt et al. (1987) who reported that human term and first trimester chorionic villi in organ culture did not respond to IFN. Perhaps this is not so much a question of species difference, but rather the use by the two laboratories of non-identical trophoblast populations. Certainly, intact pieces of chorionic villi as used by Hunt and her colleagues (1987) are likely to consist mainly of villous trophoblast, a population which is known to be devoid of MHC antigens in vivo.

Investigation is needed of the inducibility of the extravillous population, but such experiments have been difficult to perform because these trophoblast cells have not been isolated with sufficient purity. In the culture method that we have recently developed, we noticed that the cytotrophoblast cells appear to exhibit the immunocytochemical characteristics of extravillous trophoblast, and they express a certain level of Class I-like antigens (Loke and Burland, 1988). It should be of some interest to see if these trophoblast HLA antigens are increased by IFN.

Trophoblast cells which are non-inducible may have their HLA genetic control regions methylated and switched off irreversibly. This was the conclusion of Alberti and Herzenberg (1986) who observed that DNA from the HLA Class I negative JAr choriocarcinoma cell line did not transfect for HLA. However, prior treatment of JAr cells with the DNA hemimethylase inhibitor, azacytidine, resulted in the appearance of transfecting ability which lasted between two to eight weeks in culture. Whether this mechanism of gene regulation of HLA applies also to subsets of normal trophoblast remains to be seen.

If IFN is involved in normal physiological cell maturation and differentiation, its source of production is of interest. One possibility is from the target cell itself using an autocrine mechanism in which the cell secretes its own IFN and responds via its own cell surface receptors. Yarden et al. (1984) put forward such a proposal when they found that the significant increase in HLA Class I mRNA and cell surface antigen expression accompanying differentiation of the U937 lymphoma cell line towards a macrophage phenotype could be prevented by the addition of antibodies to human β-IFN to the culture medium, thus suggesting that induction of HLA expression could be in response to extracellular IFN molecules which are secreted by the differentiating cells. A similar mechanism may operate for trophoblast. Mouse fetal tissues have very low levels of IFN, but significant levels exist in the placenta and these increase progressively during gestation (Fowler et al., 1980). Data as to which placental cells are responsible for the production are not yet available. In a recent immunocytochemical study on the human placenta using a panel of monoclonal antibodies directed against α, β, and γ-IFN, we were somewhat surprised to find it was the HLA-negative villous syncytiotrophoblast which stained most strongly, although IFN was also demonstrable in a few extravillous trophoblast cells (unpublished). Thus, there seems to be an inverse relationship between intracellular IFN and cell surface Class I antigenic expression.

An additional source of IFN in the placental-uterine interface could be the unique T-lineage lymphoid population which is such a conspicuous feature of the decidua (Bulmer et al., 1989). Enhancement of Class I antigen expression has been observed in BeWo choriocarcinoma cells by using supernatants from PHA-activated lymphocytes (Anderson and Berkowitz, 1985), and in cultured renal cells with Concanavalin A-stimulated lymphocyte supernates (Halloran et al., 1985), although it has not been established that IFN is the responsible product. Maternal decidual lymphocytes would also be expected to be activated by invading trophoblast. Indeed, the local liberation of enhancing lymphokines by maternal T cells which affect placental growth and function forms the basis of the immunotropism theory enunciated by Wegmann (1987).

Expression Of Blood Group And Related Carbohydrate Antigens By Extravillous Trophoblast

Besides MHC, the other important human allogeneic antigens with transplantation potential are the ABO blood groups. Unfortunately, data on the expression of these antigens by trophoblast are really too scanty and conflicting to permit any definite conclusions. The few positive findings (Gross, 1966; Goto et al., 1980) may be criticized on the grounds that blood group substances may be adsorbed from maternal blood onto the surface of trophoblast. Conversely, the negative results (Thiede et al., 1965; Szulman, 1972) may be explained on the basis that blood group antigens are expressed in such small quantities that they are only detectable on tissue sections by the use of high titre, high affinity antibodies. Moreover, the observations relate mainly to villous syncytiotrophoblast, and not to extravillous trophoblast. The only experiments that might have contained cells from this latter population are those performed by Loke and Ballard (1973) using trophoblast disaggregated by enzymic digestion from first trimester chorionic villi. Blood group A antigens are detectable on these placental cells by the mixed antiglobulin method, but not by mixed agglutination, and this finding is interpreted to be due to antigens on trophoblast being so sparse that insufficient antibody is taken up to form stable bridges directly with indicator red cells until another link is introduced in the form of an antiglobulin, a situation that is analogous to the demonstration of Rh antigens on red cells by the Coombs test. In the preceding section, the possiblity of a similar down-regulation of MHC antigen expression on extravillous trophoblast has already been discussed.

Perhaps of greater significance in the context of trophoblast-uterine interaction are the carbohydrate molecules related to the blood group family of antigens. Many of these are found to undergo specific changes during embryogenesis, differentiation and neoplastic transformation (Feizi, 1985; Hakomori, 1987). Their exact functions are not known, but there is increasing evidence to indicate that they may be important in growth control, intercellular recognition and as components of receptor systems (Feizi and Childs, 1985). Recently an immunocytochemical study of these antigens on different trophoblast populations has been undertaken using a panel of monoclonal antibodies directed at fucosylated and sialylated carbohydrates on the Type 1 and Type 2 blood group precursor chains (King and Loke, 1988). Sialyl-Le[x] antigen is localized to extravillous trophoblast in normal and molar pregnancies and in choriocarcinoma tumor cells, with the staining being much stronger in the latter two conditions. After neuraminidase treatment, Le[x] antigen itself is identified in similar cell populations. Villous trophoblast did not react with any of the monoclonal antibodies used.

Relevance of Antigenic Expression To Trophoblast Survival

The concept of the placenta as immunologically neutral, while probably valid for villous trophoblast, needs modification in view of observations that the extravillous population, in contrast, expresses a variety of antigens which may be immunologically relevant. Most intriguing is the question regarding the functional significance of the MHC Class I-like antigen demonstrated on extravillous trophoblast. For evasion of maternal immunological attack, it would be far simpler to have all populations of trophoblast devoid of recognizable alloantigenic specificities. So why are some cells negative for these antigens while others express Class I-like determinants? It must be remembered, however, that at one stage of ontogeny during early implantation, all trophoblast populations are constitutively negative for MHC antigens, and are refractory to induction by IFN. Perhaps this is the important period when allogeneic markers need to be absent, and their subsequent appearance as trophoblast differentiates along the extravillous pathway is no longer immunologically relevant, especially as these molecules, even when they appear, are deficient in polymorphic domains. In addition, these antigens may not occur at a sufficient level of density at the cell surface to trigger alloreactivity. Preliminary experiments in this laboratory, using immunoprecipitation and immunogold staining, indicate that the Class I-like molecules on cultured cytotrophoblast cells are very scantily expressed on the cell surface, and this could be the basis for the insusceptibility of trophoblast to lysis by multispecific anti-HLA antisera (Grabowska et al., unpublished observations).

While it is possible to dismiss the presence of Class I-like antigens on extravillous trophoblast as immunologically inconsequential, there is an opposing school of thought that believes these antigens to be functionally significant in determining the success of pregnancy. Such a concept has already been proposed for the MHC-negative villous syncytiotrophoblast where the expression of a trophoblast-lymphocyte cross-reactive (TLX) antigen is thought to induce the generation of a beneficial response by the mother (McIntyre and Faulk, 1982). This forms the theoretical framework upon which a regime of treating habitually aborting women is based. A similar idea has been proposed by Wegmann (1987) for extravillous trophoblast in his `placental immunotropism' hypothesis, whereby maternal recognition of the Class I-like molecules can lead to the release by decidual lymphocytes of appropriate lymphokines which directly influence trophoblast growth and function. This view is supported by their observation that mouse placental cells in culture proliferate upon the addition of T cell-derived lymphokines like IL3 and granulocyte-macrophage colony stimulating factor, CSF-GM (Athanassakis et al., 1987).   Such a mechanism could explain the observation of Dickman and Cauchi (1978) that maternal lymphocytes, but not non-pregnant allogeneic lymphocytes, will stimulate trophoblast to secrete more hCG in vitro. Class I-inducing IFN may also be one of the substances released by trophoblast-activated decidual lymphocytes, thus providing a controlling influence over trophoblast differentiation.

It may not be necessary to have highly polymorphic determinants of a classical Class I molecule to activate maternal uterine lymphocytes. Delayed-type hypersensitivity responses have been observed in congenic mouse strains incompatible at the Qa-Tla region of the MHC (Robinson and Jordan, 1987) to which the antigens on extravillous trophoblast may belong. The benefit to implantation engendered by such a response in the uterus can be seen in the original experiments of Beer and Billingham (1984) who demonstrated that creation of a local delayed hypersensitivity reaction in one rat uterine horn to paternal alloantigens promoted blastocyst implantation within this horn, but not in the uninflamed contralateral horn. Increases in placental weights have been observed by Hamilton and Hamilton (1987) to be associated with disparity at multiple minor histocompatibility loci rather than at the MHC but, given the way these experiments were set up, the possibility must be entertained that these animals might also differ at Class I loci of the Qa-Tla type.

Bartocci and colleagues (1986) have reported that CSF-1 production in the murine uterus increases 1000-fold in pregnancy and since trophoblast has been shown to express the proto-oncogene c-fms, the product of which is the CSF-1 receptor, it is suggested that CSF-1 plays an important role in trophoblast development (Pollard et al., 1987). However, it is the uterine glandular cells which are observed to produce CSF-1. This indicates that there are alternative sources of active cytokines in the uterus secreted by cells which are subjected to hormonal regulation and not necessarily recruited by the presence of trophoblast. The need for maternal recognition of allogeneic trophoblast, therefore, becomes less attractive. Indeed, the lymphoid infiltrate characteristic of decidua (Bulmer et al., 1989) has already accumulated at the time of the secretory phase of the menstrual cycle in the non-pregnant endometrium (King et al., 1989a), again suggesting hormonal influence rather than immunological stimulation by trophoblast.

Besides the association between MHC antigens and immune reactions, there is now a great deal of interest in the role of these antigens in non-immune cellular interactions (Edidin, 1983; Klein, 1986), and it is these latter functions which could provide the explanation for the expression of Class I-like molecules by extravillous trophoblast. For example, Bodmer (1972) suggested that the true role of the MHC system is in the control of cell-cell recognition. Although this idea is formulated with the syngeneic system in mind whereby differential expression of MHC molecules by different tissues within the same organism will lead to their proper positioning, there is no reason why this concept cannot be extended to allogeneic combinations. Curtis and Rooney (1979) have reported that contact inhibition of movement of epithelial cells show MHC restriction, being determined in allogeneic combinations by mismatch in certain parts of the MHC complex. This could be relevant to migration of the allogeneic extravillous trophoblast between maternal uterine cells.

The relationship between MHC Class I molecules and cell surface receptors has also been recognized. Due (1986) reported that Mabs against monomorphic determinants of MHC Class I molecules reduced insulin binding and immunoprecipitated a 45 kD protein from ^{125}I-labeled insulin receptor preparations. A murine thymoma cell line with MHC Class I expression was found to bind more insulin than a Class I negative variant and depletion of Class I antigens on the cell surface by capping with anti-β2M or with anti-heavy chain decreased insulin binding. Together, these observations led to the conclusion that MHC Class I heavy chain and the tetrameric insulin receptor are structurally associated in the cell membrane. Thus it is possible that Class I molecules modulate receptor binding activity. This could be the mechanism for differential insulin binding according to MHC haplotype and may partly explain the association between HLA and diabetes (Phillips et al., 1986). A similar kind of interaction has also been observed between Class I molecules and the receptor for epidermal growth factor (Schreiber et al., 1984). Human trophoblast is well endowed with receptors for a variety of peptides so it is possible that the Class I-like antigen expressed by extravillous trophoblast can dictate the binding of hormones and other molecules and hence exerts a controlling influence on trophoblast growth and differentiation (Rao et al., 1985).

The regulation of Class I antigen expression by certain protooncogenes is also of some interest in the context of trophoblast. Amplification of the myc oncogene in neuroblastomas by transfection is observed to result in an increase in metastatic ability, an increase in growth rate, but a 10-15 fold reduction in Class I antigenic expression (Bernards et al., 1986). Among cells of the human placenta, c-myc product is localized predominantly to villous cytotrophoblast (Maruo and Mochizuki, 1987) and the highest amount of c-myc transcript is to be found in the cytotrophoblast shell at 4-5 weeks gestation when these cells are undergoing marked proliferative activity (Pfeifer-Ohlsson et al., 1984). However, c-myc gene transcription is not found in all rapidly proliferating embryonic cells so it is probably not simply a marker of proliferative activity per se, but could reflect some additional tissue specific gene regulation operating during the development of the placenta (Pfeifer-Ohlsson et al., 1985). The transfection experiments of Alon et al. (1987) have further demonstrated co-regulation in expression of MHC-encoded glycoproteins and the Kirsten-ras oncogene p21 protein which functions to regulate receptor-mediated transmembrane signaling. Thus, Class I molecules could play a role in intracellular and intercellular signal transduction. The expression of MHC Class I antigens by extravillous trophoblast, therefore, may be involved in a range of biological functions beyond immunological interaction. As such, these antigens may assume a pivotal role in determining trophoblast development.

As regards the blood-group related carbohydrate antigens, it is tempting to speculate that the expression of sialyl-Le^x is associated in some way with the invasive potential of extravillous trophoblast since the strongest expression of this antigen is by the infiltrative population of normal trophoblast, and also by similar cells of invasive moles and gestational choriocarcinomas. The Le^x determinant (also known as SSEA-1) is expressed during embryonic development in a stage-specific manner, and it has been suggested that it may be essential to embryonic cell-cell recognition (Feizi, 1985). Frequently, blood group antigens are reexpressed by adult tumors and this is said to be related to their state of differentiation and invasive potential (Hakomori, 1984; Coon and Weinstein, 1986). Carbohydrate antigens may also influence trophoblast survival by modifying NK cell surveillance as occurs for tumor cells (Smets and Van Beek, 1984). It is now clear that the bone marrow-derived cells present in the uterus at the time of implantation are predominantly large granular lymphocytes (LGL) with NK phenotype and function (Bulmer et al., 1989). There are very few mature T cells and no B cells. Since extravillous trophoblast expresses scarce or aberrant MHC Class I antigens, it is conceivable that maternal regulation of trophoblast invasiveness is via the non-MHC restricted NK system rather than classical allogeneic recognition, especially as carbohydrate molecules are thought to provide the targets for NK lysis. Human decidual cells do indeed show activity against cells such as the K562 cell line that act as targets for NK activity but, surprisingly, first trimester trophoblast is resistant to lysis by these decidual cells as well as by peripheral blood lymphocytes (King et al., 1989b). Increased sialylation of surface glycoconjugates has been reported to confer resistance to NK lysis in tumor cells (Smets and Van Beek, 1984) so the highly sialylated carbohydrate determinants that we observed on extravillous trophoblast could serve this function. It must be remembered, as mentioned earlier, that a certain degree of trophoblast invasion is necessary for successful implantation. Perhaps the function of decidual NK cells is to be cytostatic rather than cytolytic for trophoblast? This possibility is now being investigated with co-culture experiments.

SUMMARY

Antigens expressed by trophoblast may influence the interaction of these cells with maternal uterine tissues, although the exact mechanisms remain to be established. From observations of the nature of trophoblast antigens and of the lymphoid infiltrate in decidua, it appears that conventional allogeneic immune reactions are probably not of primary importance. Instead, the chief method of regulation may be by some para-immunological or even non-immunological mechanisms. This interaction bears a remarkable resemblance to the tumor-host relationship and provides further supportive evidence that trophoblast and cancer cells are very similar in function and behavior (Loke, 1977; Loke, 1978). Thus, the study of trophoblast antigens and implantation has relevance well beyond reproductive biology.

ACKNOWLEDGEMENTS

We are grateful for financial support from the American Friends of Cambridge University, Cancer Research Campaign, East Anglian Regional Health Authority, Medical Research Council and the Wellcome Trust. We would like to thank our obstetric colleagues and staff at Addenbrooke's Hospital for collecting the placental material. This investigation received financial support from the Special Programme of Research, Development and Research Training in Human Reproduction, World Health Organization.

REFERENCES

Alon, Y., Hammerling, G.J., Segal, S., and Bar-Eli, M. (1987) Association in the expression of Kirsten-ras oncogene and the major histocompatibility complex class I antigens in fibrosarcoma tumour cell variants exhibiting differing metastatic capabilities. *Cancer Res.* 47, 2553-2557.

Alberti, S. and Herzenberg, L.A. (1986) Transfection of DNA from choriocarcinoma cell lines and sperm cells: DNA methylation prevents the expression of genes for the Major Histocompatibility Complex (HLA) Class I and the T cell differentiation antigen, Leu-2. In: *Reproductive Immunology*, (eds.), D.A. Clark and B.A. Croy, Elsevier Science Publishers, pp. 60-66.

Anderson, D.J., and Berkowitz, R.S. (1985) γ-interferon enhances expression of Class I MHC antigens in the weakly HLA+ human choriocarcinoma cell line BeWo, but does not induce MHC expression in the HLA-choriocarcinoma cell line JAr. *J. Immunol.* 135, 2498-2501.

Athanassakis, I., Bleackley, R.C., Paetkau, U., Guildbert, L., Barr, P.J., and Wegmann, T.G. (1987) The immunostimulatory effect of T cells and T cell lymphokine on murine fetally derived placental cells. *J. Immunol.* 138, 37-44.

Barra, Y., Tanaka, K., and Davidson, W. (1985) Expression of a secreted form of the MHC Class I antigen. In: *Cell Biology of the Major Histocompatibility Complex*, (eds.), B, Pernis and H.J. Vogel, Academic Press, pp. 109-120.

Bartocci, A., Pollard, J.W., and Stanley, E.R. (1986) Regulation of colony-stimulating factor I during pregnancy. *J. Exp. Med.* 164, 856-961.

Beer, A.E. and Billingham, R.E. (1974) Host responses to intrauterine tissue, cellular and fetal allografts. *J. Reprod. Fert.* 21, 59-64.

Bernards, R., Dessain, S.K., and Weinberg, R.A. (1986) N-myc amplification causes down-modulation of MHC class I antigen expression in neuroblastoma. *Cell* 47, 667-674.

Billington, W.D. and Burrows, F.J. (1986) The rat placenta expresses paternal class I major histocompatibility antigens. *J. Reprod. Immunol.* 9, 155-160.

Bocci, U. (1985) The physiological interferon response. *Immunology Today* 6, 7-9.

Bodmer, W.F. (1972) Evolutionary significance of the HLA system. *Nature* 237, 139-145.

Bulmer, J.N., Ritson, A., and Pace, D. (1949) Endometrial leukocytes in human pregnancy. *Trophoblast Research* 4, 431-451.

Butterworth, B.H., Khong, T.Y., Loke, Y.W., and Robertson, W.B. (1985) Human cytotrophoblast populations studied by monoclonal antibodies using single and double biotin-avidin-peroxidase immunocytochemistry. *J. Histochem. Cytochem.* 33, 977-983.

Coon, J.S. and Weinstein, R.S. (1986) Blood group-related antigens as markers of malignant potential and heterogeneity in human carcinomas. *Human Pathol.* 17, 1089-1106.

Curtis, A.S.G. and Rooney, P. (1979) H-2 restriction of contact inhibition of epithelial cells. *Nature* 281, 222-223.

Dickman, W.J., and Cauchi, M.N. (1978) Lymphocyte induced stimulation of human chorionic gonadotrophin production by trophoblastic cells in vitro. *Nature* 271, 377-378.

Drake, B.L., King, N.J.C., Maxwell, L.E., and Rodger, J.C. (1987) Class I MHC antigen expression on early murine trophoblast and its induction by lymphokines in vitro. *J. Reprod. Immunol.* 10, 319-328.

Due, C., Simonsen, M., and Olsson, L. (1986) The major histocompatibility complex class I heavy chain as a structural sub-unit of the human cell membrane insulin receptor: Implications for the range of biological functions of histocompatibility antigens. *Proc. Natl. Acad. Sci. USA* 83, 6007-6011.

Earl, U., Wells, M., and Bulmer, J.N. (1985) The expression of major histocompatibility complex antigens by trophoblast in ectopic tubal pregnancy. *J. Reprod. Immunol.* 8, 13-24.

Edidin, M. (1983) MHC antigens and non-immune functions. *Immunology Today* 4, 269-270.

Ellis, S.A., Sargent, I.L., Redman, C.W.G., and McMichael, A.J. (1986) Evidence for a novel HLA antigen found on human extravillous trophoblast and a choriocarcinoma cell line. *Immunol.* 59, 595-601.

Fabre, J.W., Milton, A.D., Spencer, S., Settaf, A., and Houssin, D. (1987) Regulation of alloantigen expression in different tissues. *Transplant. Proceed.* 19, 45-49.

Faulk, W.P. and McIntyre, J.A. (1983) Immunological studies of human trophoblast: Markers, subsets and functions. *Immunol. Revs.* 75, 139-175.

Feizi, T. (1985) Demonstration by monoclonal antibodies that carbohydrate structures of glycoproteins and glycolipids are onco-developmental antigens. *Nature* 314, 53-57.

Feizi, T. and Childs, R.A. (1985) Carbohydrate structures of glycoproteins and glycolipids as differentiation antigens, tumour-associated antigens and components of receptor systems. *Trends Biochem. Sci.* 10, 24-29.

Fisher, R.A. and Lawler, S.D. (1984) The expression of major histocompatibility antigens in the chorionic villi of molar placentae. *Placenta* 5, 237-242.

Fowler, A.K., Reed, C.D., and Giron, D.J. (1980) Identification of an interferon in murine placentas. *Nature* 286, 266-267.

Ghani, A., Kunz, H.W., and Gill, T.J. (1984) Pregnancy-induced monoclonal antibody to a unique fetal antigen. *Transplantation* 37, 503-506.

Goto, S., Nishi, H., and Tomoda, Y. (1980) Blood group Rh-D factor in human trophoblast determined by immunofluorescent method. *Am. J. Obstet. Gynecol.* 137, 707-712.

Gross, S.J. (1966) Human blood group A substance in human endometrium and trophoblast localized by chromatographed rabbit antiserum. *Am. J. Obstet. Gynecol.* 95, 1149-1159.

Hakomori, S. (1984) Tumor-associated carbohydrate antigens. *Ann. Rev. Immunol.* 2, 103-126.

Hakomori, S. (1987) Glycosphingolipids as differentiation-dependent, tumour-associated markers and as regulators of cell proliferation. In: *Oncogenes and Growth Factors*, (eds.), R.A. Bradshaw and S. Prentis, Elsevier, pp.218-226.

Halloran, P.F., Jephthah-Ochola, J., Urmson, J., and Farkas, S. (1985) Systemic immunologic stimuli increases Class I and II antigen expression in mouse kidney. *J. Immunol.* 135, 1053-1060.

Halloran, P.F., Wadgymer, A., and Autenried, P. (1986) The regulation of expression of major histocompatibility complex products. *Transplantation* 41, 413-420.

Hamilton, B.L. and Hamilton, M.S. (1987) Histoincompatibility on the weight of the feto-placental unit in mice: The role of minor histocompatibility antigens. *Am. J. Reprod. Immunol. Microbiol.* 15, 153-155.

Head, J.R., Drake, B.L., and Zuckermann, F.A. (1987) Major histocompatibility antigens on trophoblast and their regulation: Implications in the maternal-fetal relationship. *Am. J. Reprod. Immunol. Microbiol.* 15, 12-18.

Ho, H.-N., Macpherson, T.A., Kinz, H.W., and Gill, T.J. (1987) Ontogeny of expression of Pa and RT1.A^a antigens on rat placenta and on fetal tissues. *Am. J. Reprod. Immunol. Microbiol.* 13, 51-61.

Hsi, B.-L., Yeh, C.-J.G., and Faulk, W.P. (1984) Class I antigens of the major histocompatibility complex on cytotrophoblast of human chorion laeve. *Immunol.* 52, 621-629.

Hunt, J.S., Andrews, G.K., and Wood, G.W. (1987) Normal trophoblasts resist induction of Class I HLA. *J. Immunol.* 138, 2481-2487.

Jones, E.A. and Bodmer, W.F. (1980) Lack of expression of HLA antigens on choriocarcinoma cell lines. *Tissue Antigens* 16, 195-202.

King, N.J.C., Drake, B.L., Maxwell, L.E., and Rodger, J.C. (1987) Class I major histocompatibility complex antigen expression on early murine trophoblast and its induction by lymphokines in vitro. II. The role of gamma interferon in the responses of primary and secondary giant cells. *J. Reprod. Immunol.* 12, 13-21.

King, A. and Loke, Y.W. (1988) Differential expression of blood-group-related carbohydrate antigens by trophoblast populations. *Placenta* 9, 513-521.

King, A., Wellings, V., Gardner, L., and Loke, Y.W. (1989a) Immunocytochemical characterization of the unusual large granular lymphocytes in human endometrium throughout the menstrual cycle. *Human Immunol.* 24, 195-205.

King, A., Birkby, C., and Loke, Y.W. (1989b) Early human decidual cells exhibit NK activity against the K562 cell line but not against first trimester trophoblast. *Cell Immunol.* 118, 337-344.

Klein, J. (1986) *Natural History of the Major Histocompatibility Complex.* New York: John Wiley and Sons.

Krangel, M.S. (1986) Secretion of HLA-A and -B antigens via an alternative RNA splicing pathway. *J. Exp. Med.* 163, 1173-1190.

Loke, Y.W. (1977) The Cancer Cell. In: *Scientific Foundations of Obstetrics and Gynaecology*, (eds.) E.E. Philipp, J. Barnes, M. Newton, London: Heinemann, pp.36-43.

Loke, Y.W. (1978) *Immunology and Immunopathology of the Human Foetal-Maternal Interaction*. Elsevier/North Holland.

Loke, Y.W., and Ballard, A.C. (1973) Blood group A antigens on human trophoblast cells. *Nature* 245, 329-330.

Loke, Y.W. and Burland, K. (1988) Human trophoblast cells cultured in modified medium and supported by extracellular matrix. *Placenta* 9, 173-182.

Loke, Y.W. and Butterworth, B.H. (1987) Heterogeneity of human trophoblast populations. In: *Immunoregulation and Fetal Survival*, (eds.), T.J. Gill and T.G. Wegmann, New York: Oxford University Press, pp. 197-209.

McIntyre, J.A. and Faulk, W.P. (1982) Allotypic trophoblast-lymphocyte cross-reactive (TLX) cell surface antigens. *Human Immunol.* 4, 27-35.

Macpherson, T.A., Ho, H.N., Kunz, H.W., and Gill, T.J. (1986) Localization of the Pa antigen on the placenta of the rat. *Transplantation* 41, 392-394.

Maruo, T. and Mochizuki, M. (1987) Immunohistochemical localization of epidermal growth factor receptor and myc oncogene product in human placenta: Implication for trophoblast proliferation and differentiation. *Am. J. Obstet. Gynecol.* 156, 721-727.

Moller, E. (1987) Recognition and response to alloantigens. *Transplant. Proceed.* 19, 40-44.

Ozato, K., Wan, Y.-J., and Orrison, B.M. (1985) Mouse major histocompatibility class I gene expression begins at mid-somite stage and is inducible in earlier-stage embryos by interferon. *Proc. Natl. Acad. Sci. USA* 82, 2427-2431.

Pfeifer-Ohlsson, S., Goustin, A.S., Rydnert, J., Wahlstrom, T., Bjersing, L., Shehelin, D., and Ohlsson R. (1984) Spatial and temporal pattern of cellular myc oncogene expression in developing human placenta: Implications for embryonic cell proliferation. *Cell* 38, 585-596.

Pfeifer-Ohlsson, S., Rydnert, J., Goustin, A.S., Larsson, E., Betsholtz, C., and Ohlsson, R. (1985) Cell-type-specific pattern of myc protooncogene expression in developing human embryos. *Proc. Natl. Acad. Sci. USA* 82, 5050-5054.

Phillips, L.M., Moule, M.L., Delovitch, T.L., and Yip, C.C. (1986) Class I histocompatibility antigens and insulin receptors: Evidence for interactions. *Proc. Natl. Acad. Sci. USA* 83, 3474-3478.

Pollard, J.W., Bartocci, A., Arceci, R., Orlofsky, A., Ladner, M.B., and Stanley, E.R. (1987) Apparent role of the macrophage growth factor, CSF-1, in placental development. *Nature* 330, 484-486.

Rao, C.V., Ramani, N., Chagini, N., Stadig, B.K., Carman, F.R., Woost, P.G., Schultz, G.S., and Cook, C.L. (1985) Topography of human placental receptors for epidermal growth factor. *J. Biol. Chem.* 260, 1705-1710.

Redman, C.W.G., McMichael, A.J., Stirrat, G.M., Sunderland, C.A., and Ting, A. (1984) Class I major histocompatibility complex antigens on human extravillous trophoblast. *Immunol.* 52, 457-468.

Robertson, W.G. (1987) Pathology of the pregnant uterus. In: *Obstetrical and Gynaecological Pathology*, (ed.), H. Fox, Churchill Livingstone, pp. 1149-1176.

Robinson, J.H. and Jordan, R.K. (1987) Delayed type hypersensitivity responses to the Qa-Tla region of the mouse major histocompatibility complex. *Immunobiol.* 174, 1-9.

Robinson, P.J. (1987) Two different biosynthetic pathways for the secretion of Qa region-associated class I antigens by mouse lymphocytes. *Proc. Natl. Acad. Sci. USA* 84, 527-531.

Schreiber, A.B., Schlessinger, J., and Edidin, M. (1984) Interaction between major histocompatibility complex antigens and epidermal growth factor receptors on human cells. *J. Cell. Biol.* 98, 725-731.

Singh, B., Raghupathy, R., Anderson, D.J., and Wegmann, T.G. (1983) The placenta as an immunological barrier between mother and fetus. In: *Immunology of Reproduction*, (eds.) T.G. Wegmann and T.J. Gill, Oxford University Press, Inc., pp. 229-250.

Smets, L.A., and Van Beek, W.P. (1984) Carbohydrates of the tumour cell surface. *Biochim. Biophys. Acta* 738, 237-249.

Spencer, S.C. and Fabre, J.W. (1987) Identification in rat liver and serum of water-soluble Class I MHC molecules possibly homologous to the murine Q10 gene product. *J. Exp. Med.* 165, 1595-1608.

Starkey, P.M. (1987) Reactivity of human trophoblast with an antibody to the HLA Class II antigen, HLA-DP. *J. Reprod. Immunol.* 11, 63-70.

Stern, P.L., Beresford, N., Friedman, C.I., Stevens, V.C., Risk, J.M., and Johnson, P.M. (1987) Class I-like MHC molecule expressed by baboon placental syncytiotrophoblast. *J. Immunol.* 138, 1088-1091.

Sunderland, C.A., Redman, C.W.G., and Stirrat, G.M. (1981) HLA A, B, C antigens are expressed on nonvillous trophoblast of the early human placenta. *J. Immunol.* 127, 2614-2615.

Sunderland, C.A., Redman, C.W.G., and Stirrat, G.M. (1985a) Characterization and localization of HLA antigens on hydatidiform mole. *Am. J. Obstet. Gynecol.* 151, 130-135.

Sunderland, C.A., Sasagawa, M., Kanazawa, K., Stirrat, G.M., and Takeuchi, S. (1985b) An immunohistological study of HLA antigen expression by gestational choriocarcinoma. *Br. J. Canc.* 51, 809-814.

Szulman, A.E. (1972) The A, B and H blood-group antigens in human placenta. *New Engl. J. Med.* 286, 1028-1031.

Thiede, H.A., Choate, J.W., Gardner, H.H., and Syntay, H. (1965) Immunofluorescent examination of the human chorionic villus for blood group A and B substance. *J. Exp. Med.* 121, 1039-1049.

Trowsdale, J., Travers, P., Bodmer, W.F., and Pattillo, R.A. (1980) Expression of HLA-A, -B, and -C and B-2 microglobulin antigens in human choriocarcinoma cell lines. *J. Exp. Med.* 152, 11s-17s.

Wegmann, T.G. (1987) Placental immunotropism: maternal T cells enhance placental growth and function. *Am. J. Reprod. Immunol. Microbiol.* 15, 67 70.

Wells, M., Hsi, B.-L., and Faulk, W.P. (1984) Class I antigens of the major histocompatibility complex on cytotrophoblast of the human placental basal plate. *Am. J. Reprod. Immunol.* 6, 167-174.

Yamashita, K., Nakamura, T., and Shimizu T. (1984) Absence of major histocompatibility complex antigens in choriocarcinoma. *Am. J. Obstet. Gynecol.* 150, 896-897.

Yarden, A., Shune-Gottlieb, H., Chebath, J., Revel, M., and Kimachi, A. (1984) Autogenous production of interferon-switches on HLA genes during differentiation of histiocytic lymphoma U937 cells. *EMBO J.* 3, 969-973.

Zuckermann, F.A. and Head, J.R. (1986) Expression of MHC antigens on murine trophoblast and their modulation by interferon. *J. Immunol.* 137, 846-853.

Zuckermann, F.A. and Head, J.R. (1987) Possible mechanisms of non-rejection of the feto-placental allograft -- trophoblast resistance to lysis by cellular immune effectors. *Transplant. Proceed.* 19, 554-556.

THE HOST TISSUE
Uterine Epithelium: Cell Biological Changes In Relation To Endometrial Receptivity

GLYCOCONJUGATE EXPRESSION AND INTERACTIONS AT THE CELL SURFACE OF MOUSE UTERINE EPITHELIAL CELLS AND PERIIMPLANTATION-STAGE EMBRYOS

Daniel D. Carson[1], Oswald F. Wilson, and Anuradha Dutt

Department of Biochemistry and Molecular Biology
The University of Texas
M. D. Anderson Cancer Center
1515 Holcombe Blvd., Box 117
Houston, Texas 77030 USA

INTRODUCTION

The periimplantation stage mammalian embryo is extraordinary with regard to its adhesive and invasive potential. At the blastocyst stage, embryos first develop the ability to attach to and displace a variety of cell types in vitro (Sherman, 1978; Sherman and Wudl, 1976; Glass et al., 1979; Van Blerkom and Chavez, 1981). Furthermore, transplanted blastocysts have the capacity to invade a number of organs and tissues in host animals (c.f. Cowell, 1969). This property contributes to the occurrence of various teratocarcinomas and ectopic pregnancies (Rubin et al., 1983). The adhesive/invasive properties of periimplantation stage embryos reside in the trophectodermal cells surrounding the embryoblast. Given the wide range of biomolecules that can serve as substrates for attachment, it may be expected that mammalian embryos express a variety of cell adhesion systems. Nonetheless, the observations that preimplantation-stage blastocysts do not display adhesive/invasive properties (Sherman and Wudl, 1976) indicate that these systems are developmentally regulated and first are expressed during development of the blastocyst.

The unique and remarkable feature of the uterus is not its ability to support embryo attachment, but rather its ability to prevent this process. It is only during a very brief interval in which the uterus is under the appropriate steroid hormone influences that the embryo is permitted to utilize its adhesive/invasive properties (Psychoyos, 1986). Indeed, implantation-competent embryos can be maintained in the uterine lumen for long periods of time if the steroid hormone influences on the uterus are manipulated properly (Bergstrom, 1978). In mice and rats, implantation can be induced rapidly in these cases of "delayed" implantation following administration of estrogen (Psychoyos, 1986; Bergstrom, 1978). The key, initial, regulated step of the implantation process is the interaction of the embryonic trophectoderm with the apical cell surfaces of uterine epithelial cells (Schlafke and Enders, 1975). The epithelial cells of the uterine lumen provide the initial barrier to embryo invasion. It is of tremendous interest to identify adhesion systems that are displayed at these cell surfaces and to understand how steroid hormones influence their expression. Such information should provide insight not only into molecular aspects of embryo implantation, in particular, but also into tissue invasion processes, in general, e.g., tumor metastasis.

---

[1]To Whom Correspondence Should Be Addressed.

Given the general distribution of glycoconjugates on cell surfaces and their putative involvement in a variety of cell adhesion/recognition phenomena (Edelman, 1983; Frazier and Glaser, 1979), the current studies have focused on the expression of these molecules at the cell surface of the mouse trophectodermal and uterine epithelial cells. A number of lines of evidence indicate that uterine synthesis and expression of various glycoconjugates is strongly influenced by estrogen and progesterone (Dutt et al., 1986b; Carson et al., 1987; Dutt et al., 1988; Munakata et al., 1985; Nelson et al., 1975). Furthermore, morphological studies indicate that the pattern of expression of both uterine and embryonic cell surface glycoconjugates changes markedly at the time of implantation (Enders and Schlafke, 1974; Anderson and Hoffman, 1984; Anderson et al., 1986). Nonetheless, it remains to be demonstrated which particular glycoconjugates play roles relevant to embryo implantation. The biology and biochemistry of heparan sulfate proteoglycans, lactosaminoglycans and their corresponding receptors with regard to implantation-related processes are being studied. Detailed reviews of the structural features and metabolism of these oligosaccharides are available (Roden, 1980; Hascall and Hascall, 1985; Fukuda, 1985). In this article, some recent studies of these molecules and their receptor systems in the mouse uterus are presented. These findings will be discussed in light of other cell adhesion systems that may operate during the implantation process.

MATERIALS AND METHODS

Materials

CF-1 mice were purchased from Harlan Sprague-Dawley (Houston, TX). [^{35}S]-Methionine (>1000 Ci/mmol), [6-^{3}H]-glucosamine (25 Ci/mmol and sodium [^{125}I] iodide (carrier-free) were from ICN Radiochemicals (Irvine, CA). Tissue culture media and reagents were from Grand Island Biological Co. (Grand Island, NY) and Irvine Scientific (Santa Ana, CA). Matrigel and laminin were supplied by Collaborative Research, Inc. (Lexington, MA). Platelet factor IV was isolated as described previously (Farach et al., 1987). Fibronectin and protein molecular weight standards were purchased from Bethesda Research Laboratories, Inc. (Gaithersburg, MD). The Arg-Gly-Asp-Ser-containing synthetic peptides were from Peninsula Laboratories (Belmont, CA). The interstitial matrix preparations were generously supplied by Dr. Steffen Gay, University of Alabama at Birmingham. The anti-fibronectin receptor antibodies have been described previously (Brown and Juliano, 1986) and were the kind gift of Dr. Pat Brown, The University of Texas Medical School at Houston. Peptide: N-glycanase was purchased from Boehinger-Mannheim Biochemicals (Indianapolis, IN). All other chemicals and reagents were purchased from Sigma Chemical Co. (St. Louis, MO).

Uterine Epithelial Cell And Embryo Culture

Primary cultures of uterine epithelial cells were prepared exactly as described by Dutt et al. (1987). The techniques used for embryo culture, monitoring embryo attachment and outgrowth, and coating tissue culture surfaces with various agents have been described in detail in previous reports (Farach et al., 1987; Carson et al., 1988a). Tissue culture surfaces were coated with basement membrane-like

(Matrigel) or interstitial matrices as follows. Concentrated solutions of these matrices, stored at 6°C, were diluted 10-fold into ice-cold, phosphate-buffered saline, and then 0.5 ml of this solution was incubated for 30 minutes at 37°C in a well of a 24-well tissue culture plate. After this interval, the solution was removed and the well washed several times with sterile phosphate-buffered saline. The well then was used immediately for embryo attachment assays. In preliminary experiments, different dilutions of the matrix solutions were tested for their ability to support embryo attachment and outgrowth.

Lactosaminoglycan-Containing Glycopeptide Analyses

Primary cultures of uterine epithelial cells were incubated overnight (16 to 18 hours) in Eagle's minimum essential medium with Earle's salts (Flow Laboratories, Inc.; McLean, VA) containing 1 mM glutamine, 50 units penicillin per ml, 50 µg streptomycin per ml and 0.5 mCi per ml [^{35}S]-methionine (50 mCi/µmol). At the end of this interval, the cell layers were washed several times with ice-cold phosphate-buffered saline. The cells were scraped from the tissue culture surface with a rubber policeman and macromolecules precipitated with ice-cold 10% (w/v) trichloroacetic acid. The precipitates were exhaustively digested with pronase, followed by digestion with nitrous acid and then chondroitinase ABC as described previously (Dutt et al., 1987; 1988). This procedure yields a glycopeptide fraction that elutes in the void volume of Sephadex G-50 that is almost entirely composed of lactosaminoglycan-bearing molecules (Dutt et al., 1987; 1988). These glycopeptides were analyzed under reducing conditions by sodium dodecyl sulfate-polyacrylamide gel electrophoresis using 10% (w/v) acrylamide as described by Laemmli (1970). When indicated, the glycopeptides were first incubated with 15 units per ml of peptide:N-glycanase in 100 µl of 0.1% (w/v) sodium dodecyl sulfate, 1% (v/v) β-mercaptoethanol, 0.1 M Tris-HCl (pH 8.0) at room temperature overnight. This procedure completely converted 50 µg of α1-acid glycoprotein used as an internal standard to a single lower M_r (deglycosylated) form.

In other cases [^{35}S]-methionine-labeled, lactosaminoglycan-bearing glycopeptides were chemically deglycosylated with trifluoromethane sulfonic acid (Edge et al., 1981). As described previously (Dutt et al., 1986a), it was found that 24 hours of such digestion at room temperature were required to convert these glycopeptides to a limit form. Glycopeptides also were analyzed before or after various digestions by molecular exclusion liquid chromatography. The column used was a 1 x 30 cm Superose 12 resin (Pharmacia, Uppsala, Sweden) eluted with 2 M guanidine-HCl, 0.01% (w/v) octylglucoside, 20 mM Tris-acetate (pH 7.0) and 0.02% (w/v) sodium azide at room temperature. The column was pumped at a flow rate of 0.7 ml per minute (back pressure = 200 psi) using a Beckman model 100A pump controlled by a model 421A controller (Beckman Instruments, Inc., Fullerton, CA). Fractions were collected every 0.5 minute. Recoveries of radioactivity from similar analyses typically ranged between 80 and 90%. In some cases, samples were incubated at 90°C for 60 minutes in the presence of 50 mM dithiothreitol immediately prior to liquid chromatographic analyses. This reduction procedure had no effect on the elution characteristics of the glycopeptides or deglycosylated peptides.

Analyses Of Cell Surface And Total Heparan Sulfate Binding Proteins

Ligand binding assays utilizing [3H]-heparin were used to identify heparan sulfate receptors on the surface of primary cultures of mouse epithelial cells. Cells were incubated with 200 nM [3H]-heparin (0.3 mCi/ng) in PBS with 2 mM Ca^{++} and Ng^{++} for 30 minutes on ice. Cells were rinsed 4 times with PBS and the amount of radioactivity associated with the cells determined. Time and heparin concentration dependence was determined as described previously (Wilson et al., submitted). Primary cultures of mouse uterine epithelial cells were prepared and metabolically labeled with [35S]-methionine as described above. To initiate a chase period, metabolically labeled cell layers were rinsed several times with prewarmed medium that did not contain isotope and was supplemented with 1 mM methionine. These cultures then were incubated for from 0 to 24 hours at 37°C and the amount of radioactive heparan sulfate binding proteins exposed at the cell surface and associated with the cell layer was determined at various intervals as described below.

The cell layers were washed several times with ice-cold phosphate-buffered saline and then incubated for 10 minutes on ice in phosphate-buffered saline that contained 50 µg trypsin per ml to release heparan sulfate binding protein fragments. Previous studies have shown that these conditions are optimal for the release of fragments of heparan sulfate receptors from the cell surface that retain heparin-binding activity (O. Wilson et al., personal communications). At the end of this incubation, the incubation medium was removed, protease inhibitors added and tryptic fragments of heparan sulfate receptors isolated by heparin-agarose affinity chromatography as described previously (O. Wilson et al., personal communications). The amount of radioactivity that bound to the resin and required more than 1 M NaCl to be eluted was used as an index of cell surface receptor. Material in the cell layer was extracted and heparan sulfate receptors isolated by heparin-agarose affinity chromatography as described previously (Wilson et al., personal communications). Briefly, medium or cell layers were extracted for 1 hour at 6°C with a solution of 4 M guanidine HCl, 1% (w/v) octylglucoside, 20 mM Tris-Acetate (pH 7.0) and a mixture of protease inhibitors (Dutt et al., 1986b). This extract was diluted 8-fold with a solution of 0.1% (w/v) octylglucoside, 20 mM Tris-Acetate (pH 7.0) and incubated with 0.5 ml of a 50% suspension of heparin-agarose at 6°C for 1 hour with constant rotary agitation. This resin was washed 5 times batchwise with 5 ml of 1 M NaCl, 20 mM Tris-Acetate (pH 7.0), 0.1% (w/v) octylglucoside. By the fifth wash, only background levels of radioactivity were eluted. The resin was then incubated with 2 ml of 4 M NaCl, 20 mM Tris-Acetate (pH 7.0), 0.1% (w/v) octylglucoside for 1 hour at 6°C with constant rotary agitation to elute the bound material. The bound material primarily consisted of the 140,000 M_r proteins described elsewhere (Wilson et al., personal communications). These studies have extensively described the specificity of the interaction of these proteins with heparin-agarose, the disposition of the 140 Kd proteins at the cell surface and the relationship of the trypsin-releasable, heparin-binding peptides of the cell surface with the 140 Kd proteins. Total heparan sulfate binding proteins were defined as the sum of the heparan sulfate binding peptides released by trypsin at the cell surface and those remaining in the cell layer following trypsin digestion.

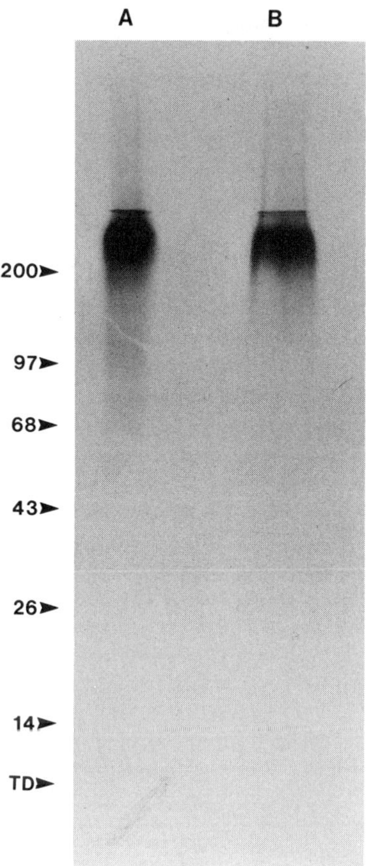

Figure 1. SDS-PAGE of Pronase LAG-Glycopeptides. Limit pronase LAG-glycopeptides were prepared from primary cultures of uterine epithelial cells labeled metabolically with [^{35}S]-methionine as described in Materials and Methods. These glycopeptides were subjected to electrophoresis on 10% (w/v) polyacrylamide gels under reducing conditions as described by Laemmli (1970) either before (lane A) or after (lane B) digestion with peptide:N-glycanase as described in Materials and Methods. The migration positions of protein standards with the indicated molecular weights (in kilodaltons) are indicated to the left of the lanes.

RESULTS AND DISCUSSION

Studies Of Lactosaminoglycan (LAG)-Bearing Molecules

Previous work has described the synthesis of LAGs by uterine epithelial cells (Dutt et al., 1987), the involvement of these molecules in epithelial cell adhesion in vitro (Dutt et al., 1987), and the marked stimulatory effects of estrogen on LAG synthesis (Dutt et al., 1988). It was of interest to determine if many or relatively few proteins synthesized by uterine epithelial cells carry these oligosaccharides. LAG-bearing glycoproteins appear to be very large both by SDS-PAGE and by molecular exclusion chromatography under dissociative conditions (see below). Therefore, we converted these molecules to limit glycopeptides by exhaustive digestion with pronase to generate fragments more amenable to analysis on the basis of size. A metabolic label was incorporated into the protein cores of these molecules so that the size of the pronase resistant peptides and glycopeptides could be estimated. Primary cultures of uterine epithelial cells were labeled with [^{35}S]-methionine and LAG-glycopeptides prepared by a series of chemical and enzymatic digestions to remove other types of molecules. As shown in Figure 1 (lane A), these limit glycopeptides exhibited very little mobility upon SDS-PAGE under reducing conditions. It should be noted that heavily glycosylated molecules migrate anomalously on SDS gels. Consequently, the apparently large M_r probably reflects the limitation of this technique more than an accurate estimate of M_r of these molecules. As a result, we employed additional techniques, our attempts to estimate the size of the LAG-glycopeptides. Previous studies have shown that LAG oligosaccharides are primarily in N-linkage to protein (Dutt et al., 1988); however, digestion with peptide:N-glycanase, an enzyme capable of releasing a wide variety of N-linked oligosaccharides from protein cores (Hirani et al., 1987), failed to shift the electrophoretic mobility of the LAG-glycopeptides (Figure 1, lane B). Consequently, it did not appear that the oligosaccharide to protein linkage region was accessible to the enzyme in these structures.

A second approach used to study the [^{35}S]-methionine-labeled glycopeptides involved their analysis by molecular exclusion column chromatography under dissociative conditions, i.e., in the presence of guanidine-HCl and detergent. As shown in Figure 2, most of the glycopeptides migrated at or near the void volume of the column suggesting that they had hydrodynamic radii similar to or greater than that of the 2,000,000 molecular weight carbohydrate, blue dextran. Preincubation of these glycopeptides with 50 mM dithiothreitol at 90°C for 1 hour did not alter this profile (data not shown). These results indicated that crosslinking of these glycopeptides by disulfide bonds could not be the sole explanation for their apparently large size. The glycopeptides then were chemically deglycosylated using trifluoromethane sulfonic acid (Edge et al., 1981) and chromatographed under the same conditions. In preliminary studies, it was found that 16 to 24 hours of incubation at room temperature was required for deglycosylation of these glycopeptides to be completed. Shorter periods of incubation generated broad distributions of products of lower apparent molecular weights than the starting material, presumably due to incomplete hydrolysis (data not shown). The chemically deglycosylated LAG core peptides were converted almost entirely to forms with an elution position similar to that of protein standards with molecular weights ranging from 9,000-12,000. SDS-PAGE of chemically deglycosylated LAG core proteins also revealed major components exhibiting M_r = 10,000-11,000 (data not shown).

The size of the protected peptide coupled with the pronase and peptide:N-glycanase resistance of the glycopeptides indicate that the LAG oligosaccharides are capable of covering or masking relatively large regions of the proteins to which they are attached. The size of the individual LAG oligosaccharides ranges from

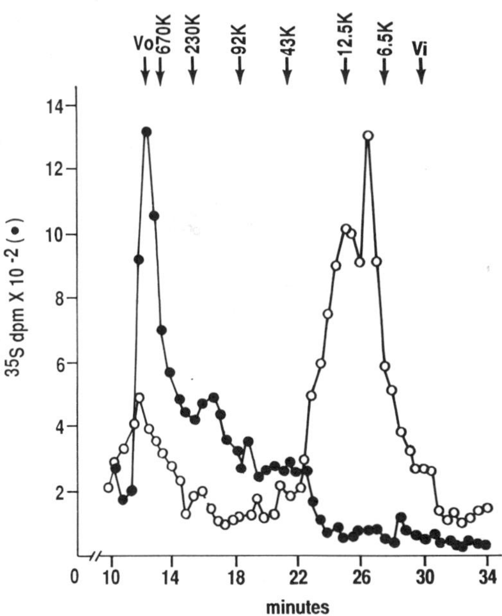

Figure 2.   Molecular Exclusion Chromatography of LAG-Glycopeptides.   [^{35}S]-Methionine-labeled LAG-glycopeptides were prepared as described in Figure 1. These glycopeptides were analyzed by liquid chromatography on a 1 x 30 cm Superose 12 column eluted with 2 M guanidine-HCl, 0.01% (w/v) octylglucoside, 20 mM Tris-acetate (pH 7.0) and 0.02% (w/v) sodium azide at 0.7 ml/minute Tris-acetate (pH 7.0) and 0.02% (w/v) sodium azide at 0.7 ml/minute at room temperature.   The elution profiles are shown of material analyzed before (•) and after (o) chemical deglycosylation with trifluoromethane sulfonic acid as described in Materials and Methods.   The elution positions of the following molecular weight standards are indicated: V$_o$, blue dextran; 670K, thyroglobulin or 530,000 MW dextran standard; 230K, catalase; 92K, phosphorylase b or 71,000 MW dextran standard; 43K, ovalbumin; 12.5K, cytochrome C; 6.5K, aprotinin; V$_i$, potassium dichromate.

7,000 to 15,000 (Dutt et al., 1988), while that of the protein cores appears to range between 9,000 to 12,000.   Amino acid analysis of the LAG glycopeptides indicates that 11% of the amino acids are in the form of Asp/Asn (Dutt and Carson, unpublished studies).   Therefore, it can be calculated that there can be no more than 8 - 11 LAG chains attached per protein core.   The corresponding glycopeptides would range in MW from 97,000 to 133,000.   This calculated MW is much smaller than the apparent MW exhibited by the glycopeptides even relative to dextran standards. Consequently, it seems quite likely that these molecules have an extreme tendency to aggregate in solution.   This may, in part, reflect their ability to support cellular adhesive functions (Dutt et al., 1987).   This property also makes it difficult to estimate the number of oligosaccharide chains actually substituted per peptide.

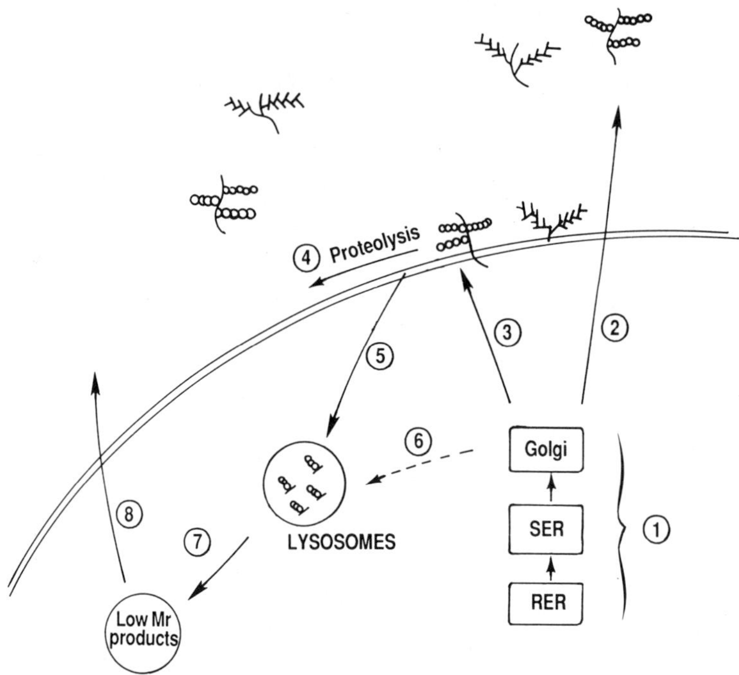

Figure 3.   Model of LAG-glycoprotein and HSPG Metabolism.   For detailed explanation see text.   Abbreviations and symbols: LAG, lactosaminoglycan; HSPG, heparan sulfate proteoglycans; RER, rough endoplasmic reticulum; SER, smooth endoplasmic reticulum; Golgi, Golgi apparatus; ⋎⋎⋏⋏⋏⋏ , lactosaminoglycan oligosaccharides; O○O○O○O, heparan sulfate oligosaccharides.

Comparative Aspects Of LAG-Glycoprotein, Heparan Sulfate Proteoglycan (HSPG) And Heparan Sulfate (HS) Receptor Metabolism

A number of features of LAG-glycoprotein (Dutt and Carson, 1990) as well as HSPG (Tang et al., 1987) metabolism in primary cultures of mouse uterine epithelial cells were described in previous studies. The overall features of these processes are summarized and compared in Figure 3 and Table 1. The kinetics by which cycloheximide inhibits LAG and HSPG assembly indicates that it takes no more than 30 to 60 minutes for these molecules to be completely glycosylated in the Golgi apparatus once their core proteins have been synthesized in the rough endoplasmic reticulum (step 2). LAG glycoproteins first become accessible to external proteases about 30 minutes after they have been synthesized. These observations suggest that this interval is the minimum time required for transit from the Golgi apparatus to the cell surface. In contrast, HSPG movement to the cell surface is significantly faster, requiring only 4 - 8 minutes (step 3).

Table 1

Metabolism of LAG-Glycoprotein and HSPGs by Mouse Uterine Epithelium

Process	LAG-GP	HSPG
Transit times		
RER ---> Golgi	30 ± 10 min	45 ± 15 min
Golgi ---> Cell surface	33 ± 7 min	6 ± 2 min
Metabolic half life	6 ± 1 h	1.5 ± 0.5 min
Proteolytic release	Yes ($\sim 30\%$)	Yes ($\sim 30\%$)
Lysosomal degradation	No	Yes
Uptake from medium	No	No

Abbreviations: HSPG, heparan sulfate proteoglycans and glycosaminoglycans; LAG-GP, LAG-bearing glycoproteins; RER, rough endoplasmic reticulum.

A large fraction of both LAG-glycoproteins and HSPGs are secreted (step 2) or released to the medium (step 4). Both classes of glycoconjugates are handled similarly in the sense that about 30% of this release can be inhibited by the inclusion of protease inhibitors in the incubation medium. Collectively, these data suggest that there are at least two systems operating in these cells to release or secrete LAG-glycoproteins and HSPGs. The first system apparently involves cell surface proteolysis (step 4) and has been described for HSPG release from mouse mammary epithelial cells (Jalkanen et al., 1988). The second system (step 2) is insensitive to protease inhibitors or chelators used. This may involve direct secretion or cell surface proteases insensitive to the battery of protease inhibitors and chelator used.

Virtually all of the LAGs are destined for secretion or release from the cell layer. In contrast, a significant fraction (30 - 40%) of the HSPGs are degraded rapidly to smaller HS chains linked to little or no protein in a chloroquine-sensitive compartment, presumably the lysosomes (step 5). Interestingly, if lysosomal degradation of HSPGs is inhibited, HSPG release or secretion is proportionately stimulated. These observations indicate that the secretory/release systems compete with the degradative systems for HSPG utilization. It also is possible that a small percentage of HSPGs may never reach the cell surface before being degraded in lysosomes (step 6). The intracellular HS chains formed via lysosomal degradation are relatively long-lived (half life = 12 hours), but ultimately are degraded to low M_r products (step 7) which are released from the cell (step 8).

At this point it is not clear if secretion of HSPGs or LAG glycoproteins is equally distributed between the apical and basolateral aspects of these cells or if there is some vectorial preference to this process. It appears that material immunologically related to HSPGs is distributed at the apical cell surface of mature mouse uterine epithelial cells in utero (Tang et al., 1987). Furthermore, it has been shown that procedures which selectively extract apical components of these cells in uterine strips also extract such molecules (Morris et al., 1988a; 1988b). Recently, procedures have been developed for the culture of polarized immature rat uterine epithelial cells (Glasser et al., 1988). This has permitted examination of patterns of vectorial secretion in vitro. Using this culture technique, we have found that rat uterine epithelial cells display a marked (80 - 90%) preference for the apical secretion of proteoglycans (Carson et al., 1988b). The development of similar procedures for the culture of polarized mature mouse uterine cells should permit direct examination of vectorial aspects of glycoprotein secretion in this system as well. Collectively, these studies strongly indicate that HSPGs are expressed and secreted at the apical cell surfaces of uterine epithelial cells. As mentioned above, LAG-glycoproteins do not appear to be significantly degraded by the cells, but rather are almost quantitatively released from the cell layer. Once released into the medium both the HSPGs and LAG-glycoproteins appear to be fairly stable with no evidence of decrease in amount for at least 24 hours. These last observations suggest that the secreted forms of the molecules are not endocytosed by the cells to a significant extent.

The biological significance for the directionality of HSPG and LAG-glycoprotein export could be the possibility that these molecules interact with embryonic cell surfaces prior to implantation. For instance, HSPGs secreted to the apical compartment could compete with HSPGs of the trophectodermal cell surface for binding to the epithelial cell surface. This point must be considered, since HS receptors not only can bind embryos in vitro (see below), but also can be blocked by HS produced by the uterine epithelial cells themselves (Wilson et al., submitted). LAG-glycoproteins appear to interact with a cell surface galactosyltransferase system in uterine epithelial cells (Dutt et al., 1987) similar to the system proposed for mouse sperm adhesion to ova (Shur, 1984). Therefore, it is possible that LAGs secreted from the apical cell surface of the uterine epithelium could interact with the galactosyltransferase system of the sperm. This interaction would diminish the number of sperm as they traverse the uterine lumen. Consequently, understanding of the directionality of LAG and HSPG secretion by uterine epithelial cells may also have biological significance beyond the particular process of implantation.

As discussed above, a significant portion of the HSPGs is degraded intracellularly, apparently in lysosomal compartments. The catabolic pathway for HSPGs involves rapid degradation of both the protein and oligosaccharide components of these molecules. The HS chains are rapidly converted from forms with a median M_r of 40,000 to forms with a median M_r of about 12,000. At this point, the process of HS degradation slows down considerably. Although the half-life of the cell-associated HSPGs is only 1 to 2 hours, that of the partially degraded HS chains is 10 to 12 hours (Tang et al., 1987). These HS chains ultimately are degraded to very low M_r products which are secreted from the cells. In contrast, the half-life of the cell-associated LAG-glycoproteins is longer, ranging from 5 to 7 hours. We recently have described 140 Kd proteins of the uterine epithelial cell surface that bind both commercial heparin and endogenous HS with high affinity (apparent Kd = 50 nM) and a high degree of specificity (Wilson et al., submitted).

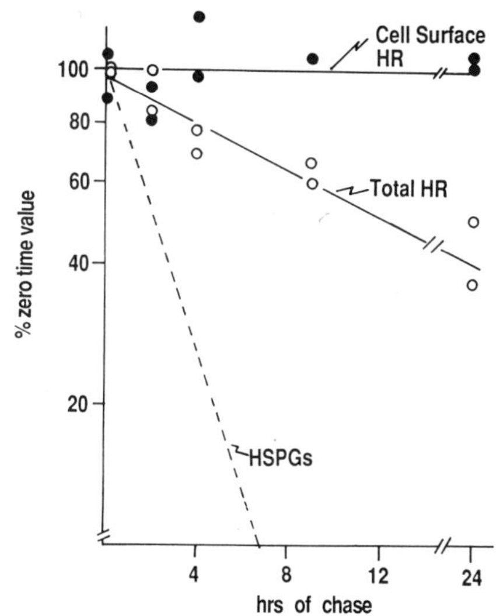

Figure 4. Turnover of Heparan Sulfate (HS) Binding Proteins. HS binding proteins were labeled metabolically with [35S]-methionine in medium containing 10 μM methionine for 16 hours in primary cultures of uterine epithelial cells as described in Materials and Methods. At time zero as indicated on the graph, the medium was replaced with prewarmed medium containing no isotope and 1 mM methionine to initiate a "chase" period. At the indicated times, duplicate cultures were removed, rinsed several times with ice-cold phosphate buffered saline, and HS binding proteins extracted and isolated as described in Materials and Methods. HS binding proteins disposed at the cell surface (Cell Surface HR) were indexed by the brief trypsinization procedure described in Materials and Methods. Total HR represents the total amount of HS binding proteins found in the trypsinate and in the trypsin-resistant, cell-associated fraction. The dashed line indicates the turnover rate of the cell-associated heparan sulfate proteoglycans (HSPGs; Tang et al., 1987) for comparison. The data are expressed as the percentage of the indicated fraction remaining at the indicated time of the chase period relative to the zero time values. The zero time values (in dpm) were $4.32 \pm 0.49 \times 10^4$ and $2.97 \pm 0.14 \times 10^5$ for cell surface and total HS binding proteins, respectively.

Other glycosaminoglycans like chondroitin sulfate, hyaluronate, keratan sulfate, and even certain forms of heparan sulfate fail to compete with heparin for binding to the cell surface or to the isolated 140 Kd proteins. As shown in Figure 4, the half-life of the heparin-binding proteins at the cell surface is quite long by comparison to that of either LAGs or HSPGs. Although the half-life of the total HS binding proteins pool in the cell layer is approximately 12 hours, no decrease in the amounts of radiolabeled heparin binding proteins is detected at the cell surface over a 24 hours period. Only 10 to 15% of the total cell-associated heparin-binding proteins appear to be expressed at the cell surface at a given time. These results suggest that the large pools of intracellular heparin-binding proteins serve to replenish those proteins at the cell surface. It should be noted that the intracellular heparin-binding proteins display the same M_r and high affinity binding to heparin-agarose as their cell surface counterparts. Consequently, the trypsin-insensitive, heparin-binding proteins appear to be very similar, if not identical, to the cell surface forms. In pulse-chase studies using trypsin accessibility as an index of exposure at the cell surface, it can be demonstrated that the half-life of "bulk" cell surface proteins exceeds 24 hours in mouse uterine epithelial cells. As is the case for the cell surface HS binding proteins, this probably reflects a combination of the actual turnover rate of these molecules and replenishment of the cell surface population with prelabeled molecules in intracellular pools. Nonetheless, it seems likely that most of the cell surface components of cultured mouse uterine epithelial cells are more stable than the major, complex glycoconjugates of their cell surface, i.e., LAG-glycoproteins and HSPGs. This property endows these cells with the capacity to alter rapidly the composition of cell surface glycoconjugates relative to other constituents. These biochemical observations are consistent with a variety of morphological studies indicating that fairly rapid alterations in various carbohydrate constituents of the lumenal surface of the uterine epithelium occur in response to early pregnancy or steroid hormones (Enders and Schlafke, 1974; Anderson and Hoffman, 1984; Anderson et al., 1986).

Attachment/Adhesion-Promoting Systems Of The Trophectoderm

As discussed above, a number of studies indicate that peri-implantation-stage embryos are capable of attaching to and invading a variety of cell types (Sherman, 1978; Glass et al., 1979), acellular substrates (Armant et al., 1986; Carson et al., 1987a; Farach et al., 1987; Carson et al., 1988a) and tissues (Cowell, 1969). Furthermore, it appears that both RGD-recognizing receptor systems (Richa et al., 1985; Armant et al., 1986; Sutherland et al., 1988; Carson et al., 1988a) and HSPGs of the trophectodermal cell surface (Farach et al., 1987) participate in aspects of these adhesion processes. The morphology of "attached" or "unattached" blastocysts in vitro is quite similar (see Figure 5A), although the blastocoelic cavity of attached embryos may partially collapse if they are maintained in implantation delay. The criterion for attachment in vitro is to determine if embryos move or remain stationary with respect to the substrate upon slight agitation of the culture vessel. This can be accomplished by gentle swirling or tapping of the side of the vessel. Outgrowth formation can become quite dramatic (see Figure 5B), although it may be difficult to observe this at very early stages. There are examples of substrates to which embryos can attach very firmly without forming significant outgrowths, even after extended periods of time. Two such substrates are platelet factor IV (Farach et al., 1987; Figure 5C) and HS-binding proteins derived from mouse uterine epithelial cells (Figure 5D, Figure 6). Both classes of molecules share the property of being able to bind HS (Handin and Cohen, 1976; Wilson et al., submitted).

The case of the epithelial HS binding proteins is of particular interest for several reasons. First, these molecules are components of the cell surface of uterine epithelial cells (Wilson et al., submitted) and, therefore, may be involved in embryo attachment in vivo. Second, their ability to support attachment without outgrowth in vitro (see below) mimics the in vivo observations indicating that attachment to the uterine epithelial cell surface much precedes penetration of the epithelial cell layer (Schlafke and Enders, 1975; Sherman and Wudl, 1976). The latter process has been suggested to be more akin to outgrowth in vitro (Sherman,

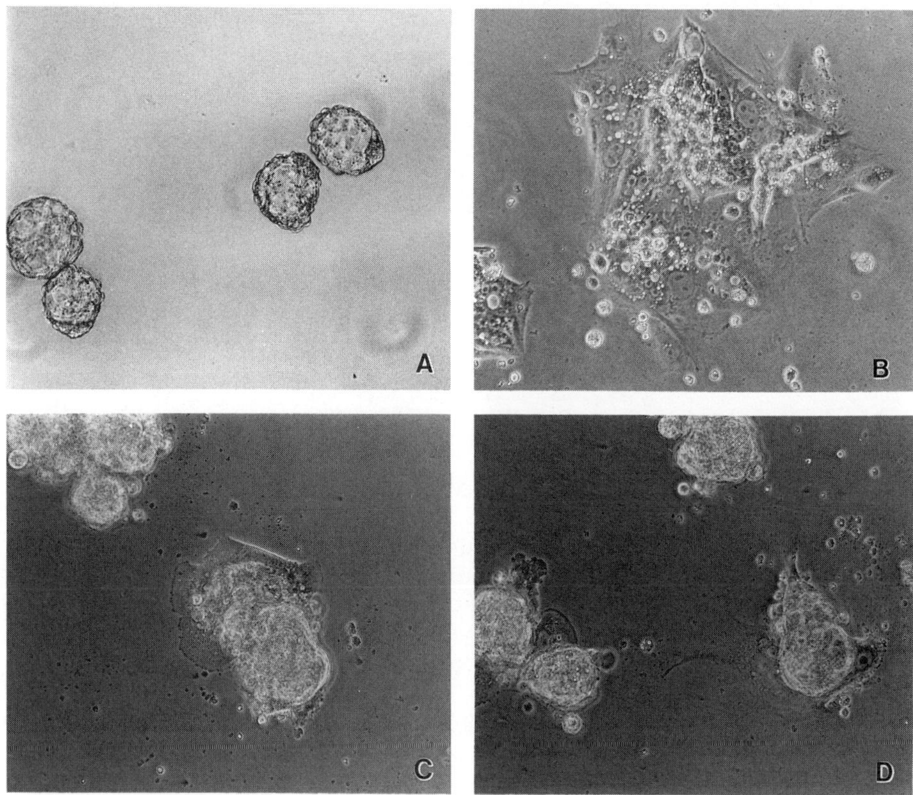

Figure 5. Blastocysts and Trophoblast Outgrowths on Fibronectin, Platelet Factor Four and Heparan Sulfate Binding Proteins. Mouse blastocysts were cultured in serum-free medium as described previously (Farach et al., 1987). Phase micrographs are at 135X magnification in Panel A) and at 270X magnification in Panels B), C) and D). Panels: A) unattached blastocysts; B) trophoblast outgrowths after 3 days of culture on a fibronectin-coated surface; C) trophoblast outgrowth after 3 days of culture on a platelet factor four-coated surface; D) trophoblast outgrowth after 3 days of culture on a heparan sulfate binding proteins-coated surface. Note the small asymmetric outgrowths formed on platelet factor four- or heparan sulfate binding proteins-coated surfaces. Most embryos cultured on these substrates fail to form any outgrowths (see Figure 6).

1978). Third, a series of studies indicate that: 1) HSPGs are expressed on the trophectodermal cell surface (Dziadek et al., 1985; Farach et al., 1987); 2) HSPG synthesis increases markedly at the time when embryos acquire attachment competence (Farach et al., 1988); 3) inhibition of HSPG synthesis inhibits embryo attachment to a variety of substrates in vitro (Farach et al., 1988); 4) digestion of the embryo cell surface with heparinase, but not other glycosidases, inhibits embryo attachment to uterine epithelial cells and other substrates in vitro (Farach et al., 1987); 5) inclusion of soluble heparin (a readily available HS analogue) in the incubation medium inhibits embryo attachment to uterine epithelial cells or other substrates in vitro, whereas other polysaccharides do not have this effect (Farach et

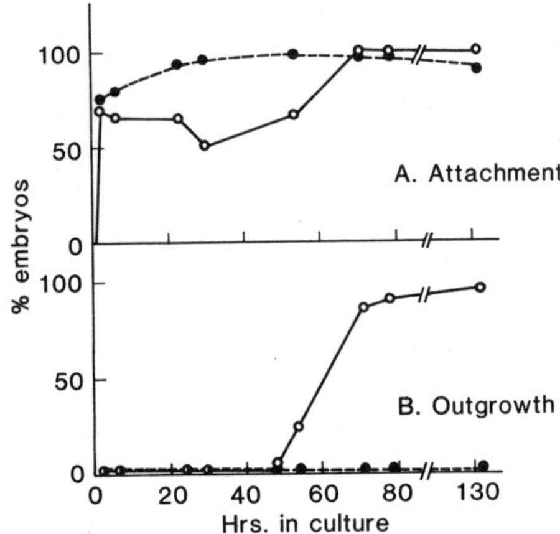

Figure 6. Kinetics of Mouse Blastocyst Attachment and Outgrowth Formation on Heparan Sulfate Binding Protein- or Fibronectin-Coated Surfaces. Mouse blastocysts were obtained and cultured as described previously in serum-free medium (Farach et al., 1987). At the indicated times, attachment, and outgrowth formation were assessed on groups of at least 45 embryos cultured on heparan sulfate binding protein-coated (●) or fibronectin-coated (○) tissue culture surface. Panels: A) attachment; B) outgrowth formation.

al., 1987). Collectively, these observations strongly argue for an important role for embryonic HSPGs in embryo adhesion phenomena. Although embryos clearly express a number of cell adhesion systems on the apical surface of the trophectoderm (see Table 2), only those involving HSPGs have been shown to influence binding to uterine epithelial cells, the biological substrate. Heparin also appears to be one of the most effective compounds that have been tested as decidualizing agents (Finn, 1986). Consequently, HS-dependent interactions must be considered as potentially important components of implantation-related cell adhesion processes.

HS binding proteins expressed at the cell surface of uterine epithelial cells have been isolated by affinity chromatography on heparin-agarose and used to coat tissue culture surfaces. As shown in Figure 6, periimplantation-stage mouse blastocysts rapidly attach to these matrices in vitro and maintain their attachment for several days. In contrast to fibronectin, another embryo attachment-supporting molecule (Armant et al., 1986), HS binding proteins fail to support subsequent outgrowth of the attached embryos (Figure 5D). A number of molecules now have been shown to support both embryo attachment and outgrowth in vitro. The list includes fibronectin (Armant et al., 1986; Farach et al., 1987; Sutherland et al., 1988), laminin (Armant et al., 1986; Farach, et al., 1987; Sutherland et al., 1988), vitronectin (Armant et al., 1986), hyaluronic acid (Carson et al., 1987), and collagen types I - VI (Carson et al., 1988a). In addition to simple matrices, embryos attach and outgrow rapidly on complex substrates containing various mixtures of extracellular matrix components. Figure 7 compares embryo attachment and outgrowth on a commercially available basement membrane-like complex matrix [Matrigel; EHS tumor matrix (Kleinman et al., 1986)] as well as a reconstituted preparation of interstitial extracellular matrix (prepared and provided by Dr.

Table 2

Occurrence of Various Adhesion-Promoting Molecules on the Cell
Surfaces of Periimplantation Mouse Trophectoderm and Uterine Epithelium

Adhesion-Promoting Molecule	Trophectoderm	Uterine Epithelium	References
HSPGs	+	+	Tang et al., 1987; Morris et al., 1988; Dziadek et al., 1985; Farach et al., 1987
Laminin	+	-	Leivo et al., 1980; Dziadek and Timpl, 1985; Wu et al., 1983; Glasser et al., 1987;
Nidogen/ entactin	+	+	Dziadek and Timpl, 1985; Wu et al., 1983
Collagens	+	-	Leivo et al., 1980; Sherman et al., 1980; Minamoto et al., 1987; Leivo et al., 1980
Fibronectin	-	-	Wartiovaara et al., 1979; Grinnell et al., 1982; Glasser et al., 1987
CAM-105	-	-	Svalander et al., 1987; Svalander et al., in press
Uvomorulin cadherins, GP 120/80	-	-	Damjanov et al., 1986; Vestweber et al., 1987; Yoshida-Noro et al., 1984; Damjanov et al., 1986; Vestweber and Kemler, 1984
N-CAM	?	-	Wu et al., 1983; Carson and Julian, unpublished observations
Hyaluronate	?	-	Carson et al., 1987
Galactosyl-transferase	+	+	Sato et al., 1984; Dutt et al., 1987 Bayna et al., 1988
Lactosamino-glycans	+/-	+	Sato and Muramatsu, 1986; Marticorena et al., 1983; Damjanov et al., 1986; Dutt et al., 1987
RGD-dependent receptors	+	?	Richa et al., 1985; Armant et al., 1986; Carson et al., 1988; Sutherland et al., 1988

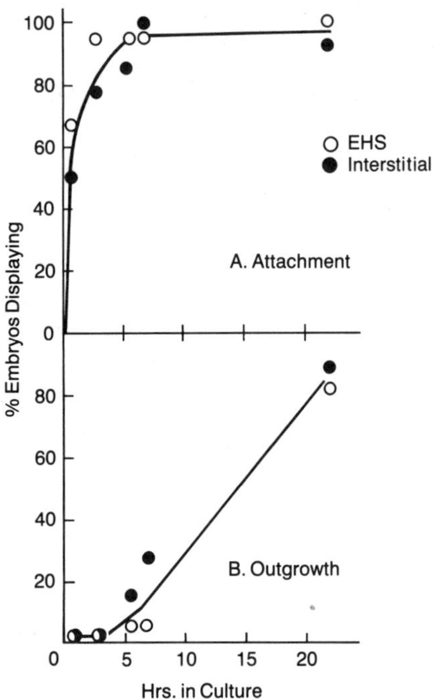

Figure 7.   Kinetics of Attachment and Outgrowth Formation on Complex Extracellular Matrices. Mouse blastocysts were obtained and cultured in serum-free medium as described previously (Farach et al., 1987). Tissue culture surfaces were precoated with either a commercially available, basement membrane-like matrix (EHS, Matrigel; Kleinman et al., 1986) or a reconstituted interstitial matrix (Interstitial) provided by Dr. Steffan Gay (University of Alabama-Birmingham). At the times indicated, attachment and outgrowth formation were assessed on groups of at least 40 embryos. Panels: A) attachment; B) outgrowth formation.

Steffen Gay, University of Alabama at Birmingham). Collectively, these studies demonstrate the promiscuous ability of embryos to utilize extracellular matrix components for attachment.

It is clear that a large number of proteins and polysaccharides found in the extracellular matrix do not support embryo attachment or outgrowth in vitro (Carson et al., 1987a; Sherman, 1978). Therefore, the embryo is not totally indiscriminate with regard to attachment substrates. In spite of the variety of molecules that has been shown to support embryo attachment in vitro, none of these molecules have been shown to be expressed at the apical cell surface of uterine epithelial cells in utero (see Table 2). Consequently, although these molecules may act as secondary promoters of implantation, i.e., embryo invasion of the basal lamina and stroma, they are unlikely mediators of initial embryo attachment. Uterine LAGs do not support embryo attachment in vitro nor do they inhibit embryo attachment and outgrowth on uterine epithelial cells cultured in vitro (J.P. Tang, personal communications). Thus, it seems unlikely that LAGs of the epithelial cell surface are essential for embryo adhesion. Galactosyltransferase, a putative receptor for LAGs (Shur, 1984), has been implicated as an adhesion-promoting molecule in mouse uteri (Dutt et al., 1987) and has been found abundantly expressed on microvilli in human placentae (Nelson et al., 1977). One report indicates that embryonic galactosyltransferase is not expressed at the apical cell surface of the mouse trophectoderm (Sato et al., 1984); however, the antibody used in these studies was directed at a form of the enzyme isolated from bovine milk. It must be considered that the murine cell surface form of galactosyltransferase may not be identical to that found in milk and/or expressed by other species. A more recent study indicates that this activity is present on blastocyst cell surfaces (Bayna et al., 1988). Consequently, it seems worthwhile to reexamine the cell surface distribution of galactosyltransferase both at the embryo and uterine epithelial cell surface using antibodies directed at the murine, cell surface molecule(s). Potential ligands for cell surface galactosyltransferase appear to exist at the trophectodermal surface in the form of N-acetylglucosaminyl residues (Anderson et al., 1986) and at the epithelial cell surface in the form of LAGs (Dutt et al., 1987). Thus, it remains possible that this system supports certain aspects of embryo attachment and/or outgrowth.

It is interesting to note that a number of components of basal lamina are expressed at the apical surface of trophectoderm at the periimplantation stage (Table 2). Multiple independent investigations have described expression of HSPGs, laminin, collagens, and nidogen/entactin. From this standpoint, it appears that the apical trophectodermal cell surface expresses a remarkable degree of "basal" character. It is tempting to speculate that this atypical expression of extracellular matrix components may promote interaction between the apical surfaces of trophectoderm and uterine epithelium. In any event, it is clear that the embryonic and uterine epithelial cell surfaces that interact during embryo attachment display unusual characteristics relative to most apical cell surfaces (Simons and Fuller, 1985).

Several studies have indicated that cell adhesion molecules associated with many adult epithelia, i.e., uvomorulin (Vestweber et al., 1987), cadherins (Yoshida-Noro et al., 1984), and GP 120/80 (Damjanov et al., 1986), are not found on the apical cell surface of mouse trophectoderm. It also appears that at least two types of receptors for Arg-Gly-Asp sequences (RGD sequences; Ruoslahti and Pierschbacher, 1987) are expressed at these surfaces of the mouse embryo (Armant et al., 1986; Sutherland et al., 1988; Carson et al., 1988a); however, polyclonal antisera directed against an adult RGD-receptor proposed to function as a fibronectin receptor (Brown and Juliano, 1986) neither inhibit the rate of embryo attachment and outgrowth to fibronectin in vitro (Figure 8) nor do they react with the surface of the trophectoderm above levels observed with preimmune antisera (Tang et al., unpublished observations). The same antisera has been shown to inhibit cell adhesion processes in both rodent and human cell lines (Brown and Juliano, 1986). In contrast, antisera directed at a similar group of extracellular matrix receptors also failed to inhibit embryo attachment, although they markedly inhibited trophoblast outgrowth on both fibronectin and laminin. These antisera

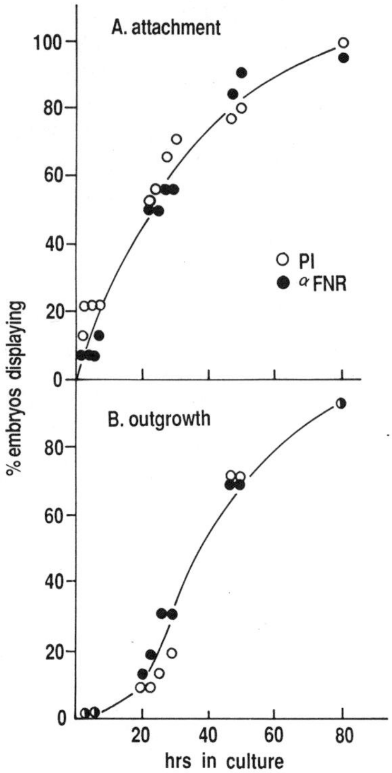

Figure 8.   Effects of Fibronectin Receptor Antibodies on Mouse Blastocyst Attachment and Outgrowth on Fibronectin.   Mouse blastocysts were obtained and cultured in serum-free medium on fibronectin-coated tissue culture surfaces as described previously (Farach et al., 1987).   The medium also contained 1 mg/ml of either preimmune rabbit IgG (○) or IgG from rabbits immunized with fibronectin receptors (●; Brown and Juliano, 1986).   At the indicated times, attachment and outgrowth formation were assessed on groups of at least 40 embryos.

also do not appear to react with blastocyst cell surfaces nor do they inhibit embryo attachment (Sutherland et al., 1988).   Taken together, these observations indicate that these extracellular matrix receptors do not function in the initial events of implantation, i.e., embryo attachment; however, these molecules may participate in secondary events such as outgrowth formation.

It has been shown previously that RGD-containing peptides as well as heparin specifically inhibit the rate of embryo attachment and outgrowth on fibronectin (Armant et al., 1986; Farach et al., 1987).   The concentratons required for these inhibitory effects are on the order of 1 mM for RGDS peptides and 4 - 40 μM for heparin.   Importantly, similar concentrations of related compounds, e.g., RGES peptides or chondroitin sulfate, have no such inhibitory effects on embryos (Armant et al., 1986; Farach et al., 1987; Carson et al., 1988).   Consequently, the inhibition does not appear to be due to toxic side effects of such compounds, but rather

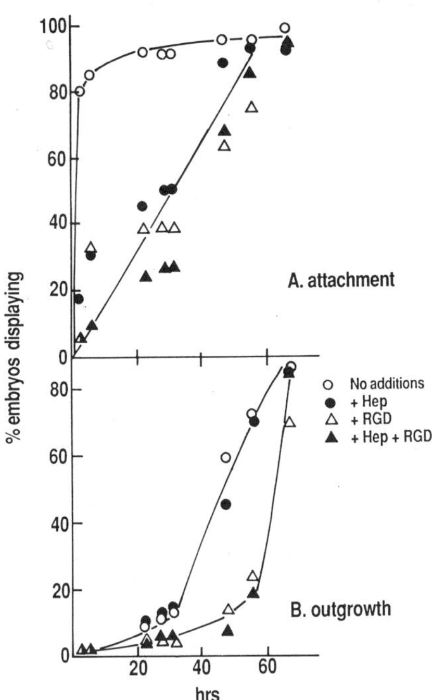

Figure 9. Recovery of "Synchronized" Embryos from Inhibition of Attachment and Outgrowth Formation on Fibronectin by Heparin and RGD-Containing Peptides. Mouse blastocysts were obtained and cultured on fibronectin-coated tissue culture surfaces in serum-free medium as previously described (Farach et al., 1987). Blastocysts were "synchronized" by retaining them in serum-free medium on a non-attachment-supporting surface (bovine serum albumin-coated) for 24 - 48 hours prior to transfer to the fibronectin-coated surface (Farach et al., 1987). In addition, the medium was supplemented with either no further additions (O), or 0.5 mg/ml heparin (●), or 0.5 mg/ml RGDS peptide (△) or 0.5 mg/ml heparin plus 0.5 mg/ml RGDS peptide (▲). At the times indicated, attachment and outgrowth formation were determined on groups of at least 35 embryos in each case.

specific interference with cell adhesion systems. The effects of heparin on embryo attachment and outgrowth are somewhat different than those of RGD-containing peptides. This point can be demonstrated if embryos are synchronized in vitro (see legend to Figure 9) so that they rapidly attach to an appropriate substrate (Farach et al., 1987). This procedure allows one to obtain a population of embryos of which virtually all are in an attachment-competent state, i.e., at a slightly later stage of development. For the studies described below fibronectin was chosen as an attachment substrate, since this molecule expresses a number of attachment domains with different specificities (Yamada, 1985). This property allows the study of several adhesion processes on a single, well-defined substrate. The initial attachment reaction to fibronectin is inhibited to a similar degree by RGD-containing peptides, heparin or the combination of the two (Figure 9A). The

observation that combining the two competitors does not produce a greater effect
than either one alone suggests that each agent affects this process maximally and
acts at the same level. It seems unlikely that both agents compete for the same sites
given the structural dissimilarities between these compounds; however, it is
possible that both agents bind to different sites of the same molecule and, thereby,
mutually interfere with the actions of the other. Since attachment proceeds, albeit at
a slower rate, in the presence of both agents, it appears that one of the other
attachment-promoting systems of the trophectoderm (see Table 3) supports this
process.

Although heparin substantially delays the rate of embryo attachment to
fibronectin under these conditions, it has virtually no effect on the rate of embryo
outgrowth (Figure 9B). In contrast, embryos incubated in the presence of RGD-
containing peptides also display a retarded initial rate of outgrowth formation. In
this case as well, embryos overcome the RGD-dependent inhibition of outgrowth
formation and "catch-up" to embryos not treated with inhibitors. The outgrowths
ultimately formed by embryos are morphologically similar in all cases (data not
shown). The fact that embryos recover from inhibition indicates that these
compounds are not toxic to the embryos. Furthermore, these observations indicate
that although some embryonic attachment systems may predominate at earlier
stages of in vitro development, additional systems arise later that can support
attachment or outgrowth. These additional systems appear to be distinct from those
expressed at earlier stages, since they are insensitive to the same agents that
interfere with "early" attachment/outgrowth. For example, as shown above,
heparin does not inhibit the rate of outgrowth formation on fibronectin by embryos
that have been "synchronized" by implantation delay in vitro; however, this agent
does inhibit outgrowth formation on this substrate by non-delayed embryos (Farach
et al., 1987). This apparently reflects the expression of additional, heparin-
insensitive adhesion systems by delayed embryos. These additional systems do
not seem to be expressed as early in development as the heparin-sensitive ones.

The suggestion that embryos sequentially express various adhesion
systems is drawn not only from observations of the kinetics of embryo
attachment/outgrowth on single, multidomain, adhesion-promoting molecules
like fibronectin, but also from studies of embryo attachment/outgrowth on different
adhesion-promoting molecules (Carson et al., 1988a). In these cases, it again
appears that embryos acquire the ability to attach/outgrow on particular matrices at
different stages of development in vitro. Assuming that the program of
development in vitro follows the program used in vivo, it seems that the
periimplantation stage of embryo development is characterized by the successive
acquisition of a number of adhesion and invasion-promoting systems. Table 3
presents a summary of the types of adhesion-promoting systems mouse embryos
have been shown to display in vitro along with the time at which these systems are
expressed relative to each other and during periimplantation development in vitro.

As discussed above, it is not clear which adhesion systems are expressed at
the apical cell surface of the uterine epithelium. Any adhesion system proposed to
operate at the embryo cell surface to initiate embryo attachment to the uterine wall
must be shown to have a complementary receptor system on the opposing surface of
the uterine lumenal epithelium. Furthermore, this uterine system must display

Table 3

Expression of Cell Adhesion Systems by Mouse Trophectoderm
During the Periimplantation Stages

Cell Adhesion System[b]	Stage of Periimplantation Development [a]		
	Early	Middle	Late
For Uterine Epithelium	+	+	+
HSPGs	+	+	+
HA receptors	+	+	+
Collagen type VI receptors	+	+	+
RGD receptor-1	+	+	+
RGD receptor-2	-	+	+
Other collagen receptors	-	-	+

[a]The staging refers to the sequence of in vitro development during which these adhesion systems are expressed. Early attachment typically occurs within 15 - 25 hours of embryo hatching from zonae pellucidae in vitro. Middle is defined as occurring later than early systems (typically 30 - 45 hours after hatching from zonae pellucida in vitro). Late is defined as occurring after the middle systems are expressed (typically 70 - 90 hours after hatching from zonae pellucidae in vitro). It is assumed for the sake of simplicity that each system continues to be expressed once it has appeared, although, in most cases, this has not been rigorously demonstrated.

[b]The abbreviations are: HSPG, heparan sulfate proteoglycans; RGD receptor 1, molecules that recognize Arg-Gly-Asp sequences and are not related immunologically to fibronectin receptors (Brown and Juliano, 1986; Sutherland et al., 1988); RGD receptor-2, molecules that recognize Arg-Gly-Asp sequences and are related immunologically to fibronectin receptors (Brown and Juliano, 1986; Sutherland et al., 1988); HA receptors, molecules that bind hyaluronate; Collagen type VI receptors, molecules that bind collagen type VI, but not collagen types I-V; Other collagen receptors, molecules that recognize collagens by mechanisms not inhibitable specifically with Arg-Gly-Asp containing peptides or heparin. Studies comparing the times of in vitro acquisition of these various adhesion systems are reported in Farach et al. (1987) and Carson et al. (1988).

certain properties in order to explain the biology of implantation. It is expected that uterine receptors for embryos would be: 1) preferentially available at the antimesometrial side of the uterus where implantation almost invariably occurs (Schlafke and Enders, 1975; Sherman and Wudl, 1976); 2) not expressed functionally in the uterine lumen throughout most of the estrous cycle (Schlafke and Enders, 1975; Sherman and Wudl, 1976); and 3) tightly regulated by steroid hormones in terms of their functional expression (Psychoyos, 1986).

Models Of Embryo Implantation

Figure 10 depicts three models of receptor regulation that might explain conversion from a non-receptive to a receptive uterine state. In all three cases, it is considered that it is the actions of steroid hormones that bring about this conversion. For the purposes of this discussion, production of uterine receptivity is defined as the expression of functional embryo receptors at the apical cell surface of the epithelial cells lining the uterine lumen. It should be noted that non-receptive surfaces may express a low or subthreshhold level of receptors, since cell adhesion processes, in general, seem to require a critical concentration of such ligands (Hoffman and Edelman, 1983). The first model is one of simple induction whereby steroid hormones stimulate the de novo synthesis of receptors. A nonreceptive state would be regained by attenuation of this synthesis and receptor turnover. As discussed above, the synthesis of certain glycoproteins of the uterine cell surface is markedly stimulated by steroid hormones. Moreover, these molecules appear to turn over rapidly enough to fit this scheme. However, other lines of evidence do not support this model. First, it appears that transcription in the uterus decreases markedly under the conditions that produce a receptive uterine state (Glasser and McCormack, 1979). Second, inhibitors of RNA or protein synthesis fail to inhibit conversion from a nonreceptive to a receptive uterine state (Finn and Bredl, 1973; Tarachand et al., 1982). Consequently, it seems that de novo receptor synthesis is an unlikely means of regulation of uterine receptivity.

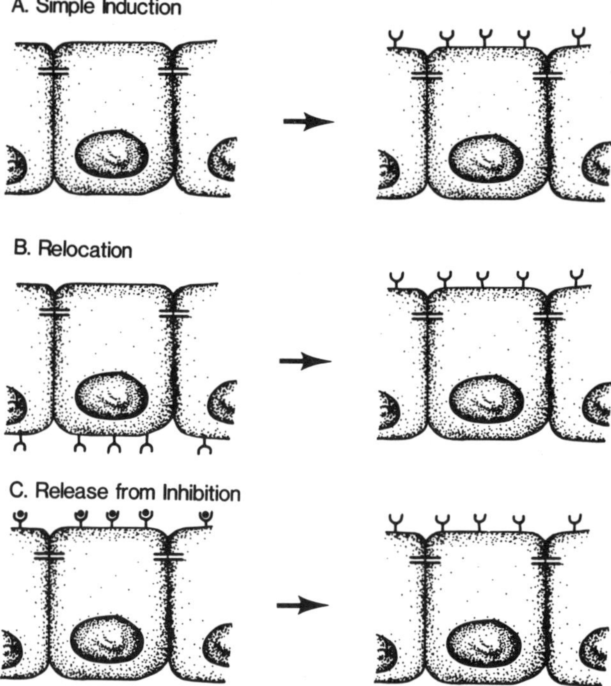

Figure 10. Models of Receptor Regulation. The apical surface is directed upward in the diagrams. For detailed explanation see text. Symbols: Y, embryo receptors; ♥ endogenous inhibitors of embryo receptors.

The second model does not require de novo synthesis to attain a receptive state. Rather, the situation may be that the receptors are expressed constitutively by the epithelial cells, but are not displayed at the apical cell surface. In the cartoon, the receptors in the nonreceptive case are situated at the basal cell surface; alternatively, these molecules also could be situated laterally and/or stored intracellularly. Under the appropriate hormonal environment, receptors would be mobilized to the apical cell surface. A nonreceptive state would reflect redistribution of these molecules to a nonapical site. This model takes into account the knowledge that the uterine lumenal epithelial cell is a polarized cell type and, as such, expresses distinct apical and basolateral cell surface compositions (Simons and Fuller, 1985). Recent evidence indicates that certain aspects of the secretory pathway may be influenced by steroid hormones (Firestone et al., 1987). Furthermore, Parr (1980, 1982) has presented morphological evidence that intracellular vesicular trafficking is augmented in uterine epithelial cells under certain steroid hormone influences. It will be necessary to use specific probes to cell surface components of the uterine epithelium to determine if redistribution of these molecules actually occurs during production of a receptive uterine state. In addition, the recent development of techniques for the culture of polarized uterine epithelial cells (Glasser et al., 1988) should provide an excellent opportunity to study the ability of these cells to mobilize and redistribute cell surface components.

The third model proposes that embryo receptors are physically present at the apical cell surface at all times; however, they are inhibited functionally in the non-receptive case. This might be accomplished by a covalent modification that modulates the activity of the receptor, e.g., phosphorylation. Another possibility is that the receptor is active but is "blocked" by an endogenous inhibitor. As discussed above, HS binding proteins not only can bind embryos, but also can bind HS expressed by these cells (O. Wilson, personal communication). Consequently, endogenous HSPGs may act as repressors of embryo binding to HS binding proteins. It is interesting that proteoglycan turnover by these cells is stimulated not only by estrogen, but also on the day of embryo implantation (Morris et al., 1988a; 1988b). This process would tend to reduce cell surface HS expression. Since HS binding proteins have long half lives compared to HSPGs, these events should result in increased expression of unoccupied HS binding proteins. In any event, it must be considered that maintenance of a nonreceptive uterine state may require sustained expression of an inhibitor. A transient, hormonally-mediated interruption in the expression of this inhibitor would thereby generate a receptive state. Resumed synthesis of the inhibitor would regenerate a nonreceptive state.

A model of mouse embryo implantation that incorporates our working hypotheses as well as a number of suggestions and observations of others (Schlafke and Enders, 1975; Sherman and Wudl, 1976) is presented in Figure 11. Following hatching from the zona pellucida, the embryo acquires the ability to attach to the uterine wall. It is believed that expression of HSPGs by the embryo is one important aspect of acquisition of attachment competence. At approximately the same time, the uterus expresses complementary receptor molecules at the apical surfaces of the lumenal epithelium. In the proposed model, some of these receptors would be those for HSPGs. A period follows during which the embryo remains attached without penetrating the epithelial cell layer. During this interval it appears that the decidual cell response is triggered. Given that the trigger for this response is generated prior to penetration of the epithelial cell layer and that it can be produced

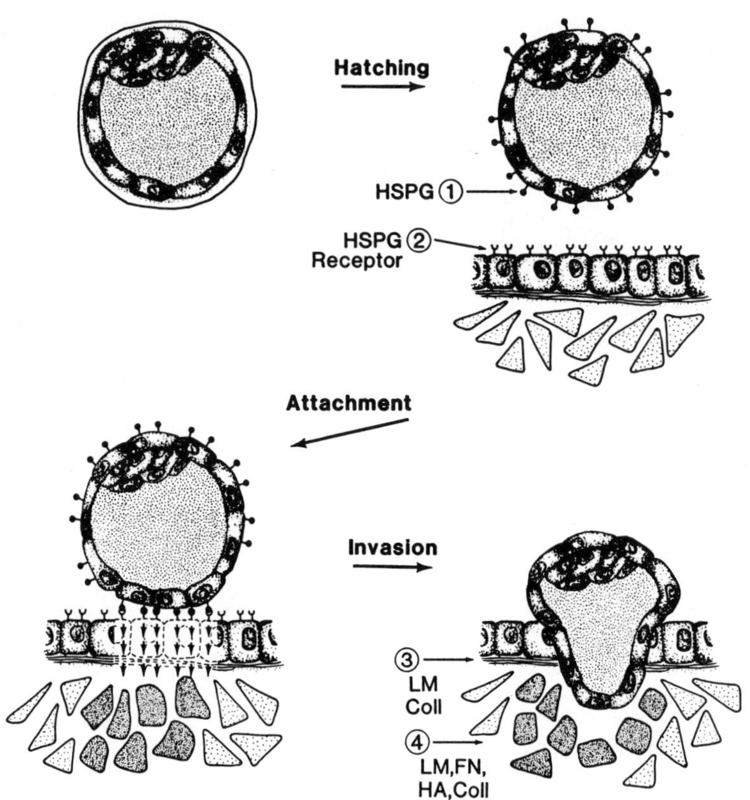

Figure 11. Model of Embryo Implantation. Upon hatching from the zona pellucida, the embryo expresses cell surface components that mediate attachment to the uterine epithelium (♦) designated as HSPG in this model (1). Coordinately, the uterine epithelium expresses corresponding embryo receptors (Ψ) designated as HSPG Receptor in this model (2). Following attachment, the initial events of the decidual response are triggered by signals emanating from the site of embryo attachment. These signals lead to localized differentiation of underlying stromal cells (shaded). Once the embryo breaches the epithelium, it encounters a number of molecules in the basement membrane (3) and in the stromal tissue (4) that can support further invasion. Symbols: Coll, collagens; FN, fibronectin; HA, hyaluronate; HSPG, heparan sulfate proteoglycans; HSPG receptor, heparan sulfate receptor; LM, laminin.

in the absence of embryos (Finn, 1986), it seems likely that the intervening epithelial cells transmit signals to the stromal cells. Upon penetration of the epithelial cell layer, the embryo encounters a host of molecules that can support its further invasive potential. These molecules include hyaluronate, laminin, fibronectin, and several types of collagen. Subsequently, embryo invasion is limited to the stromal tissue of the implantation site. The molecular basis by which the uterus exerts its control over the invasive potential of the embryo at these later phases of invasion represents another interesting aspect of early mammalian development.

SUMMARY

The cell surface of mouse uterine epithelial cells is characterized by the presence of large molecular weight, heavily glycosylated molecules, notably LAG-bearing proteins and HSPGs. In the case of the LAG-bearing proteins, the oligosaccharide chains are capable of protecting large regions of the protein core from proteolytic attack. Both types of glycoconjugate are metabolized rapidly relative to many other cell surface proteins. This property coupled with the marked effects of steroid hormones on uterine glycoprotein assembly is likely to contribute to the ability of these cells to modify rapidly the composition of cell surface glycoconjugates under certain steroid hormone influences. These cells also express complementary receptors for LAGs and HSPGs in the form of galactosyltransferase and HS binding proteins. At least in the case of HSPGs and HS binding proteins, there is evidence for a role in embryo-uterine epithelial cell interactions. In contrast, the LAG-galactosyltransferase system appears to function in adhesion among uterine epithelial cells, although it is not clear if this system is involved in embryo interactions as well. The trophectodermal cell surface is characterized by the developmental acquisition of multiple adhesion systems. Expression of these systems in vitro roughly parallels the point at which embryos would encounter complementary uterine molecules in vivo. Collectively, it appears that glycoconjugate-dependent cell adhesion systems play important roles in interactions that take place between cells of the embryo and uterus.

ACKNOWLEDGMENTS

The authors are grateful for the critical readings and helpful discussions with Drs. M.C. Farach, S.R. Glasser, A. Jacobs, N. Raboudi, Mr. D. Farrar, Mrs. JoAnne Julian, and Mrs. J.-P. Tang. The excellent typing of Ms. Ellen Madson is most appreciated. We are particularly indebted to Drs. Pat Brown and Steffen Gay for their gifts of anti-fibronectin receptor and interstitial matrices, respectively. This work was supported by grants from the March of Dimes (1-958) and American Cancer Society (BC-503) awarded to D.D.C. and National Institutes of Health training grant (HD 07325) supporting O.W.

REFERENCES

Anderson, T.L., Olson, G.E., and Hoffman, L. (1986) Stage-specific alterations in the apical membrane glycoproteins of endometrial epithelial cells related to implantation in rabbits. *Biol. Reprod.* 34, 701-720.

Anderson, T.L. and Hoffman, L.H. (1984) Alterations in epithelial glycocalyx of rabbit uteri during early pseudopregnancy and pregnancy, and following ovariectomy. *Am. J. Anat.* 171, 321-334.

Armant, D.R., Kaplan, H.A., Mover, H., and Lennarz, W.J. (1986) The effect of hexapeptides on attachment and outgrowth of mouse embryos in vitro: Evidence for the involvement of the cell recognition tripeptide Arg-Gly-Asp. *Proc. Natl. Acad. Sci. USA* 83, 6751-6755.

Bayna, E.M., Shaper, J., and Shur, B. (1988) Temporally specific involvement of cell surface β-1,4 galactosyltransferase during mouse embryo morula compaction. *Cell* 53, 145-157.

Bergstrom, S.I. (1978) Experimentally delayed implantation. In: *Methods In Mammalian Reproduction*, (ed.) J.C. Daniel, Jr., Academic Press, NY, pp. 419-435.

Brown, P.J. and Juliano, R.L. (1986) Expression and function of a putative cell surface receptor for fibronectin in hamster and human cell lines. *J. Cell Biol.* 103, 1595-1603.

Carson, D.D., Dutt, A., and Tang, J.-P. (1987a) Glycoconjugate synthesis during early pregnancy: Hyaluronate synthesis and function. *Dev. Biol.* 120, 228-235.

Carson, D.D., Tang, J.-P., and Gay, S. (1988a) Collagens support embryo attachment and outgrowth in vitro: Effects of the Arg-Gly-Asp sequence. *Dev. Biol.* 127, 368-375.

Carson, D.D., Tang, J.-P., and Hu, G. (1987b) Estrogen influences dolichyl phosphate distribution among glycolipid pools in mouse uteri. *Biochem.* 26, 1598-1606.

Carson, D.D., Tang, J.-P., Julian, J., and Glasser, S.R. (1988b) Vectorial secretion of proteoglycans by polarized rat uterine epithelial cells. *J. Cell Biol.*, 107, 2425-2435.

Cowell, T.P. (1969) Implantation and development of mouse eggs transferred to the uteri of non-progestational mice. *J. Reprod. Fertil.* 19, 239-245.

Damjanov, I., Damjanov, A., and Damsky, C.H. (1986) Developmentally regulated expression of the cell-cell adhesion glycoprotein cell-CAM 120/80 in periimplantation mouse embryos and extraembryonic membranes. *Dev. Biol.* 116, 194-202.

Dziadek, M., Fujiwara, S., Paulsson, M., and Timpl, R. (1985) Immunological characterization of basement membrane types of heparan sulfate proteoglycan. *EMBO J.* 4, 905-912.

Dziadek, M. and Timpl, R. (1985) Expression of nidogen and laminin in basement membranes during mouse embryogenesis and in teratocarcinoma cells. *Dev. Biol.* 111, 372-382.

Dutt, A. and Carson, D.D. (1990) Lactosaminoglycan assembly, cell surface expression and release by mouse uterine epithelial cells. *J. Biol. Chem.* 265, in press.

Dutt, A., Tang, J.-P., and Carson, D.D. (1987) Lactosaminoglycans are involved in uterine epithelial cell adhesion in vitro. *Dev. Biol.* 119, 27-37.

Dutt, A., Tang, J.-P., and Carson, D.D. (1988) Estrogen preferentially stimulates lactosaminoglycan-containing oligosaccharide synthesis in mouse uteri. *J. Biol. Chem.* 263, 2270-2279.

Dutt, A., Tang, J.-P., Chu, R., Stewart, S., and Carson, D.D. (1986a) Isolation and characterization of a lactosaminoglycan-glycoprotein from mouse uteri. *J. Cell Biol.* 103, 382a.

Dutt, A., Tang, J.-P., Welply, J.K., and Carson, D.D. (1986b) Regulation of N-linked glycoprotein assembly in uteri by steroid hormones. *Endocrinol.* 118, 661-673.

Edelman, G. (1983) Cell adhesion molecules. *Science* 219, 450-457.

Edelman, G. (1985) Expression of cell adhesion molecules during embryogenesis and regeneration. *Exp. Cell Res.* 161, 1-16.

Edge, A.S.B., Faltynek, C.R., Hof, L., Reichert, L.E., Jr., and Weber, P. (1981) Deglycosylation of glycoproteins by trifluoremethane-sulfonic acid. *Anal. Biochem.* 118, 131-137.

Enders, A.C. and Schlafke, S. (1974) Surface coats of the mouse blastocyst and uterus during the preimplantation period. *Anat. Rec.* 180, 31-46.

Farach, M.C., Tang, J.-P., Decker, G.L., and Carson, D.D. (1987) Heparin/heparan sulfate is involved in attachment and spreading of mouse embryos in vitro. *Dev. Biol.* 123, 401-410.

Farach, M.C., Tang, J.-P., Decker, G.L., and Carson, D.D. (1988) Differential effects of p-nitrophenyl-D-xylosides on mouse blastocysts and uterine epithelial cells. *Biol. Reprod.* 39, 443-455.

Finn, C.A. (1986) Implantation, menstruation and inflammation. *Biol. Rev.* 61, 313-328.

Finn, C.A. and Bredl, J.C.S. (1973) Studies on the development of the implantation reaction in the mouse uterus: Influence of actinomycin D. *J. Reprod. Fertil.* 34, 247-253.

Firestone, G.L., John, N.J., Haffar, O.K., and Cook, P.W. (1987) Genetic evidence that the steroid-regulated trafficking of cell surface glycoproteins in rat hepatoma cells is mediated by glucocorticoid-inducible cellular components. *J. Cell. Biochem.* 35, 271-284.

Frazier, W. and Glaser, L. (1979) Surface components and cell recognition. *Ann. Rev Biochem.* 48, 491-523.

Fukuda, M. (1985) Cell surface glycoconjugates as onco-differentiation markers in hematopoietic cells. *Biochim. Biophys. Acta* 780, 119-150.

Glass, R.H., Spindle, A.I., and Pedersen, R.A. (1979) Mouse embryo attachment to substratum and interaction of trophoblast with cultured cells. *J. Exp. Zool.* 208, 327-336.

Glasser, S.R., Julian, J., Decker, G.L., Tang, J.-P., and Carson, D.D. (1988) Development of morphological and functional polarity in primary cultures of immature rat uterine epithelial cells. *J. Cell Biol.* 107, 2409-2423.

Glasser, S.R., Lampelo, S., Munir, M.I., and Julian, J. (1987) Expression of desmin, laminin and fibronectin during in situ differentiation (decidualization) of rat uterine stromal cells. *Different.* 35, 132-142.

Glasser, S.R. and McCormack, S.A. (1979) Estrogen-modulated uterine gene transcription in relation to decidualization. *Endocrinol.* 104, 1112-1118.

Goldstein, J.L., Brown, M.S., Anderson, R.G.W., Russell, D.W., and Schneider, W.J. (1985) Receptor-mediated endocytosis: Concepts emerging from the LDL receptor system. *Ann. Rev. Cell Biol.* 1, 1-39.

Grinnell, F., Head, J.R., and Hoffpauir, J. (1982) Fibronectin and cell shape in vivo: Studies on the endometrium during pregnancy. *J. Cell Biol.* 94, 597-606.

Handin, R.I. and Cohen, H.J. (1976) Purification and binding properties of human platelet factor four. *J. Biol. Chem.* 251, 4272-4282.

Hascall, V.C. and Hascall, G.K. (1985) Proteoglycans. In: *Cell Biology Of The Extracellular Matrix*, (ed.) E.D. Hay, Plenum Press, pp. 39-63.

Hirani, S., Bernasconi, R.J., and Rasmussen, J.R. (1987) Use of N-glycanase to release asparagine-linked oligosaccharides for structural analyses. *Anal. Biochem.* 162, 485-492.

Hoffman, S. and Edelman, G.M. (1983) Kinetics of homophilic binding by embryonic and adult forms of the neural cell adhesion molecule. *Proc. Natl. Acad. Sci. USA* 80, 5762-5766.

Jalkanen, M., Rapraeger, A., Saunders, S., and Bernfield, M. (1988) Cell surface proteoglycan of mouse mammary epithelial cells is shed by cleavage of its matrix-binding ectodomain from its membrane-associated domain. *J. Cell Biol.* 105, 3087-3096.

Kleinman, H.K., McGarvey, M.L., Hassell, J.R., Star, V.L., Cannon, F.B., Laurie, G.W., and Martin, G.R. (1986) Basement membrane complexes with biological activity. *Biochem.* 25, 312-318.

Laemmli, U.K. (1970) Cleavage of structural proteins during the assembly of the head of bacteriophage T4. *Nature* 227, 680-685.

Leivo, I., Vaheri, A., Timpl, R., and Wartiovaara, J. (1980) Appearance and distribution of collagens and laminin in the early mouse embryo. *Dev. Biol.* 76, 100-114.

Marticorena, P., Hogan, B., DiMeo, A., Artzt, K., and Bennett, D. (1983) Carbohydrate changes in pre- and peri-implantaton mouse embryos as detected by a monoclonal antibody. *Cell Different.* 12, 1-10.

Minamoto, T., Arai, K., Hirakawa, S., and Nagai, Y. (1987) Immunohistochemical studies on collagen types in the uterine cervix in pregnant and non-pregnant states. *Am. J. Obstet. Gynecol.* 156, 138-144.

Morris, J.E., Potter, S.W., and Gaza-Bulseco, G. (1988a) Estradiol induces an accumulation of free heparan sulfate glycosaminoglycan chains in uterine epithelium. *Endocrinol.* 122, 242-253.

Morris, J.E., Potter, S.W., and Gaza-Bulseco, G. (1988b) Estradiol-stimulated turnover of heparan sulfate proteoglycan in mouse uterine epithelium. *J. Biol. Chem.* 263, 4712-4718.

Munakata, H., Isemura, M., and Yosizawa, Z. (1985) Effects of female hormones on the activity of 3'-phosphoadenylylsulphate: desulphated heparan sulphate sulphotransferase in the endometrium of rabbit uterus. *Int. J. Biochem.* 17, 1077-1083.

Muramatsu, T. (1988) Developmentally regulated expression of cell surface carbohydrates during mouse embryogenesis. *J. Cell. Biochem.* 36, 1-14.

Nelson, D.M., Enders, A.C., and King, B.F. (1977) Galactosyltransferase activity of the microvillus surface of human placental syncytial trophoblast. *Gynecol. Invest.* 8, 267-281.

Nelson, J.D., Jato-Rodriguez, J.J., and Mookerjea, S. (1975) Effect of ovarian hormones on glycosyltransferase activities in the endometrium of ovariectomized rats. *Arch. Biochem. Biophys.* 169, 181-191.

Parr, M. (1982) Apical vesicles in the rat uterine epithelium during early pregnancy: A morphometric study. *Biol. Reprod.* 26, 915-924.

Parr, M.B. (1980) Endocytosis at the basal and lateral membranes of rat uterine epithelial cells during early pregnancy. *J. Reprod. Fertil.* 60, 95-99.

Psychoyos, A. (1986) Uterine receptivity for nidation. *Ann. N.Y. Acad. Sci.* 476, 36-42.

Richa, J., Damsky, C.H., Buck, C.A., Knowles, B.B., and Solter, D. (1985) Cell surface glycoproteins mediate compaction, trophoblast attachment, and endoderm formation during early mouse development. *Dev. Biol.* 108, 513-521.

Roden, L. (1980) Structure and metabolism of connective tissue proteoglycans. In: *The Biochemistry Of Glycoproteins and Proteoglycans*, (ed.), W.J. Lennarz, Plenum Press, NY, pp. 167-371.

Rubin, G.L., Peterson, H.B., Dorfman, S.F., Layde, P.M., Maze, J.M., Ory, H.W., and Cates, W. (1983) Ectopic pregnancy in the United States - 1970 through 1978. *JAMA* 249, 1725-1729.

Ruoslahti, E. and Pierschbacher, M.D. (1987) New perspectives in cell adhesion: RGD and integrins. *Science* 238, 491-497.

Sato, M., Muramatsu, T., and Berger, E.G. (1984) Immunological detection of cell surface galactosyltransferase in preimplantation mouse embryos. *Dev. Biol.* 102, 514-518.

Sato, M. and Muramatsu, T. (1986) Oncodevelopmental carbohydrate antigens: Distribution of ECMA 2 and 3 antigens in embryonic and adult tissues of the mouse and in teratocarcinomas. *J. Reprod. Immunol.* 9, 123-135.

Schlafke, S. and Enders, A.C. (1975) Cellular basis of interaction between trophoblast and uterus at implantation. *Biol. Reprod.* 12, 41-65.

Sherman, M.I. (1978) Implantation of mouse blastocysts in vitro. In: *Methods In Mammalian Reproduction*, (ed.) J.C. Daniel, Academic Press, NY, pp. 247-257.

Sherman, M.I., Gay, R., Gay, S., and Miller, E.J. (1980) Association of collagen with preimplantation and peri-implantation mouse embryos. *Dev. Biol.* 74, 470-478.

Sherman, M.I. and Wudl, L.R. (1976) The implanting mouse blastocyst. In: *The Cell Surface In Animal Development*, (eds.) G. Poste and G.R. Nicolson, North Holland Publishers: Amsterdam, pp. 81-125.

Shur, B.D. (1984) The receptor function of galactosyltransferase during cellular interactions. *Mol. Cell. Biochem.* 61, 143-158.

Simons, K. and Fuller, S.D. (1985) Cell surface polarity in epithelia. *Ann.Rev. Cell Biol.* 1, 243-288.

Sutherland, A.E., Calarco, P.G., and Damsky, C.H. (1988) Expression and function of cell surface extracellular matrix receptors in mouse blastocyst attachment and outgrowth. *J. Cell Biol.* 106, 1331-1348.

Svalander, P.C., Odin, P., Nilsson, B.O., and Obrink, B. (1987) Trophectoderm surface expression of the cell adhesion molecule cell-CAM 105 on rat blastocysts. *Develop.* 100, 653-660.

Svalander, P.C., Odin, P., Nilsson, B.O., and Obrink, B. (1989) Expression of cell-CAM 105 in the apical surface of rat uterine epithelium is controlled by ovarian steroid hormones. *Develop.*, in press.

Tang, J.-P., Julian, J., Glasser, S.R., and Carson, D.D. (1987) Heparan sulfate proteoglycan synthesis and metabolism by mouse uterine epithelial cells in vitro. *J. Biol. Chem.* 262, 12832-12842.

Tarachand, U., Sivabalan, R., and Eapen, J. (1982) Effect of cycloheximide on deciduoma morphogenesis and implantaton in rats. *Int. J. Fertil.* 27, 238-

Turley, E.A. (1984) Proteoglycans and cell adhesion. *Canc. Metastasis Rev.* 3, 325-339.

VanBlerkom, J. and Chavez, D.J. (1981) Morphodynamics of outgrowths of mouse trophoblast in the presence and absence of a monolayer of uterine epithelium. *Am. J. Anat.* 162, 143-155.

Vestweber, D., Gossler, A., Boller, K., and Kemler, R. (1987) Expression and distribution of cell adhesion molecule uvomorulin in mouse preimplantation embryos. *Dev. Biol.* 124, 451-456.

Vestweber, D. and Kemler, R. (1984) Rabbit antiserum against a purified surface glycoprotein decompacts mouse preimplantation embryos and reacts with specific adult tissues. *Expt. Cell Res.* 152, 169-178.

Wartiovaara, J., Leivo, I., and Vaheri, A. (1979) Expression of the cell surface-associated glycoprotein, fibronectin, in the early mouse embryo. *Dev. Biol.* 69, 247-257.

Wu, T.-C., Wan, Y.-J., Chung, A.E., and Damjanov, I. (1983) Immunohistochemical localization of entactin and laminin in mouse embryos and fetuses. *Dev. Biol.* 100, 496-505.

Yamada, K.M. (1985) Fibronectin and other structural proteins. In: *Cell Biology Of The Extracellular Matrix*, (ed.) E.D. Hay, Plenum Press, NY, pp. 95-114.

Yoshida-Noro, C., Suzuki, N., and Takeichi, M. (1984) Molecular nature of the calcium-dependent cell-cell adhesion system in mouse teratocarcinoma and embryonic cell studied with a monoclonal antibody. *Dev. Biol.* 101, 19-27.

UTERINE RECEPTIVITY TO IMPLANTATION IN THE RABBIT: EVIDENCE FOR A 42 kDa GLYCOPROTEIN AS A MARKER OF RECEPTIVITY

Loren H. Hoffman, Virginia P. Winfrey,
Ted L. Anderson, and Gary E. Olson

Department of Cell Biology
Vanderbilt University School of Medicine
Nashville, Tennessee 37232, USA

INTRODUCTION

The requirement for synchronous development of the mammalian embryo and uterus was evidenced by the failure of asynchronous embryo transfer to result in implantation except at the time of normal receptivity for each species (Chang, 1950; Finn, 1977). Evidence linking embryo-uterine synchrony to hormonal balance in the rabbit was based on steroid hormone-induced delay (Beier, 1976) or advancement (Adams, 1971; McCarthy et al., 1977) of uterine preparation relative to embryonic development; ovariectomy was directly lethal to contained blastocysts (Adams, 1958). That the epithelium constituted a major component of the "hostile uterine environment" resulting from inadequate hormone preparation was argued by Nilsson (1967) and Potts and Psychoyos (1967). Further support for this view was reported by Cowell (1969) who obtained "implantation" of mouse blastocysts in otherwise unreceptive uteri only after scraping of the epithelium to expose the underlying stroma. The uterine epithelium may be involved in a variety of implantation-related processes (absorption, secretion, barrier to trophoblast, message transduction to underlying tissues; Martin, 1980).

While other components of the uterus (myometrium, vasculature, stromal-decidual cells, glands) obviously play important roles in the implantation process, it has been proposed that the epithelial surface, being the initial site at which embryo attachment must occur, is likely the primary barrier to attachment/implantation in the non-receptive phase. It is essential that the surface factors associated with the "receptive" uterus be recognized in order to design appropriate experimental models (e.g., in vitro systems) for mechanistic studies and, ultimately, to understand fully the implantation process. A number of properties for the surface membrane of rabbit uterine epithelial cells have been reported to change with the acquisition of receptivity. Such changes include alterations in surface charge and lectin affinities (Nalbach and Denker, 1983; Anderson and Hoffman, 1984) and in membrane protein composition (Ricketts et al., 1984; Lampelo et al., 1985; Anderson et al., 1986a). It is reported here that one such membrane constituent from uterine epithelium of rabbits during the receptive phase, a glycoprotein of M_r 42,000 has been isolated (Anderson et al., 1986a). Antisera were prepared against this glycoprotein and used to follow its appearance during the pre- and peri-implantation periods and to determine its specific localization in endometrial cells.

MATERIALS AND METHODS

Animals And Tissue Processing

New Zealand White rabbits were used for the study. Pseudopregnancy was induced by intravenous injection of 50 I.U. human chorionic gonadotropin (hCG, Ayerst, New York, NY). For early stages of pregnancy, females were mated twice followed by an injection of hCG to ensure ovulation. The time of mating or of hCG injection was designated as the beginning of day 0 of pregnancy or pseudopregnancy. Tissue samples were obtained at estrus, and at days 2, 4, and 7 of pseudopregnancy. Uteri from pregnant females were obtained on days 4, 6, 6 1/2, 6 3/4, 7, and 8. Animals were euthanatized by administration of an overdose of anesthetic (Nembutal, Abbott Laboratories, N. Chicago, IL) and uterine horns removed and flushed with 15 - 20 ml of phosphate-buffered saline (PBS, pH 7.3) prior to removing samples for fixation. For 7 day pregnant females, implantation sites were marked prior to flushing and the uteri cut into separate implant and non-implant regions. The non-uterine tissues collected for light microscopic immunolocalization experiments included oviduct, jejunum, and liver. Tissue was preserved in either Bouin's solution or in buffered picric acid-formaldehyde (Zamboni and Martino, 1967), dehydrated through an alcohol series and embedded in paraffin wax. Sections 6 μm in thickness were obtained and used for immunolocalization studies.

Isolation Of 42 kDa Glycoprotein And Production Of Antiserum

Extracts of luminal epithelial membrane were prepared from estrous or pseudopregnant females using a detergent extraction procedure modified slightly from that previously employed (Anderson et al., 1986a). Briefly, uteri were excised, flushed with warm PBS to minimize contamination due to secretory material, and then infused with the detergent solution (0.1% Triton X-100 in 140 mM NaCl, 25 mM Tris-HCl, pH 7.5, containing 0.5 mM phenylmethyl-sulfonyl fluoride, 5 mM benzamidine, 20 μg/ml pepstatin A, and 20 μg/ml leupeptin). Pepstatin A and leupeptin were obtained from Calbiochem (LaJolla, CA), the other reagents were from Sigma Chemical Company (St. Louis, MO). Horns were ligated and incubated with gentle shaking at 37°C for 15 minutes. Pooled extracts from the four uterine horns were centrifuged at 1000 X g for 15 minutes to remove particulates, then desalted on Sephadex G-25 (Pharmacia Inc., Piscataway, NJ) and collected in imidazole buffer (20 mM, pH 6.5) containing 0.02% Triton X-100. The desalted extract was placed on a DEAE-Sephacel column (LKB Instruments, Gaithersburg, MD), washed with imidazole buffer and sequentially eluted with 0.1, 0.3, and 0.5 M NaCl in the imidazole buffer. The appropriate fractions, as determined by gel electrophoresis (see below) were equilibrated with lectin buffer (25 mM Tris-HCl, 140 mM NaCl, 1 mM $MgCl_2$, 1 mM $CaCl_2$ and 0.02% Triton X-100, pH 7.2) by chromatography on Sephadex G-25. The extract was placed on a *Ricinus communis* (RCA-I)-agarose affinity column (Vector Laboratories, Burlingame, CA), the column was exhaustively washed with lectin buffer, and then the specifically bound material was eluted with lectin buffer containing 0.2 M D-galactose. After final desalting on PD-10 columns (Pharmacia), aliquots were removed for protein assay (Bradford, 1976) and the remainder frozen in liquid nitrogen and lyophilized.

Aliquots of detergent extracts and column fractions were suspended in sample buffer, boiled for 5 minutes, and analyzed by SDS-polyacrylamide gel electrophoresis (SDS-PAGE, Laemmli, 1970). Equivalent protein loadings along with protein standards (Bio-Rad Laboratories, Richmond, CA) were electrophoresed onto 10 - 15% gradient gels (Boehringer-Mannheim Biochemicals, Indianapolis,IN) with 4% acrylamide stacking gels. Polypeptides were visualized by staining with Coomassie Blue or by the silver staining procedure of Wray et al. (1981). Western blot analysis was utilized to confirm appropriate lectin binding of polypeptides and to test the specificity of antiserum preparations. For this, polypeptides were electrophoretically transferred from SDS-PAGE gels onto nitrocellulose sheets by the method of Towbin et al. (1979). The transfer buffer consisted of 25 mM Tris-HCl, 192 mM glycine, and 20% methanol, pH 8.3. Detection of lectin-binding components on blots utilized biotin-conjugated lectins (RCA-I, succinylated wheat germ agglutinin [sWGA]; Vector), followed by streptavidin-peroxidase (Zymed Laboratories Inc., San Francisco, CA) and a diaminobenzidine solution containing hydrogen peroxide. Controls included saccharide inhibitors at 0.1 M concentration (D-galactose for RCA-I; N-acetyl-D-glucosamine for sWGA). Lectin reagents were prepared in PBS and the nitrocellulose blots were blocked using 0.5% bovine serum albumin in PBS for 30 minutes prior to incubation with the lectin reagents.

Antigen-antibody interaction on blots was conducted using a similar procedure. Biotin-conjugated goat anti-guinea pig IgG (Kirkegaard and Perry Laboratories, Inc., Gaithersburg, MD) was utilized following incubation with diluted guinea pig sera, and streptavidin-peroxidase/DAB-H_2O_2 was employed as above. An equivalent dilution of pre-immune guinea pig serum was used as a control and PBS solutions and the primary antiserum were prepared with 1% normal goat serum.

For generation of antisera to the 42 kDa glycoprotein, the pooled membrane extract from two 7 day pseudopregnant rabbits was fractionated on DEAE-Sephacel followed by RCA-I affinity separation as detailed above. The total specifically-eluted fraction from the RCA-agarose column was subjected to SDS-PAGE electrophoresis by dividing it into 8 aliquots of 50 µg protein and running separate lanes on preparative gels (1.5 mm in thickness). After staining the gels with Coomassie Blue, the individual bands representing the 42 kDa glycoprotein were excised, and an appropriate aliquot reduced to a slurry by homogenization (glass/glass) in Freund's complete adjuvant. Densitometer scanning (LKB Ultroscan XL) of the gel revealed that approximately 12% of the total protein per lane was in the 42 kDa band; the pooled bands contained 48 - 50 µg of this glycoprotein. The slurry was employed as immunogen for two guinea pigs. Animals were anesthetized using intraperitoneal methoxyfluorane injections and a small amount of blood obtained for pre-immune serum by cardiac puncture. Primary immunization was carried out using an estimated 12 µg protein/guinea pig. Injections were given in the Freund's complete adjuvant into multiple dermal sites and a single intramuscular site. Booster injections, estimated at 6 µg protein per female and prepared by homogenization of acrylamide bands in Freund's incomplete adjuvant, were administered at 22 and 36 days. Blood was collected by cardiac puncture 8 days after the second booster injection and used to prepare serum.

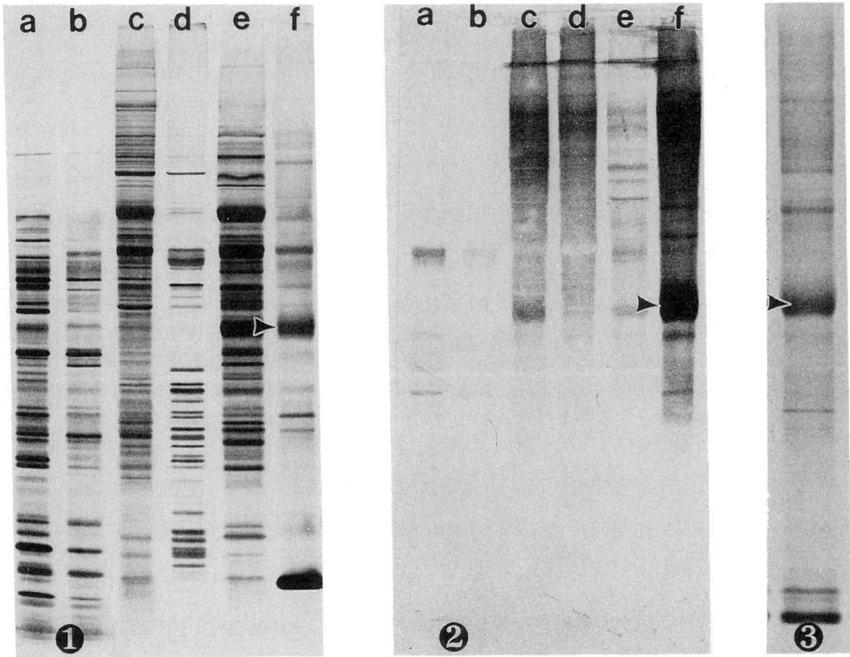

Figures 1 and 2. Detergent extracts from 7 day pseudopregnant females. Silver stained SDS-PAGE gel of column fractions from DEAE Sephacel and RCA-Sepharose (Figure 1), and Western blot of similar gel reacted with RCAI-biotin-avidin-peroxidase (Figure 2). Lanes a and b, buffer eluates of DEAE column; lanes c and d, fractions eluted from DEAE column with 0.3 and 0.5 M NaCl, respectively; lane e, 0.1 M NaCl fraction from DEAE column subsequently applied to RCA-Sepharose and eluted with buffer; lane f, same DEAE fraction as in "e", but eluted from RCA-Sepharose with D-galactose. Arrowheads indicate 42 kDa band. A silver-stained band in lane e has a similar M_r (~42,000), but has little or no affinity for the lectin (2e).

Figure 3. Sample lane of preparative gel stained with Coomassie Blue. Pooled extracts were subjected to DEAE fractionation and RCA-Sepharose affinity steps. Bands at 42 kDa (arrowhead) were excised and used as immunogen.

Light Microscopic Immunolocalization

Paraffin sections of Bouin- or Zamboni (buffered formaldehyde/picric acid)-fixed tissue were treated for immunolocalization using the indirect biotin-avidin-peroxidase method (Hsu et al., 1981). After deparaffinizing the sections, they were reacted with methanol containing 0.2% H_2O_2, hydrated, and incubated in PBS containing 2% normal goat serum. Guinea pig antisera to the 42 kDa glycoprotein were employed at dilutions of 1:100 to 1:5000, and biotinylated goat anti-guinea pig IgG (secondary antiserum) at a 1:100 dilution in PBS. PBS rinse solutions contained 0.02% Tween 20. After the reaction with streptavidin-peroxidase (1:200 dilution) and diaminobenzidine-H_2O_2, some sections were counterstained with Gill's hematoxylin before dehydration and mounting. For photographic purposes, uncounterstained sections were employed. In this case, $NiCl_2$ (Hsu and Soban,1982) was added to the diaminobenzidine-H_2O_2 solution to enhance contrast on black and white film. Control procedures included 1) the use of pre-immune guinea pig serum instead of specific antiserum, 2) elimination of primary antiserum incubation, and 3) elimination of both primary and secondary (biotinylated) antibody steps to establish the level of endogenous peroxidase activity remaining after methanolic peroxide treatment.

To minimize possible cross reaction of biotinylated secondary (goat anti-guinea pig IgG) antiserum with rabbit immunoglobulins endogenous to the tissue sections, samples of the secondary antiserum were absorbed. Absorption was carried out using whole rabbit serum coupled to CNBr-agarose (Sigma). Testing of the absorbed and non-absorbed serum against guinea pig immunoglobulin (Sigma) by dot blot assay on nitrocellulose strips (Jahn et al., 1984) revealed only a slight loss of reactivity due to absorption. Thus, absorbed secondary serum was employed at slightly higher concentrations than non-absorbed serum when used for immunostaining.

RESULTS

Antigen Isolation And Generation Of Antiserum

Experiments using detergent extracts of uteri from 7 day pseudopregnant females demonstrate that the 42 kDa glycoprotein is enriched when the appropriate fraction (0.1 M NaCl elution) from DEAE-Sephacel columns is subsequently subjected to affinity isolation using RCA-I-agarose. Strong staining of the 42 kDa band is present in silver-stained PAGE gels (Figure 1), while only the fraction eluted from the RCA-agarose with galactose reacted positively with biotinylated RCA on Western blots from the gels (Figure 2). The pooled extract from two females, subjected to this procedure and run as multiple lanes on preparative SDS-PAGE gels results in a clean separation of the 42 kDA glycoprotein which represents a significant fraction (~12%) of the total protein loaded. A single lane of the Coomassie Blue-stained preparative gel is illustrated in Figure 3.

Immunostaining of Western blots of SDS-PAGE-separated polypeptides from estrous and 7 day pseudopregnant detergent extracts with the guinea pig antisera reveals a positive reaction at the location of the 42 kDa glycoprotein (Figure 4). No comparable staining is observed in the lane loaded with an equivalent quantity of protein from an extract prepared from the uterus of an estrous female. The reaction is much stronger and more specific with antiserum from one of the guinea pigs, and this serum is used for all immunolocalization experiments.

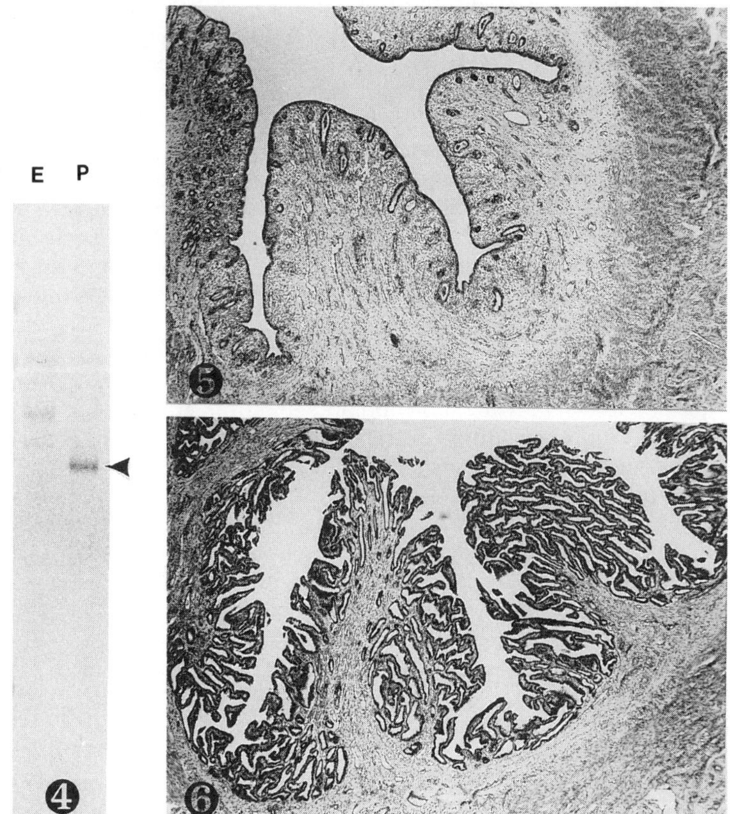

Figure 4. Western blot of detergent extracts from estrous (lane E) and 7 day pseudopregnant (lane P) uteri. The blot was immunostained with guinea pig antiserum against the 42 kDa glycoprotein. Arrowhead indicates reaction of the antiserum with a band (M_r 42,000) in pseudopregnant extracts, but no comparable reaction with estrous extract.

Figures 5 and 6. Paraffin sections of uteri from estrous (5) and 7 day pseudopregnant (6) females stained with hematoxylin and eosin. Note the presence of numerous minor folds of the mucosa in the pseudopregnant (receptive) stage. X22.

Immunolocalization Of 42 kDa Glycoprotein

The endometrium of the rabbit undergoes dramatic changes in cell proliferation, epithelial cell shape, and luminal topography during the pre-implantation period (Davies and Hoffman, 1973, 1975; Busch et al., 1981a). When viewed in cross-sectional profile, the endometrium at estrus presents a relatively smooth surface interrupted only by short tubular glands penetrating into the stroma (Figure 5). By days 6 - 7 of pseudopregnancy, the surface area of endometrium has increased many fold due to the appearance of numerous "minor folds" of mucosal tissue which project above the glands (Figure 6). Non-implantation regions of uteri from 6 - 7 day females appear identical to the pseudopregnant uterus, while in implantation sites the endometrial folds in contact with the blastocyst undergo some flattening (Segalen and Chambon, 1983; Anderson and Hoffman, 1984). The epithelium at estrus is simple columnar with a small number of ciliated cells in most females. During the phase of rapid proliferation (days 3 - 5), epithelial cells become somewhat more elongated, presenting an almost pseudostratified appearance, and return to a more typical simple columnar shape by days 6 - 7. Cells lining the minor folds ("cryptal epithelium") have a similar appearance to the luminal epithelium by days 6 - 7, but cells in the glands do not undergo comparable changes. Lateral fusion of luminal epithelial cells begins after day 6 and occurs irrespective of the presence of blastocysts. Larger aggregates of epithelial syncytium ("symplasma") are seen on or after day 8 of pregnancy.

Preliminary testing of the guinea pig antiserum employed dilutions ranging from 1:100 to 1:5000 on both Bouin- and Zamboni-fixed tissue sections. Optimal results were obtained using Bouin-fixed tissue and an antiserum dilution of 1:500 or 1:1000. Background staining of rabbit tissue, especially vascular structures, was completely eliminated by absorbing the biotinylated goat anti-guinea pig serum with rabbit serum and, thereafter, this procedure was used routinely.

Sections of uteri from estrous or 2 day pseudopregnant females exhibited no staining of any cell types (Figure 7); experimental slides were essentially comparable to control preparations treated with pre-immune serum. By day 4 of pregnancy or pseudopregnancy reaction product was apparent on the apical aspect of some luminal epithelial cells (Figure 8). Slight reactivity was seen along cryptal epithelium as well, in which case the reaction often appeared to be localized to the supranuclear region of epithelial cells more than to the apical membrane (Figure 9). Glandular epithelium was non-reactive as were other cell types in the endometrium (stromal cells, endothelium). Localization of the 42 kDa glycoprotein was identical on day 6 or 7 of pseudopregnancy. In these an intense line of reaction product was localized to the apical surface of luminal and cryptal epithelial cells (Figure 10). No difference in staining intensity was noted between mesometrial and antimesometrial surfaces. Ciliated cells, when encountered, were negative as were glandular epithelial cells and control sections treated with pre-immune guinea pig serum (Figure 11). A similar pattern of immunoreactivity was seen in non-implantation regions of uteri from 6, 6 1/2, 6 3/4, and 7 day pregnant females. At implantation sites on days 6 - 7 the pattern of immunostaining was similar to non-implantation regions; however, on day 8, some decrease in staining was apparent on the antimesometrial surface of implantation chambers. The specimens on day 8 had not been flushed to remove conceptuses; neither embryonic cells nor trophoblastic tissues were immunostained. Uterine symplasm formation was observed in these 8 day pregnant samples, and a redistribution of the reaction product was seen. In symplasmic aggregates, the reaction was present in the cortical cytoplasm surrounding the centrally-located nuclei, and was less intense on the cell surface (Figure 12). Immunostaining of non-uterine tissues (liver, oviduct, jejunum) using comparable dilutions of the guinea pig antiserum yielded negative results.

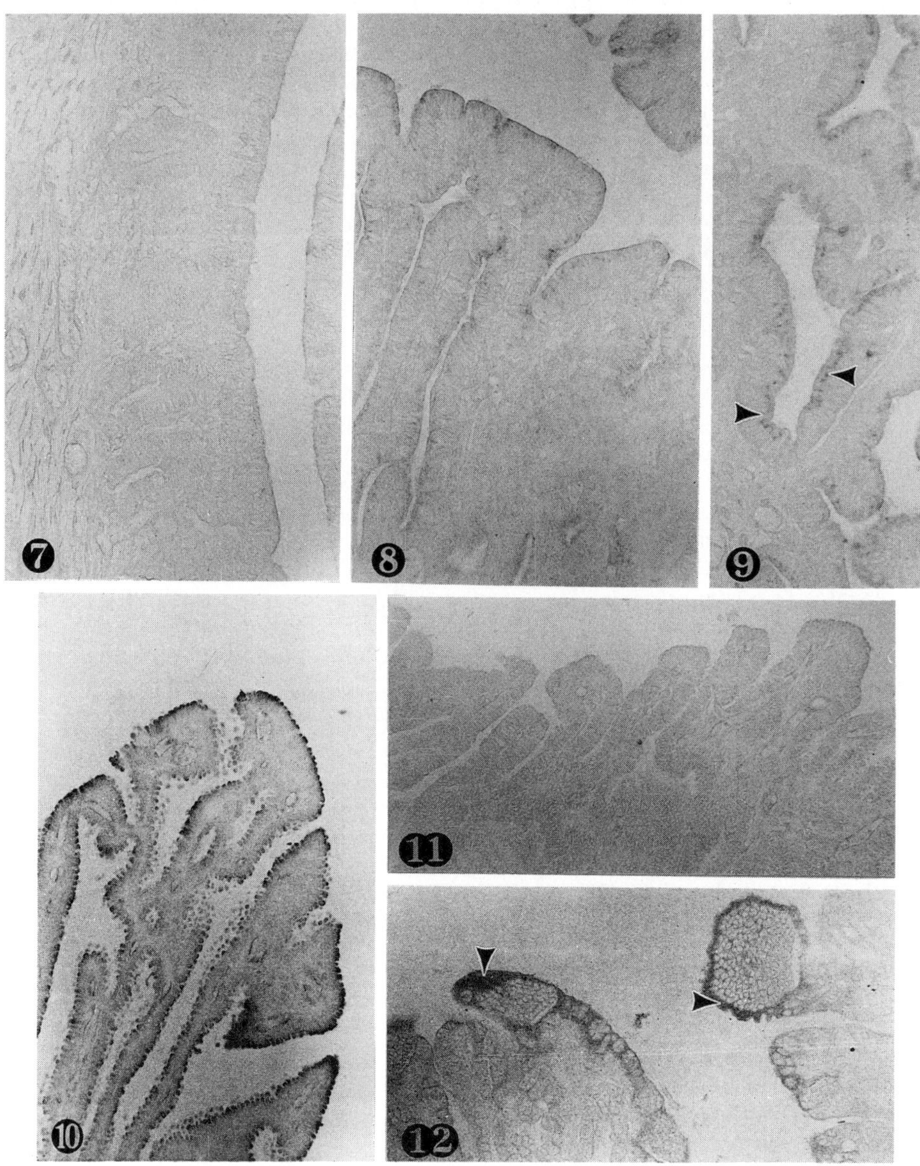

DISCUSSION

Implantation in the rabbit is initiated between 6 1/2 and 7 days p.c. (Denker, 1977). The duration or "window" of receptivity in this species is believed to be 48 - 72 hours, encompassing days 5 - 7 p.c. based on rates of successful embryo transfers to uteri of pseudopregnant recipients (Chang, 1950) as well as the optimal period required for the blastocyst to trigger vascular and decidual reactions associated with implantation (Hoos and Hoffman,1980). For purposes of this discussion, these days (5 -7 p.c.) will represent the period of receptivity to an implanting blastocyst, while the estrous state and days 1 - 4 of pregnancy or pseudopregnancy will be considered as nonreceptive.

As described under Results, the epithelial cells lining the uterine lumen of the rabbit undergo marked changes observable by light and electron microscopy during the pre-implantation period. Other alterations in the epithelium during this time include changes in the pattern of apical membrane enzyme activities (Classen-Linke et al., 1987). Alkaline phosphatase, aminopeptidase M, gamma-glutamyl transferase, and dipeptidyl peptidase IV all localize to the apical surface of epithelial cells and reach maximal activities by day 5, but decline in activity during the receptive phase. Additional changes in the pattern of enzyme activity are noted on days 7 - 8 p.c. in the endometrium directly apposed to implanting blastocysts. Alterations in tight junctional morphology and reorganization of intramembranous particles of epithelial plasma membranes during the pre-implantation period have been reported by Winterhager and Kuhnel (1982) and Winterhager et al. (1984). Increased communication via gap junctions is also associated with epithelial cells of the implantation chamber in the rabbit (Brummer et al., 1985). These events are of particular interest since they may correlate with the lateral fusion process resulting in multinucleated epithelial cells which occurs during the receptive phase, days 7 - 8 p.c. (Davies and Hoffman, 1973, 1975; Busch et al., 1981b). Furthermore, such membrane events may predispose to fusion of luminal epithelial cells with trophoblastic knobs at implantation (Larsen, 1961; Enders and Schlafke, 1971).

Figures 7 - 12. Paraffin sections of uteri at various reproductive stages, immunostained with guinea pig antiserum against the 42 kDa glycoprotein. Bouin fixation.

Figure 7. 2 day pseudopregnant; no immunoreaction observed. X165.

Figure 8. 4 day pregnant; moderate immunostaining present at apices of luminal epithelium. X165.

Figure 9. Section through minor folds on day 4 of pregnancy, cryptal epithelium. In this location the first appearance of immunoreaction was in the supranuclear cytoplasm (arrowheads) rather than at the apical surface of epithelial cells. X250.

Figure 10. 7 day pseudopregnant; intense immunostaining of apical surface of luminal and cryptal epithelium overlying the minor folds. X165.

Figure 11. 7 day pseudopregnant; same tissue as in Fig. 10 but reacted with pre-immune guinea pig serum as control. X165.

Figure 12. 8 day pregnant uterus; from a site immediately adjacent to an implantation chamber. Large symplasmic aggregates of luminal epithelial cells are forming and in these the immunoreaction appears to be in the cortical cytoplasm (arrowheads) rather than on the apical surface. X165.

Anderson and Hoffman (1984) examined the epithelial surface coat or glycocalyx of the rabbit endometrium during the pre-implantation period and following steroid hormone treatment of ovariectomized females. Luminal epithelial cells of the non-receptive (estrous) uterus had a prominent glycocalyx as demonstrated by periodic acid-Schiff (PAS) staining at the light microscopic level and the periodic acid-alkaline bismuth (PABi) reaction at the electron microscopic level. The glycocalyx had a net negative charge as determined by its affinity for polycationic ferritin (PCF), although ciliated cells bound little or no PCF. Loss of affinity for PCF following pretreatment of uteri with neuraminidase or trypsin suggested that sialoglycoproteins were a likely contributor to the surface negativity. Experiments employing saccharide-specific lectins (see below) and previous histochemical studies (Denker, 1977) likewise suggest the presence of sialic acid in the glycocalyx of these cells. A gradual decrease in the anionic property of the surface was noted during the pre-implantation period and in ovariectomized females receiving progesterone treatment (Anderson and Hoffman, 1984). By days 6 - 7, the cationic probe did not bind to the epithelial surface. PABi staining of the glycocalyx revealed qualitative changes coincident with loss of negativity, but the change was not due simply to removal of the luminal glycocalyx. Evidence that reduction of repulsive charges on apposing membranes could play a role in enhancement of cellular adhesion was discussed in light of these findings (Anderson and Hoffman, 1984).

Changes in the nature of saccharide components in membrane glycoconjugates of luminal epithelial cells have been examined using fluorescent-labeled lectins (Anderson et al., 1986a). Of the battery of lectins tested in that study, only RCA-I appeared to exhibit increased binding which correlated with the acquisition of receptivity. Comparable experiments employing electron microscopic analysis of ferritin-conjugated lectin binding confirmed the appearance of RCA-I affinity in the receptive phase and also revealed increased binding of succinylated wheat germ agglutinin (sWGA). Binding of these lectins was inhibited by the appropriate saccharides, D-galactose, and N-acetyl-D-glucosamine, respectively. Likewise, a reduction in binding of these lectins was apparent at implantation sites of 7 1/2 day pregnant females. Nalbach and Denker (1983) have also reported stage-dependent changes in affinity for various lectins. Of the lectins examined in that study, those with the greatest affinity for luminal and cryptal epithelium were RCA and WGA.

Biochemical characterization of endometrial membranes of rabbit uteri has been reported by Bullock and co-workers (Ricketts et al., 1984; Lampelo et al., 1985) as well as by this laboratory (Anderson et al., 1986a). Ricketts et al. (1984) analyzed surface proteins of endometrial epithelial cells radioiodinated after enzymatic dissociation or after 3 days in culture. Changes in surface iodination patterns were observed which correlated with the transition of the uterus from a nonreceptive (day 4) to a receptive (day 6.5) state. The most prominent changes in labeling were noted for proteins of 38 and 42 kDa as determined by SDS-PAGE. Lampelo et al. (1985) employed concanavalin A (Con A)-affinity chromatography and SDS-PAGE separation of solubilized proteins from rabbit endometrium to characterize the membranes of receptive (day 6) and non-receptive (day 3) females. Radioiodination of epithelial cells and plasma membrane marker enzyme activities were used to ensure that appropriate membrane fractions were analyzed.

Membrane components which characterized the receptive phase included proteins of 38, 55 and 84 kDa (Con A non-bound) and 30 kDa (Con A bound). A single protein of 78 kDa (Con A bound) was specific for the non-receptive phase.

A different approach to obtaining luminal membrane constituents was taken in our study (Anderson et al., 1986a). Membrane polypeptides and proteins were extracted by luminal incubation of intact uteri in a detergent-containing solution. Silver-stained SDS-PAGE gels of such extracts revealed several stage-specific changes in polypeptide patterns. The most notable were three polypeptides of 24, 42, and 58 kDa which appeared specific to membranes from receptive uteri. They were present in epithelial membrane preparations from day 7 of pseudopregnancy and pregnancy (reduced slightly in implantation sites) and were not detectable in extracts of estrous uteri. These bands were not seen on similar gels of uterine secretory proteins from receptive or nonreceptive uteri. Electrophoretic transfer of these polypeptides to nitrocellulose (Western blotting) was used to examine affinities for the lectins sWGA, RCA-I, and Con A. Of the three stage-specific polypeptides cited above, only the 42 kDa band bound sufficient lectin to be detected; both RCA-I and sWGA had affinity for this glycoprotein. Lectin staining of Western blots also revealed additional bands in the extracts from receptive uteri which were not present in estrous uteri nor in serum. These included a wide band at 80 - 86 kDa (sWGA affinity), multiple bands between 90 and 245 kDa (RCA-I affinity) and a prominent glycoprotein of 145 kDa which bound Con A. The fact that the 42 kDa protein was present in sufficient quantity in extracts from receptive stage uteri to be visualized by silver or Coomassie Blue staining of acrylamide gels made it a logical candidate for isolation and further characterization as was carried out in the current study. Its affinity for lectins (RCA-I or sWGA) presented the opportunity for affinity purification of the glycoprotein for use as an immunogen.

The antiserum prepared against this glycoprotein appeared to be highly specific for a single band (42 kDa) on immunoblots of membrane components from receptive uteri and did not react with polypeptides on blots from nonreceptive (estrous) females. Using immunolocalization procedures, specific staining of apical surfaces of epithelial cells in the receptive stage was observed. The glycoprotein appeared to be confined to the apical membrane, no staining was observed on lateral or basal surfaces of these epithelial cells. It was of interest that on day 4, the first stage at which the glyoprotein was detected, the antiserum also bound to material in the supranuclear cytoplasm of cryptal epithelium. This could be interpreted as Golgi processing of the protein prior to its appearance on the apical surface. At later times, through day 8, the antigen appeared to be specifically localized to the apical surface of luminal and cryptal epithelia. Although stages beyond day 8 of pregnancy were not examined, the finding that symplasmic epithelia demonstrated an altered distribution of the antigen may be an indicator of imminent disappearance of the 42 kDa glycoprotein as a surface constituent. It was not present in non-endometrial tissues or organs included in this study (myometrium, endometrial stroma or glands, oviduct, small intestine, liver, day 8 conceptus). The glycoprotein has the requisite properties of a membrane marker for the progestational stage coincident with implantation (presence on days 5 - 7), i.e., a marker of receptivity in the rabbit uterus. Lampelo et al. (1985) have also prepared an antibody (monoclonal) which reacted with luminal epithelium of receptive uteri of the rabbit. The antigen against which this antibody reacted was not characterized, although it was reported that the antibody bound to sites in several non-uterine organs (liver, lung, kidney, oviduct) in addition to uterine epithelium.

Of obvious relevance to implantation is the possibility that membrane constituents identified as markers of receptivity could be involved in the recognition and attachment of the trophoblast; changes in epithelial adhesive properties are most certainly involved in the acquisition of receptivity. Cell adhesion molecules have received considerable attention with regard to both development processes and cell-cell interactions in numerous cell types (Subtelny and Wessells, 1980; Obrink, 1986). Evidence that galactose-containing membrane glycoconjugates are increased in the rabbit uterine epithelium during the receptive phase has been confirmed for other species as well, including the mouse and primates (Chávez and Anderson, 1985; Anderson et al., 1986b). This finding could reflect increased activity of galactosyltransferase, a cell surface enzyme which appears to play a role in cell migration and adhesion in a variety of cell types, and in the fertilization process (Shur, 1982, 1983). Preliminary experiments suggest that αlactalbumin, which binds to galactosyltransferase, reduces the implantation rate in rabbits (T. Anderson, unpublished observations) and mice (Chávez, 1986a) when infused into the uterine lumen during the peri-implantation period. This line of reasoning is given added credence by the observation that N-acetylglucosamine (detected by sWGA affinity) serves as the site for galactosylation by this enzyme and sWGA binding was decreased on "adhesive" trophectoderm of the mouse (Chávez, 1986b) and on uterine epithelial cells apposed to implanting blastocysts of the rabbit (Anderson et al., 1986a). Some uterine proteins, such as β-glycoproteins in the rabbit, make their appearance during the peri-implantation period and may well serve as markers of the receptive phase (Thie et al., 1984). Immunolocalization of β-glycoprotein places it on or in ciliated cells of the endometrium, however, and these cells seem unlikely to be involved in initial trophoblastic adhesion and fusion with the uterine surface (Enders and Schlafke, 1971). Further progress in relating membrane markers of receptivity in the rabbit to trophoblast recognition and attachment will come from experiments employing in vitro cell or blastocyst adhesion assays (e.g., Morris et al., 1983, mouse; Hohn and Denker, 1985, rabbit). The current study supports, and adds to, previous findings suggesting that acquisition of uterine receptivity to implantation is accompanied by changes in protein composition and/or alterations in protein glycosylation of the epithelial plasma membrane. The 42 kDa, RCA-I-binding glycoprotein described in this report is the only stage-specific membrane protein or glycoconjugate to be characterized and localized to the cells most directly involved in trophoblast-endometrial interaction in the rabbit. It would appear to be an obvious candidate for further investigation focusing on embryo attachment in this species.

SUMMARY

Changes in uterine epithelial cell properties during the pre-implantation period are required to prepare the endometrium for implantation. A number of changes in apical membrane properties of luminal epithelium in the rabbit have been described, some of which appear to correlate with the acquisition of the receptive state, although none has been conclusively demonstrated to play a direct role in blastocyst attachment or invasion. In this study, one such stage-specific membrane constituent, a 42 kDa glycoprotein, was isolated and used as an immunogen for preparation of specific antiserum. The antiserum reacted with the appropriate glycoprotein on immunoblots of membrane extracts from receptive (day 7 pseudopregnant) uterine epithelium separated by SDS-PAGE. No comparable binding was observed with similar extracts from non-receptive (estrous) uteri. Light microscopic immunolocalization demonstrated the presence of the glycoprotein on the apical surface of luminal epithelial cells from receptive phase uteri (days 6 - 7). The 42 kDa glycoprotein was first detectable on day 4 of pregnancy or pseudopregnancy; it was not present in earlier (non-receptive) stages nor in other uterine cell types or non-uterine tissues examined. The results confirm the hypothesis that this glycoconjugate is a marker of receptivity in the rabbit, although its possible role in trophoblast attachment has yet to be demonstrated.

ACKNOWLEDGMENTS

This research was supported by Research Grant HD 18123 and Center Grant HD 05797 from the National Institutes of Health, USA.

REFERENCES

Adams, C.E. (1958) Egg development in the rabbit: The influence of post coital ligation of the uterine tube and of ovariectomy. *J. Endocrinol.* 16, 283-293.

Adams, C.E. (1971) The fate of fertilized eggs transferred to the uterus or oviduct during advancing pseudopregnancy in the rabbit. *J. Reprod. Fert.* 26, 99-111.

Anderson, T.L. and Hoffman, L.H. (1984) Alterations in epithelial glycocalyx of rabbit uteri during early pseudopregnancy and pregnancy, and following ovariectomy. *Am. J. Anat.* 171, 321-334.

Anderson, T.L., Olson, G.E., and Hoffman, L.H. (1986a) Stage-specific alterations in the apical membrane glycoproteins of endometrial epithelial cells related to implantation in rabbits. *Biol. Reprod.* 34, 701-720.

Anderson, T.L., Simon, J.A., and Hodgen, G.D. (1986b) Molecular evidence for the window of uterine receptivity to implantation in the cycling nonhuman primate. *J. Cell Biol.* 103, 463a.

Beier, H.M. (1976) Uteroglobin and related biochemical changes in the reproductive tract during early pregnancy in the rabbit. *J. Reprod. Fert.* Suppl. 25, 53-69.

Bradford, M.M. (1976) A rapid and sensitive method for the quantitation of microgram quantities of protein utilizing the principle of protein-dye binding. *Anal. Biochem.* 72, 248-254.

Brummer, F., Winterhager, E., Denker, H.-W., and Hulser, D.F. (1985) Cell-cell communication via gap junctions in rabbit uterine epithelium related to implantation. *Eur. J. Cell Biol.* 36 (Suppl. 7), 11.

Busch, L.C., Kuhnel, W., and Mootz, U. (1981a) Scanning electron microscopical studies of the rabbit endometrium during estrus and preimplantation. In: *Three Dimensional Microanatomy Of Cells And Tissue Surfaces*, (eds.), J.A. DiDio, P.M. Motta, and D.J. Allen, Amsterdam: Elsevier North Holland, pp. 267-278.

Busch, L.C., Winterhager, E., and Kuhnel, W. (1981b) Symplasmatische Umwandlung des Cavumepithels im Uterus pseudogravider Kaninchen. *Acta Anat.* 111, 22.

Chang, M.C. (1950) Development and fate of transferred rabbit ova or blastocyst in relation to the ovulation time of recipients. *J. Exp. Zool.* 114, 197-226.

Chávez, D.J. (1986a) Inhibition of mouse embryo implantation by perturbation of galactose metabolism. *J. Cell Biol.* 103(5), pt.2, 488a.

Chávez, D.J. (1986b) Cell surface of mouse blastocysts at the trophectoderm-uterine interface during the adhesive stage of implantation. *Am. J. Anat.* 176, 153-158.

Chávez, D.J. and Anderson, T. (1985) The glycocalyx of the mouse uterine luminal epithelium during estrus, early pregnancy, the peri-implantation period, and delayed implantation. I. Acquisition of <u>Ricinus Communis</u> I binding sites during pregnancy. *Biol. Reprod.* 32, 1135-1142.

Classen-Linke, I., Denker, H.-W., and Winterhager, E. (1987) Apical plasma membrane-bound enzymes of rabbit uterine epithelium. Pattern changes during the periimplantation phase. *Histochem.* 87, 517-529.

Cowell, T.P. (1969) Implantation and development of mouse eggs transferred to the uteri of non-progestational mice. *J. Reprod. Fert.* 19, 239-245.

Davies, J. and Hoffman, L.H. (1973) Studies on the progestational endometrium of the rabbit: I. Light microscopy, day 0 to 13 of gonadotrophin-induced pseudopregnancy. *Am. J. Anat.* 137, 423-445.

Davies, J. and Hoffman, L.H. (1975) Studies on the progestational endometrium of the rabbit. II. Electron microscopy, day 0 to 13 of gonadotrophin-induced pseudopregnancy. *Am.J. Anat.* 142, 335-366.

Denker, H.-W. (1977) Implantation. The role of proteinases, and blockage of implantation by proteinase inhibitors. *Adv. Anat. Embryol. Cell Biol.* 53, (Part 5) 1-123.

Enders, A.C. and Schlafke, S. (1971) Penetration of the uterine epithelium during implantation in the rabbit. *Am. J. Anat.* 132, 219-240.

Finn, C.A. (1977) The implantation reaction. In: *Biology Of The Uterus*, (ed.), R.M. Wynn, New York: Plenum Press, pp. 259-267.

Hohn, H.-P. and Denker, H.-W. (1985) Attachment and invasion of rabbit blastocysts confronted with endometrium in organ culture. *Eur. J. Cell Biol.* 36 Suppl. 7, 28.

Hoos, P.C. and Hoffman, L.H. (1980) Temporal aspects of rabbit uterine vascular and decidual responses to blastocyst stimulation. *Biol. Reprod.* 23, 453-459.

Hsu, S.-M., Raine, L., and Fanger, H. (1981) The use of avidin-biotin-peroxidase complex (ABC) in immunoperoxidase techniques: A comparison between ABC and unlabeled antibody (PAP) procedures. *J. Histochem. Cytochem.* 29, 577-580.

Hsu, S.-M. and Soban, E. (1982) Color modification of diaminobenzidine (DAB) precipitation by metallic ions and its application for double immunohistochemistry. *J. Histochem. Cytochem.* 30, 1079-1082.

Jahn, R., Schiebler, W., and Greengard, P. (1984) A quantitative dot-immunobinding assay for proteins using nitrocellulose membrane filters. *Proc. Natl. Acad. Sci. USA* 81, 1684-1687.

Laemmli, U.K. (1970) Cleavage of structural proteins during the assembly of the bacteriophage T4. *Nature* 227, 680-685.

Lampelo, S.A., Ricketts, A.P., and Bullock, D.W. (1985) Purification of rabbit endometrial plasma membranes from receptive and non-receptive uteri. *J. Reprod. Fert.* 75, 475-484.

Larsen, J.F. (1961) Electron microscopy of the implantation site in the rabbit. *Am. J. Anat.* 109, 319-334.

Martin, L. (1980) What roles are fulfilled by uterine epithelial components in implantation? In: *Blastocyst-Endometrium Relationships. Prog. Reprod. Biol.*, Vol 7, (eds.), F. Leroy, C.A. Finn, A. Psychoyos, and P.O Hubinont, Basel: S.Karger, pp. 54-69.

McCarthy, S.M., Foote, R.H., and Maurer, R.R. (1977) Embryo mortality and altered uterine luminal proteins in progesterone-treated rabbits. *Fertil. Steril.* 28, 101-107.

Morris, J.E., Potter, S.W., Rynd, L.S., and Buckley, P.M. (1983) Adhesion of mouse blastocysts to uterine epithelium in culture: A requirement for mutual surface interactions. *J. Exp. Zool.* 225, 467-479.

Nalbach, B.P. and Denker, H.-W. (1983) Stage-dependent changes in lectin binding patterns in rabbit uterus and blastocyst during the pre-implantation period and implantation. *Eur. J. Cell Biol. Suppl.* 4, 13.

Nilsson, O. (1967) Attachment of rat and mouse blastocysts onto uterine epithelium. *Int. J. Fertil.* 12, 5-13.

Obrink, B. (1986) Epithelial cell adhesion molecules. *Exp. Cell Res.* 163, 1-21.

Potts, D.M. and Psychoyos, A. (1967) L'ultrastructure des relations ovoendometriales au cours du retard experimental de nidation chez la souris. *C.R. Acad. Sci.[D] (Paris)* 264, 956-958.

Ricketts, A.P., Scott, D.W., and Bullock, D.W. (1984) Radioiodinated surface proteins on separated cell types from rabbit endometrium in relation to the time of implantation. *Cell Tissue Res.* 236, 421-429.

Segalen, J. and Chambon, Y. (1983) Ultrastrutural aspects of the antimesometrial implantation in the rabbit. *Acta Anat.* 115, 1-7.

Shur, B.D. (1982) Cell surface glycosyltransferase activities during fertilization and early embryogenesis. In: *The Glycoconjugates*, Vol. 3, (ed.), M.I. Horowitz, New York: Academic Press, pp. 145-185.

Shur, B.D. (1983) Embryonal carcinoma cell adhesion: The role of surface galactosyltransferase and its 90K lactosaminoglycan substrate. *Dev. Biol.* 99, 360-372.

Subtelny, S. and Wessells, N. (1980) *The Cell Surface: Mediator Of Developmental Processes*, New York: Academic Press.

Thie, M., Bochskanl, R., and Kirchner, C. (1984) Purification and immunohistology of a glycoprotein secreted from the rabbit uterus before implantation. *Cell Tissue Res.* 237, 155-160.

Towbin, H., Staehlin, T., and Gordin, J. (1979) Electrophoretic transfer of proteins from polyacrylamide gels to nitrocellulose sheets; procedure and some applications. *Proc. Natl. Acad. Sci. USA* 76, 4350-4354.

Winterhager, E. and Kuhnel, W. (1982) Alterations in intercellular junctions of the uterine epithelium during the preimplantation phase in the rabbit. *Cell Tissue Res.* 224, 517-526.

Winterhager, E., Busch, L.C., and Kuhnel, W. (1984) Membrane events involved in fusion of uterine epithelial cells in pseudopregnant rabbits. *Cell Tissue Res.* 235, 357-363.

Wray, W.T., Boulikas, T., and Wray, V.P. (1981) Silver staining of proteins on polyacrylamide gels. *Anal. Biochem.* 118, 197-203.

Zamboni, L. and Martino, C.D. (1967) Buffered picric acid-formaldehyde: A new, rapid fixative for electron microscopy. *J.Cell Biol.* 35, 148A.

POSSIBLE INVOLVEMENT OF D-GALACTOSE IN
THE IMPLANTATION PROCESS

Daniel J. Chávez

Department of Anatomy
School of Medicine
Southern Illinois University
Carbondale, Illinois 62901 USA

INTRODUCTION

Embryo implantation involves the attachment of embryonic trophoblast cells to the luminal epithelium of the uterus. Attachment occurs in all eutherian mammals, regardless of whether or not subsequent implantation is interstitial. Adhesion of embryonic trophoblast to uterine luminal epithelium presents an apparent paradox in cell biology for it represents apical adhesion of two presumably polarized epithelia. The acquisition of adhesiveness between these two cell types has been presumed to be an intrinsic property of trophectoderm because trophectoderm cells readily adhere to extrauterine sites. However, examination of mouse uteri during the peri-implantation period has revealed that the apposed walls of the uterine lumen become adherent to one another at the same time that trophoblast becomes adherent to epithelium. Apical adhesion of epithelium is an uncommon phenomenon or else the patency of the uterus (and any other hollow organ in the body) would not be maintained. The molecular basis for the cell to cell interactions which result in adhesion is of importance to those interested in developmental biology, and tumor metastasis and invasion, as well as implantation. Embryo implantation as it occurs in the mouse presents a convenient model for studying this fundamental phenomenon because the temporal regulation of the acquisition of adhesiveness is well understood. In this respect, both adhesive and non-adhesive cells are accessible for analysis.

Although informational molecules which could be synthesized and secreted to pass from cell to cell have been postulated as mediators of intercellular communication, their isolation and characterization has thus far been equivocal. On the other hand, surface molecules which are integral components of the cell membrane have been implicated in providing cells with markers for identification to the environment, recognition for adjacent cells, receptors for hormones and other macromolecules, and adhesion factors which maintain the integrity of tissue (Bennett et al., 1971; Jenkinson and Billington, 1976; Phillips and Gartner, 1980; Frazier and Glaser, 1979; Sharon, 1984a; 1984b).

Mouse embryos have been shown to synthesize a variety of glycoconjugates. Modifications of cell surface glycoproteins have been reported during preimplantation (Pinsker and Mintz, 1973; Brownell, 1977; Johnson and Calarco, 1980) and peri-implantation (Wu and Chang, 1978; Carollo and Weitlauf, 1979; Chávez and Enders, 1981, 1982; Jenkinson and Searle, 1977; Hakansson, Heyener,

Sundqvist, and Bergstrom, 1975; Hakansson and Sundqvist, 1975; Searle et al., 1976; Chávez, 1986) stages of embryogenesis. Studies involving the mouse uterus show similar changes to occur at the surface of the luminal epithelium during pregnancy (Holmes and Dickson, 1973; Chávez and Anderson, 1985; Hewitt, Beer and Grinnell, 1979).

In the mouse, and many other mammals, the onset of implantation begins with an initial stage of apposition of the trophoblast to the uterine luminal epithelium. This is followed by a stage of adhesion of the trophoblast to the epithelial surface. The stage of apposition is brought about by closure of the uterine lumen and is under control of normal hormones of pregnancy. The stage of adhesion involves adhesion of mural trophoblast, but not polar trophoblast, to the uterine luminal epithelium (Kirby et al., 1967; Chávez et al., 1984; Chávez, 1986). The adhesive stage also appears to involve adhesion of the apposed luminal epithelial cells to one another (Chávez and Shriver, 1987). The stage of adhesion has been shown to coincide with subtle qualitative changes in the glycocalyx of adhering trophectoderm (Holmes and Dickson, 1973; Chavez, 1986) and qualitative changes in the thickness and amount of glycocalyx present at the apical surface of the luminal epithelium (Chávez and Anderson, 1985).

The data reported come from an investigation of the saccharides present in plasma membrane glycoconjugates of trophectoderm and uterine luminal epithelium. Non-pregnant and pregnant uteri were examined at a variety of stages including estrus, days 4 and 5 of pregnancy, and delayed implantation. The aim of these studies was to determine whether specific molecules could be temporally associated with the series of events that occurs during early implantation. Blastocysts were also examined in situ. The results obtained from these observations were used to design experiments with the aim of determining whether there is a sequence of expression of cell surface molecules and to investigate their possible regulatory mechanisms. Further experiments were done to determine whether perturbation of the normal sequence of glycoconjugate expression would result in inhibition of the implantation process. The experimental data are summarized in this paper.

MATERIALS AND METHODS

The observations described were obtained from a colony of random-bred ICR mice kept under a controlled photoperiod of 14h light: 10h dark. Mice were caged five per box (four females and a single male). Females were checked daily for the presence of a vaginal plug, and the day the plug was found was designated day 1 of pregnancy. Females were removed and replaced as they were mated; each male was in constant company of four females.

Electron Microscopy

During estrus (determined by vaginal cytology), day 4, and delayed implantation, mice were perfused under metofane anesthesia via a cannula inserted into the thoracic aorta. Perfusion effulent was allowed to escape via a cut made distal to the renal veins. Phosphate buffered saline (PBS) was allowed to flow through the vascular system until the effulent became clear and then a glutaraldehyde/paraformaldehyde fixative (Karnovsky, 1965) was allowed to flow until the uterine horns became fixed and hard. On day 5 of pregnancy, at 1100h (before blastocyst adhesion) and 1500h (after blastocyst adhesion) 0.1 ml of 1% Evans blue in PBS was injected either via a femoral vein or via a tail vein. Fifteen minutes after Evans blue injection, the animals were perfused as described. Either implantation sites (identified by decidual swellings and blue spots) or uterine segments were isolated and finely trimmed. The apposed sites of uterine epithelium were gently separated either by splitting or by gentle spreading with the fine points of microforceps under a dissecting microscope. Fixation was continued by immersion for 1-2 hours at 4°C. After fixation, the tissue was rinsed overnight in cacodylate buffer. The tissue was rinsed again for 1 hour in cacodylate buffer containing glycine in order to neutralize any unreacted aldehydes.

Ferritin-conjugated lectins were purchased from E-Y Laboratories (San Mateo, CA). Polycationized ferritin (PCF) was purchased from Miles Laboratories (Elkhart, IN). Immediately before use, the lectins and PCF were diluted to 0.5 mg/ml with PBS. Uterine segments were exposed to 10 ml drops of lectin or PCF for 10 minutes at ambient temperature. Care was taken to separate the apposed sides of uterine epithelium while in the lectin or PCF drops in order to expose the entire surface of the implantation chamber or intersite to the reagent. After exposure to lectin or PCF, the tissue was rinsed in PBS. Care was taken to separate gently the apposed tissue to allow thorough rinsing of unbound reagent. The tissue was postfixed in OsO_4, dehydrated in a graded series of increasing ethanol concentrations, and embedded in epoxy resin. As a control, to demonstrate specificity, tissue was exposed to each lectin in the presence of 0.25 mM of the sugar to which it specifically binds.

Embedded sites were hand-trimmed with a razor blade. Each block was sectioned with a glass knife until the blastocyst was encountered and then thin-sectioned with a diamond knife. Sections were collected on copper grids, stained with lead citrate only, and examined in a Hitachi H500 electron microscope operated at an accelerating voltage of 75 kV.

Radioautography

Radiolabeled sugars were purchased from New England Nuclear (Boston, MA). Radioactivity was diluted to 25 µCi/ml in PBS and 5 µl of PBS with radiolabelled sugar was injected directly into the uterine lumen of day 5 pregnant mice at 1000h. Access to uterine horns was via a mid-flank incision under metofane anesthesia. Fluid was injected at the proximal (oviductal) end of the uterus. The incisions were secured by wound clips and the animals were allowed to recover from anesthesia. Exactly 1 hour and 45 minutes after injection of radioactivity, the animals received Evans blue injections and were perfused as described above. The uteri were excised; implantation sites were isolated and finely trimmed. Fixation was continued as above except that an additional exhaustive rinse with three changes of 5% trichloroacetic acid (TCA) was inserted into the protocol in order to remove all unincorporated radiolabel. Animal carcasses, cages, bedding, perfusion effluent, and fixation rinses were treated as contaminated waste. The sites were post-fixed and embedded as above. After trimming and facing of the blocks, semithin (1 µm) sections were cut and mounted on glass slides. The slides were dipped in Kodak NTB-2 Nuclear Track Emulsion (Rochester, NY) and exposed at 4°C for 2 to 6 weeks.

Histochemistry

In situ labeling of the galactosyl transferase (Galtase) enzyme was done on ethanol-fixed cryostat sections of fresh mouse uteri. On the appropriate day of pregnancy, mice were killed, their uteri excised and immediately frozen on dry ice. Cryostat sections (6-10 mm) were air-dried and fixed in ethanol. Sections were exposed to a reaction mixture containing 1 mg ovalbumin, 10 mM sodium cacodylate (pH 7.35), 10 mM $MnCl_2$, and 20 nM UDP-(^{3}H)galactose (specific activity 17.3 Ci/mM) per 100 ml of assay mixture. The reaction was allowed to proceed 15 minutes at 37°C and then stopped by immersing the slides in ice cold 5% TCA. After being rinsed, the slides were dipped in photographic emulsion (above) and exposed at 4°C for 2 to 8 weeks. These radioautograms were stained with fast green and photographed with phase contrast optics.

Intrauterine Injections

In order to determine whether the presence of specific carbohydrate moieties in the glycocalyx was necessary for successful implantation, a series of enzymes, inhibitors, and antimetabolites was injected directly into the uterine lumen of mice on either day 4 or day 5 of pregnancy. Injections were performed transcervically after visualization of the cervix through an operating otoscope inserted into the vagina. A specially designed microliter syringe was made to facilitate injections into only one uterine horn. The syringe was fitted with a truncated needle which had a 10 degree bend approximately 15 mm from the end. After gentle dilation of the cervical os, the needle was inserted through the cervix. A slight rotation of the needle was sufficient to guide it into a specific horn. The needle was inserted 10-15 mm into the desired horn, exactly 5 μl of fluid was injected, and the needle was slowly withdrawn. The cervix could be monitored during withdrawal; normally, very little fluid leaked out of the dilated cervix following withdrawal of the needle. Mice were laparotomized on day 9 of pregnancy. The presence of fetuses within well-developed decidual swellings and comparable to the non-injected contralateral horn was considered to indicate successful implantation.

An additional control involved in vitro culture of peri-implantation blastocysts in the same concentrations of the specific compounds injected into the uterine lumen. Incubation was done in the standard manner using microdrops of Dulbecco's modification of Eagle's Minimum Essential Medium (DMEM), supplemented with 10% fetal bovine serum, under oil at 37°C, in an atmosphere of 5% CO_2 and 95% room air at saturated humidity. Although some concentrations may have increased the osmotic pressure of the medium, only non-lethal concentrations were injected into the uterus.

Although the strain of mice used had only a single cervix and the lumina of the uterine horns were continuous with one another, transfer of injected fluid from one horn to the other did not occur. Implantation inhibition, when it occurred, only affected the injected horn. Mice which were not pregnant in either horn were considered to be non-pregnant at the time of injection and were not included in the data base. Non-pregnant mice were encountered as often in control treatments as in experimental treatments. Sterile mating is a scourge of all reproductive biologists.

RESULTS

The glycocalyx of the uterine luminal epithelium was seen to be reduced markedly in thickness during early pregnancy and the peri-implantation period (Figures 1 and 2). The glycocalyx was 0.06-0.10 mm thick during estrus and the first three days of pregnancy. By day 4 the glycocalyx became reduced to 0.06-0.08 mm thick. This thickness was reduced further by the time of trophectoderm attachment on day-5, at which time precise measurement became difficult (Chávez and Anderson, 1985).

Reduction in glycocalyx thickness was determined to be primarily due to loss of the thick sialic acid coat which is the normal terminal portion of the glycopeptide sequence. Desialylation had the coincident result of reducing the net electronegative charge of the apical surface of the epithelium. Desialylation was inferred from the reduction or loss of binding of Limulus polyphemus agglutinin (LPA) through days 3, 4, and 5 of pregnancy. Reduction in electronegativity was manifested by reduced binding of PCF during the same time period (Figures 1,2).

Desialylation of the epithelial glycocalyx also resulted in the uncovering or unmasking of subterminal sugar moieties that are not otherwise expressed. Examination of the surface of the epithelium using lectin markers indicated that D-galactose (Gal) was subterminal to sialic acid and that the Gal was in a β-linkage to penultimate N-acetyl glucosamine (GlcNAc) residues (Chávez and Anderson, 1985; Chávez, 1986). This observation was confirmed by exposing fixed and unfixed segments of uteri to sialidase (Sigma, St. Louis, MO) and subsequent esposure to ferritin-conjugated Ricinus communis Agglutinin I (RCA-I). Gal was revealed abundantly after the sialic acid had been enzymatically removed.

Radioautography was used to ascertain whether the expression of Gal residues was solely the result of desialylation or new incorporation of glycopeptides into the plasma membrane and revealed that Gal was incorporated by neither the uterine luminal epithelium nor the blastocyst. The nucleotide derivative, however, was actively incorporated into the blastocyst, as shown by the presence of dense reduced silver grains in the radioautograms (Figure 8). The luminal and glandular epithelium of the uterus showed minimal incorporation of the sugar and very little transport to the cell surface during the two hour labeling period. Therefore, it appeared likely that the majority of the Gal expressed after desialytion was already present in the glycocalyx.

On day 5 of pregnancy, the proliferation and transformation of uterine stroma created a crypt surrounding each blastocyst. Although implantation crypts were formed and blastocysts were tightly clasped by apposing sides of the uterine lumen, it was possible to separate gently trophoblast from epithelium before midday on day 5. However, by 1500h on day 5, blastocysts had established foci of adhesion with the uterine luminal epithelium. Such sites of adhesion always involved mural trophectoderm and never polar trophectoderm. It was not possible to separate the trophectoderm from epithelium at sites of attachment without damaging the trophectoderm cells. Nor was it possible to separate epithelium from epithelium without damaging the cells (Figure 4). By examining more than two dozen sites of adhesion and comparing lectin binding data with non-adhesive sites it was

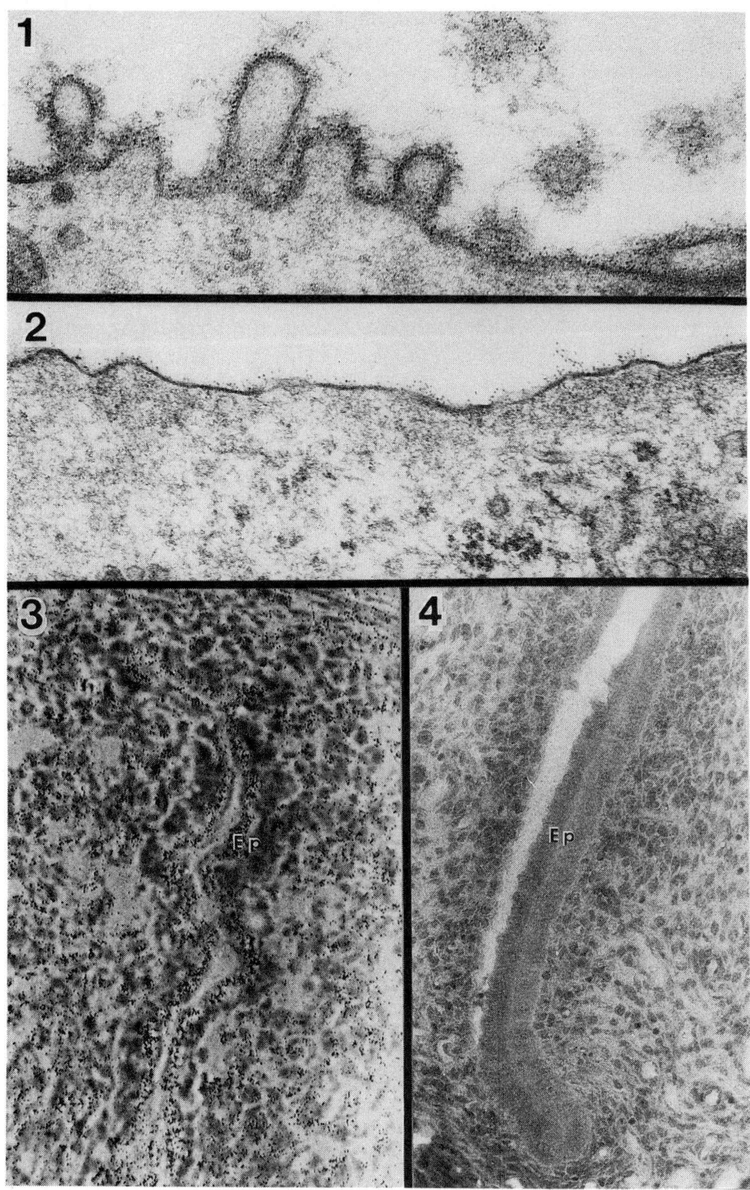

possible to determine that specific changes did occur in the biochemistry of uterine and trophectoderm glycocalyces in conjunction with the acquisition of adhesiveness by these two cell types.

Qualitative differences in lectin binding that correlated with adhesion of trophectoderm to luminal epithelium of the uterus were found with succinylated wheat germ agglutinin (S-WGA), which binds to GlcNAc. S-WGA bound uniformly, albeit rather sparsely, to the surface of preadhesive blastocysts but did not bind at all to trophectoderm at sites of adhesion. At the free nonadhered surface of adhesive blastocysts (e.g., polar trophectoderm) the lectin did bind to the cell surface. These results indicate that at the time of adhesion terminal GlcNAc residues become either masked or lost from the cell surface (Chávez, 1986).

Figure 1. Electron micrograph of a section of uterine luminal epithelium of mouse uterus during estrus. The uterus has been exposed to polycationized ferritin. The presence of ferritin particles demonstrates the electronegativity of the glycocalyx. The glycocalyx is thick and filamentous in appearance. The electronegativity is located adjacent to the plasma membrane and not in the filamentous terminus (Magnification X64500).

Figure 2. This section came from an implantation site on day 5 of pregnancy. The uterus was exposed to PCF. Although ferritin particles are seen bound to the surface of the epithelium, the net electronegativity is significantly reduced from that seen in Figure 1. Thickness is also reduced and the filamentous appearance of the glycocalyx is not apparent at this time. (Magnification X64500).

Figure 3. An autoradiogram of cryostat section of day 5 pregnant mouse uterus demonstrating the fixation of [3H]UDP-Gal by the galactysyltransferase enzyme. Silver grains are distributed throughout the uterine stroma and over the epithelium (Ep) although there do not appear to be any located specifically over the plasma membrane of either.

Figure 4. Light micrograph of a section from a day 5 pregnant mouse uterus. During the apposition stage of implantation the opposite walls of the uterine lumen become tightly apposed to one another. At this stage it is not possible to inject fluid into the uterine lumen or to recover blastocysts. This figure demonstrates the result of attempting to split open the uterus manually. The plane of fracture extends from the uterine lumen through the epithelium (Ep) and separates the epithelium from the underlying stroma. This demonstrates the apex to apex adhesion of the uterine epithelium.

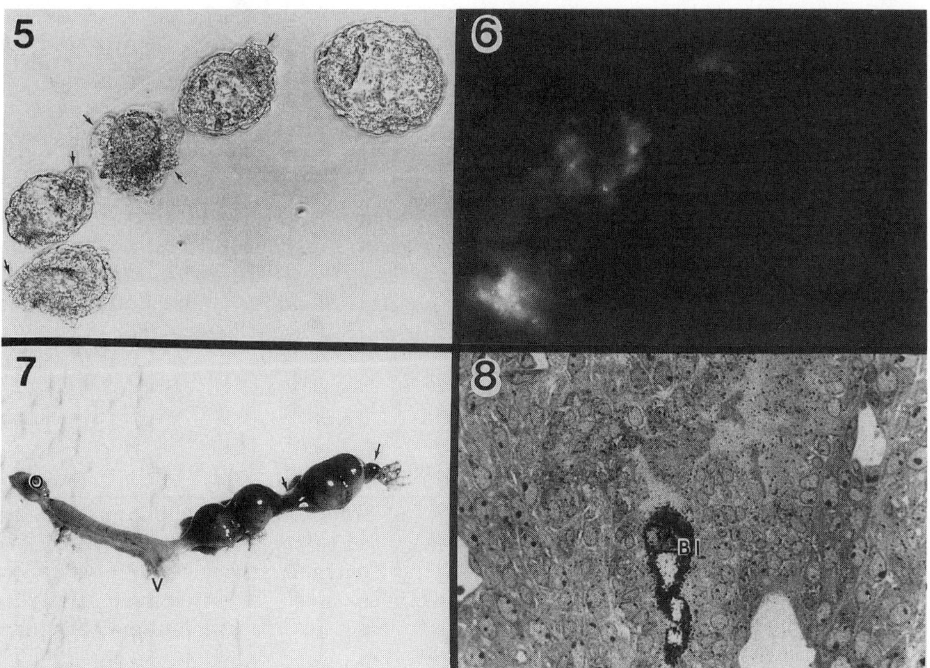

Figure 5. Phase contrast micrograph of mouse blastocysts recovered from the uterus during the peri-implantation period (1200h, day 5). Four of the five blastocysts had begun to attach to the uterine luminal epithelium and damaged fragments can be seen adhering to abembryonic trophectoderm (arrows).

Figure 6. The same blastocysts shown in Figure 5 were exposed to polyclonal rabbit anti-mouse galactosyltransferase antibody and then exposed to FITC-conjugated protein A. Under fluorescence microscopy, it can be seen that no antibody was bound to the surface of peri-implantaiton blastocysts. Abundant antibody can be seen in the damaged uterine fragments.

Figure 7. Photograph of a day 9 pregnant mouse uterus. The horn on the left had been treated with an intraluminal injection of UDP-Gal on day 5 of pregnancy. The horn on the right contains four normal fetuses and two resorbing sites (arrows). Embryo resorption is common in laboratory mice. O, ovary; V, vagina.

Figure 8. Radioautogram of an implantation site exposed to an intraluminal injection of UDP-[^{3}H]Gal. The presence of reduced silver grains demonstrates abundant incorporation by the blastocyst (B) but minimal incorporation by either the uterine epithelium or stroma.

Because Gal becomes linked to GlcNAc by the action of galactosyltransferase (Galtase) it became of interest to ascertain the activity of this enzyme during the peri-implantation period. Although Sato et al. (1984) reported the presence of Galtase at the surface of preimplantation embryos, the use of a polyclonal rabbit anti-Galtase antibody did not confirm its presence in peri-implantation blastocysts. However, the antibody did bind to fragments of uterine epithelial cells adherent to flushed blastocysts. This suggests that the enzyme is present in uterine epithelium (Figures 5 and 6); however, the fluorescent label did not show whether the enzyme was intracellular or at the cell surface. These results, in conjunction with the data indicating active incorporation of Gal by blastocysts and not by uterine epithelium (Figure 8), stimulated consideration of the Roseman hypothesis (Roseman, 1970), a mechanism by which cells may recognize and adhere to one another via binding between enzymes located on one cell surface and substrates located on an adjacent cell surface.

To test this hypothesis and examine the possible involvement of the sugar, a series of experiments was performed by injecting compounds which interfere with galactose metabolism directly into the uterine lumen during pre- and peri-implantation stages. Inhibition of implantation (Figure 7) was interpreted to indicate that the pathway which was interfered with, by a specific compound, was necessary for the normal sequence of events related to growth and/or differentiation of the blastocyst or uterus and, therefore, was directly or indirectly involved in the implantation process. Parallel in vitro (control) experiments with the same compound indicated that the concentrations used were non-lethal to the blastocyst. Partial data are presented in Table 1. The data clearly indicate that interference with galactose metabolism did result in interference with the implantation process.

DISCUSSION

The presumed biological importance of cell surface glycoconjugates is that they may be involved in cell-cell recognition. Recognition appears to be associated with selective adhesion between cells. Although adhesion of cells to one another must involve physicochemical interactions such as calcium bridging, hydrogen bonding, disulfide bonding and others (Grinnell, 1978), selective adherence may require binding between specific cell surface molecules. Adherence mediated by surface carbohydrates could involve selective interaction between the saccharide moieties themselves or via cell surface lectins or soluble extracellular lectins acting as intermediates (Sharon, 1984a,b).

In the case of embryo implantation, consideration may be given independently to the uterine luminal surface and trophectoderm surface, as well as to their interaction. Examination of epithelium at sites distant from the blastocyst has revealed that these changes are not site-specific. Although blastocyst adhesion occurs at the antimesometrial aspect of the uterine lumen, and blastocysts become evenly spaced within the uterus, there is no evidence to support site-specific attachment.

Histochemical examination of the uterine luminal epithelium with the lectins thus far employed has indicated that the surface coat throughout the entire uterine lumen is homogeneous. Identical lectin binding characteristics are observed from the oviductal to the cervical end. There is no difference between mesometrial or antimesometrial aspects of the uterine lumen. Although individual epithelial cells did present heterogeneity in surface projections, e.g., lobopodia and microvilli, there is no variation in lectin binding among the projections.

Table 1

**Effect On Blastocyst Implantation Of Intrauterine Injection Of
Carbohydrate Antimetabolites On Day 5 Of Pregnancy**

Compound	Implantation
α-D-Galactosidase[1]	+
α-D-N-Acetylgalactosaminidase[2]	-
Sialidase[3]	-
α-Lactalbumin[4]	+/-
UDP-Galactose[5]	-
UDP-Glucose[6]	+
Tunicamycin[7]	-
β-D-Galactosidase[8]	+
N-Acetylneuraminic acid[9]	+/-
(given day 4)	-
PBS (control)	+

1. Bethesda Research Laboratories (Gaithersburg, MD) 0.4 units
2. Bethesda Research Laboratories (Gaithersburg, MD) 0.004 units
3. Bethesda Research Laboratories (Gaithersburg, MD) 0.1 unit
4. Type I, Sigma (St. Louis, MO) 20 mg
5. Sigma (St. Louis, MO) 5 mM
6. Sigma (St. Louis, MO) 5 mM
7. Calbiochem (La Jolla, CA) 20 mg
8. Sigma (St. Louis, MO) 200 units
9. Type VI Sigma (St. Louis, MO) 20 mg

Blastocysts present two distinct subpopulations of trophectoderm cells (Chávez et al., 1984). Polar trophectoderm, overlying the inner cell mass, remains non-adhesive; the mural trophectoderm adheres to the uterine epithelium (Chávez, 1986). Following the luminal closure reaction, the uterine luminal epithelial cells appeared to adhere to apposed cells from the opposite wall of the lumen as well as adhering to the blastocyst. Blastocysts adhere to the uterine luminal epithelium but never to other blastocysts.

The temporal sequence of glycocalyx modification during the peri-implantation period of both uterine epithelium and trophectoderm begins with desialylation. Removal of non-reducing terminal sialic acid from the apical surface of uterine luminal epithelium is coincident with arrival of the blastocysts in the uterus on day 4 of pregnancy. Removal of the sialic acid moiety a) reduces the thickness of the surface coat, b) reduces the electronegativity of the cell surface, and c) exposes or unmasks Gal and GlcNAc residues. The net effect of a and b may be to allow closer apposition of cells and thus facilitate cell-to-cell contact. The ultimate effect of c is equivocal; however, it is possible that these two molecules are somehow involved in the early events of the implantation process.

In the normal state, the apex of any polarized epithelium must be non-adhesive, whether one considers luminal epithelium of a hollow organ (uterus) or superficial epithelium (trophectoderm). Apical nonadhesiveness may be required to maintain the structure and function of the hollow organs. The acquisition of adhesiveness by uterine luminal epithelium may be considered to be an abnormal situation, although it is necessary in order for implantation to occur. Epithelium in adhesion with blastocysts appears degenerative when viewed by electron microscopy. The cells lose their columnar appearance, microvilli and basal cell processes become retracted, and apico-basal cell polarity appears to be lost (Denker, 1986; Chávez and Shriver, 1987).

Whether these observations reflect the normal progression of apoptotic epithelial cell death, which occurs during embryo invasion of the epithelium, or normal modification coincident with acquisition of adhesiveness by epithelium cannot be stated at this time. The important consideration is that there is a progression of events which render normally non-adhesive cells adhesive. These events involve a series of morphological changes within the trophoblast cells, as well as at the cell surface.

Treatment of pregnant mouse uteri with either sialidase or N-acetyl-neuraminic acid resulted in inhibition of implantation in the treated horn. It is possible that sialic acid present at the surface of cells prevents them from adhering to one another either by increased electronegativity, masking of sub-terminal adhesive groups, or maintaining a critical intercellular distance. The results of the present study do not provide evidence for or against this hypothesis. Nevertheless, it remains clear that desialylation is one in a series of physiochemical events that leads to cell adhesion.

Treatment of pregnant mouse uteri with a variety of compounds that perturb galactose metabolism resulted in inhibition of implantation. Specifically, it appears that the enzyme galactosyltransferase may be necessary for implantation to occur. Both UDP-Gal and α-lactalbumin are Galtase substrates. Introduction of either compound into a uterine horn during the peri-implantation period resulted in inhibition of implantation in the injected horn. The appropriate control was UDP-Glc, which has identical molecular weight to UDP-Gal but different configuration at ring carbon 3.

It has been nearly 20 years since Roseman (1970) proposed that cells may recognize and selectively adhere to one another via binding of enzymes located on one cell surface to substrates located on an adjacent cell surface. Such a mechanism does exist for Galtase in a variety of cell types (Roth et al., 1971; Roth and White, 1972; Shur, 1983; Shur and Hall, 1982a, b). Although Sato et al. (1984) did report the presence of cell surface Galtase on preimplantation mouse embryos, studies using a polyclonal antibody failed to confirm its presence at the surface of peri-implantation blastocysts. The results indicated that the enzyme was abundantly present in the uterine tissue although neither the antibody used nor the in situ histochemical assay could detect it at the cell surface. In this regard, the activity of the enzyme, as judged by an in vitro bioassay, was found to increase eight-fold in uterine tissue during the peri-implantation period of pregnancy (Chávez, 1986; Moran et al., 1986). Perhaps more specific monoclonal antibodies are necessary to localize precisely this enzyme.

SUMMARY

During the peri-implantation stages of pregnancy, the glycocalyces of both trophoblast (trophectoderm) and uterine luminal epithelium lost their sialic acid terminus. Desialylation resulted in decreased thickness of glycocalyx and thus permitted closer apposition of cell membranes. Electronegativity of the glycocalyx was also diminished. D-Galactose molecules, which were normally sub-terminal to sialic acid became exposed in a terminal position in the glycocalyx. Interference with galactosyltransferase resulted in inhibition of implantation. The mechanism by which the galactose residue is involved in the implantation process remains unknown.

ACKNOWLEDGEMENTS

I am grateful to Ms. Elaine Shriver for technical assistance during much of this work. Candida Morris and Jeanette Robinson typed the manuscript. The majority of this work was supported by NIH Grant HD-16703.

REFERENCES

Bennett, D., Boyse, E.A., and Old, L.J. (1971) *Cell Interactions*, (ed). L.G. Silvestri, North-Holland, Amsterdam, p. 247.

Brownell, A.G. (1977) Cell surface carbohydrates of preimplantation embryos as assessed by lectin binding. *J. Supramol. Struct.* 7, 223-234.

Carollo, J.R. and Weitlauf, H.M. (1979) A comparison of changes in Con A binding and surface area of mouse blastocysts at implantation. *Anat. Rec.* 193, 477A.

Chávez, D.J. and Enders, A.C. (1981) Temporal changes in lectin binding of peri-implantation mouse blastocysts. *Devel. Biol.* 87, 267-276.

Chávez, D.J. and Enders, A.C. (1982) Lectin binding of mouse blastocysts: Appearance of Dolichos biflorus agglutinin binding sites on the trophoblast during delayed implantation and their subsequent disappearance during implantation. *Biol. Reprod.* 26, 545-552.

Chávez, D.J., Enders, A.C., and Schlafke, S. (1984) Trophectoderm cell subpopulations in the peri-implantation mouse blastocyst. *J. Exp. Zool.* 231, 267-271.

Chávez, D.J. and Anderson, T.L. (1985) The glycocalyx of mouse uterine luminal epithelium during estrus, the peri-implantation period, and delayed implantation. I. Acquisition of RCA-I binding sites during pregnancy. *Biol. Reprod.* 32, 1135-1142.

Chávez, D.J. (1986) Cell surface of mouse blastocysts at the trophectoderm-uterine interface during the adhesive stage of implantation. *Am. J. Anat.* 176, 153-158.

Chávez, D.J. and Shriver, E.M. (1987) Modification of mouse uterine luminal epithelium during apposition and adhesion of early implantation. *Anat. Rec.* 218(1), 22-23A.

Denker, H.-W. (1986) Epithel-Epithel-Interaktionen bei der Embryo-Implantation: Ansätze zur Lösung eines zellbiologischen Paradoxons. *Verh. Anat. Ges. (Anat. Anz. Suppl.* 160) 80, 93-114.

Frazier, W. and Glaser, L. (1979) Surface components and cell recognition. *Ann. Rev. Biochem.* 48, 49-52.

Grinnell, F. (1978) Cellular adhesiveness and extracellular substrata. *Int. Rev. Cytol.* 53, 65-144.

Hakansson, S., Heyener, S., Sundqvist, K.G., and Bergstrom, S. (1975) The presence of paternal H-2 antigens on hybrid mouse blastocysts during experimental delay of implantation and the disappearance of these antigens after onset of implantation. *Int. J. Fertil.* 20, 137-140.

Hakansson, S. and Sundqvist, K.G. (1975) Decreased antigenicity of mouse blastocysts after activation for implantation from experimental delay. *Transplantation* 19, 475-484.

Hewitt, K., Beer, A.E., and Grinnell, F. (1979) Disappearance of anionic sites from the surface of the rat endometrial epithelium at the time of blastocyst implantation. *Biol. Reprod.* 21, 691-707.

Holmes, P.V. and Dickson, A.D. (1973) Estrogen-induced surface coat and enzyme changes in the implanting mouse blastocyst. *J. Embryol. Exp. Morph.* 29, 639-645.

Jenkinson, E.J. and Billington, W.D. (1976) Cell surface properties of early mammalian embryos. In: *The Cell Surface In Animal Embryogenesis And Development*, (eds.), G. Poste and G.L. Nicolson, Elsevier/North-Holland, Amsterdam, pp. 235-266.

Jenkinson, E.J. and Searle, R.F. (1977) Cell surface changes on the mouse blastocyst at implantation. *Exp. Cell Res.* 106, 386-390.

Johnson, L.V. and Calarco, P.G. (1980) Mammalian preimplantation development. The cell surface. *Anat. Rec.* 196, 201-219.

Karnovsky, M.J. (1965) A formaldehyde-glutaraldehyde fixative of high osmolarity for use in electron microscopy. *J. Cell Biol.* 27, 137A.

Kirby, D.R.S., Potts, D.M., and Wilson, L.B. (1967) On the orientation of the implanting blastocyst. *J. Embryol. Exp. Morph.* 17, 527-532.

Martin, L. (1984) On the source of uterine 'luminal fluid' proteins in the mouse. *J. Reprod. Fert.* 71(1), 73-80.

Moran, G., Kurilla, G., and Babiarz, B. (1986) Identification of galactosyltransferase activity and its substrate during murine decidual cell differentiation. *J. Cell Biol.* 103(5), 487a.

Phillips, D.R. and Gartner, D.K. (1980) Cell recognition systems in eukaryotic cells. In: *Receptors And Recognition*, (ed.), E.H. Beachey, Chapman and Hall, London, pp. 400-438.

Pinsker, M.C. and Mintz, B. (1973) Change in cell-surface glocoproteins of mouse embryos before implantation. *Proc Nat. Acad. Sci. USA* 70, 1645-1648.

Roseman, S. (1970) The synthesis of complex carbohydrates by multi-glycosyltransferase systems and their potential function in intercellular adhesion. *Chem. Phys. Lipids* 5, 270-297.

Roth, S., McGuire, E.J., and Roseman, S. (1971) Evidence for cell-surface glycosyltransferase. *J. Cell Biol.* 51, 536-547.

Roth, S. and White, D. (1972) Intercellular contact and cell-surface galactosyl transferase activity. *Proc. Natl. Acad. Sci. USA.* 69, 485-489.

Sato, M., Muramatsu, T., and Berger, E.G. (1984) Immunological detection of cell surface galactosyltransferase in preimplantation mouse embryos. *Devel. Biol.* 102, 514-518.

Searle, R.F., Sellens, H.H., Elson, J., Jenkinson, E.J., and Billington, W.D. (1976) Detection of alloantigens during preimplantation development and early trophoblast differentiation in the mouse by immunoperoxidase labeling. *J. Exp. Med.* 143, 348-359.

Sharon, N. (1984a) Surface carbohydrates and surface lectins are recognition determinants in phagocytosis. *Immunol. Today* 5, 143-147.

Sharon, N. (1984b) Carbohydrates as recognition determinants in phagocytosis and in lectin-mediated killing of target cells. *Biol. Cell* 51, 239-246.

Shur, B.D. and Hall, N.G. (1982) Sperm surface galactosyltransferase activities during in vitro capacitation. *J. Cell Biol.* 95, 567-573.

Shur, B.D., and Hall, N.G. (1982) A role for mouse sperm surface galactosyltransferase in sperm binding to the egg zona pellucida. *J. Cell Biol.* 95, 574-579.

Shur, B.D. (1983) Embryonal carcinoma cell adhesion: The role of surface galactosyltransferase and its 90K lactosaminoglycan substrate. *Devel. Biol.* 99, 360-372.

Wu, T.J. and Chang, M.C. (1978) Increase in Concanavalin A binding sites in mouse blastocysts during implantation. *J. Exp. Zool.* 105, 447-453.

HISTOCHEMICAL CHARACTERISTICS OF THE ENDOMETRIAL SURFACE RELATED TEMPORALLY TO IMPLANTATION IN THE NON-HUMAN PRIMATE (Macaca fascicularis)

Ted L. Anderson[1,3], James A. Simon[2], and Gary D. Hodgen

The Jones Institute for Reproductive Medicine
Laboratory for Cellular and Molecular Biology
Department of Obstetrics and Gynecology
Eastern Virginia Medical School
Norfolk, Virginia 23507 USA

INTRODUCTION

Appropriate differentiation of the trophoblast is surely requisite for blastocyst invasion. However, there is overwhelmingly compelling evidence to suggest that differentiation of the endometrium into a "receptive state" is the primary determinate of successful nidation. The highly invasive trophoblast of certain species appears able to "implant" almost without discrimination in a variety of ectopic sites, regardless of the host tissue or the hormonal milieu (Kirby, 1971). Conspicuously, the only tissue in which trophoblastic invasion *does not* occur indiscriminately is the natural implantation site, the uterine lining. The uterus is permissive to blastocyst implantation for only a brief interval or "receptive window" following appropriate hormonal conditioning (Casimiri and Psychoyos, 1981), as evidenced by the failure of transferred blastocysts to attach to endometrium except at the time of normal receptivity for that species (Finn, 1977). Further evidence is derived from observations that peri-ovulatory administration of progesterone leads to implantation failure due to embryo-uterine asynchrony in rats, rabbits and (albeit debated) humans (Psychoyos, 1976; Schacht and Foote, 1978; Trounson et al., 1986).

Additionally, in human in vitro fertilization programs where ovarian hyperstimulation is used to obtain multiple ova for fertilization, excessively high serum levels of estradiol and progesterone often result in advanced endometrial (histologic) maturation. This could contribute to embryo-uterine asynchrony, and could be partially responsible for the relatively low success rate (compared with the animal science industry) seen in these programs (Garcia et al., 1983; Jones, 1985). These relatively rigid constraints on implantation appear to be mediated through the endometrial epithelium (Nilsson, 1967; Potts and Psychoyos, 1967; Cowell, 1969; Martin, 1980).

The initial attachment of blastocysts to the endometrium is not only limited temporally, but it also requires heterotypic cell adhesion between the apical membranes of these two cell populations; such an event is usually not a

---

[1]Present address: Department of Obstetrics and Gynecology, Endocrine Laboratory, Vanderbilt Medical School, Nashville, TN 37232 USA

[2]Present address: Department of Obstetrics and Gynecology, Georgetown University Medical School, Washington, DC 20007, USA

[3]To Whom Correspondence Should Be Addressed.

characteristic of polarized epithelial cells. Thus, stage-specific alterations in these epithelial surfaces appear necessary to permit this event. Indeed, alterations in surface components of preimplantation embryos (charge, lectin affinity, and glycoproteins) have been reported by a number of investigators, particularly in relation to development of "adhesive" properties (Nilsson et al., 1973; Johnson and Calarco, 1980; Magnuson and Epstein, 1981; Chávez and Enders, 1982). Additionally, studies in a variety of species have identified stage-specific alterations in the apical (luminal) membrane of endometrial epithelial cells related temporally to uterine receptivity. Salient features common among studies reported include: 1) decreased luminal surface negativity; 2) changes in glycocalyx thickness and/or "morphology"; 3) changes in lectin affinity; and 4) stage-specific expression of epithelial membrane proteins and glycoproteins (Enders and Schlafke, 1972; Hewitt et al., 1979; Murphy and Rogers, 1981; Guillomot et al., 1982; Anderson and Hoffman, 1984; Whyte and Robson, 1984; Chávez and Anderson, 1985; Lee and Damjanov, 1985; Anderson et al., 1986; Lampelo et al., 1986).

This study has examined specific membrane properties of the luminal epithelium of regularly-cycling primates throughout the normal menstrual cycle. Cynomolgus monkeys were chosen for these studies because of their extensive mimicry of fundamental anatomic, physiologic, and temporal aspects of the human hypothalamic-pituitary-ovarian-uterine axis (Bartelmez et al., 1951; Healy and Hodgen, 1983; Padykula et al., 1984; Enders et al., 1985; Brenner et al., 1983). Reported here are the results of histochemical studies that have identified membrane properties unique to the interval of receptivity to implantation.

MATERIALS AND METHODS

Animals

Cynomolgus monkeys (*Macaca fascicularis*) used during these studies were caged individually in a 12:12 hour light:dark environment under the direction of a full-time veterinarian. All monkeys in the colony maintained by this institute were checked daily for onset of menses (cycle day 1) by the animal husbandry core laboratory; historical records were kept. Additionally, all monkeys assigned to protocols had blood drawn daily (if necessary) between 7:00 and 9:00 AM to monitor serum hormone levels. For these studies, endometrial biopsies were obtained during laparotomy from normally-cycling cynomolgus monkeys during the following cycle intervals: mid-proliferative (days 4-6), peri-ovulatory (days 13-16), peri-implantation (days 6-8 post-ovulation), and late secretory (1-2 days prior to expected onset of menses, based on menstrual history). At least four biopsies were examined from each of these intervals, and analysis was performed as described below.

Chemicals

Polycationic ferritin (PCF, pI_{av} 8.4) and native ferritin (NF, $pI < 7.0$) were obtained from Miles Laboratories (Elkhart, IN). Bismuth subnitrate was obtained from Polysciences (Warrington, PA) and ferritin-conjugated *Ricinus communis-I* agglutinin (F-RCA-I) was purchased from E-Y Laboratories (San Mateo, CA).

All other chemicals were obtained from Sigma Chemical Company (St. Louis, MO).

Surface Charge

The initial goal was to determine whether generalized luminal surface changes occur during the menstrual cycle. To this end, portions of biopsies were washed under a stream of phosphate buffered saline (PBS) from a syringe to remove any secretory material. Unfixed tissue was exposed to PCF or NF for 2 minutes (0.5 mg/ml in PBS) as described previously (Anderson and Hoffman, 1984). Incubation of uterine tissues with these probes was accomplished at physiological pH (7.4) to avoid induced alterations in native surface charge. Tissue was then rinsed with PBS, fixed with glutaraldehyde, and processed routinely for electron microscopy (EM). Ultrathin sections were mounted on copper grids and examined, without counterstain, using a Philips 301 electron microscope. Additionally, uterine luminal surface charge was examined histochemically using alcian blue (AB, pH 2.5) on 5 µm sections of formalin-fixed, paraffin-embedded tissue (Lillie and Fulmer, 1976).

Glycocalyx Histochemistry

Periodic acid-Schiff (PAS) histochemistry was performed on 5 µm paraffin sections of formalin-fixed tissue using well-established procedures (Lillie and Fulmer, 1976) to examine the uterine luminal glycocalyx. Additionally, portions of biopsies were examined using the periodic acid - alkaline bismuth (PABi) technique, an EM correlate of PAS, as described previously (Anderson and Hoffman, 1984). Briefly, ultrathin sections of tissue processed routinely for EM were mounted on nickel grids and exposed to 1% aqueous periodic acid for 15 minutes. After a subsequent 45 minutes exposure to an alkaline bismuth subnitrate solution, grids were washed well with water and examined, without counterstain, as described above.

Lectin Binding

Ferritin-conjugated *Ricinus communis-I* agglutinin (F-RCA-I) was chosen for initial studies of glycoconjugate alterations in the luminal epithelial membrane because increased binding of this lectin, recognizing D-galactose residues, has been observed in several other species related to uterine receptivity to implantation (Murphy and Rogers, 1981; Chávez and Anderson, 1985; Lee and Damjanov, 1985; Whyte and Robson, 1984; Anderson et al., 1986). Biopsies were flushed with PBS and fixed with glutaraldehyde for 30 minutes. Free aldehydes were blocked by rinsing tissue with Tris-buffered saline (TBS) containing 1% ammonium chloride; tissue was subsequently rinsed (3 x 10 minutes) with lectin buffer (TBS containing 2 mM $CaCl_2$, and 2 mM $MgCl_2$).

Biopsies were incubated 45 minutes with F-RCA-I (250 µg/ml in lectin buffer), rinsed briefly, re-fixed with glutaraldehyde, and processed for electron microscopy. Similar exposure of tissues to lectins in the presence of 0.1 M galactose served as control incubations.

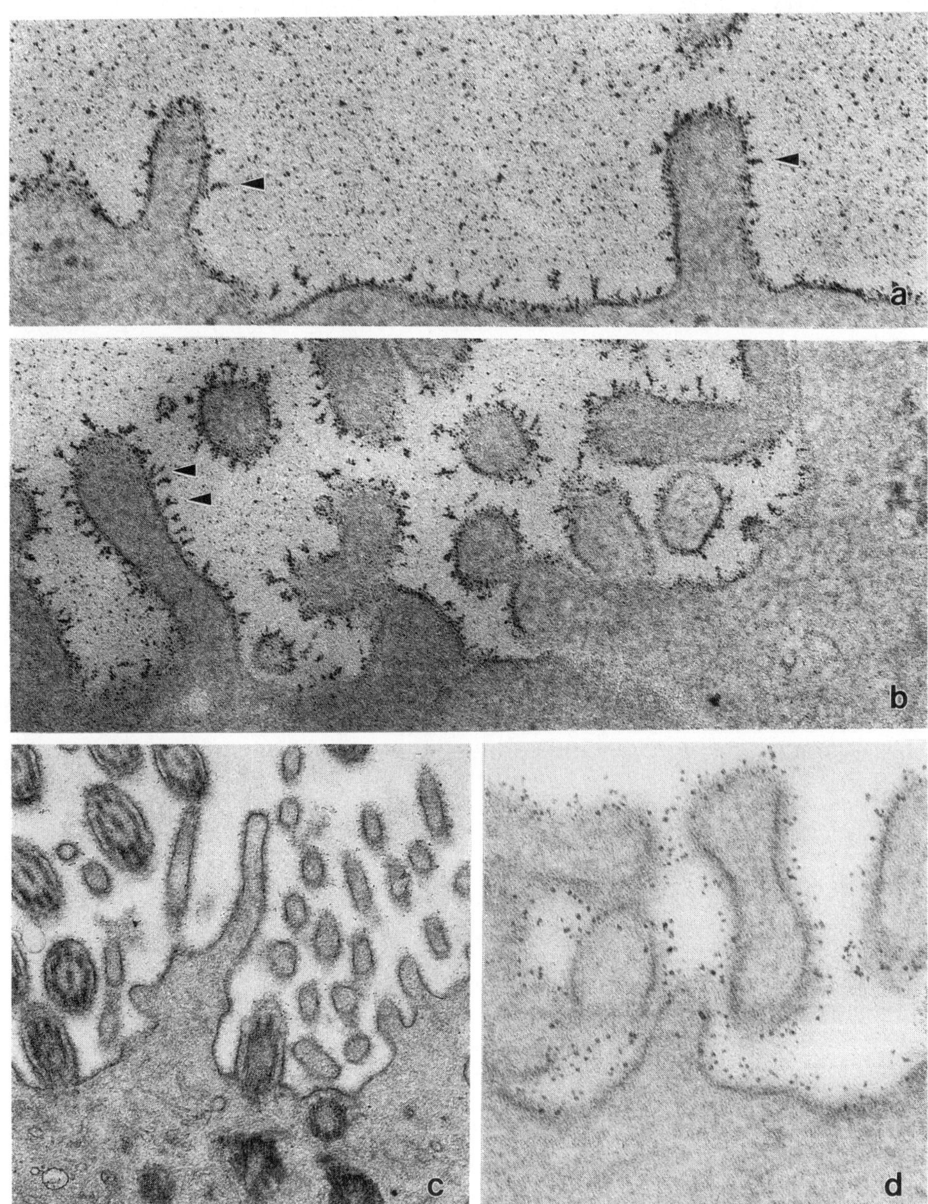

Table 1

Histochemical evaluation of luminal epithelial cells of the
cynomolgus monkey throughout the normal menstrual cycle

Technique *	Mid-proliferative	Peri-ovulatory	Peri-implantation	Late secretory
PCF	+++	++	-	-
AB	+++	++	+	+
PAS	++	++	+++	++
PABi**	+++ (60 nm)	+++ (60 nm)	+++ (90 nm)	+++ (90 nm)
F-RCA-I	-	-	+++	-

(-) not detected; (+) weak but detectable; (++) moderate and uniform, or strong and
patchy; (+++) strong and uniform; *see text for details; **glycocalyx thickness in
parentheses; PCF = Polycationic Ferritin; AB = Alcian Blue; PAS = Periodic Acid-
Schiff; PABi = Periodic Acid - Alkaline Bismuth; F-RCA-I = Ferritin-Conjugated
Ricinus Communis-I Agglutinum

RESULTS

The results of these histochemical studies were summarized in Table 1.
Binding of PCF to the luminal surface of biopsies from mid-proliferative uteri was
uniform and extensive, but no binding was observed during either the peri-
implantation or late secretory intervals. Similar examination of biopsies obtained
during the peri-ovulatory interval revealed occasional PCF binding; no binding
was observed on most cells in these biopsies, and only patchy binding was detected
on a few others. No NF binding to the surface of biopsies from any cycle interval
was observed. Similarly, AB (pH 2.5) staining was most intense during the mid-
proliferative interval, moderate during the peri-ovulatory interval, and weak but
discernable during the peri-implantation and late secretory intervals.

Figure 1. Histochemical properties of the luminal surface of uteri from
cynomolgus monkeys. The fine granular reaction product of the PABi technique
demonstrates a simple glycocalyx during the mid-proliferative interval (a), which
extends up to 60 nm from the plasma membrane (arrows). During the peri-
implantation interval (b), a branched glycocalyx morphology is appreciated
(arrows), which extends up to 90 nm from the plasma membrane. Stage-specific
binding of ferritin-conjugated RCA-I is noted on ciliated (c) and non-ciliated (d)
epithelial cells during the peri-implantation interval. (Magnification: a,b,d
X60,000; c X20,000)

Surface staining by PAS was moderate, but relatively uniform, during the mid-proliferative and peri-ovulatory intervals, but was appreciably more intense and uniform during the peri-implantation interval. However, during the late secretory interval, PAS surface staining appeared reduced to a level comparable to that of the mid-proliferative interval. The PABi technique resulted in deposition of a fine granular reaction product at the surface of all biopsies examined, which allowed analysis of "glycocalyx morphology". In all cases, PABi-positive material was identified extending from the apical plasma membrane and microvilli. During the mid-proliferative and peri-ovulatory intervals, the PABi technique revealed a uniform but simple glycocalyx, which did not appear to be branched, and extended an average of 60 nm from the plasma membrane (Figure 1a). Similar examination of the glycocalyx during the peri-implantation interval revealed a uniform glycocalyx that appeared highly branched and extended an average of 90 nm from the plasma membrane (Figure 1b). The glycocalyx morphology of biopsies obtained during the late secretory interval was not appreciably different from that of the peri-implantation interval.

Uniform binding of F-RCA-I to the luminal surface of biopsies was observed only during the peri-implantation interval. Such lectin binding was noted on both ciliated cells (Figure 1c) and non-ciliated cells (Figure 1d), but binding to the latter was more uniform and intense. Interestingly, no lectin binding was detected during either the mid-proliferative nor the late secretory intervals, and only rare patchy binding to some non-ciliated cells was noted in biopsies obtained from the peri-ovulatory interval. No lectin binding was detected upon examination of control experiment tissues exposed to F-RCA-I in the presence of 0.1 M galactose.

DISCUSSION

These results demonstrate that glycocalyx material is present on the uterine luminal surface throughout the normal primate menstrual cycle, and that the saccharide composition of the membrane glycoconjugates is altered toward the time of implantation under the influence of a changing hormonal milieu. Specifically, the electronegativity of this surface is reduced (PCF, AB), and the glycocalyx appears to thicken and assume a branched conformation (PAS, PABi). These membrane alterations correlate with increased abundance of D-galactose residues at the luminal surface (F-RCA-I), or with the unveiling of such residues already present in the glycocalyx, or both. Observations reported here compare favorably with similar examination of uteri from other species, including one reported study of human biopsies (Hewitt et al., 1979; Chávez and Anderson, 1985; Jansen et al., 1985; Anderson et al., 1986). It is believed that appearance of F-RCA-I binding to the uterine luminal surface 6-8 days post-ovulation, and its subsequent disappearance before the onset of menses, provides the first molecular evidence for what is thought to be the "window of uterine receptivity" in nonhuman primates.

That binding of PCF to the uterine luminal surface is related specifically to surface electronegativity is demonstrated by the lack of native (electroneutral) ferritin binding to this surface. Reduction of luminal surface negativity prior to the time of implantation has been described previously in the rat and rabbit (Hewitt et al., 1979; Anderson and Hoffman, 1984); evidence is provided that electronegativity of the non-receptive uterus is related to the presence of sialic acid.

Though specific examination of sialic acid involvement in surface negativity is not a goal of this study, one speculates that a similar relationship between sialic acid removal and reduced negativity is present in the primate uterus.

Removal of surface charge toward the time of implantation is consistent with the polymer exclusion theory of cell-cell interactions (Maroudas, 1975), which asserts that events such as adhesion and fusion are enhanced as repulsive charges on the respective cell surfaces are diminished. Furthermore, removal of sialic acid to reveal D-galactose residues on glycoproteins is a recognition mechanism common to a variety of biological systems. Certain serum glycoproteins (transferrin, fetuin, and ceruloplasmin) contain terminal sialic acid - D-galactose units (Uhlenbruck et al., 1978); removal of sialic acid from these is required for recognition by the liver asialoglycoprotein receptor (Ashwell and Harford, 1982). Skutelsky et al. (1974) proposed that sialic acid may play a "protective role" during differentiation by masking antigenic or recognition sites. This notion is particularly attractive in relation to the acquisition of uterine receptivity to implantation, and is consistent with results presented here.

The biological significance of an increased thickness of glycocalyx on the uterine luminal surface at the time of implantation, as revealed by PAS and PABi techniques, is not clear. In fact, the initial impression had been that a *reduction* of this glycocalyx would better facilitate apposition of trophoblastic and endometrial epithelial surfaces. Such a reduction of surface coat material at prospective implantation sites in the rat uterus has been reported (Salizar-Rubio et al., 1980), but Enders et al. (1980) did not detect direct effects of blastocysts on the uterine epithelial glycocalyx in that model. Reduction of glycocalyx material specifically at sites of embryo-uterine apposition has been demonstrated in the ferret (Enders and Schlafke, 1972) and rabbit (Anderson et al., 1986). Identification and isolation of areas in potential trophoblast-endometrial interaction prior to implantation in the primate have not been successful; thus, whether local modification of the luminal surface by implanting embryos occurs in this model remains uncertain. The changes in glycocalyx thickness and morphology described here reflect a "decondensation" due to reduced negativity, as well as increased saccharide complexity of glycoproteins synthesized under specific hormonal stimulation, which may be indicative of a carbohydrate-based cell recognition system during the initial phase of embryo-uterine interactions.

The stage-specific binding of F-RCA-I to the uterine luminal surface during the peri-implantation period suggests an embryo recognition/adhesion system involving D-galactose. Recognition of galactose-containing glycoproteins by apical membrane "endogenous lectins" mediates adhesion of chick blastoderm cells (Milos and Zalik, 1982), as well as cell-cell and cell-substratum adhesion of some tumor cells in relation to metastatic potential (Meromsky et al., 1986). Furthermore, similar endogenous lectins have been identified in several rabbit tissues (Harrison et al., 1984), and have been postulated as mediators of both fertilization (Shur and Hall, 1982) and embryo-uterine interactions (Chávez, 1986) in the mouse. Though F-RCA-I binding to the uterine luminal surface appears limited to the peri-implantation interval, this does not necessarily mean that galactose residues are present in the glycocalyx only during this time. It is possible that galactose residues are present in glycoconjugates at other times of the cycle, but are only accessible to the large ferritin-conjugated probe (400 kDa) during a specific interval due to changes in conformational organization of the glycocalyx.

The molecular mechanism by which luminal epithelial cells manifest stage-specific glycocalyx alterations in the primate endometrium is not clear. On one hand, regulation may be at the level of post-translational modification so that constitutively synthesized proteins are differentially glycosylated depending on the hormonal environment. This notion is supported by the demonstration that estradiol stimulation increases cellular levels of mannosylphosphoryldolichol in the mouse uterus, which enhances N-linked protein glycosylation (Carson et al., 1987). Alternatively, these luminal surface changes may be mediated through differential gene transcription under the influence of a changing hormonal environment, resulting in the synthesis and glycosylation of a variety of stage-specific membrane proteins.

Evidence has been provided that both mechanisms are operative in the rabbit uterus under the influence of progesterone (Anderson et al., 1986), and preliminary evidence for similar dual action in the primate uterus as well has been developed.

Despite differences in the mechanism of implantation and placentation in a variety of species, remarkably similar molecular alterations in the uterine surface appear to occur. As such, many of these molecular changes represent alterations common in a variety of molecular recognition and adhesion systems, and are not specific for the process of implantation per se. Rather, these membrane alterations facilitate the apposition and adhesion of trophoblast with uterine luminal epithelium (regardless of species) in such a way that other, more implantation-specific and/or species-specific, molecular interactions may be facilitated.

SUMMARY

Properties of the endometrial luminal surface in relation to uterine receptivity to implantation in cynomolgus monkeys was examined using a variety of histochemical techniques. The results demonstrated that glycocalyx material was present on the uterine luminal surface throughout the normal primate menstrual cycle, but that alterations in saccharide composition of this surface occured toward the time of implantation. Specifically, the electronegativity was reduced (examined by polycationic ferritin binding and alcian blue histochemistry), and the glycocalyx appeared to thicken and assumed a branched conformation (examined using PAS and periodic acid - alkaline bismuth histochemistry). These alterations correlated with the increased presence of D-galactose residues at the luminal surface, or with the unveiling of such residues already presented in the glycocalyx, or both (detected by ferritin-conjugated *Ricinus communis*-I agglutinin). The appearance of this lectin binding to the uterine luminal surface 6-8 days post-ovulation, and its subsequent disappearance before the onset of menses, provided the first molecular evidence for what was thought to be the "window of uterine receptivity" in nonhuman primates. The membrane alterations described here mediated apposition and adhesion of trophoblast with uterine luminal epithelium in such a way that other, more implantation-specific and/or species-specific, molecular interactions may be facilitated.

ACKNOWLEDGEMENTS

The authors express appreciation to Ms. Lynn Sharpe for assistance with animal husbandry and surgical aspects of these studies, and Mr. Jim Slusser for excellent technical assistance with electron microscopy. This work was supported, in part, by a grant from the Thomas F. and Kate Miller Jeffress Memorial Trust.

REFERENCES

Anderson, T.L. and Hoffman, L.H. (1984) Alterations in the epithelial glycocalyx of rabbit uteri during early pseudopregnancy and pregnancy, and following ovariectomy. *Am. J. Anat.* 171, 321-334.

Anderson, T.L., Olson, G.E., and Hoffman, L.H. (1986) Stage-specific alterations in the apical membrane glycoproteins of endometrial epithelial cells related to implantation in rabbits. *Biol. Reprod.* 36, 599-618.

Ashwell, G. and Harford, J. (1982) Carbohydrate-specific receptors of the liver. *Ann. Rev. Biochem.* 51, 531-554.

Bartelmez, G.W., Corner, G.W., and Hartman, C.G. (1951) Cyclic changes in the endometrium of the rhesus monkey (*Macaca mulatta*). *Contr. Embryol. Carnegie Inst.* 34, 99-146.

Brenner, R.M., Carlisle, K.S., Hess, D.L., Sandow, B.A., and West N.B. (1983) Morphology of the oviducts and endometria of cynomolgus macaques during the menstrual cycle. *Biol. Reprod.* 29, 1289-1302.

Carson, D.D., Tang, J.-P., and Hu, G. (1987) Estrogen influences dolichol phosphate distribution among glycolipid pools in mouse uteri. *Biochem.* 26, 1598-1606.

Casimiri, V. and Psychoyos, A. (1981) Embryo-endometrial relationships during implantation. In: *The Endometrium* (ed.), J. DeBrux, R. Mortel, and J.P. Gautray, New York: Plenum Press, pp. 63-79.

Chávez, D.J. (1986) Cell surface of mouse blastocysts at the trophectoderm-uterine interface during the adhesive stage of implantation. *Am. J. Anat.* 176, 153-158.

Chávez, D.J. and Anderson, T.L. (1985) The glycocalyx of the mouse luminal epithelium during estrus, early pregnancy, the peri-implantation period, and delayed implantation. I. Acquisition of *Ricinus communis*-I binding sites during pregnancy. *Biol. Reprod.* 32, 1135-1142.

Chávez, D.J. and Enders, A.C. (1982) Lectin binding of mouse blastocysts: Appearance of Dolichos bifloris binding sites on the trophoblast during delayed implantation and their subsequent disappearance during implantation. *Biol. Reprod.* 21, 545-552.

Cowell, T.P. (1969) Implantation and development of mouse eggs transferred to the uterus of non-progestational mice. *J. Reprod. Fert.* 19, 239-245.

Enders, A.C. and Schlafke, S. (1972) Implantation in the ferret: Epithelial penetration. *Am. J. Anat.* 133, 291-313.

Enders, A.C., Schlafke, S., and Welsh, A.O. (1980) Trophoblastic and uterine epithelial surfaces at the time of blastocyst adhesion in the rat. *Am. J. Anat.* 159, 59-72.

Enders, A.C., Welsh, A.O., and Schlafke, S. (1985) Implantation in the rhesus monkey: Endometrial response. *Am. J. Anat.* 173, 147-169.

Finn, C.A. (1977) The implantation reaction. In: *Biology of the Uterus*, (ed.), R.M. Wynn, New York: Plenum Press, pp. 245-303.

Garcia, J.E., Acosta, A.A., and Jones, H.W. (1983) Advanced endometrial maturation after ovulation induction with hMG/hCG for in vitro fertilization. *Fertil. Steril.* 39, 426.

Guillomot, M., Flechon, J.-E., and Wintenberger-Torres, S. (1982) Cytochemical studies of uterine and trophoblastic surface coats during blastocyst attachment in the ewe. *J. Reprod. Fert.* 65, 1-8.

Harrison, F.L., Fitzgerald, J.E., and Catt, J.W. (1984) Endogenous beta-galactoside-specific lectins in rabbit tissues. *J. Cell Sci.* 72, 147-162.

Healy, D.L. and Hodgen, G.D. (1983) The endocrinology of human endometrium. *Obstet. Gynecol. Survey* 38, 509-530.

Hewitt, K., Beer, A.E., and Grinnell, F. (1979) Disappearance of anionic sites from the surface of the rat endometrial epithelium at the time of implantation. *Biol. Reprod.* 21, 691-707.

Jansen, R.P.S., Turner, M., Johannisson, E., Landgren, B.-M., and Diczfalusy, E. (1985) Cyclic changes in human endometrial surface glycoproteins: A quantitative histochemical study. *Fertil. Steril.* 44, 85-91.

Johnson, L.V. and Calarco, P.G. (1980) Mammalian preimplantation development: The cell surface. *Anat. Rec.* 196, 201-219.

Jones, H.W. (1985) Factors affecting the luteal phase in IVF and ET programs. In: *Human In Vitro Fertilization*; INSERM Symposium No. 24, (ed.), J. Testart and R. Frydman, Amsterdam: Elsevier, pp. 219-235.

Kirby, D.R.S. (1971) The transplantation of mouse eggs and trophoblast to extra-uterine sites. In: *Methods in Mammalian Embryology*, (ed.), J.C. Daniel, San Francisco: W.H. Freeman and Company, pp. 146-156.

Lampelo, S.A., Anderson, T.L., and Bullock, D.W. (1986) Monoclonal antibodies recognize a cell surface marker of epithelial differentiation in the rabbit reproductive tract. *J. Reprod. Fert.* 78, 663-672.

Lee, M.-C. and Damjanov, I. (1985) Pregnancy-related changes in the human endometrium revealed by lectin histochemistry. *Histochem.* 82, 275-280.

Lillie, R.D. and Fullmer, H.M. (1976) *Histopathologic Technic and Practical Histochemistry*, New York: McGraw-Hill, pp. 611-678.

Magnuson, T. and Epstein, C.J. (1981) Characterization of Concanavalin A precipitated proteins from early mouse embryos: A 2-dimensional gel electrophoresis study. *Dev. Biol.* 81, 193-199.

Maroudas, N.G. (1975) Polymer exclusion, cell adhesion, and membrane fusion. *Nature* 254, 695-696.

Martin, L. (1980) What roles are fulfilled by uterine epithelial components in implantation? In: *Blastocyst-Endometrium Relationships; Prog. Reprod. Biol.*, Vol. 7, (eds.), F. Leroy, C.A. Finn, A. Psychoyos, and P.O. Hubinot, Basel: S. Karger, pp. 54-69.

Meromsky, L., Lotan, R., and Raz, A. (1986) Implications of endogenous tumor cell surface lectins as mediators of cellular interactions and lung colonization. *Cancer Res.* 46, 5270-5275.

Milos, N. and Zalik, S.E. (1982) Mechanisms of adhesion among the cells of the early chick blastoderm. Role of the beta-D-galactosamine-binding lectin in the adhesion of extra-embryonic endoderm cells. *Differentiation* 21, 175-182.

Murphy, C.R. and Rogers, A.W. (1981) Effects of ovarian hormones on cell membranes in the rat uterus. III. The surface carbohydrates at the apex of the luminal epithelium. *Cell Biophys.* 3, 305-320.

Nilsson, O. (1967) Attachment of rat and mouse blastocysts onto uterine epithelium. *Int. J. Fertil.* 12, 5-13.

Nilsson O., Lindkvist, I., and Ronqvist, G. (1973) Decreased surface charge of mouse blastocysts at implantation. *Exp. Cell Res.* 83, 421-423.

Padykula, H.A., Coles, L.G., McCracken, J.A., King, N.W., Longcope, C., and Kaiserman-Abramof, I.R. (1984) A zonal pattern of cell proliferation and differentiation in the rhesus endometrium during the estrogen surge. *Biol. Reprod.* 31, 1103-1118.

Potts, D.M. and Psychoyos, A. (1967) L'ultrastructure des relations ovoendometriales au cours du retard experimental de nidation chez la souris. *C.R. Acad. Sci. [D] (Paris)* 264, 956-958.

Psychoyos, A. (1976) Hormonal control of uterine receptivity for nidation. *J. Reprod. Fert. (Suppl.)* 25, 17-28.

Salizar-Rubio, M., Gil-Recansens, M.E., Hicks, J.J., and Gonzalez-Angulo, A. (1980) High resolution cytochemical study of the uterine epithelial surface of the rat at identified sites previous to blastocyst-endometrial contact. *Arch. Invest. Med. (Mex.)* 11, 117-127.

Schacht, C.J. and Foote, R.H. (1978) Progesterone-induced asynchrony and embryo mortality in rabbits. *Biol. Reprod.* 19, 534-539.

Shur, B.D. and Hall, N.G. (1982) A role for mouse sperm surface galactosyl-transferase in sperm binding to the egg zona pellucida. *J. Cell Biol.* 95, 574-579.

Skutelsky, E., Marikovsky, Y., and Danon, D. (1974) Immunoferritin analysis of membrane antigen density: A. Young and old human blood cells. B. Developing erythroid cells and extruded erythroid nuclei. *Eur. J. Immunol.* 4, 512-518.

Trounson, A., Howlett, D., Rogers, P., and Hoppen, H.-O. (1986) The effect of progesterone supplementation around the time of oocyte recovery in patients superovulated for in vitro fertilization. *Fertil. Steril.* 45, 532-535.

Uhlenbruck, G., Baldo, B.A., Steinhausen, G., Schwick, H.G., Chatterjee, B.P., Horejsi, V., Krajhanzl, A., and Kocourek, J. (1978) Additional precipitation reactions of lectins with human serum glycoproteins. *J. Clin. Chem. Clin. Biochem.* 16, 19-23.

Whyte, A. and Robson, T. (1984) Saccharides localized by fluorescent lectins on trophectoderm and endometrium prior to implantation in pigs, sheep, and equids. *Placenta* 5, 533-540.

CHANGES IN LECTIN BINDING PATTERNS IN RABBIT ENDOMETRIUM DURING PSEUDOPREGNANCY, EARLY PREGNANCY AND IMPLANTATION

A. Bükers, J. Friedrich[2], B.P. Nalbach, and H.-W. Denker[1,2]

Institut für Anatomie
RWTH Aachen, Melatener Str. 211
D-5100 Aachen, West Germany

ABBREVIATIONS

apm, apical plasma membrane; ConA, concanavalin A; DBA, Dolichos biflorus agglutinin; FITC, fluorescein isothiocyanate; Gal, galactose; GalNAc, N-acetyl galactosamine; Glc, glucose; GlcNAc, N-acetyl glucosamine; GS-I-B4, Griffonia simplicifolia isolectin B4; hCG, human chorionic gonadotropin; PNA, peanut agglutinin; RCA-I, Ricinus communis agglutinin I (120); RIP, relative inhibitor potency; SBA, soybean agglutinin; UEA I, Ulex europeus agglutinin I; WGA, wheat germ agglutinin.

INTRODUCTION

In the initial phase of embryo implantation, adhesion of the trophoblast to the uterine epithelium is thought to be a critical first step to be followed in invasive types of implantation by penetration through the uterine epithelium into the stroma. Adhesion to the apical plasma membrane of an epithelium is an intriguing phenomenon of largely an unknown mechanism.

Based on recent concepts of cell adhesion and recognition, e.g., during differentiation (Hakomori, 1985), in fertilization (Wassarman, 1987) or tumor metastasis (Nicolson, 1984) carbohydrate groups can be assumed to play an important role in trophoblast-endometrial interaction. Some reports have been published describing stage-dependent glycoconjugate expression on the surfaces of trophoblast and uterine epithelium during this initial phase of attachment (Wu and Chang, 1978; Chávez and Enders, 1981; Guillomot et al., 1982; Nalbach and Denker, 1983; Anderson and Hoffman, 1984; Chávez and Anderson 1985; Nalbach, 1985; Thie et al., 1986). Lectins have proved to be particularly useful for such studies because of their specific binding to defined carbohydrate molecules allowing 1) identification of these molecules at the cell surfaces involved and study of changes in their density during the pre- and post-implantation phases; and 2) blocking of such molecules by application of lectins to the uterus in order to analyze their physiological significance (Guillomot et al., 1982; Sato and Muramatsu, 1985; Hicks and Guzman-Gonzales, 1979; Wu and Gu, 1981; Sretarugsa et al., 1987; Friedrich, in press).

[1]To Whom Correspondence Should Be Addressed.
[2]Present address: Institut für Anatomie, Universitätsklinikum, Hufelandstrasse 55, D-4300 Essen, West Germany

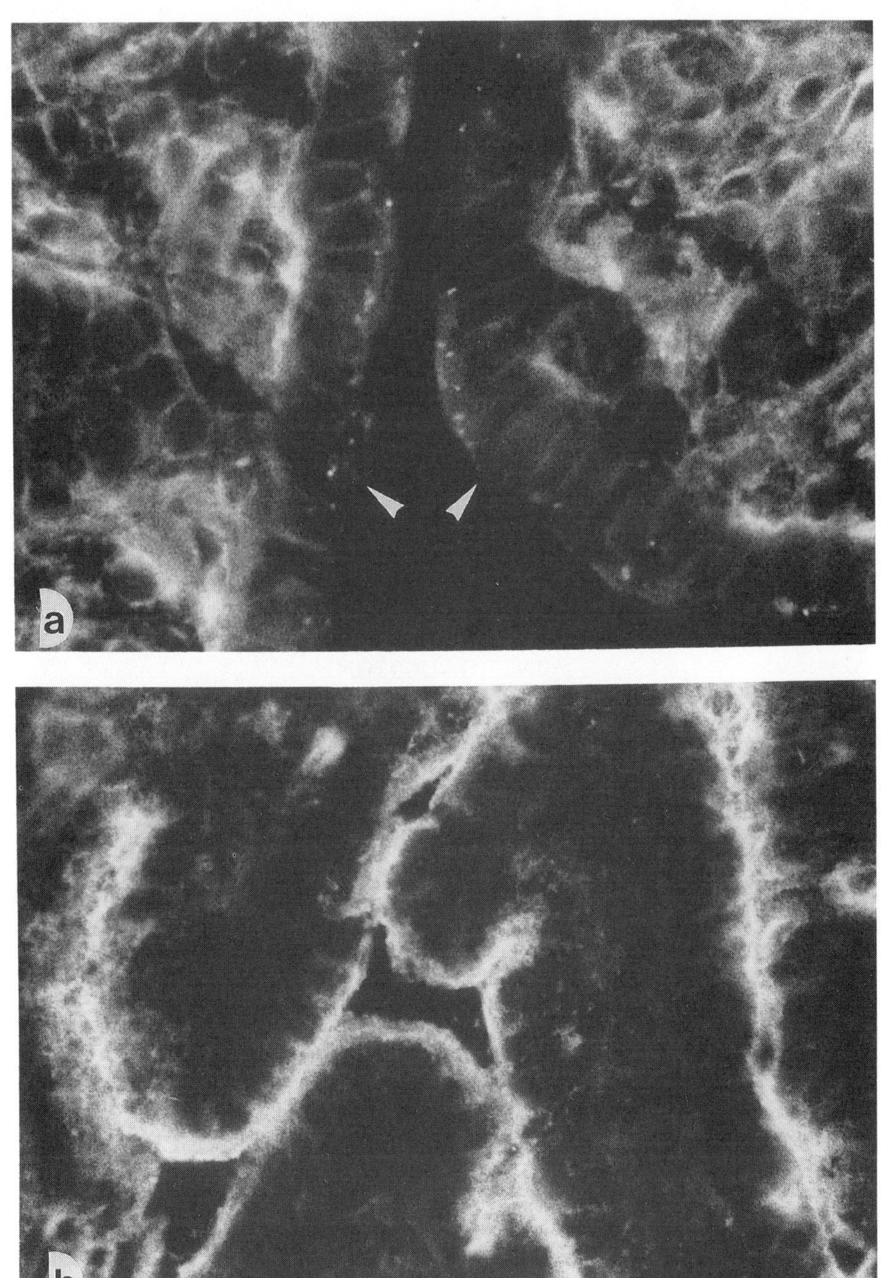

The present communication reports on histochemical lectin binding studies performed in rabbit implantation stages in continuation of previous work by Nalbach, 1985 (cf. Nalbach and Denker, 1983). Unfixed cryostat sections were used in order to minimize artifacts produced by fixation and embedding. In contrast to some previous experiments (Anderson et al., 1986), the use of histologic sections was preferred to the application of lectins via the uterine lumen so that binding to structures located more deeply in the endometrium (e.g., basolateral membranes of the epithelium) could also be studied.

MATERIALS AND METHODS

Adult female rabbits were housed individually in air-conditioned rooms on a 12 hour-light/12 hour-dark cycle, as described previously (Denker, 1977). Before the experiment, rabbits were quarantined for at least 14 days to exclude pseudopregnancy induced by stress ovulation during transport.

Pseudopregnancy was induced by injection of 75 IU hCG (Prolan®, Bayer, Leverkusen, FRG) into the marginal ear vein. Pregnancies were obtained by mating does with males of proven fertility. Does were sacrificed by stunning and exsanguination at desired stages of pregnancy (days post coitum = d p.c.) or pseudopregnancy (days after hCG injection = d p.hCG). Uterine horns were cut into pieces of 1 cm length, quenched in liquid nitrogen, and stored in air-tight bags at -25°C to -30°C until used. At least two animals were used for each stage.

Lectin Histochemistry

FITC-conjugated lectins were obtained from E & Y Laboratories (San Mateo, California, USA). Preliminary experiments determined the optimal time of incubation and lectin dilution from the stock solution (original protein concentration 1 mg/ml in 0.01 M phosphate buffered saline (PBS) pH 7.3): WGA 0.05 mg/ml PBS 30 minutes, RCA I 0.1 mg/ml PBS 30 minutes, GS-I-B4 0.1 mg/ml PBS 60 minutes, PNA 0.5 mg/ml PBS 60 minutes, SBA 0.1 mg/ml PBS 30 minutes, DBA 1 mg/ml PBS 60 minutes, UEA I 0.5 mg/ml PBS 30 minutes. For binding specificity of the lectins see Goldstein and Hayes (1978) and Goldstein and Poretz (1986).

Figure 1. WGA-binding (Sections not rinsed before lectin treatment). a) Non-pregnant, mesometrial endometrium. Low binding to the apical plasma membrane (apm) of luminal epithelium (arrowheads) with a stronger reaction of the apical cytoplasm. X560. b) 8 d p.c., mesometrial endometrium. Strong reaction at the apm of upper parts of crypts. (This reaction is lost at this stage at the cavum epithelium, not shown in this field.) X360.

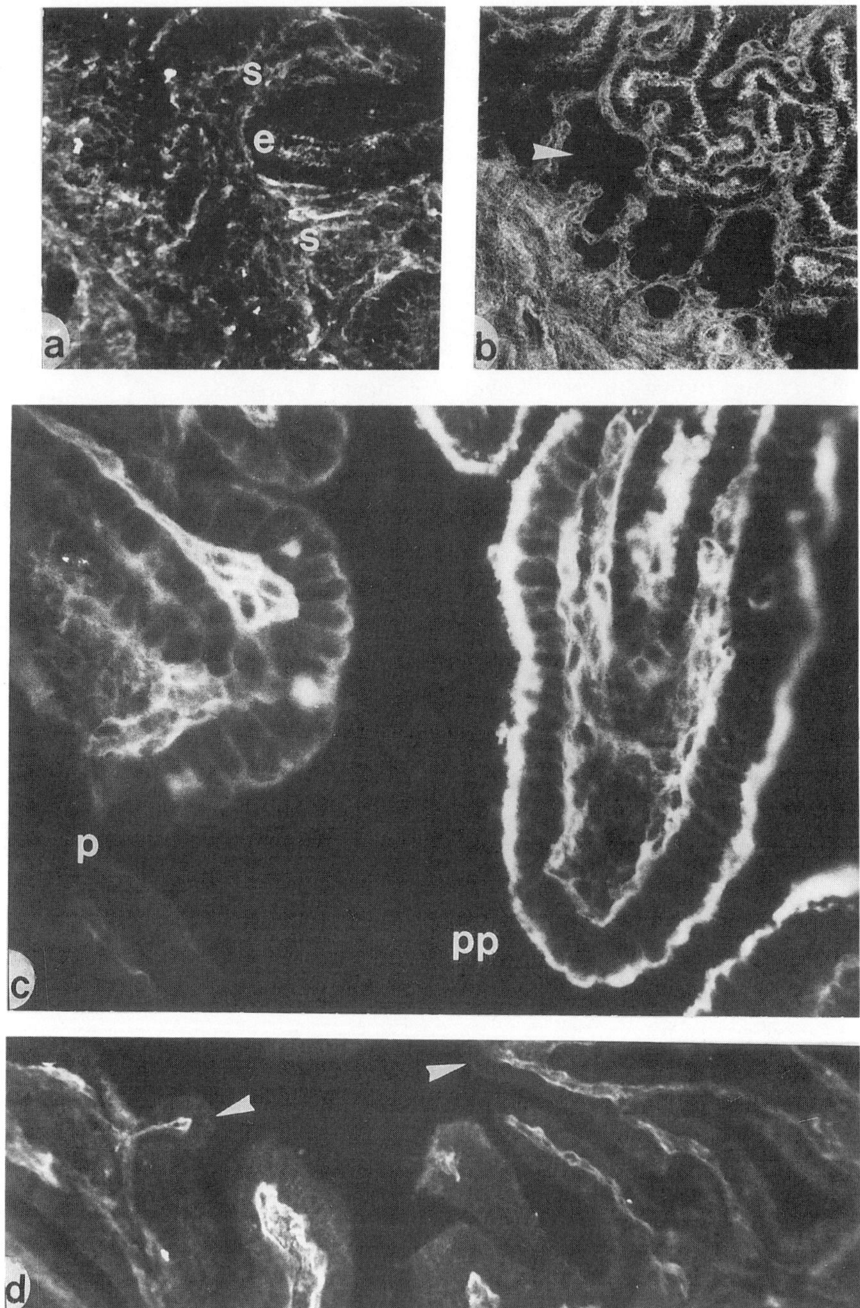

Stock solutions were kept at -25°C to -30°C and working dilutions were always freshly prepared before use. Sections of about 14 μm were cut on a cryostat (Dittes-Duspiva, Heidelberg, FRG) at -25°C, mounted on glass slides at room temperature, and air dried for 1-2 minutes. Sections taken from one piece of a 5 d p.hCG uterus served as a standard and were included in all series of other stages. If not indicated otherwise, slides were rinsed in PBS for 10 minutes at room temperature. Examples from non-rinsed sections are given in Figures 1, 2b, 3, and 5. This was followed by incubation for 30 minutes with 50 μl lectin/section in a moist chamber at room temperature and by washing in PBS. Mounting in glycerol/PBS resulted in non-specific fluorescence which made evaluation difficult, so that PBS alone was preferred as mounting medium. Sections were immediately studied on a Zeiss Photomikroskop II equipped with epifluorescent illumination (filter combination 09, excitation 450-490 nm, barrier filter LP 520). Photodocumentation was performed on Kodak Ektachrome 400 film using standardized exposure times.

Pictures were taken from different locations in the uterus: mesometrial area with placental and paraplacental fold; antimesometrial area with obplacental fold (for definition of these various regions of the rabbit uterus and for illustrations see Böving, 1963 and Denker, 1977).

Controls

To examine the specificity of lectin binding the following inhibitory sugars were used in controls: WGA: N-acetyl glucosamine, 44.2 mg/ml PBS, or N, N', N"-triacetylchitotriose, 0.033 mg/ml PBS; RCA-I: lactose, 14.4 mg/ml PBS; GS-I-B4: α-methyl galactoside, 10.6 mg/ml PBS, β-methyl galactoside, 41.3 mg/ml PBS; PNA: D-galactosamine, 19.4 mg/ml PBS; SBA: N-acetyl-D,L-galactosamine, 44.2 mg/ml PBS; DBA: N-acetyl-D,L-galactosamine, 44.2 mg/ml PBS; UEA-I: α-L-fucose, 32.8 mg/ml PBS. Different sugar concentrations were chosen since inhibitory potencies are known to differ. The employed sugar concentration was calculated by dividing the widely used 0.2 M value by the RIP (relative inhibitor potency; Goldstein and Poretz, 1986).

Figure 2. RCA-I binding. a) 3 d p.hCG (pseudopregnancy), placental fold. Moderate to low binding of the lectin to the epithelium (e), stronger reaction with subepithelial stroma (s). b) 5 d p.c., paraplacental fold (Section not rinsed before lectin treatment). Strong apical reaction of uterine crypts (top right), except for the deepest parts of the crypts (dark, bizarre structures, arrowhead). Moderate reaction in endometrial stroma, stronger in the myometrium (lower left). X112. c) 8 d p.c., implantation chamber. Differential binding to the apm regions of the placental (p) and paraplacental (pp) fold. The strongly reacting paraplacental apm domain exhibits also a significant difference to the basolateral one whereas this difference is less at the placental fold due to the decreased staining of the apm region. X370. d) Control with lactose. 5 d p.c., placental fold. Staining of the apm of the luminal epithelium (arrowheads) is blocked effectively. X148.

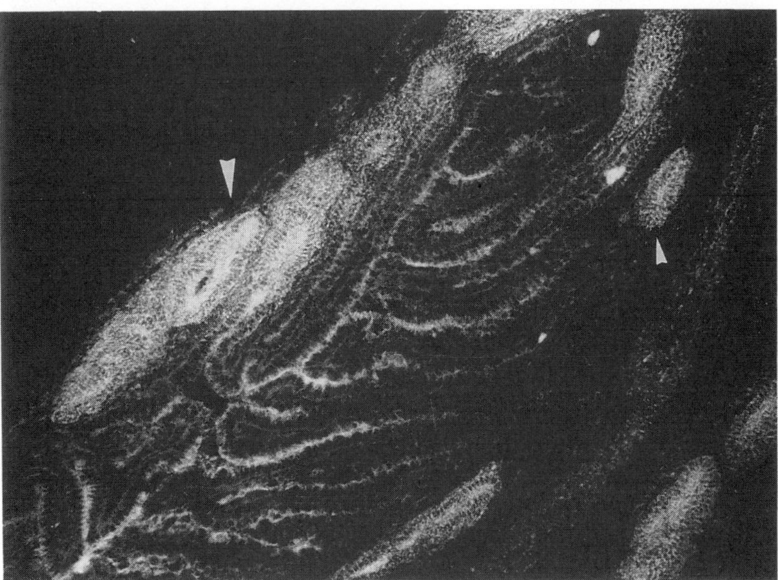

Figure 3.   SBA binding. (Section not rinsed before lectin treatment).   7 d p.c., overview over the mesometrial endometrium. Strong reaction of the deeper parts (arrowheads) of the crypts including both the apm region and the whole cytoplasm and moderate to strong binding to the apm region of upper parts of crypts and the cavum epithelium. X70.

RESULTS

Results are reported predominantly with regard to lectin binding to the apical plasma membrane region (apm) of uterine epithelial cells (see also Tables 1-7) while binding to other structures is described briefly in the text and only where it reached considerable degrees in relation to the reaction of the epithelium. Lectin binding was generally stronger in non-rinsed sections compared to those that were rinsed before lectin treatment.

WGA (primary specificity: D-GlcNAcβ1-4D-GlcNAcβ1-4D-GlcNAc, D-GlcNAcβ1-4D-GlcNAc, D-GlcNAc; in sialic acid-rich structures, binding to this sugar is also possible) bound moderately to strongly to the apical plasma membrane region of the luminal epithelium (Table 1). Uteri from pseudopregnant and non-pregnant rabbits (Figure 1a) tended to react less intensely than those of pregnant animals. A marked difference between the intensity of reaction on the placental and paraplacental fold (Figure 1b) could be seen starting on the 8th day of pregnancy. Such a difference was not detected in interblastocyst segments or in pseudopregnancy. Basal lamina and cell membranes reacted moderately with WGA. Inhibition with triacetylchitotriose and isopotential solutions of N-acetyl glucosamine caused a moderate decrease in binding in all listed sites.

RCA-I (primary specificity: β-D-Gal, D-Galβ1-4D-Glc) binding was weak to moderate at the apical and basolateral membrane of the pseudopregnant rabbit uterine epithelium and of interblastocyst segments (mesometrial endometrium =

antimesometrial side) (Table 2, Figures 2a and b) with a peak at day 7 and 8 (++). Starting on day 8, implantation chambers yielded a marked differential reaction at the apm of the placental (+) and paraplacental fold (+++) (Figure 2c). The mesometrial and antimesometrial endometrium showed significant differences in staining of the apical (+ - +++) versus the basolateral (+) domains of the epithelial membrane. On the placental fold, however, this difference disappeared, beginning on day 8. RCA-I also stained the basement membrane of the epithelium and the elastica interna of the arteries. This could not be suppressed by lactose whereas the reaction on the apical side of the luminal epithelium was blocked effectively (Figure 2d). The deep parts of cryptal epithelium reacted weaker than the upper parts and the luminal epithelium but the pattern was similar.

GS-I-B4 (primary specificity: α-D-Gal, α-D-GalNAc) binding to luminal epithelium showed an increase and decrease over the time period examined (Table 4). Maximum binding was found from day 5 to 7. At the apm of the placental fold the reaction was again less intense than at the paraplacental fold including the antimesometrial area (Figure 4). Binding in the deep crypts paralleled the distribution and strength of reaction of the upper parts. In all parts of the uterus examined, especially at day 7 and 8, significant differences were obvious between the apical and basolateral domain of the membrane. This difference was strong in the luminal epithelium (++-+++ apical and 0-+ basolateral) but not detectable in the deep crypts because of the low apical reaction here. Due to the described decrease of apical staining at the placental fold (in contrast to the paraplacental fold) beginning on day 8 the difference was reduced. Tissues incubated with GS-I-B4 plus α-methyl galactoside or β-methyl galactoside were invariably unlabeled.

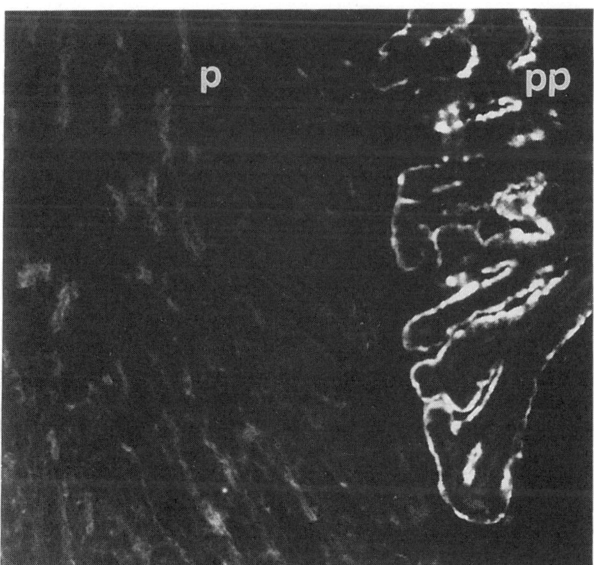

Figure 4. GS-I-B4 binding. 8 d p.c., implantation chamber. Strong binding to the apm region of the paraplacental fold (pp) in contrast to the decreased, very low staining at the placental fold (p). X370.

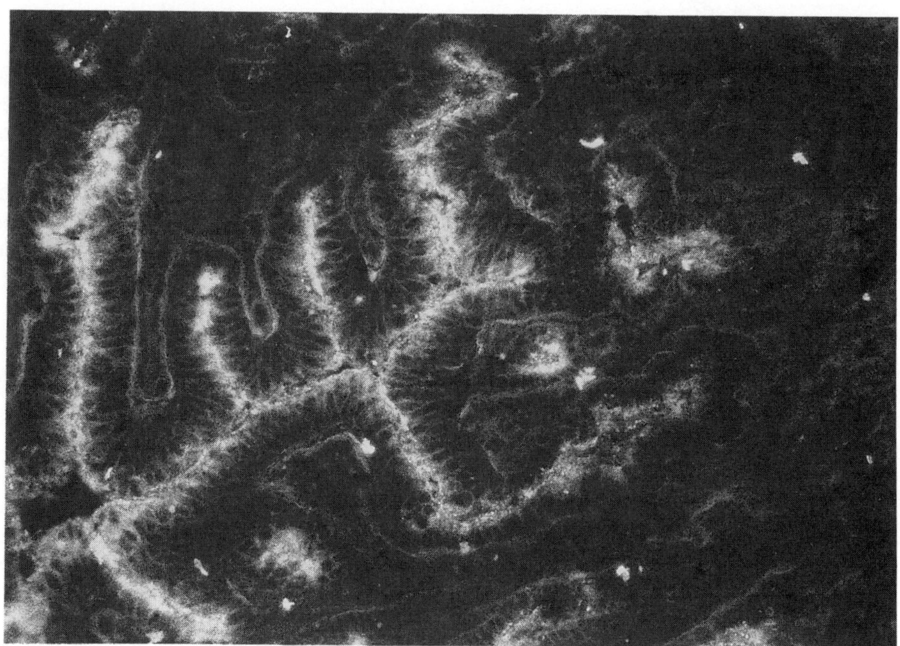

Figure 5. PNA-binding. 3 d p.c., placental fold. (Section not rinsed before lectin treatment). Strong apical reaction of the luminal epithelium, decreasing towards the basolateral membrane domain. Basal lamina moderately stained. Lower binding in the region of the deep crypts, shown in the right corner of this field the placental fold. X180.

PNA (primary specificity: D-Galβ1-3D-GalNAc, β-D-GalNH$_2$, α-D-Gal) showed (Table 5) two phases of increasing reaction at the apical side of the luminal epithelium. The first rise was seen at 3 d p.hCG (Figure 5) and p.c. This was followed by a slight decrease after 5 d. The second increase started at 8 d and only appeared in pregnant rabbits. Interesting results were also found in the deep crypts. Beginning on day 8 there was always (as with SBA) a slightly higher reaction detectable in the deep crypts versus the upper parts of the crypts and the luminal epithelium. An impressive difference was seen between a weak reaction at the placental fold and a strong reaction at the paraplacental fold especially at day 8 and 9 only in the implantation chamber. This strong paraplacental reaction was inhibited specifically by galactosamine. Differences between the apical and basolateral membrane domain as seen with GS-I-B4 could also be detected with PNA. However, in contrast to GS-I-B4, this difference was stronger in the deeper crypts (apical ++-+++ and basolateral only +) than in the luminal epithelium and the upper parts of the crypts (+-++ apical and 0-+ basolateral). This observation was most impressive at day 5 of pregnancy in the interblastocyst segment (+++ apical and + basolateral) and was inhibited by galactosamine in all stages and locations.

TABLE 1

WGA Reaction Pattern of the Uterine Epithelium

	Cavum Epithelium and Upper Parts of the Crypts								Deep Crypts			
	Pseudopregnancy*		Pregnancy						Pregnancy and Pseudopregnancy			
			Interblastocyst segment*		Implantation**chamber				All Locations			
					Paraplacental fold		Placental fold		Cytoplasm		Cell Membrane	
	(a)	(b)	(a)	(b)	(a)	(b)	(a)	(b)	(a)	(b)	(a)	(b)
0d	0-+	0-+							0-+	0-+	0-+	0-+
3d	+	+							0-+	0-+	+	+
5d	0-+	0-+	+-++	+-++					+	+	+	+
7d	+	+	+++	+-++	++	++	++	++	0-+	0-+	0-+	0-+
8d	+	+	++	+-++	++	0-+	+-++	+-++	+	+	+	+
9d	0-+	0-+	++	+-++	+++	+++	++	++	0-+	0-+	+	+

*   Identical reaction of the placental and paraplacental fold
**   First detectable from the 6th day on
(a) =   apical
(b) =   basolateral

TABLE 2

RCA I Reaction Pattern of the Uterine Epithelium

	Cavum Epithelium and Upper Parts of the Crypts								Deep Crypts			
	Pseudopregnancy*		Pregnancy						Pregnancy and Pseudopregnancy			
			Interblastocyst segment*		Implantation**chamber				All Locations			
					Paraplacental fold		Placental fold					
									Cytoplasm		Cell Membrane	
	(a)	(b)	(a)	(b)	(a)	(b)	(a)	(b)	(a)	(b)	(a)	(b)
0d	0-+	0-+							0-+	0-+	+	+
3d	0-+	0-+							0-+	0-+	+	+
5d	+-++	+	+-++	+					0-+	0-+	+	+
7d	++	+	++	+	+++	+	+++	+	0-+	0-+	+	+
8d	++	++	++	+	+++	+	+	+	0-+	0-+	+	+
9d	0-+	0-+	+-++	+	+-++	+	+	+	0-+	0-+	+	+

*   Identical reaction of the placental and paraplacental fold
**   First detectable from the 6th day on
(a) = apical
(b) = basolateral

TABLE 3

SBA Reaction Pattern of the Uterine Epithelium

	Cavum Epithelium and Upper Parts of the Crypts								Deep Crypts			
	Pseudopregnancy*		Pregnancy						Pregnancy and Pseudopregnancy			
			Interblastocyst segment*		Implantation**chamber				All Locations			
					Paraplacental fold		Placental fold		Cytoplasm		Cell Membrane	
	(a)	(b)	(a)	(b)	(a)	(b)	(a)	(b)	(a)	(b)	(a)	(b)
0d	++	++							++	++	++	++
3d	+-++	+-++							0-+	0-+	0-+	0-+
5d	+-++	+-++	0-+	0-+					+	+	+-++	+-++
7d	+-++	+-++	+	+	++-+++	++-+++	++-+++	++-+++	+++	+++	++-+++	++-+++
8d	+	+	0-+	+	++	0-+	0-+	0-+	++	++	+++	+++
9d	+-++	+-++	0	0	+++	++	+	+	++	++	++	++

*   Identical reaction of the placental and paraplacental fold
**   First detectable from the 6th day on
(a) = apical
(b) = basolateral

TABLE 4

GS-I-B4 Reaction Pattern of the Uterine Epithelium

	Cavum Epithelium and Upper Parts of the Crypts								Deep Crypts			
	Pseudopregnancy*		Pregnancy						Pregnancy and Pseudopregnancy			
			Interblastocyst segment*		Implantation**chamber				All Locations			
					Paraplacental fold		Placental fold					
									Cytoplasm		Cell Membrane	
	(a)	(b)	(a)	(b)	(a)	(b)	(a)	(b)	(a)	(b)	(a)	(b)
0d	0	0							0	0	0	0
3d	+	+							0	0	0-+	0-+
5d	+-++	+-++	+++	+					0	0	0-+	0-+
7d	+-++	+	+-++	+	+-++	+	+-++	+	0	0	0-+	0-+
8d	+-++	+	+-++	+	+-++	+	+	+	0-+	0-+	0-+	0-+
9d	0	0	+	0-+	++	+	+	+	0	0	0-+	0-+

*   Identical reaction of the placental and paraplacental fold
**   First detectable from the 6th day on
(a) =   apical
(b) =   basolateral

TABLE 5

PNA Reaction Pattern of the Uterine Epithelium

	Cavum Epithelium and Upper Parts of the Crypts								Deep Crypts			
	Pseudopregnancy*		Pregnancy						Pregnancy and Pseudopregnancy			
			Interblastocyst segment*		Implantation**chamber				All Locations			
					Paraplacental fold		Placental fold					
									Cytoplasm		Cell Membrane	
	(a)	(b)	(a)	(b)	(a)	(b)	(a)	(b)	(a)	(b)	(a)	(b)
0d	0	0							0	0	0	0
3d	++	0-+							0	0	0-++	0-+
5d	++	0-+	+-++	+					+	+	0-++	0
7d	0	0-+	0	0-+	0-+	0-+	0-+	0-+	+	+	+-++	+-++
8d	0	0-+	0-+	0	++	0-+	0-+	0-+	++	+	++	+
9d	0	0	+-++	0	+-++	0-+	0-+	0-+	++	+	++	+

*   Identical reaction of the placental and paraplacental fold
**   First detectable from the 6th day on
(a) = apical
(b) = basolateral

TABLE 6

PNA Reaction Pattern of the Uterine Epithelium

	Cavum Epithelium and Upper Parts of the Crypts								Deep Crypts			
	Pseudopregnancy*		Pregnancy						Pregnancy and Pseudopregnancy			
			Interblastocyst segment*		Implantation**chamber				All Locations			
					Paraplacental fold		Placental fold		Cytoplasm		Cell Membrane	
	(a)	(b)	(a)	(b)	(a)	(b)	(a)	(b)	(a)	(b)	(a)	(b)
0d	0-+	0-+							0-+	0-+	+-++	+-++
3d	0-+	0-+							0-+	0-+	+-++	+-++
5d	0	0	0	0					+-++	+-++	0-+	0-+
7d	0	0	0	0	+-++	+-++	+	+	+-++	+-++	+-++	+-++
8d	0	0	0	0	0	0	0	0	0-+	0-+	0-+	0-+
9d	0	0	0	0	0	0	0	0	0-+	0-+	0-+	0-+

*   Identical reaction of the placental and paraplacental fold
**   First detectable from the 6th day on
(a) =   apical
(b) =   basolateral

TABLE 7

UEA I Reaction Pattern of the Uterine Epithelium

	Cavum Epithelium and Upper Parts of the Crypts								Deep Crypts			
	Pseudopregnancy*		Pregnancy						Pregnancy and Pseudopregnancy			
			Interblastocyst segment*		Implantation**chamber				All Locations			
					Paraplacental fold		Placental fold		Cytoplasm		Cell Membrane	
	(a)	(b)	(a)	(b)	(a)	(b)	(a)	(b)	(a)	(b)	(a)	(b)
0d	0-+	0-+							0-+	0-+	0-+	0-+
3d	0-+	0-+							0-+	0-+	0-+	0-+
5d	+	+	0-+	0-+					0-+	0-+	0-+	0-+
7d	+	+	0-+	0-+	+	+	+	+	0-+	0-+	0-+	0-+
8d	0-+	0-+	0-+	0-+	+	+	0-+	0-+	0-+	0-+	0-+	0-+
9d	0-+	0-+	0-+	0-+	+	+	0-+	0-+	0-+	0-+	0-+	0-+

*   Identical reaction of the placental and paraplacental fold
**   First detectable from the 6th day on
(a) =   apical
(b) =   basolateral

SBA (primary specificity: α-D-GalNAc, β-D-GalNAc) was the only lectin that bound strongly to estrous endometrium (apm of cavum epithelium) (Table 3). At later stages, reactivity decreased (3d, 5d) and rose again towards time of implantation when a differential reaction of placental (+) and paraplacental fold (+++) could be noted similar to RCA-I. This difference was not seen in interblastocyst segments or pseudopregnancy. SBA-reaction intensity pattern in the deep parts of crypts was always roughly one point higher than the reaction in the luminal epithelium (Figure 3) but followed the differential distribution within the implantation chamber in basically the same manner. SBA always exhibited a granular reaction in contrast to the homogeneous distribution in case of all other lectins. N-acetyl glucosamine was an inhibitor of striking potency with total inhibition of SBA binding sites in all parts of the samples even in the strongly reacting deep crypts thus confirming specificity.

DBA (primary specificity: α-D-GalNAc, D-GalNAcα1-3-D-Gal) binding was unique in various aspects (Table 6). There was no or only a very weak reaction at the luminal epithelium in any stage and location. Only a minor peak occurred at the 7th day in the implantation chamber. In the deep crypts, however, binding was moderate also in other stages and interesting changes were seen. On day 0 and 3, DBA bound here predominantly to the cell membrane whereas on the 5th and 7th day, reaction could be clearly localized in the cytoplasm. The distribution switched back to a membrane dominance on day 8 and 9. In all stages and regions, terminal bars were stained but this was resistant to blocking with N-acetyl galactosamine. Therefore those structures are suspected to be stained unspecifically.

UEA-I (primary specificity: α-L-fucose) presented only a weak reaction in the different stages and parts of the uterus even at the apm (Table 7). Incubation with α-L-fucose gave no significant change in binding intensity.

DISCUSSION

Galactosyl end groups of surface-bound glycoproteins (or glycolipids) have previously been proposed to play a role in trophoblast attachment to the uterine epithelium, based primarily on histochemical (and limited biochemical) data (Chávez, 1986; Anderson et al., 1986). This appears indeed very suggestive, since a wealth of data from more global studies on cell surface carbohydrates in these structures documents that changes do occur here in the periimplantation phase (Jenkinson and Searle, 1977; Wu and Chang, 1978; Nilsson et al., 1980; Carollo and Weitlauf, 1981; Chávez and Enders, 1981; Guillomot et al., 1982; Marticorena et al., 1983; Anderson and Hoffman, 1984; Whyte and Robson, 1984; Chávez and Anderson, 1985; Sato and Muramatsu, 1985; Thie et al., 1986). However, a detailed analysis including a comparison of various regions of the implantation chamber, interblastocyst segments and the appropriate stages of pseudopregnancy reveals that there are conflicting observations which need to be discussed.

Anderson et al. (1986) studied the binding of FITC-conjugated WGA, RCA I, UEA, and SBA to estrous and pseudopregnant uterine epithelium. It was concluded that acquisition of terminal galactosyl groups at the apical cell membrane of the uterine epithelium is positively correlated with "receptivity". Only FITC-ConA was employed for binding studies to uteri of 6 1/2 day pregnant rabbits and revealed no differential reaction of the antimesometrial and mesometrial region of the implantation chamber. This was confirmed by studies using ferritin-labeled lectins (RCA-I, succinylated ConA and WGA, UEA) in uteri of 7 1/2 day pregnant rabbits. Those authors did not find any differences in reaction pattern of pseudopregnant uteri and interblastocyst segments of the corresponding pregnant stages. However, they describe a reduction in binding of both ferritin-labeled lectins at implantation sites compared to nonimplantation sites in pseudopregnant uteri.

In contrast, the current investigation documented a decrease of the RCA I reaction only on the placental fold but an increase at the obplacental and paraplacental uterine epithelium of the implantation chamber at day 7 and 8 of pregnancy. This would be in agreement with Anderson's hypothesis about the role of Gal groups (see above) only as far as trophoblast attachment to the antimesometrial epithelium is concerned. Unfortunately, however, the same stages were not examined by those authors (Anderson et al., 1986). In pseudopregnancy and early stages - before 7 days p.c. - of pregnancy, major differences between the different folds were likewise never found in the current study.

Some of the differences between the cited study and the present one may be due to the different experimental approaches. Anderson et al. (1986) did not perform the lectin binding reaction on sections but rather introduced the lectins into the uterine lumen of segments of flushed and fixed uteri so that only the apical (uterine) epithelial plasma membrane was able to react. In the current study, lectins were applied to sections of uteri so that binding could also occur to basolateral plasma membranes and to cells/matrix located more deeply in the endometrium. In order to prevent blurring of the pattern by redistribution of uterine (particularly secretory) glycoconjugates, sections - except for some examples given in the figures - were gently rinsed in PBS before applying the lectins.

A difficulty in interpreting the histochemical lectin binding data is posed by the importance of two facts: 1. primary sugar specificity of the lectin; 2. steric configuration of the sugar part of the glycoconjugate. Lectins with the same primary specificity for sugar moieties (Goldstein and Hayes, 1978; Goldstein and Poretz, 1986) may differ strongly in their reaction pattern due to differences in complex molecular structure of the optimal ligand (for a discussion see Nalbach, 1985). SBA and DBA both exhibit great affinity for non-reducing end groups of N-acetyl galactosamine. A comparison of the reaction pattern found for those lectins (see Table 3 and 6) presents a stronger binding of SBA in nearly all regions and phases of implantation. Another difference exists in the granular reaction product of SBA. Furthermore, SBA shows a stronger inhibition by N-acetyl galactosamine in contrast to DBA. The molecular weight of both lectins ranges at about 120000, but there are four carbohydrate binding sites in SBA and only two in DBA (Goldstein and Poretz, 1986). RIP of galactose to SBA is 0.04 and to DBA is 0.01.

As far as Gal-binding lectins are concerned, RCA I prefers the β1-4 linkage whereas PNA binds 50-fold stronger to Galβ1-3GalNAc than to galactose. We found that α-methyl galactoside displayed the same inhibitory potency. It must be left open whether the observed effects reflect an abundance of βGal at the epithelium.

The finding that may be most important for implantation physiology is the differential reaction of various parts of endometrium within the implantation chamber. In this series, RCA-I, SBA, PNA, GS-I-B4, and WGA showed intense binding to the antimesometrial region and very low binding to the mesometrial region from the 8th day on. Such a difference was never seen in interblastocyst segments or in pseudopregnancy. All of these lectins except for WGA, recognizing D-GlcNAc and eventually sialic acids, have primary specificity for D-Gal or D-GalNAc terminal groups. Definitive chorioallantoic placentation of the rabbit blastocyst takes place selectively on the placental fold of the mesometrial region. One would assume, therefore, that a loss of D-Gal or D-GalNAc at the mesometrial uterine epithelium may be a precondition for trophoblast attachment at least in this region (although another D-GalNAc-binding lectin, DBA, did not react differentially on the placental and paraplacental fold).

The reduced binding of the placental fold uterine epithelium is in contrast to the previously proposed concept (Anderson et al., 1986; Chávez and Anderson, 1985). It postulates that expression of galactosyl groups at the uterine epithelial surface is a precondition for trophoblast attachment.

Our histochemical findings concerning the antimesometrial part of the endometrium would be in agreement with such a concept. However, the high density of galactosyl groups is most striking here after the uterine epithelium has been transformed into a symplasm, a stage which has possibly already passed receptivity (Böving, 1963). In contrast, the current observations made on the lectin binding patterns at the placental fold, where chorioallantoic placentation takes place in the rabbit, are not in agreement with such a mechanism. This appears quite relevant since the situation met here can be expected to be particularly suited for the histochemical evaluation so that results should be meaningful: Attachment and fusion of trophoblast and uterine epithelium occur over long membrane stretches comprising many cells, i.e., membrane changes that do occur cannot easily be overlooked, whereas antimesometrially trophoblastic knobs attach first only punctually to single uterine epithelial cells. The generalized loss of binding sites for a variety of lectins at the placental folds would indeed fit better with an alternative hypothesis, i.e., that a partial loss of apical-type characteristics prepares the apical plasma membrane of the uterine epithelium for trophoblast attachment in the "receptive phase" (Denker, 1986). Obviously, more work on the histochemical patterns of membrane-bound carbohydrate groups, including additional lectins and localization at the EM level, will be needed before the question about the correct model can be definitely answered.

SUMMARY

Previous investigations have already provided some data on the composition of the surface coats of the uterine epithelium and the trophoblast using lectin binding studies at both light and electron microscopical levels. The present investigation seeks to compare, in the rabbit, a larger number of different reproductive states as well as different regions within the implantation chamber including sites where the trophoblast normally attaches and those where it does not.

Thus, binding of 7 plant lectins (WGA, RCA I, SBA, GS-I-B4, PNA, DBA, and UEA I) was studied histochemically at the light microscopical level on cryostat sections taken from non-pregnant, pseudopregnant (3-9 days after hCG injection), and pregnant (5-9 days post coitum) rabbits. Differences between binding at the antimesometrial (obplacental and paraplacental fold) and the mesometrial (placental fold) parts of the endometrium and, in pregnant animals, between implantation chamber and interblastocyst regions were recorded.

Generally strong reactivity was found at the uterine apical epithelial cell surface for D-Gal- or D-GalNAc-binding lectins (RCA I, SBA, GS-I-B4, and PNA). Binding intensity increased, at the obplacental and paraplacental fold of the uterine epithelium in the implantation chamber, from 7 to 9 days of pregnancy or pseudopregnancy. Interestingly, reactivity decreased at the apical membranes at the placental fold, from 8 days post coitum on, in the implantation chamber. Thus, there are remarkable differences between the lectin binding behavior of the two parts of the rabbit endometrium, the obplacental/paraplacental region (where localized adhesion of trophoblastic knobs of the abembryonic hemisphere of the blastocyst occurs during the first phase of attachment) and the placental fold (where trophoblast of the embryonic pole attaches, one day later, on a large front). It is proposed that reduction of certain sugar residues at the glycocalyx of the apical plasma membrane (as measured by plant lectins) may precondition the uterine epithelium for trophoblast attachment (as an element of "receptivity") at the placental fold. The contradiction, however, that the antimesometrial endometrium does not show this type of behavior cannot be resolved at present. Thus a comparative investigation of the various reproductive states and uterine regions puts into doubt the previously proposed hypothesis that the expression of galactosyl groups at the uterine epithelial cell surface is positively correlated with "receptivity" for implantation. The results are, however, consistent with the concept that reduction of apical-type characteristics of the apical plasma membrane of the uterine epithelium may be important in the acquisition of "receptivity" at least at the placental fold.

ACKNOWLEDGEMENTS

The authors like to thank Ms. Lisa Hölscher and Ms. Gerda Helm for excellent technical assistance, Ms. Gisela Mathieu for secretarial help and Ms. Marina Luerkens for darkroom work. These investigations were supported by grants from the Deutsche Forschungsgemeinschaft.

REFERENCES

Anderson, T.L. and Hoffman, L.H. (1984) Alterations in epithelial glycocalix of rabbit uteri during early pseudopregnancy and pregnancy, and following ovariectomy. *Am. J. Anat.* 171, 321-334.

Anderson, T.L., Olson, G.E., and Hoffman, L.H. (1986) Stage-specific alterations in the apical membrane glycoproteins of endometrial epithelial cells related to implantation in rabbits. *Biol. Reprod.* 34, 701-720.

Böving, B.G. (1963) Implantation mechanisms. In: *Conference On Physiological Mechanisms Concerned With Conception*, (ed.), C.G. Hartmann, Pergamon Press, pp. 321-396.

Chávez, D.J. (1986) Cell surface of mouse blastocysts at the trophectoderm-uterine interface during the adhesive stage of implantation. *Am. J. Anat.* 176, 153-158.

Chávez, D.J. and Enders, A.C. (1981) Temporal changes in lectin binding of peri-implantation mouse blastocysts. *Develop. Biol.* 87, 267-276.

Chávez, D.J. and Anderson, T.L. (1985) The glycocalyx of the mouse uterine luminal epithelium during estrous, early pregnancy, the peri-implantation period, and delayed implantation. I. The aquisition of Ricinus communis I binding sites during pregnancy. *Biol. Reprod.* 32, 1135-1142.

Carollo, J.R. and Weitlauf, H.M. (1981) Regional changes in the binding of (^{3}H) Concanavalin A to mouse blastocysts at implantation: An autoradiographic study. *J. Exp. Zool.* 218, 247-251.

Denker, H.-W. (1977) Implantation: The role of proteinases, and blockage of implantation by proteinase inhibitors. *Adv. Anat. Embryol. Cell Biol.* 53, Fasc. 5.

Denker, H.-W. (1986) Epithel-Epithel-Interaktionen bei der Embryo-Implantation: Ansätze zur Lösung eines zellbiologischen Paradoxons. *Verh. Anat. Ges.* 80 (*Anat. Anz. Suppl.* 160), 93-114.

Goldstein, I.J. and Hayes, C.E. (1978) The lectins: Carbohydrate-binding proteins of plants and animals. *Adv. Carbohydr. Chem. Biochem.* 35, 127-340.

Goldstein, I.J. and Poretz, R.D. (1986) Isolation, physicochemical characterization, and carbohydrate-binding specificity of lectins. In: *The Lectins; Properties, Functions, And Applications In Biology And Medicine*, (eds.) I.E. Liener et al., Academic Press, Inc., pp. 33-214.

Guillomot, M., Fléchon, J.-E., and Winterberger-Torres, S. (1982) Cytochemical studies of uterine and trophoblastic surface coats during blastocyst attachment in the ewe. *J. Reprod. Fert.* 65, 1-8.

Hakomori, S. (1985) Glycosphingolipids as differentiation and tumor markers and as regulators of cell proliferation. In: *Molecular Biology of Tumor Cells*, (eds.) B. Wahren et al., New York, Raven Press, pp. 139-156.

Hicks, J.J. and Guzman-Gonzales, A.M. (1979) Inhibition of implantation by intraluminal administration of Concanavalin A in mice. *Contraception* 20, 129-136.

Jenkinson, E.J. and Searle, R.F. (1977) Cell surface changes on the mouse blastocyst at implantation. *Exp. Cell Res.* 106, 386-390.

Marticorena, P., Hogan, B., DiMeo, A., Artzt, K., and Bennett, D. (1983) Carbohydrate changes in pre- and peri-implantation mouse embryos as detected by a monoclonal antibody. *Cell Differ.* 12, 1-10.

Nalbach, B.P. (1985) Lektinbindungsmuster in Uterus und Blastozyste des Kaninchens während der Präimplantationsphase und der frühen Implantation. Histochemie und Methodenkritik. Dissertation, Medizinische Fakultät der RWTH Aachen.

Nalbach, B.P. and Denker, H.-W. (1983) Stage-dependent changes in lectin binding patterns in rabbit uterus and blastocyst during the preimplantation period and implantation. *Europ. J. Cell Biol. Suppl.* 4, 13.

Nicolson, G.L. (1984) Cell surface molecules and tumor metastasis. Regulation of metastatic phenotypic diversity. *Exp. Cell Res.* 150, 3-22.

Nilsson, B.O., Naeslund, G., and Curman, B. (1980) Polar differences of delayed and implanting mouse blastocysts in binding of Alcian blue and Concanavalin A. *J. Exp. Zool.* 214, 177-180.

Sato, M. and Muramatsu, T. (1985) Reactivity of five N-acetyl galactosamine-recognizing lectins with preimplantation embryos, early post-implantation embryos, and teratocarcinoma cells of the mouse. *Differentiation* 29, 29-38.

Sretarugsa, P., Sobhon, P., Bubpaniroj, P., and Yodyingyuad, V. (1987) Inhibition of implantation of hamster embryos by lectins. *Contraception* 35, 507-515.

Thie, M., Bochskanl, R., and Kirchner, C. (1986) Glycoproteins in rabbit uterus during implantation. Differential localization visualized using ^{3}H-N-acetyl-glucosamine labelling and FITC-conjugated lectins. *Histochem.* 84, 73-79.

Wassarman, P.M. (1987) The biology and chemistry of fertilization. *Science*, 235, 553-560.

Whyte, A. and Robson, T. (1984) Saccharides localized by fluorescent lectins on trophectoderm and endometrium prior to implantation in pigs, sheep and equids. *Placenta* 5, 533-540.

Wu, J.T. and Chang, M.C. (1978) Increase in Concanavalin A binding sites in mouse blastocysts during implantation. *J. Exp. Zool.* 205, 447-453.

Wu, J.T. and Gu, Z. (1981) The effect of intrauterine injection of Concanavalin A on implantation in mice and rats. *Contraception* 23, 667-675.

PREPARATION OF RABBIT UTERINE EPITHELIUM FOR TROPHOBLAST ATTACHMENT: HISTOCHEMICAL CHANGES IN THE APICAL AND LATERAL MEMBRANE COMPARTMENT

I. Classen-Linke[1] and H.-W. Denker[2]

Institut für Anatomie
Medizinische Fakultät der RWTH Aachen
Melatener Strasse 211
D-5100 Aachen, Federal Republic of Germany

INTRODUCTION

Embryo implantation is initiated by an interaction of the trophoblast with the uterine epithelium. In the preimplantation phase the uterine epithelium undergoes extensive changes that are detectable by morphological and histochemical methods. It is usually assumed that these changes 'are somehow related to the acquisition of a state of "receptivity" (Psychoyos, 1976) for trophoblast adhesion and invasion. The morphological transformation is already well documented in the rabbit, one of the most widely used models for implantation studies (Böving, 1963; Denker, 1970, 1977; Enders and Schlafke, 1971; Beier, 1973; Davies and Hoffman, 1973, 1975; Suzuki and Tsutsumi, 1980, 1981; Busch, 1982; Winterhager, 1985).

In the present investigation changes in the composition of the uterine epithelial plasma membrane were monitored. Such studies sought to explain the cell biological paradox (Denker, 1986) that, at implantation, interaction occurs between the otherwise non-adhesive apical surfaces of the trophoblast and the uterine epithelium. So-called brush border enzymes (alkaline phosphatase, aP; aminopeptidase M, APM; gamma-glutamyl transferase, GGT; dipeptidyl peptidase IV, DPP IV) were studied as markers for the apical plasma membrane of a typical epithelium with polarized organization, and their changes with the stage of pregnancy or pseudopregnancy were noted. Previously data were reported on the light microscopical (LM) histochemistry of the enzymes (Classen-Linke et al., 1987). These data are now augmented by immunohistochemical and electron microscopical (EM) investigations to elucidate whether the reported pattern changes are due to: 1) activation/inactivation phenomena; 2) changes in density of microvilli; or 3) changes in the concentration of the membrane-bound proteins. In addition, the desmosomal plaque proteins desmoplakin I and II were investigated as markers for the basolateral membrane compartment.

MATERIALS AND METHODS

Rabbits were kept and bred as described previously (Denker, 1977). Pseudopregnancy was induced by injecting 75 I.U. of human chorionic

[1]To Whom Correspondence Should Be Addressed.
[2]Present address: Institut für Anatomie, Universitätsklinikum, Hufelandstrasse 55, D-4300 Essen 1, Federal Republic of Germany

gonadotropin (hCG) intravenously.   Does were euthanatized at defined stages post coitum (days post coitum = d p.c., pregnancy) or after hCG injection (days post hCG injection = d p.hCG, pseudopregnancy) by stunning and exsanguinating.

Light Microscopical Enzyme Histochemistry

The uterus was quickly removed, cut into pieces, frozen unfixed in liquid nitrogen and stored at -25°C to -30°C in airtight plastic bags for about 6 to 12 months without loss of enzyme activity. Two to five rabbits were used for the non-pregnant state and for each of the stages 3, 5, 6, 7, 7 1/2, 8, and 9 days p.c. and of 3, 5, 6, 7, 8, and 9 days after hCG injection.   In pregnancy, implantation chambers and interblastocyst segments were investigated separately. Native longitudinal or cross sections, 8 - 14 µm thick were cut on a cryostat (Dittes-Duspiva, Heidelberg, FRG) at -25°C, mounted at room temperature on glass slides and air-dried for 1-2 minutes. Enzyme activity was demonstrated histochemically by a metal salt (lead) procedure (aP) or by a simultaneous azo-coupling procedure (APM, GGT, DPP IV) (for further details, see Classen-Linke et al., 1987).

Light Microscopical Immunohistochemistry

Monoclonal antibodies (1D8.1 and 4H7.1) against the purified intestinal aminopeptidase M of the rabbit were produced and kindly provided by J.-P. Gorvel and S. Maroux, C.N.R.S., Marseille, France. These antibodies are mouse IgG1. The production and specificity is described by Gorvel et al. (1989).

The immunohistochemical studies were performed as follows: 8-14 µm thick cryostat sections were mounted on gelatin-coated glass slides, air-dried for about 1-2 minutes, collected in PBS (phosphate buffered saline, Biochrom, Berlin, FRG) and then pre-incubated for 20 minutes in non-immune rabbit serum (1:20 diluted in PBS/BSA (bovine serum albumin, Sigma, Deisenhofen, FRG) 1.5%, pH 7.4). Serum was tapped off and excess solution was wiped away from the slides. The sections were incubated with the first antibody (diluted in PBS/BSA 1.5%, pH 7.4; 1D8.1 - 1:16, 4H7.1 - 1:17) for 1 hour at room temperature in a humid chamber. After rinsing in 5 changes of PBS, the second antibody (FITC-conjugated rabbit anti-mouse IgG, Dakopatts, Hamburg, FRG) was applied in a dilution of 1:20 for 1/2 hour. After further rinsing in PBS, the sections were mounted under coverslips in a 9:1 mixture of glycerol:PBS.

The monoclonal antibodies with the epitope designation 2.15, 2.17, and 2.20 against the desmosomal plaque proteins desmoplakin I and II (Cowin et al., 1985) were produced and kindly provided by W. Franke, German Cancer Research Center, Heidelberg, FRG. The production and specificity is described by Cowin et al. (1985) and Moll et al. (1986). For the immunohistochemical procedure see above. The sections were incubated with a mixture of the three monoclonal antibodies (undiluted supernatant). The second antibody (FITC-conjugated rabbit anti-mouse IgG, Dakopatts, Hamburg, FRG) was applied in a dilution of 1:20 for 1/2 hour. After rinsing in PBS, the sections were rinsed in aqua bidest and in 100% ethanol, air-dried, and were mounted in Mowiol (Hoechst, Frankfurt, FRG).

Sections were studied and photographed in a Zeiss Photomikroskop II equipped with epifluorescence (excitation at 450-490 nm, emission wavelength: 500-550 nm).

Electron Microscopy

For electron microscopical investigations uterine pieces of about 4 mm thickness were fixed in 1.25% paraformaldehyde + 0.1% glutaraldehyde for about 5 hours, rinsed overnight in a 5% saccharose-phosphate buffer pH 7.4, incubated for 4-5 hours in phosphate buffer containing 15% saccharose (cryoprotectant) and frozen in liquid propane. The histochemical or immunohistochemical reaction was performed on 40 μm cryostat sections.

For the demonstration of alkaline phosphatase activity according to Oledzka-Slotwinska et al. (1967) (cf. Lewis and Knight, 1977) the sections were incubated in a medium containing cytidine monophosphate as substrate and $CeCl_3$ instead of $Pb(NO_3)_2$ in Tris/maleate buffer pH 8.5 for 1/2 hour at room temperature. After rinsing in buffer, the sections were fixed for 1 hour in 2.5% glutaraldehyde, rinsed again, postfixed for 1 hour in 2% OsO_4, dehydrated in graded ethanol solutions, and embedded in araldite for ultrathin sectioning. Some of the thin sections were stained with uranyl acetate and lead citrate.

For the immunohistochemical localization of aminopeptidase M at the EM level the 40 μm sections were incubated overnight with the first antibody 1D8.1 (see above) at a dilution of 1:16 PBS/BSA 1.5%, pH 7.4 in a humid chamber at 4°C. After rinsing in PBS, the second antibody (peroxidase-conjugated rabbit anti-mouse IgG, Dakopatts, Hamburg, FRG) was applied at a dilution of 1:20 PBS/ BSA 1.5% pH 7.4 for 4 hours. After rinsing in PBS, 0.05% diamino benzidine + 0.01% H_2O_2 was applied for 10-15 minutes. The sections were rinsed in PBS, fixed for 1/2 hour in 2.5% glutaraldehyde, rinsed again in PBS, postfixed in 2% OsO_4, dehydrated in graded ethanol solutions and embedded in araldite. Ultrathin sections were viewed in a Zeiss EM 10.

RESULTS

Whereas enzyme reactions were absent or very weak at the uterine epithelium of the non-pregnant state, all enzymes investigated (aP, APM, GGT, DPP IV) demonstrated a strong reaction at the apical plasma membrane of the epithelium in the pre-implantation phase (3-6 d p.c./p. hCG) (Table 1). A positive enzyme reaction was already clearly visible at 3 d p.c./p.hCG (Figure 8) and maximum activity could be demonstrated at 5 d p.c./p. hCG for aP and APM (Figures 1 and 10) and for GGT and DPP IV at 5 and 6 d p.c. /p. hCG (Figures 4 and 6). Activity of aP and APM started to decline at 6 d p.c./p. hCG and for GGT and DPP IV one to two days later, i.e., at 7-8 d p.c./p. hCG (Table 1). At 8 d p.c./p. hCG (Figures 5 and 7) the enzyme activity at the apical plasma membrane was nearly completely abolished in uterine epithelium of pseudopregnant animals as well as in interblastocyst segments of pregnant animals. At the time of implantation (about 7 d p.c.) a difference in activity of aP and DPP IV between implantation chamber and the interblastocyst segments first became detectable. There were also differences between the antimesometrial region, where the first attachment of trophoblastic knobs occurs, the paraplacental fold and the placental fold, i.e., the

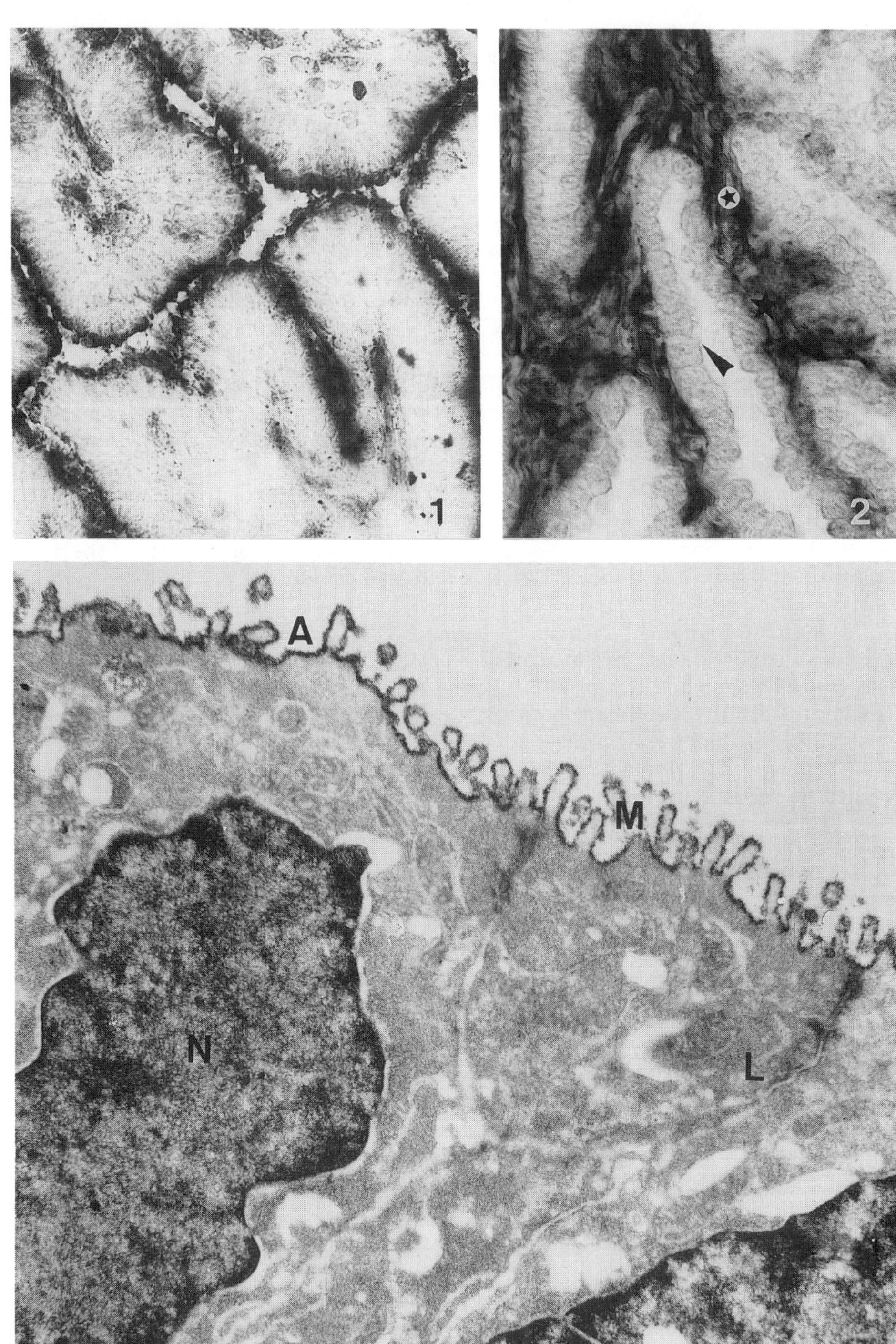
1
2
3
A
M
L
N

Table 1

Activity of marker enzymes at the apical plasma membrane.
Rabbit uterine epithelium during pseudopregnancy

	non-pregnant state	days post hCG					
		3	5	6	7	8	9
APM	0-+	++	+++	(+)-++	0-+	0-(+)	0
aP	0-(+)	++	++(+)	+(+)	0	0	0
GGT	0-(+)	++	++(+)	+++	+	(+)	0-(+)
DPP IV	0-+	++	+++	+++	+++	(+)	

Table 2

Activity of marker enzymes at the apical plasma membrane. Rabbit
uterine epithelium during pregnancy, implantation chamber[1]

	days post coitum									
	5		6		7		7 1/2		8	
	I	II	I	II	I	II	I	II	I	II
APM	+++	+++	+	+	0	0	0	0	0	0
aP	+++	+++	+(+)	+(+)	++	+(+)	+++	+(+)	+++	+
GGT			++(+)	++(+)	++	+(+)	+	(+)	(+)	0
DPP IV	+++	+++	+++	+++	+++	++	+++	++	+++	+

I = paraplacental fold and antimesometrial region; II= placental fold
[1]At 5 d p.c., the blastocysts have not yet reached their final position within the uterus so that data given are for the part of the uterine epithelium that happened to be adjacent to the blastocyst at this stage.

Figure 1. Alkaline phosphatase, LM. Pre-implantation phase (5 d p.c.). Strong reaction at the apical plasma membrane of uterine epithelium; there is no staining of the lateral membranes; X350.

Figure 2. Alkaline phosphatase, LM. Implantation phase (7 1/2 d p.c.). The main reaction of aP is found in the subepithelial stroma (*); the apical plasmalemma (arrow) of epithelial cells is nearly negative; X350.

Figure 3. Alkaline phosphatase, EM. 5 d p.hCG. The strong and distinct staining of the apical plasma membrane and of the microvillar area is evident; the lateral membranes are not stained; X17600. N = Nucleus; M = Microvilli; A = Apical membrane; L = Lateral membrane.

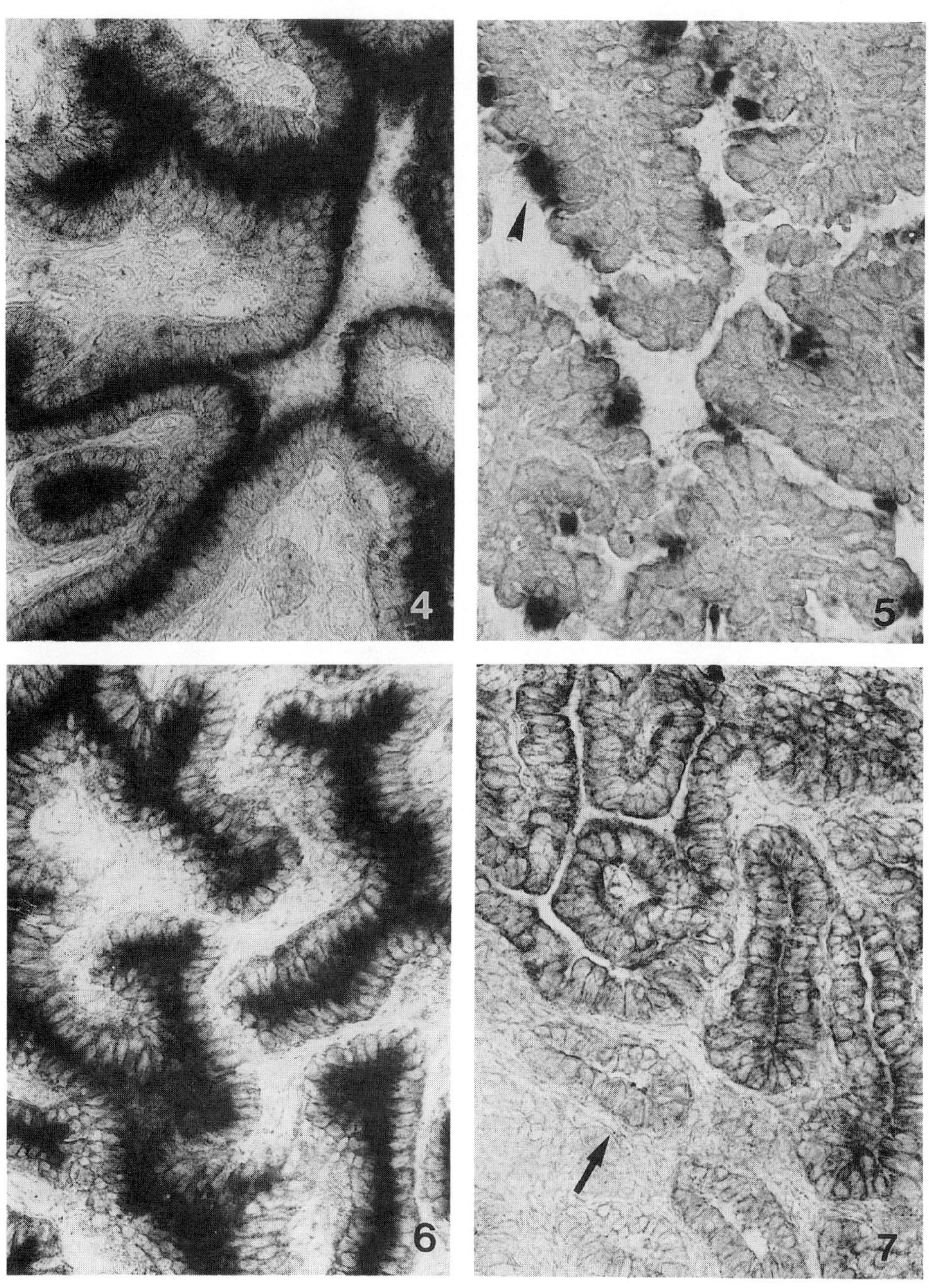

region of definitive chorioallantoic placentation (Table 2; for more details see Classen-Linke et al., 1987).

In the cases of aP, APM, and DPP IV, no reaction in the deepest parts of crypts compared to middle and upper crypts could be discerned (Figure 7), whereas GGT showed no difference between deep, middle and upper parts of crypts.

In the immunohistochemical approach with the two monoclonal antibodies 1D8.1 and 4H7.1 the aminopeptidase M pattern was similar to the reaction based on the catalytic function of the enzyme. Again a strong reaction was obtained in the pre-implantation phase (3-6 d p.c./p. hCG) and a marked reduction at 7-9 d p.c./p. hCG. The staining was largely restricted to the apical cell membrane in contrast to the basolateral membrane area (Figures 8 and 9).

These patterns were confirmed electron microscopically for aP and APM. It could be clearly demonstrated that the enzymes were localized at the apical membrane of the uterine epithelium at 5 d p.c./p. hCG (Figures 3 and 10) without any staining of the lateral membranes becoming apparent. In order to exclude artifactual lack of staining at the lateral membranes due to hampered diffusion at this site, exposed lateral membranes at the surfaces of the thick sections were carefully evaluated and were found to be negative or only very weakly stained. At 7 d p.c./p. hCG the reaction of APM at the apical cell membrane was strongly reduced (Figure 11) confirming the light microscopical findings (see above). The basolateral membrane domain remained negative.

Desmoplakin I and II immunohistochemistry was evaluated, in these studies, only for pseudopregnancy and interblastocyst segments. A strong staining was seen at the lateral membranes in the pre-implantation phase (3-6 d p.c./p. hCG) (Figure 12). This staining had a distinct maximum in the subapical region, i.e., the junctional zone. At 7-9 d p.c./p. hCG the strong subapical localization was lost and the desmosomal plaque protein appeared more evenly distributed over the whole lateral membrane (Figure 13).

Figure 4. gamma-Glutamyl transferase, LM. 6 d p.hCG. Strong reaction of the apical membranes of the uterine epithelium; X350.

Figure 5. gamma-Glutamyl transferase, LM. Implantation phase (8 d p.c.). Clear reduction of enzyme activity at the apical plasma membrane; there is a persistence of activity at some single cells (arrowhead); X350.

Figure 6. Dipeptidyl peptidase IV, LM. Pre-implantation phase (6 d p.c.). The apical part of the uterine epithelial cells is strongly stained; X350.

Figure 7. Dipeptidyl peptidase IV, LM. Implantation phase (8 d p.c.). The staining of the apical region of the uterine epithelium is strongly reduced, while the weak staining of the basolateral membranes becomes more apparent. The deepest crypts (arrow) are not stained at all; X350.

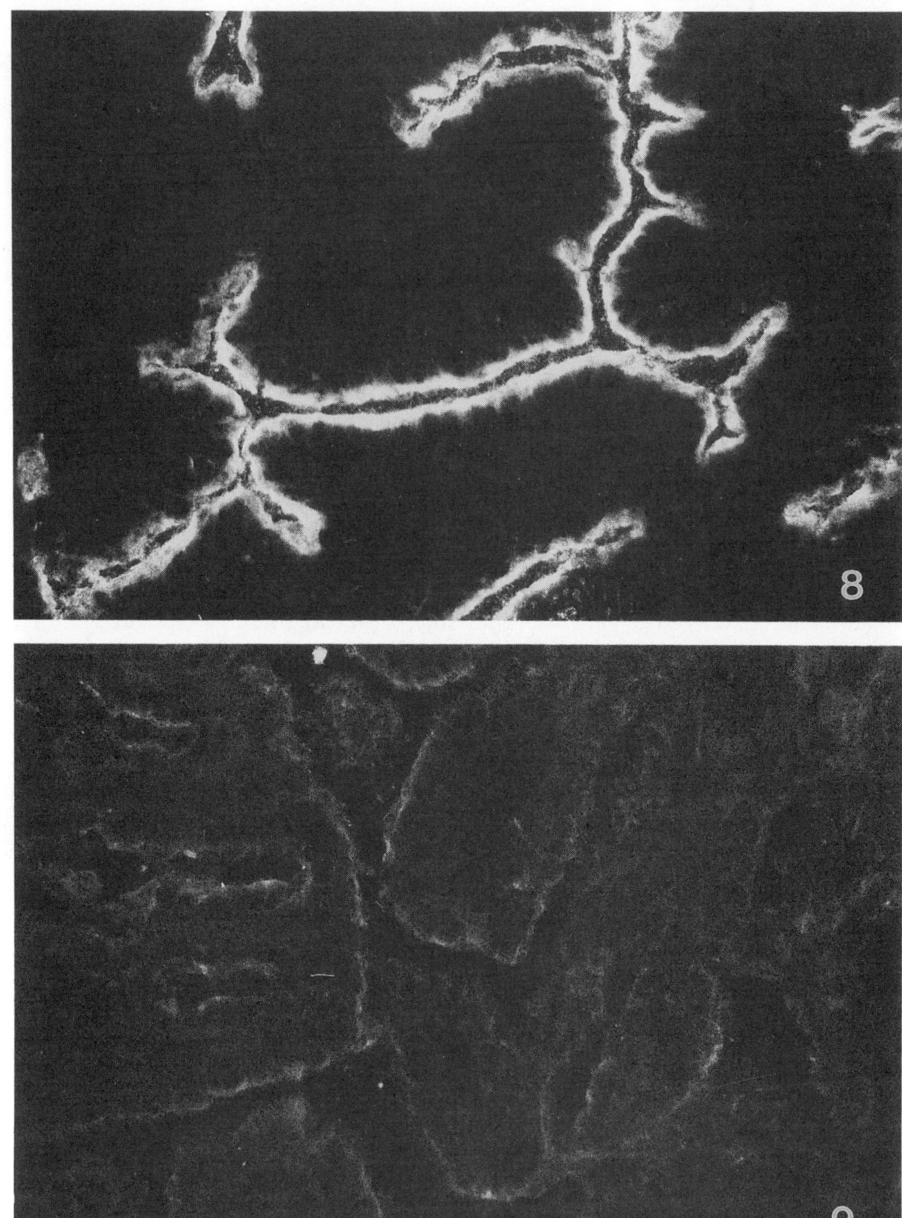
8
9

DISCUSSION

In the present investigations, a histochemical approach was used in order to monitor changes that the apical and basolateral plasma membranes of the rabbit uterine epithelium might undergo in preparation for trophoblast attachment. In the first part, four enzymes (aP, APM, GGT, and DPP IV) were used as cytochemical markers, which are typically expressed at the apical domain in polarized epithelia (Kinne, 1976; Feracci et al., 1981; Louvard, 1980; Simons and Fuller, 1985). The strong apical expression of these enzymes at the uterine epithelium in pseudopregnancy or in pregnancy in the pre-implantation phase (3-6 d p.c./p. hCG) and the loss of membrane enzyme activity at the time of trophoblast attachment and invasion in the implantation phase (7-8 d p.c./p. hCG) reflect a general and dramatic change of apical plasma membranes of the uterine epithelium. In a functional sense, it remains to be found out what the loss of these enzymes may imply for the properties of the apical plasma membrane. In particular, as discussed before (Classen-Linke et al., 1987), there is no known reason to connect the four enzymes in any sense (positively or negatively) with attachment processes. Therefore, one tends to regard the described changes as an indication of profound alterations of the composition of this membrane, which will certainly also involve adhesion-related molecules. Interestingly, however, it is not only specific adhesion-related molecules that change but apical markers of a more general type.

The present data for aminopeptidase M obtained by the immunohistochemical approach address the question of whether the loss of enzyme activity (Classen-Linke et al.,1987) is due to loss of enzyme proteins or to inactivation. It could be shown that the reaction with the two monoclonal antibodies 1 D8.1 (directed against the catalytic site of the aminopeptidase M) and 4 H7.1 (not directed against the catalytic site and with broad cross-reactivity) is similar to the reaction pattern obtained with the enzyme histochemical method. Thus, inactivation of the enzyme towards implantation time seems to be rather unlikely and it appears more probable that this integral membrane protein is being lost by a process of shedding and/or by a reduced rate of insertion into the apical membrane.

Figure 8. Aminopeptidase M, LM immunohistochemistry. 3 d p. hCG. Monoclonal antibody 1 D8.1, indirect immunofluorescence. Staining at the apical surface of the epithelium. The pattern is similar to that obtained by the enzyme histochemical method based on the catalytic function (van Hoorn and Denker, 1975; Classen-Linke et al., 1987); X350.

Figure 9. Aminopeptidase M, LM immunohistochemistry. 8 d p. hCG. The immunofluorescent staining is strongly reduced; X350.

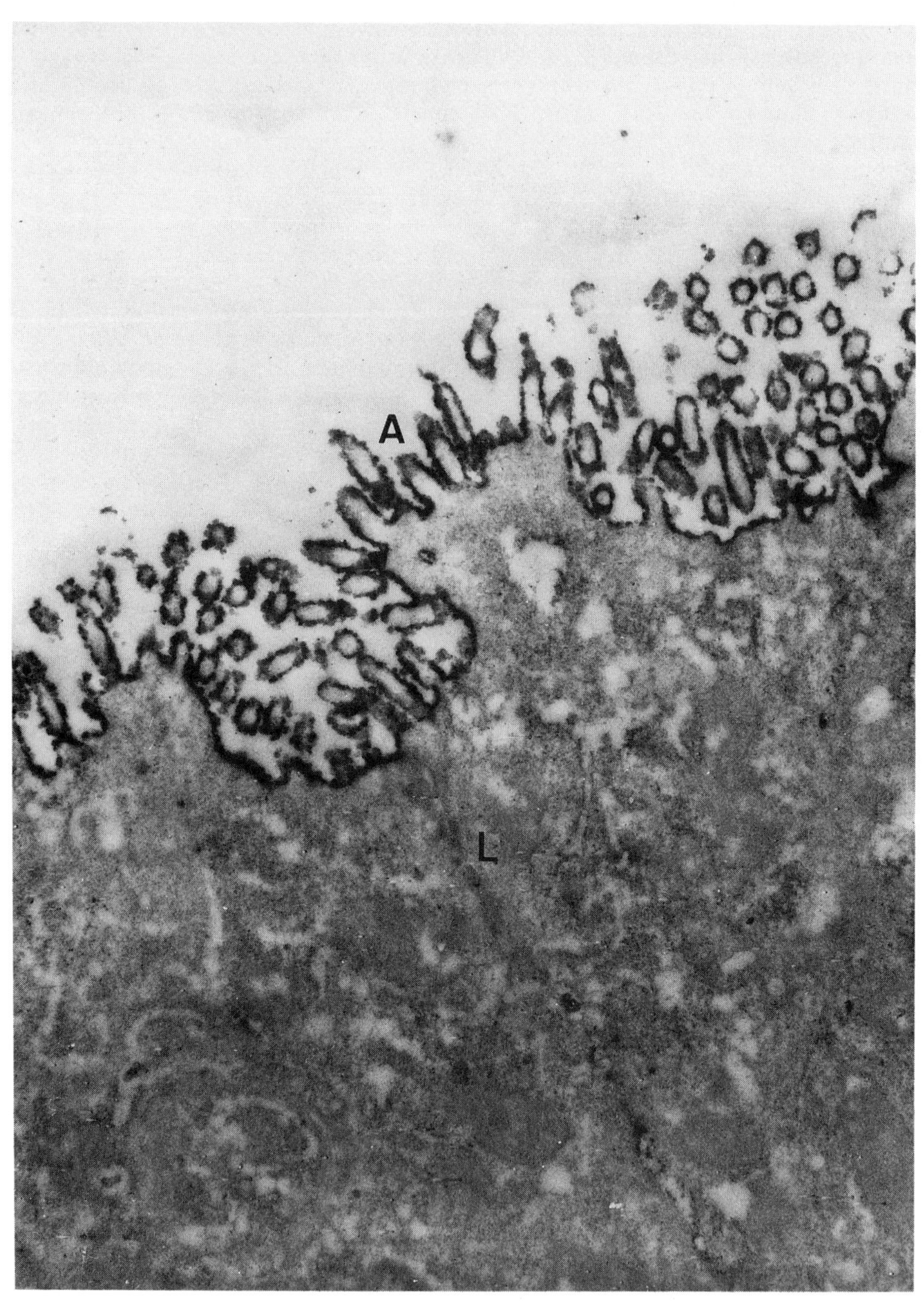
A
L

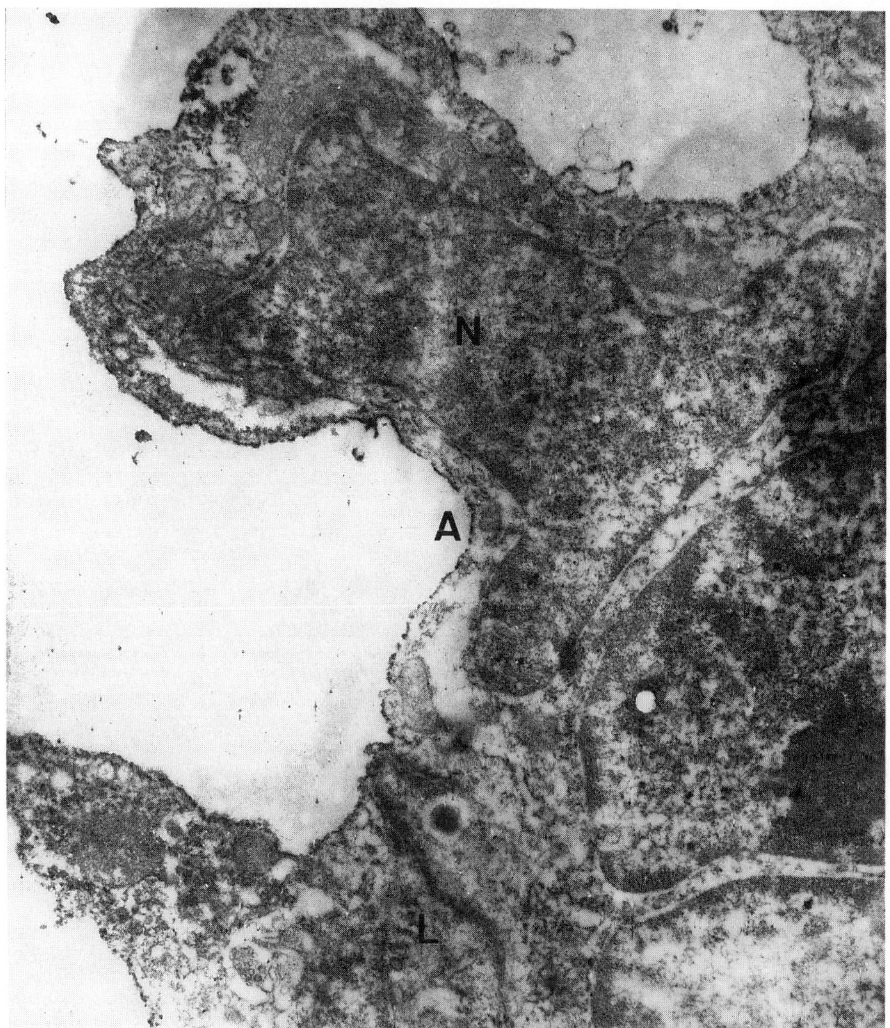

Figure 11. Aminopeptidase M, EM immunohistochemistry. Implantation phase (7 d p.c.). Monoclonal antibody 1 D8.1. Immunoperoxidase staining. The apical reaction is strongly reduced; the lateral membranes are not stained. The nucleus is in part positioned in an apical protrusion of the epithelial cell as often found at this stage. X16000. N = Nucleus; A = Apical membrane; L = Lateral membrane.

Figure 10:   Aminopeptidase M, EM immunohistochemistry. 5 d p.hCG. Monoclonal antibody 1 D8.1.   Immunoperoxidase staining at the apical microvillar area of the uterine epithelial cells. No staining at the lateral membrane. X22000. A = Apical membrane; L = Lateral membrane.

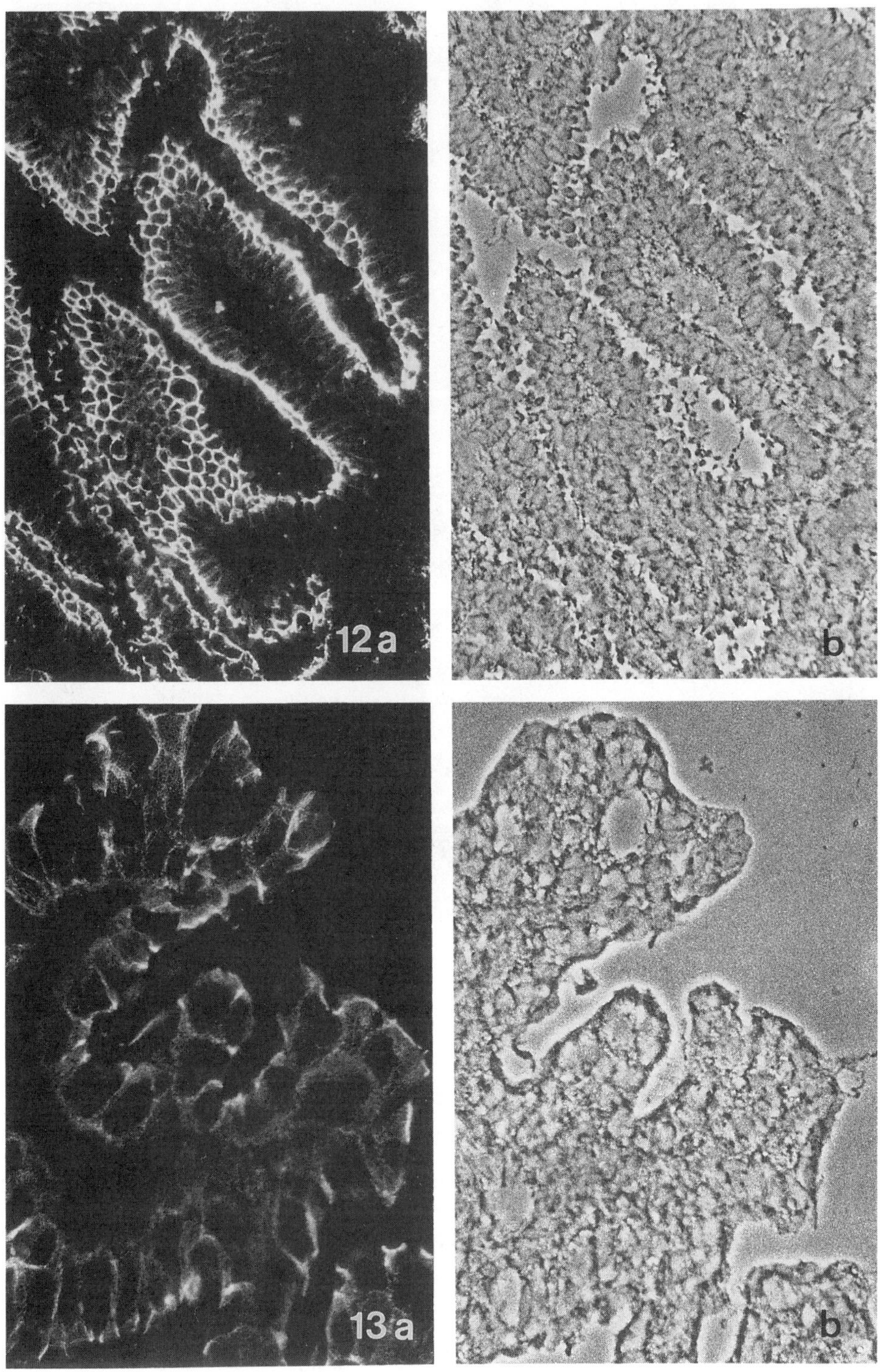

Localization of the enzymes aP and APM at higher resolution could be obtained by electron microscopical investigation. This demonstrated that they are found exclusively at the apical membrane of the uterine epithelium without any staining of lateral membranes. At 7 d p.c./p. hCG the reaction of APM at this site was strongly reduced, while no staining of lateral membranes could be detected, thus confirming the light microscopical findings. Clearly, the changes of enzyme reactivity at the apical cell pole described previously at the light microscopical level are not due to e.g., a change in density of microvilli.

In the implantation chamber, the reduction of enzyme activity was found typically at the placental fold, i.e., the region of definitive chorioallantoic placentation. In contrast, the behavior of the antimesometrial uterine epithelium is more difficult to understand. For aP and DPP IV no obvious loss of enzyme activity was observed here until 8 d p.c. However, it cannot be excluded that at the narrow sites of attachment of trophoblastic knobs the enzyme activity is reduced (compare Figure 14, Classen-Linke et al., 1987).

It is interesting that in correlation with these changes, proteins that are associated with the lateral plasma membrane were also found to change their pattern of distribution. For desmoplakin I and II, the strong subapical localization at the lateral membranes, a characteristic feature of polarized epithelia (Moll et al., 1986), was found to be lost at implantation time (7-9 d p.c./p. hCG). The protein was now discerned along the whole lateral membrane. These findings suggest that the described loss of typical markers at the apical plasma membrane is not an isolated phenomenon but is somehow connected with a redistribution of marker proteins associated with the lateral membrane domain. This is consistent with the assumption that the uterine epithelium shows reduced expression of some elements of apico-basal polarity during the "receptive phase" (Denker, 1986; Denker, 1988). It should be of interest to investigate the behavior of a wider spectrum of marker proteins for both membrane domains, during this phase, in order to find out whether the changes in composition of the membranes are more general or more specific with respect to molecules related to adhesion events.

Figure 12a. Desmoplakin I and II, LM immunohistochemistry. Pre-implantation phase (6 d p.c.). Indirect immunofluorescence using a mixture of the monoclonal antibodies 2.15, 2.17 and 2.20. Strong subapical staining in the region of the lateral membranes; X350.
Figure 12b. Phase contrast of the same field.

Figure 13a. Desmoplakin I and II, LM immunohistochemistry. Implantation phase (9 d p.c.). Indirect immunofluorescence using a mixture of the monoclonal antibodies 2.15, 2.17, and 2.20. As compared to 6 d p.c. (Figure 12), the confinement of staining to the subapical regions of the lateral membranes is lost and the desmosomal plaque proteins can now be detected rather evenly distributed in the whole lateral membrane region; X350.
Figure 13b. Phase contrast of the same field.

SUMMARY

Changes in the apical and lateral membrane compartment of rabbit uterine epithelium were investigated via histochemical techniques at the light and electron microscopical level. As markers for the apical plasma membrane domain so-called brush border enzymes (alkaline phosphatase, aminopeptidase M, gamma-glutamyl transferase, and dipeptidyl peptidase IV) were examined. Activity was maximal at about 5 d p.c./p. hCG and declined towards implantation time (7-8 d p.c.). Desmosomal plaque proteins desmoplakin I and II were studied immunohistochemically as examples of proteins associated with the lateral membrane domain. In the pre-implantation phase (3-6 d p.c./p. hCG) there was a distinct staining of desmoplakin I and II in the subapical region of the junctional zone, which is typical for a polarized epithelial cell. At 7-9 d p.c./p. hCG the strong subapical localization was lost and the desmosomal plaque protein was more evenly distributed over the whole lateral membrane. The observed changes in the apical and lateral membrane compartments of the uterine epithelium are discussed to indicate a loss of typical elements of apico-basal polarity, which may be a precondition for trophoblast adhesion and invasion.

ACKNOWLEDGEMENTS

We thank G. Helm for her excellent and expert technical assistance, G. Bock for photographic work, and G. Mathieu for typing the manuscript. We are very grateful to Dr. J.-P. Gorvel (C.N.R.S., Marseille, France) who provided the monoclonal antibodies against aminopeptidase M and gave technical advice. We thank Prof. W.W. Franke (German Cancer Research Center, Heidelberg, FRG) for providing the monoclonal antibodies against desmoplakin I and II and C. Kuhn for her technical assistance. These investigations were supported by Deutsche Forschungsgemeinschaft Postdoktorandenstipendium to I. Classen-Linke and by grant De 181/9-6.

REFERENCES

Beier, H.M. (1973) Die hormonelle Steuerung der Uterussekretion und frühen Embryonalentwicklung des Kaninchens. Habilitationsschrift, Universität Kiel.

Böving, B.G. (1963) Implantation mechanisms. In: *Conference On Physiological Mechanisms Concerned With Conception*, (ed.), C.G. Hartman, Oxford, New York, London, Paris, Pergamon Press, pp. 321-396.

Busch, L.C. (1982) Ultrastrukturelle Untersuchungen zur Transformation des Endometriums. Habilitationsschrift, Medizinische Fakultät, RWTH Aachen.

Classen-Linke, I., Denker, H.-W., and Winterhager, E. (1987) Apical plasma membrane-bound enzymes of rabbit uterine epithelium. Pattern changes during the periimplantation phase. *Histochem.* 87, 517-529.

Cowin, P., Kapprell, H.-P., and Franke, W.W. (1985) The complement of desmosomal plaque proteins in different cell types. *J. Cell. Biol.* 101, 1442-1454.

Davies, J. and Hoffman, L.H. (1973) Studies on the progestational endometrium of the rabbit. I. Light microscopy, day 0 to day 13 of gonadotrophin-induced pseudopregnancy. *Am. J. Anat.* 137, 423-446.

Davies, J. and Hoffman, L.H. (1975) Studies on the progestational endometrium of the rabbit. II. Electron microscopy, day 0 to day 13 of gonadotrophin-induced pseudopregnancy. *Am. J. Anat.* 142, 335-366.

Denker, H.-W. (1970) Topochemie hochmolekularer Kohlenhydratsubstanzen in Frühentwicklung und Implantation des Kaninchens, I. *Zool. Jahrb. Abt. Allgem. Zool. Physiol.* 75, 141-245.

Denker, H.-W. (1977) Implantation. The role of proteinases, and blockage of implantation by proteinase inhibitors. *Adv. Anat. Embryol. Cell Biol.* 53, Fasc. 5.

Denker, H.-W. (1986) Epithel-Epithel-Interaktionen bei der Embryo-Implantation: Ansätze zur Lösung eines zellbiologischen Paradoxons. *Verh. Anat. Ges. (Anat. Anz. Suppl.* 160) 80, 93-114.

Denker, H.-W. (1988) Implantation: Recent approaches to understand a cell biological paradox. In: *Human Reproduction. Current Status/Future Prospect*, (eds.), R. Iizuka and K. Semm, Excerpta Medica, Amsterdam, New York, Oxford, Elsevier Publishers B.B., pp. 237-240.

Enders, A.C. and Schlafke, S. (1971) Penetration of the uterine epithelium during implantation in the rabbit. *Am. J. Anat.* 132, 219-240.

Feracci, H., Bernadac, A., Hovsépian S., Fayet, G., and Maroux, S. (1981) Aminopeptidase N is a marker for the apical pole of porcine thyroid epithelial cells in vivo and in culture. *Cell Tissue Res.* 221, 137-146.

Gorvel, J.-P., Mishal, Z., Liegey, F., Rigal, A., and Maroux, S. (1989) Conformational change of rabbit aminopeptidase N into enterocyte plasma membrane domains analyzed by flow cytometry fluorescence energy transfer. *J. Cell Biol.* 108, 2193-2200.

Kinne, R. (1976) Membrane-molecular aspects of tubular transport. *Int. Rev. Physiol.* 11, 169-210.

Lewis, P.R. and Knight, D.R. (1977) Staining methods for sectioned materials. In: *Practical Methods In Electron Microscopy*, (ed.), A.M. Glauert, Vol. 5, Amsterdam, Elsevier/North-Holland, pp. 184-185.

Louvard, D. (1980) Apical membrane aminopeptidase appears at site of cell-cell contact in cultured kidney epithelial cells. *Proc. Natl. Acad. Sci. USA*, 77, 4132-4136.

Moll, R., Cowin, P., Kapprell, H.-P., and Franke, W.W. (1986) Biology of disease. Desmosomal proteins: New markers for identification and classification of tumors. *Lab. Invest.* 54 (1), 4-25.

Oledzka-Slotwinska, H., Creemers, J., and Desmet, V. (1967) Cytidine monophosphate as substrate for the electron microscopic visualization of alkaline phosphatase activity. *Histochemie* 9, 320-326.

Psychoyos, A. (1976) Hormonal control of uterine receptivity for nidation. *J. Reprod. Fertil. Suppl.* 25, 17-28.

Simons, K. and Fuller, S.D. (1985) Cell surface polarity in epithelia. *Ann. Rev. Cell Biol.* 1, 243-288.

Suzuki, H. and Tsutsumi, Y. (1980) Morphological changes of uterine and cervical epithelium during early pregnancy in rabbits. *J. Fac. Agr. Hokkaido Univ.* 60, 47-61.

Suzuki, H. and Tsutsumi, Y. (1981) Morphological studies of uterine and cervical epithelium in pseudopregnant rabbits. *J. Fac. Agr. Hokkaido Univ.* 60, 123-132.

van Hoorn, G. and Denker, H.-W. (1975) Effect of the blastocyst on a uterine amino acid arylamidase in the rabbit. *J. Reprod. Fertil.* 45, 359-362.

Winterhager, E. (1985) Dynamik der Zellmembran: Modellstudien während der Implantationsreaktion beim Kaninchen. Habilitationsschrift, Medizinische Fakultät, RWTH Aachen.

CHANGES IN LIPID ORGANIZATION OF UTERINE EPITHELIAL CELL MEMBRANES AT IMPLANTATION IN THE RABBIT

E. Winterhager[1] and H.-W. Denker[1]

Institut für Anatomie der RWTH Aachen
Melatener Strasse 211
D-5100 Aachen, West Germany

INTRODUCTION

Interaction of the blastocyst with the endometrium at implantation starts with adhesion of the apical membrane of the trophoblast to the apical membranes of the uterine epithelium (Enders and Schlafke, 1972; Schlafke and Enders, 1979). In the rabbit this event is followed by apical membrane fusion of both partner tissues thus leading to the formation of a mixed symplasm, and by invasion into the endometrium (Enders and Schlafke, 1971). At the same time, widespread fusion of the lateral cell membranes of the uterine epithelial cells starts. This uterine epithelial cell fusion occurs in two overlapping phases, in the antimesometrial portion of the endometrium from 7 days post coitum (7 d p.c.) associated with attachment and invasion of the abembryonic trophoblast (yolk sac placentation), and mesometrially on the endometrium of the placental folds from 8 d p.c. (associated with chorioallantoic placentation). Uterine symplasms formed in this way are found exclusively in the implantation chamber, i.e., in the vicinity of a blastocyst, and are very large, containing hundreds of nuclei.

Fusion of the lateral (formation of endometrial symplasms) as well as the apical membranes (mixed symplasm of trophoblast plus uterine epithelium) could depend on alterations in membrane composition leading to changes in membrane fluidity. In general, membrane fusion is accompanied by an increase in membrane fluidity which is often thought to be its prerequisite, as reported by Chandler (1984) for exocytosis and by Wolf et al. (1981) for fertilization. Anionic phospholipids have been proposed to be involved in increasing membrane fluidity so favoring membrane fusion (Papahadjopoulos, 1978). On the other hand cholesterol is known to increase membrane rigidity (Shinitzky, 1984).

For rabbit uterine epithelium, the concept of a functional interrelationship between the two types of lipids is tested here using freeze-fracture histochemistry with probes for both lipids that may permit detection of membrane microdomains. The peptide antibiotic polymyxin B (PXB), has been reported to bind specifically to anionic phospholipids in membranes (Hartmann et al., 1978; Sixl and Galla, 1981). Complexes of polymyxin B with anionic phospholipids can be detected as characteristic lesions on freeze-fracture replicas. For the localization of cholesterol filipin was used as a probe. Filipin interacts specifically with 3-β-hydroxy sterols (Elias et al., 1978; 1979) and leads to well defined membrane protrusions which can be detected on freeze-fracture replicas as well as in ultrathin sections.

[1]Present address: Institut für Anatomie, Universitätsklinikum, Hufelandstr. 55, D-4300 Essen 1, West Germany

Interest was focused on possible differences in lipid composition of the uterine epithelial membranes in the implantation chamber where fusion processes are extremely widespread, or in the blastocyst-free segments as well as in the uterine epithelium of pseudopregnant animals where lateral cell fusion is limited (Busch et al., 1981; Winterhager et al., 1984).

MATERIALS AND METHODS

Sexually mature female rabbits (mixed breeds) were caged individually on a 12 hour light/12 hour dark cycle and were fed a standardized pellet diet. Pseudopregnancy was induced by injection of 75 I.U. human chorionic gonadotropin (hCG) i.v. (Prolan ®, Bayer, Leverkusen, FRG). Mating was performed with two bucks of proven fertility. The day of mating and day of hCG injection were designated as day 0. Material for transmission electron microscopy was obtained from pseudopregnant (6, 7 days post hCG injection, p.hCG) as well as pregnant (6, 7 d p.c.) animals. In the case of pregnancy the endometrium of the implantation chamber and the blastocyst-free segments were investigated separately. Freeze-fracture histochemistry was exclusively performed on day 7 p.c./p.hCG.

Polymyxin B Treatment

Rabbits were anaesthetized by sodium pentobarbitone (Nembutal®, 0.4 ml/kg body weight i.v.), the uterus was exposed and polymyxin B (Upjohn Co.) was applied by injection of a 1% to 7% solution in physiological saline into the uterine cavity. Before injection the uteri were ligated into short segments (about 1 cm). Care was taken not to impair blood supply. After different times of incubation (1, 2, 5, 10, 20, and 30 minutes) uteri were fixed by vascular perfusion as described previously (Winterhager and Kühnel, 1982).

Filipin Treatment

Filipin (Sigma) was solubilized in dimethylsulfoxide (DMSO) (1 mg filipin/50 µl DMSO) (= filipin stock solution). Incubation with filipin was performed with pieces of endometrium that had been immersion fixed (2.5% glutaraldehyde in 0.1 M cacodylate buffer for 4 - 6 hours). After opening the uterine cavity, uterine segments were treated with filipin at different concentrations of the stock solution (0.05% and 0.02%) and different incubation times (1, 2, 4, 6, 8, and 12 hours) at room temperature in the dark. Controls were treated with DMSO alone.

Transmission Electron Microscopy Ultrathin Sections And Freeze-Fracture

Specimens from perfusion-fixed uteri were routinely embedded in araldite. The PXB and filipin-treated endometria as well as controls, were processed for freeze-fracturing after several washings with cacodylate buffer as described previously (Winterhager and Kühnel, 1982). Replicas were produced in a Bioetch apparatus 2005 (Leybold Heraeus). Replicas and ultrathin sections were examined in an EM 10 (Zeiss) electron microscope.

RESULTS

Morphology

At the time of trophoblast invasion, the apical membrane of uterine epithelial cells undergoes dramatic changes in its configuration in the implantation chamber. The microvilli which are thin and short in the preimplantation phase transform into numerous swollen and irregularly shaped cell projections in the implantation chamber beneath the adjacent trophoblast prior to and during attachment (Figure 1). These cell projections seem to collapse and fuse with one another forming ring-like structures. The deformation of the microvilli into cell organelle-free projections is accompanied by an umbrella-like distension of the apical cell region (Figure 1), thereby overlapping the neighboring cell like a tile. As a consequence, the lateral membranes are tilted in the apical region and this part which contains the broad tight junctional belt may now run nearly parallel to the cell apex.

In the crypts, these blunt projections of the apical plasma membrane probably fuse with those of opposite cells (Figure 2). This process appears to play a role in the formation of the exceedingly large uterine epithelial symplasms which proceed to cover virtually the whole antimesometrial surface of the endometrium, smoothing out many former depressions.

Distribution Of Cholesterol

Pseudopregnancy And Blastocyst-Free Segments In Pregnancy

Cholesterol complexes are detected in the plasma membrane at 7 d p.hCG/p.c. mainly in the form of typical lesions, i.e., circular protuberances of approximately 20 - 25 nm in diameter on the P-face and depressions on the E-face (Figure 3, inset). In the uterine epithelial cells of pseudopregnant rabbits as well as of the blastocyst-free segments of pregnant does, these lesions seem to be homogeneously distributed in the apical as well as lateral membranes. Apical membranes exhibit more filipin-cholesterol complexes than the lateral membranes (compare Figure 3 with Figure 4). On the lateral membranes, however, the number of lesions decreases from the apical to the basal portion of the membrane (Figure 4). This fact does not seem to be due to hampered diffusion of the reagent into the tissues as indicated by endothelial cells of underlying capillaries which are highly decorated with lesions. In contrast to the observations on the reaction with PXB (see below), the pattern of filipin lesions is homogeneous on all epithelial cells throughout the whole length of the uterus. Microdomains on the lateral or apical membranes, i.e., regions with specialized reaction patterns, are restricted to the membrane region of the tight junctional belt. Within this area a reaction with filipin is not detectable. A small strip of membrane below the belt shows a reduced number of lesions.

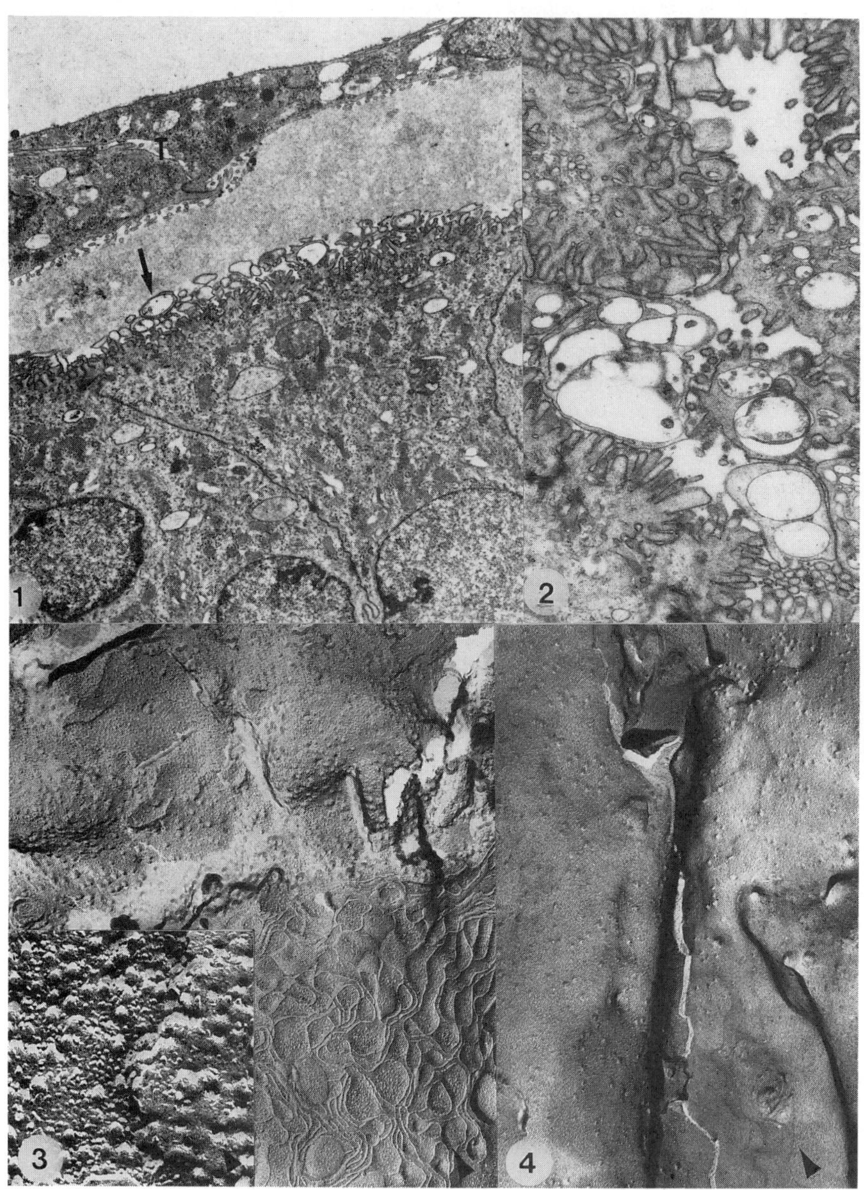

Pregnancy: Implantation Chamber

Most lateral membranes of the uterine epithelium in the implantation chamber exhibit many more cholesterol-filipin complexes than those of blastocyst-free segments of the same stage (7 d) of pseudopregnancy. Alternatively, a few lateral membranes can be found which are nearly free of cholesterol-filipin complexes. In that case the complexes which are seen, appear in small clusters (Figure 5). All apical membranes are decorated extensively so that the membrane does not show any smooth area (Figures 5 and 6). This pattern is found in the apical membranes of the whole epithelium surrounding the blastocyst.

In one respect, however, the localization of membrane cholesterol-filipin complexes in the implantation chamber is similar to that in the blastocyst-free segments and in pseudopregnancy: a gradient of lesions, i.e., a number that decreases towards the basal region, is seen on the lateral membrane in all cases.

The tight junctional area is again nearly free, the numerous gap junctions which are exclusively found between the epithelial cells of the implantation chamber at this reproductive stage (Winterhager et al., 1988) are totally free of filipin-cholesterol protrusions (Figure 7). Even in membrane areas where gap junctions seem to be formed (formation plaques, indicated by loosely aggregating particles) no reaction with filipin is obtained.

Figure 1. Uterine epithelium of the implantation chamber (7 d p.c.) with confronting trophoblast (T). The microvilli of the uterine epithelium are transformed into irregularly shaped cell projections forming partly ring-like structures (arrow). The cell apex is distended, shifting thereby the upper parts of the lateral membranes into a nearly surface-parallel position. X5,300

Figure 2. Apical cell surfaces of uterine epithelial cells in the implantation chamber (7 d p.c.) located vis-à-vis in the crypts. The nearly organelle-free, ectoplasmic cell projections interdigitate, obliterating the lumen. X10,000

Figures 3 - 4. Freeze fracture cytochemistry with filipin as a probe for cholesterol, pseudopregnancy.

Figure 3. Freeze fracture replica of uterine epithelium of a pseudopregnant animal (7 d p. hCG) incubated with filipin (0.2%, 3 hours). The apical membranes are decorated with typical protrusions whereas the region of the tight junctional belt is devoid of such reaction products. X24,500. The arrowhead indicates the direction of shadowing in all freeze fracture micrographs. Inset: Higher magnification of filipin/cholesterol complexes reveals circular protrusions (20-25 nm) on the P-face and mainly depressions on the E-face. X65,000

Figure 4. Lower portion of the lateral cell membranes of the same cell as in Figure 3. Membrane lesions are less frequent than in the apical membrane and their number decreases towards the basal portion. X24,500

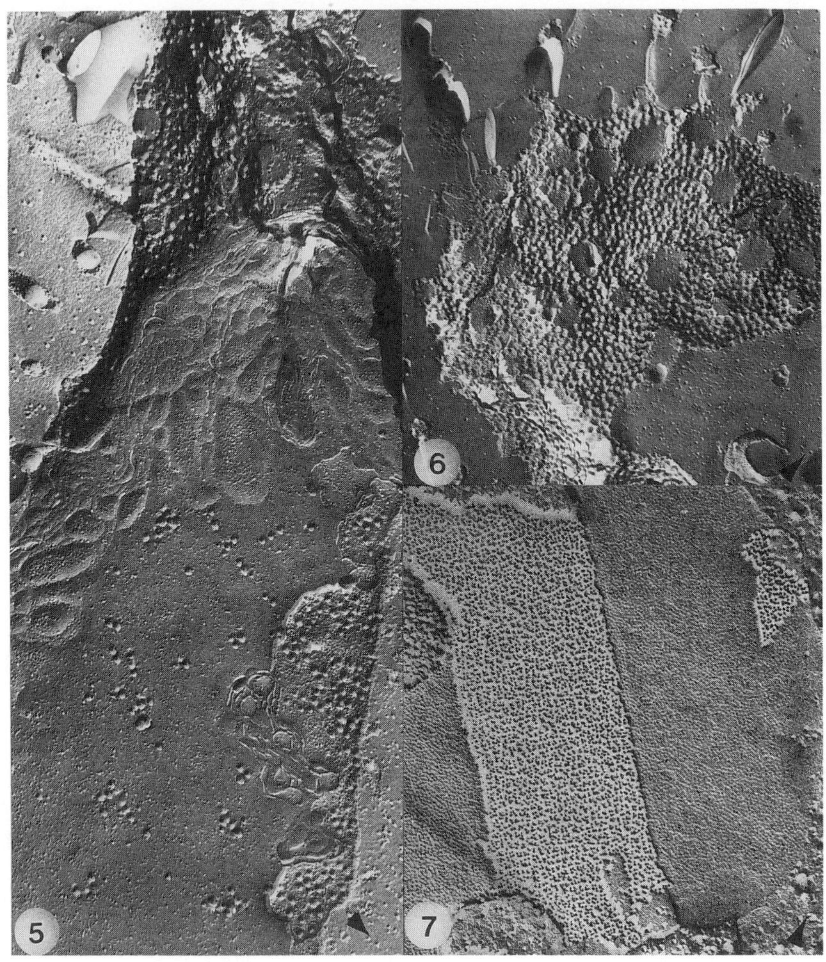

Figures 5 - 7. Freeze fracture immunocytochemistry with filipin as a probe for cholesterol: pregnancy. Uterine epithelial cells of the implantation chamber (7 days p.c.); incubation with 0.2% filipin for 4 hours.

Figure 5. Labeling with filipin is enhanced   on the apical membranes (a) as well as on most lateral membranes (see right side of the figure). Some lateral membranes, however, exhibit only few lesions arranged in clusters (lower left). X25,200

Figure 6. The apical membranes of the uterine epithelial cells in the implantation chamber (7 d p.c.) exhibit abundant circular protrusions indicative for cholesterol. X20,600

Figure 7. The membrane area with numerous gap junctional proteins is free of filipin/cholesterol complexes. X57,600

Distribution Of Anionic Phospholipids

Pseudopregnancy And Blastocyst-Free Segments In Pregnancy

Uterine epithelial cell membranes of pseudopregnant rabbits and those of the blastocyst-free segments of pregnant rabbits show the same reaction pattern of negatively charged phospholipids with PXB, at implantation time (7 d p.hCG/p.c.). Figures 8a, b, c demonstrate different types of membrane lesions that can be observed. This polymorphism includes an aggregation of particles combined with circular particle-free protrusions on the P-face (Figure 8a) or elongated protrusions on the P-face associated with intramembranous particles (IMPs) which seem to be associated with depressions of similar shape on the E-face (Figure 8b). With increasing incubation time membranes are more often seen to decompose and seem to form vesicles (Figure 8c). Results are evaluated using 7% PXB and an incubation time of 10 minutes. This methodological approach is chosen based upon an obvious balance between labeled, unlabeled, and destroyed cells. In addition to the polymorphism of lesions just described, the reaction to PXB is quite different between adjacent uterine epithelial cells. Cells without any reaction to PXB can be found that are located next to cells with perturbations of their membranes (Figure 9). However, the type of reaction with PXB is homogeneous within each single cell. Apical as well as lateral membranes of a given epithelial cell always exhibit the same kind of membrane lesions. The cell-to-cell differences in reaction pattern to PXB do not disappear after longer incubation times (up to 30 minutes). The number of cells which are destroyed by the detergent action of PXB, however, increases but even then unaltered cell membranes (i.e., without lesions) are detectable adjacent to those that have disintegrated into vesicles (Figure 9).

Except for the tight junctional membrane area (Figure 8a) (which is never decorated) compartments or microdomains on the lateral or apical membranes are not detectable.

Thus, a detailed evaluation of the results suggests that there are methodological problems that need to be discussed and which may result from the fact that treatment with PXB has been performed on native membrane material (see Discussion).

Pregnancy: Implantation Chamber

To obtain results with PXB on the epithelial membranes of uterine cells in the vicinity of the blastocyst, the concentration of PXB as well as the incubation time have to be considerably reduced, otherwise total cell destruction is observed. Concentrations between 1.5 to 3.5% and an incubation time between 1 and 2 minutes are found sufficient to get well defined lesions on the membranes. In contrast to the results obtained in pseudopregnancy and in the blastocyst-free segments (pregnancy) at day 7, the type of membrane lesions due to PXB is relatively homogeneous. These appear as particle aggregations or as elongated protrusions associated with the IMPs as illustrated for pseudopregnancy in Figure 8b. The reaction on the lateral as well as apical membranes is the same, whether cells are still non-fused or are already transformed into symplasms. Some of the lateral membranes, however, appear more damaged as indicated by a wavy structure and

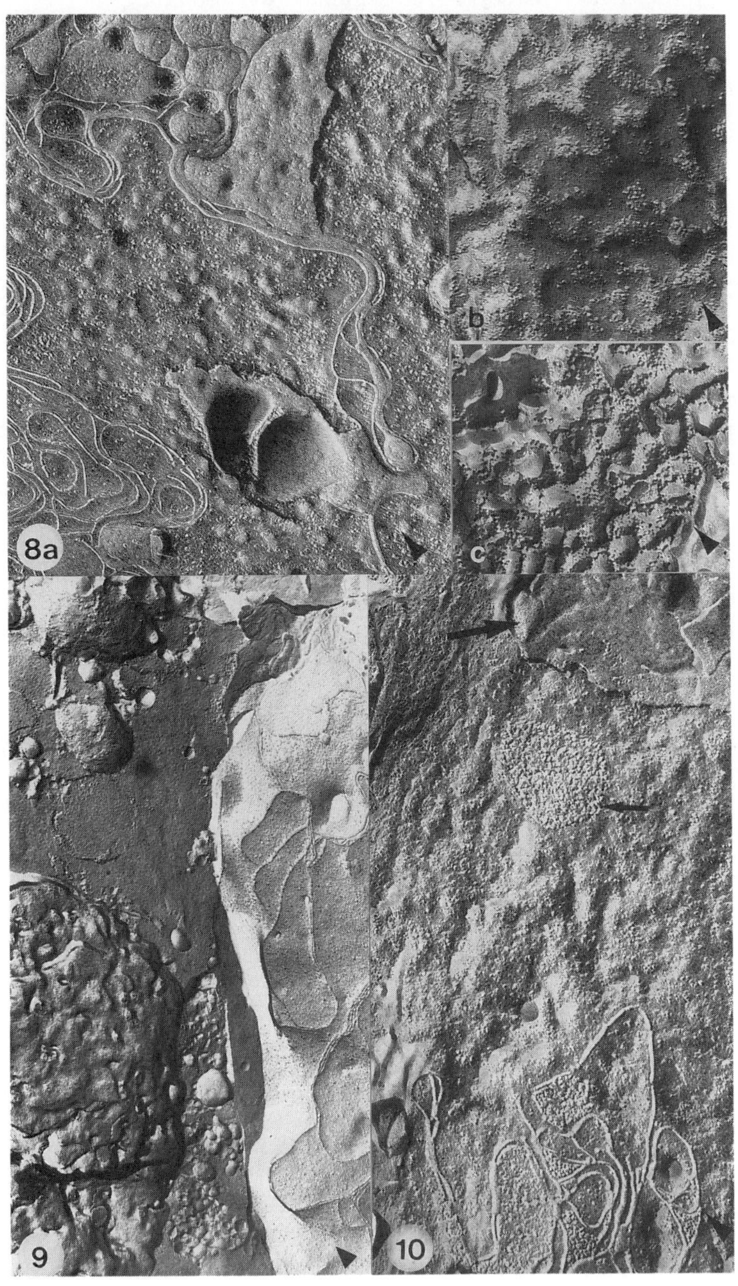

segregation of proteins and lipids (Figure 10). Microdomains (except the tight junctional region, as already described) are not detectable on either membrane compartment. Gap junctions, which are frequently found do seem to be affected by PXB treatment. This means some gap junctions exhibit agglomerated junctional particles (Figure 10), and in other plaques, particles are dispersed.

DISCUSSION

Changes in the morphology of the apical plasma membrane of the uterine epithelium in relation to embryo implantation have been seen not only in the rabbit (as described briefly in the present communication) but also in a number of other species and may indeed be a general phenomenon (Leiser, 1979, 1980; Enders and Schlafke, 1967, 1972; Enders, 1975; Wooding et al., 1980; Murphy and Martin, 1985). Particularly obvious in the vicinity of the blastocyst, microvilli become transformed into blunt cell projections of considerably varying shape and size. This phenomenon is accompanied by flattening of the apical plasma membrane and tilting of the apical portion of the lateral membranes. The described morphological changes in the apical plasma membrane have to be discriminated from the well known dome-shaped cell protrusions related to maternal progesterone serum levels. Traditionally, such apical protrusions have been viewed as a sign of apocrine secretion (Beier and Kühnel, 1973) or as specializations for endocytosis in the rat (Enders and Nelson, 1973). In the rabbit, the apical protrusions are already considerably reduced in number in the implantation chamber when trophoblast attachment commences.

Figures 8 - 10. Freeze fracture cytochemistry with polymyxin B as a probe for anionic phospholipids.

Figures 8a,b,c. Different membrane reactions obtained with polymyxin B treatment (7% for 10 minutes) on uterine epithelial cells 7 d p. hCG (pseudopregnancy). a) Circular protrusions are similar to those obtained with filipin and are devoid of particles; the tight junctional area is free of reaction products, X33,000; b) Elongated protrusions with agglomerated particles, X50,000; c) Membranes decomposed into vesicle-like structures, X50,000.

Figure 9. Uterine epithelial cells of a blastocyst-free segment at pregnancy (7 d p.c.) incubated with 7% polymyxin B for 30 minutes. Two neighboring cells with different reaction to PXB: the membrane of the left cell is totally decomposed whereas the adjacent cell demonstrates a membrane without any reaction product. X15,800

Figure 10. Lateral membrane of a uterine epithelial cell of the implantation chamber (7 d p.c.) treated with 3.5% polymyxin for 1 minute. The membrane shows a wavy structure with segregated lipid areas (big arrow) and agglomerated gap junctional proteins (small arrow), whereas the tight junctional area with intercalated gap junctions does not appear affected. X31,700

All these phenomena suggest that the physico-chemical properties of the membranes and of underlying cytoskeletal elements may undergo alterations at this phase. Consequently it appears reasonable to ask whether the lipid composition of the apical membrane with respect to the phospholipid and/or cholesterol content has changed. This conclusion has already previously been drawn for the rat (Murphy and Martin, 1984, 1985; Murphy and Dwarte, 1987) using a similar methodological approach. This study likewise suggests that both types of lipids investigated, i.e., cholesterol and anionic phospholipids, undergo changes in density and distribution in the uterine epithelial membranes at implantation also in the rabbit.

However, these observations must be interpreted with caution since artefacts may be generated. The heterogenous reaction of the uterine epithelial cells to PXB (in pseudopregnancy as well as in the blastocyst-free segments in pregnancy) is an example. The non-reaction is surely not due to a total absence of anionic phospholipids. This current series of differing treatment times suggests that some of the anionic phospholipid molecules may indeed be masked, since prolonged incubation increases the amount of membranes with alterations. Severs and Robenek (1983) have suggested that PXB may be adsorbed by negatively charged glycoproteins, or that the anionic phospholipids may alternatively be masked by calcium. Murphy and Martin (1984) observed non-reactivity to PXB of the lateral epithelial cell membranes of the endometrium in the rat. They interpret these observations as evidence for differences in anionic lipid content but without excluding other unknown causes. Nevertheless, it appears to be of interest that this method does reveal differences in reactivity between membranes of uterine epithelial cells, of, on one hand, pseudopregnant animals and the blastocyst-free segments of pregnant animals, and, on the other hand, cells surrounding the blastocyst. The cells of pseudopregnant animals and the blastocyst-free segments are characterized by mosaic-like reactions to PXB whereas in the implantation chamber a homogeneous reaction of apical as well as lateral membranes indicates a more homogeneous composition.

In the implantation chamber, epithelial cells have previously been shown to be coupled by newly formed gap junctions (Winterhager et al., 1988). This must be expected to give rise to greater similarity due to metabolic cooperation. In addition to greater uniformity of reaction, the density of characteristic lesions in the implantation chamber probably indicates an increased concentration of anionic phospholipids and/or better accessibility. The latter could be due to a reduction of the thickness of the glycocalyx of the apical cell membrane particularly in the implantation chamber (Anderson and Hoffman, 1984).

The selectivity of filipin for binding to cholesterol is well established for simplified membrane model systems (Kinsky et al., 1967; Norman et al. 1972; Verkleij et al., 1973). However, it has been suggested that the complexity of cell membranes may in certain cases lead to false negative results due to a stabilizing influence of proteins (in the case of coated pits,the clathrin coat) (Feltkamp et al., 1982; Severs and Simons, 1983; Steer et al., 1984). Consistent with this stabilizing influence is the current finding that filipin plaques are absent in gap junctions of the uterine epithelium. The lower number of lesions in tight junctional membrane areas is difficult to interpret in this context since the role of proteins integral to or associated with the tight junctions is still incompletely understood (Citi et al.,

1988). The observed increase in the number of plaques in the apical membrane in the implantation chamber, in contrast, cannot be explained on the basis of interference of proteins since in freeze fracture particle density is found to be increased (Winterhager, 1985) a phenomenon which also holds true for the rat (Murphy and Swift, 1983). If not due to changes in the glycocalyx as discussed above for PBX-binding one would therefore assume that the apical plasma membranes of uterine epithelial cells show an increased content of cholesterol in the implantation chamber. Since diffusion problems cannot be excluded completely for the lateral membranes, it appears less certain whether the observed lower number of lesions does indeed show that the cholesterol content is less here than in the apical membranes. The same reservation may apply for the gradient seen in the lateral membrane with most lesions being formed in the apical portion.

It is remarkable that in the implantation chamber the filipin binding shows more diversity on the lateral membranes than in other parts whereas the reactivity of the apical cell membrane is more uniform.

With regard to implantation physiology, the reported results are in many respects unexpected for the distribution of anionic phospholipids as well as for cholesterol. Keeping in mind that many fusion processes are going on at the lateral membranes (uterine symplasm formation) as well as at the apical membranes (fusion implantation), it is expected to find a decrease in membrane cholesterol content and an increase in anionic lipids based on results obtained in simplified model systems (Yanagimachi et al., 1972; Bearer and Friend, 1981; 1982). Not withstanding the methodological problems discussed above, the current results would suggest that the concentration of anionic lipids increases - or they have better access to the probe - in the implantation chamber, combined with an increase in the cholesterol content, particularly in the apical membranes. About the same is found for the cholesterol content in the apical membranes of the uterine epithelium in the rat which seems to be increased at implantation stage as compared to a non-pregnant animal (Murphy and Dwarte, 1987). In the rat, however, fusion does not occur, neither for lateral membranes or between trophoblast and uterine epithelium. Alternatively, studies performed in other systems indicate that fusion can be independent of membrane rigidity and sterol content. Apical membranes of villous goblet cells, for example, exhibit abundant filipin-sterol complexes during secretion, i.e., more than nonsecretory cells (Trier and Madara, 1984). Conversely, enhanced membrane fluidity is not necessarily a prerequisite for cell fusion since photolabeling experiments show that virus-induced fusion in a variety of cell types is not accompanied by increased lateral mobility of cell membrane components (Aroeti and Harris, 1986).

In conclusion, the results presented on freeze fracture histochemistry using filipin and PXB as probes suggest that uterine epithelial plasma membranes do undergo changes in lipid (cholesterol and anionic phospholipids) composition in preparation for trophoblast attachment and implantation. In spite of the fact that some details need to be clarified due to certain problems inherent to the application of this methodology to complex tissues, the results suggest that the cholesterol content of the apical epithelial cell membranes increases at implantation in the vicinity of the blastocyst. This is in contrast to what one could expect when a simple fusion model as derived from liposome studies would be applied (White and Helenius, 1968; Papahadjopoulos, 1973) and also to findings obtained with sperm

(Bearer and Friend, 1981, 1982). Obviously the uterine epithelium/trophoblast system is very complex and components other than the two classes of lipids studied here may come into play, including e.g., integral membrane proteins and the cytoskeleton. Cell fusion related to embryo implantation is not a peculiarity of the rabbit. Fusion between uterine epithelial cells occurs in carnivores (Leiser, 1979; Enders, 1972), while fusion of trophoblast with uterine epithelium occurs in ungulates (Wooding et al., 1980) and a group of marsupials (peramelids, cf., Padykula, 1976). The processes of fusion that occur in the rabbit implantation chamber should provide a useful model system to study the cell biology of this phenomenon.

SUMMARY

The uterine epithelium of the rabbit shows conspicuous morphological changes of the cell membranes during the preimplantation and implantation phases, and finally undergoes cell fusion of both homologous (uterine epithelial cells with each other) and heterologous type (uterine epithelium with trophoblast). In the present investigations, it is asked whether these membrane alterations may be due to changes in lipid composition of the membranes.

Polymyxin B as well as filipin are used as histochemical probes to detect anionic phospholipids and cholesterol, respectively, in the membranes of the uterine epithelium at implantation. Typical lesions produced by these probes are visualized on freeze fracture replicas. The implantation chamber is being separately investigated and compared to the blastocyst-free segments and the endometrium of pseudopregnant animals of the same stage.

Filipin-induced lesions increase enormously in the apical and in most of the lateral membranes of the epithelial cells of the implantation chamber suggesting that the content of cholesterol is higher here than in the blastocyst-free segments and in pseudopregnant animals. The reaction with polymyxin B (anionic phospholipids) is found to be very heterogeneous with respect to the reaction products, in both the epithelium of pseudopregnant animals and the blastocyst-free segments. This heterogeneity disappears in the epithelium of the implantation chamber resulting in a homogeneous and strong reaction of apical and lateral membranes.

The results are discussed in the light of model experiments that have previously been done e.g., with liposomes, suggesting that negatively charged phospholipids increase membrane fluidity and that cholesterol is responsible for membrane rigidity. It is concluded that the described membrane alterations and fusion processes in the uterine epithelium represent a more complex system. Although considerable changes in lipid composition are indeed found, these do not fulfill predictions from model experiments in detail. Additional components (e.g., integral membrane proteins) may play a decisive role in this biological system.

ACKNOWLEDGEMENTS

These investigations form part of a Habilitation thesis (Winterhager, 1985). We thank Mrs. Diana Seelis for excellent technical help, Mrs. Gabriele Bock for photographic work. We also appreciate Mrs. Gisela Mathieus help with typing the manuscript. The work was supported by the Deutsche Forschungsgemeinschaft (Wi 774/1-1).

REFERENCES

Anderson, T.L. and Hoffman, L.H. (1984) Alterations in epithelial glycocalyx of rabbit uteri during early pseudopregnancy and pregnancy, and following ovariectomy. *Am. J. Anat.* 171, 321-334.

Aroeti, B. and Henis, Y.I. (1986) The lateral mobility of cell membrane components is not altered following cell fusion induced by Sendai virus. *Exp. Cell Res.* 162, 243-254.

Bearer, E.L. and Friend, D.S. (1982) Modifications of anionic-lipid domains preceding membrane fusion in guinea pig sperm. *J. Cell Biol.* 92, 604-615.

Beier, H.M. and Kühnel, W. (1973) Pseudopregnancy in the rabbit after stimulation by human chorionic gonadotropin. *Hormone Res.* 4, 1-27.

Busch, L.C., Winterhager, E., and Kühnel, W. (1981) Symplasmatische Umwandlung des Cavumepithels im Uterus pseudogravider Kaninchen. *Acta Anat.* 111, 22.

Chandler, D.E. (1984) Comparison of quick-frozen and chemically fixed sea urchin eggs: Structural evidence that cortical granule exocytosis is preceded by a local increase in membrane mobility. *J. Cell Sci.* 72, 23-36.

Citi, S., Sabanay, H., Jakes, R., Geiger, B., and Kendrick-Jones, J. (1988) Cingulin, a new peripheral component of tight junctions. *Nature* 333, 272-276.

Elias, P.M., Friend, D.S., and Goerke, J. (1979) Membrane sterol heterogeneity. Freeze-fracture detection with saponins and filipin. *J. Histochem. Cytochem.* 27, 1247-1260.

Elias, P.M., Goerke, J., and Friend, D.S. (1978) Freeze-fracture identification of sterol-digitonin complexes in cell and liposome membranes. *J. Cell Biol.* 577-596.

Enders, A.C. (1975) The implantation chamber, blastocyst and blastocyst imprint of the rat: A scanning electron microscope study. *Anat. Rec.* 182, 137-150.

Enders, A.C. and Nelson, D.M. (1973) Pinocytotic activity of the uterus of the rat. *Am. J. Anat.* 138, 277-300.

Enders, A.C. and Schlafke, S. (1967) A morphological analysis of the early implantation stages in the rat. *Am. J. Anat.* 120, 185-226.

Enders, A.C. and Schlafke, S. (1971) Penetration of the uterine epithelium during implantation in the rabbit. *Am. J. Anat.* 132, 219-240.

Enders, A.C. and Schlafke, S. (1972) Implantation in the ferret: Epithelial penetration. *Am. J. Anat.* 133, 291-316.

Feltkamp, C.A. and van der Waerden, A.W.M. (1982) Membrane associated proteins affect the formation of filipin-cholesterol complexes in viral membranes. *Exp. Cell Res.* 140, 289-297.

Hartmann, W., Galla, H.-J., and Sackmann, E. (1978) Polymyxin binding to charged lipid membranes an example of cooperative lipid-protein interaction. *Biochim. Biophys. Acta* 510, 124-139.

Kinsky, S.C., Luse, S.A., Sopf, D., van Deenen, L.L.M., and Haxby, J. (1967) Interactions of filipin and derivatives with erythrocyte membranes and lipid dispersions. *Biophys. Acta* 135, 844-861.

Leiser, R. (1979) Blastocystenimplantation bei der Hauskatze. Licht- und elektronenmikroskopische Untersuchung. Zbl. Vet. Med. C. *Anat. Histol. Embryol.* 8, 79-96.

Leiser, R. (1980) Funktionelle Morphologie der Implantation und der frühen Plazentation bei der Hauskatze. Licht- und elektronenmikroskopische Untersuchung mit histocytochemischer Ergänzung. Habilitationsschrift, Institut für Tieranatomie der Universität Bern.

Murphy, C.R. and Martin, B. (1984) Freeze-fracture cytochemistry with polymyxin B. A study on the plasma membrane of uterine epithelial cells. *Histochem.* 80, 327-331.

Murphy, C.R. and Martin B. (1985) Cholesterol in the plasma membrane of uterine epithelial cells: A freeze-fracture cytochemical study with digitonin. *J. Cell Sci.* 78, 163-172.

Murphy, C.R. and Dwarte, D.M. (1987) Increase in cholesterol in the apical plasma membrane of uterine epithelial cells during early pregnancy in the rat. *Acta Anat.* 128, 76-79.

Murphy, C.R. and Swift, J.G. (1983) Pronase reduces intramembranous particle density in uterine epithelial cells in vivo. *Acta Anat.* 116, 174-179.

Norman, A.V., Demel, R.A., de Kruijff, B., and van Deenen, L.L.M. (1972) Studies on the biological properties of polyene antibiotics. Evidence for the direct interaction of filipin with cholesterol. *J. Biol. Chem.* 247, 1918-1929.

Padykula, H.A. and Taylor, J.M. (1976) Ultrastructural evidence for loss of the trophoblastic layer in the chorioallantoic placenta of Australian bandicoots (Marsupialia: Peramelidae) *Anat. Rec.* 186, 357-386.

Papahadjopoulos, D. (1978) Calcium-induced phase changes and fusion in natural and model membranes. In: *Membrane Fusion*, (eds.) G. Poste and G.L. Nicolson, Elsevier: North-Holland Biomedical Press, pp. 765-790.

Schlafke, S. and Enders, A.C. (1975) Cellular basis of interaction between trophoblast and uterus at implantation. *Biol. Reprod.* 12, 41-65.

Severs, N.J. and Robenek, H. (1983) Detection of microdomains in biomembranes. An appraisal of recent developments in freeze-fracture cytochemistry. *Biochim. Biophys. Acta* 737, 373-408.

Severs, N.J. and Simons, H.L. (1983) Failure of filipin to detect cholesterol rich domain in smooth muscle plasma membrane. *Nature (Lond.)* 303, 637-638.

Shinitzky, M. (1984) Membrane fluidity and cellular functions. In: *Physiology of Membrane Fluidity*, (ed.) M. Shinitzky, Vol. 1., Boca Raton, Florida: CRC Press Inc.

Sixl, F. and Galla, H.-J. (1981) Polymyxin interaction with negatively charged lipid bilayer membranes and the competetive effect of Ca^{++}. *Biochim. Biophys. Acta* 643, 626-635.

Steer, J.C., Bisher, M., Blumenthal, R., and Steven, A.C. (1984) Detection of membrane cholesterol by filipin in isolated rat liver coated vesicles is dependent upon removal of the clathrin coat. *J. Cell Biol.* 99, 315-319.

Trier, J.S. and Madara, J.L. (1984) Distribution of filipin-sterol complexes in villus goblet cell membranes of rat small intestine. *Lab. Invest* 50, 673-682.

Verkleij, A.J., de Kruijff, B., Gerritsen, W.F., Demel, R.A., van Deenen, L.L.M., and Ververgaert, P.H.J. (1973) Freeze-etch electron microscopy of erythrocytes, Acholeplasma laidlawii cells and liposomal membranes after the action of filipin and amphotericin B. *Biochim. Biophys. Acta* 291, 577-581.

White, J. and Helenius, A. (1980) pH-dependent fusion between the Semliki forest virus membrane and liposome. *Proc. Natl.Acad. Sci. USA* 77, 3273-3277.

Winterhager, E. (1985) Dynamik der Zellmembran: Modellstudien während der Implantationsreaktion beim Kaninchen. Habilitationsschrift, *Medizinische Fakultät*, RWTH Aachen.

Winterhager, E. and Kühnel, W. (1982) Alterations in intercellular junctions of the uterine epithelium during the preimplantation phase in the rabbit. *Cell Tissue Res.* 224, 517-526.

Winterhager, E., Busch, L.C., and Kühnel, W. (1984) Membrane events involved in fusion of uterine epithelial cells in pseudopregnant rabbits. *Cell Tissue Res.* 235, 357-363.

Winterhager, E., Brümmer, F., Dermietzel, R., Hülser, D.F., and Denker, H.-W. (1988) Gap junction formation in rabbit uterine epithelium in response to embryo recognition. *Dev. Biol.* 126, 203-211.

Wolf, D.E., Kinsey, W., Lennarz, W., and Eddin, M. (1981) Changes in the organization of the sea urchin egg plasma membrane upon fertilization: Indications from the lateral diffusion rates of lipid-soluble fluorescent dyes. *Dev. Biol.* 81, 133-138.

Wooding, F.B.P., Chambers, S.G., Perry, J.S., George, P.M., and Heap, R.B. (1980) Migration of binucleate cells in the sheep placenta during normal pregnancy. *Anat. Embryol.* 158, 361-370.

Yanagimachi, R., Noda, Y.D., Fujimoto, M., and Nicolson, G.I. (1972). The distribution of negative surface charges on mammalian spermatozoa. *Am. J. Anat.* 135, 497-520.

CELL SURFACE COMPONENTS OF HUMAN ENDOMETRIAL EPITHELIUM: MONOCLONAL ANTIBODY STUDIES

Mourad W. Seif and John D. Aplin[1]

Departments of Obstetrics, Gynaecology,
Biochemistry and Molecular Biology
University of Manchester
Manchester, United Kingdom

INTRODUCTION

During the preimplantation phase, the endometrium undergoes a variety of histological (Dallenbach-Hellweg, 1981) and compositional (Bell, 1986; Aplin, 1989) alterations. These are believed to create an appropriate milieu for the implanting blastocyst. The changes include the production of secretory material for the implanting blastocyst; the alteration of cell surface composition at the luminal epithelium; and changes in the stromal environment. We (Seif et al., 1989) and others (Bell, 1986) have reported previously studies of hormonally regulated secretory components of human endometrium and decidua. Here we describe an approach to the characterization of endometrial cell surfaces during the menstrual cycle which may be useful both in furthering understanding of reproductive function and in diagnosis of reproductive failure. We have focused on endometrial epithelium since this forms the site of initial interaction of maternal and fetal cells.

The endometrial cycle is controlled by the ovarian cycle, and evidence from several animal species suggests the probable existence of hormone-dependent changes in cell surface composition of endometrial epithelium in the secretory phase when implantation takes place (cf. Aplin, 1989; Anderson et al., 1986). In humans there is also evidence, especially from lectin histochemistry, that hormone-dependent compositional changes occur at the endometrial epithelial cell surface (Lee and Damjanov, 1985; Bychkov and Toto, 1986; Yen et al., 1986). While these changes represent solely the maternal factor operating during the process of implantation, and occur irrespective of the presence of an embryo, other data suggest a possible role of embryonic signals in implantation (Kosasa, 1973; Morton et al., 1977; Denker, 1982; 1983; Classen-Linke et al., 1987; Imakawa et al., 1987).

In analyzing molecular changes in the endometrial epithelial cell surface, two considerations evolve. All epithelial cells show morphlogical and compositional variations between apical, lateral and basal domains of the cell surface (Van Meer and Simons, 1986; Matlin and Simons, 1984) and this is clearly the case in endometrium (Lee and Damjanov, 1985; Bychkov and Toto, 1986; Classen-Linke et al., 1987; Kimber et al., 1988; Dockery et al., 1988; Smith et al., 1989). Secondly, the endometrium contains two epithelial cell populations, surface (or luminal) and glandular. Though both seem to originate in a common precursor cell following menstruation and endometrial shedding (Ferenczy,

[1]To Whom Correspondence Should Be Addressed: Research Floor, St. Mary's Hospital, Manchester, M13 0JH, United Kingdom

1976a; 1976b), SEM studies (Ferenczy, 1976a; Ferenczy, 1977) have demonstrated differences between the two in morphology and sensitivity to cyclic hormonal stimuli. Immunochemical studies in human and other species have also demonstrated differences in the secretory activity (Fazleabas et al., 1985) and in the expression of cytoplasmic components (Ciocca et al., 1983) between these two cell types.

The current approach to analyze changes in endometrial cell surface involved raising monoclonal antibodies (mcAbs) against endometrial epithelial cells obtained in the secretory phase (Aplin and Seif, 1985). Epithelial aggregates and isolated cells were used in alternating doses as immunogen thus exposing the epithelial component of the endometrium as well as the cell surface of individual cells to the immune system of the recipient mouse. A panel of mcAbs has been obtained to cell surface components of endometrial epithelium, and four of these markers are described herein.

MATERIALS AND METHODS

Isolation Of Endometrial Epithelium

Endometrial tissue was obtained by dilatation and curettage in the secretory phase of the cycle based on the patients' menstrual history and date of the last menstrual cycle. Tissue was assessed retrospectively by histological examination.

Curettings were cut into 1-2 mm cubes then washed with PBS containing antibiotics. Tissue digestion was carried out with crude collagenase (Sigma type 1A) at a concentration of 160 U/ml (Kirk et al., 1978) in a 1:1 mixture of Dulbecco's Modified Eagle's Medium and Ham's F12, either for short period (3-4 hours at 37°C) or by long exposure (15-20 hours) in medium supplemented with 5% (v/v) fetal calf serum. Digestion was followed by vigorous shaking to disaggregate epithelial fragments which were then allowed to settle under gravity. The pellet was filtered through a nylon mesh (30 μm) to remove single stromal cells and the retained epithelial aggregates were back washed and recovered. Single epithelial cell populations were obtained by dispersal of aggregates with trypsin-EDTA (1:250, Flow Labs., Paisley, U.K.) for 3 minutes at 37°C followed by gentle pipetting, addition of Soy trypsin inhibitor and then centrifugation. These fragments consisted largely (>90%) of epithelial cells (Aplin and Seif, 1985) and formed colonies of typically epithelioid morphology in culture. Further fractionation of the cell population was avoided since the pattern of expression for these antigens was unknown at the start.

Antibody Preparation

Each mouse received 7 intraperitoneal injections (300 μl) alternating between glandular fragments and single cell suspensions (approx. 10^7 cells). A final boost was given intravenously 7 days after the penultimate injection and contained single trypsinized epithelial cells. After 3 days the mouse was killed, the spleen removed, cells prepared and fused with NS1 myeloma cells using polyethylene glycol-mediated fusion in a standard protocol (Kohler and Milstein, 1975; Milstein, 1981). Hybridoma culture media were screened using indirect immunofluorescence on acetone-fixed cytospin preparations of endometrial epithelial cells (Figure 1b) and subsequently on tissue cryosections. A critical step in the screening for McAbs CC25 and F51 was to compare staining of proliferative and secretory phase tissue to make a crude initial selection on the basis of menstrual cycle variation of expression. Antibody-producing cultures were selected and cloned by limiting dilution. Single clones were selected and screened by indirect immunofluorescence on cryosections of snap frozen acetone-fixed endometrial tissue (Figure 1a) to identify the distribution of the antigens on endometrial cell surfaces. Selected cultures were recloned in the same way.

Immunofluorescence

Tissue sections were prepared from endometrial curettings collected in PBS. The curettings were washed several times in the buffer, cut into cubes and snap frozen in liquid nitrogen. Sections of 10 μm thickness were cut using a cryostat. The sections were allowed to dry in air at room temperature for 30 minutes before being stored at -20°C in dry, closed boxes. Fixation was carried out, just before staining, for 3 minutes using acetone; unfixed tissue gave similar results. All four new mcAbs recognized determinants that were sensitive to procedures involved in wax embedding and so routinely processed tissue could not be studied. The optimal dilution of the first antibody (pooled culture supernatant) was determined for each mcAb (v/v) as follows: G71 1:50, DH71 1:200, F51 1:20, CC25 1:100; clone 80 anti-cytokeratin (Sanbio, Cambridge, U.K), 1:10. These dilutions provided strong specific staining with a minimum of non-specific background staining.

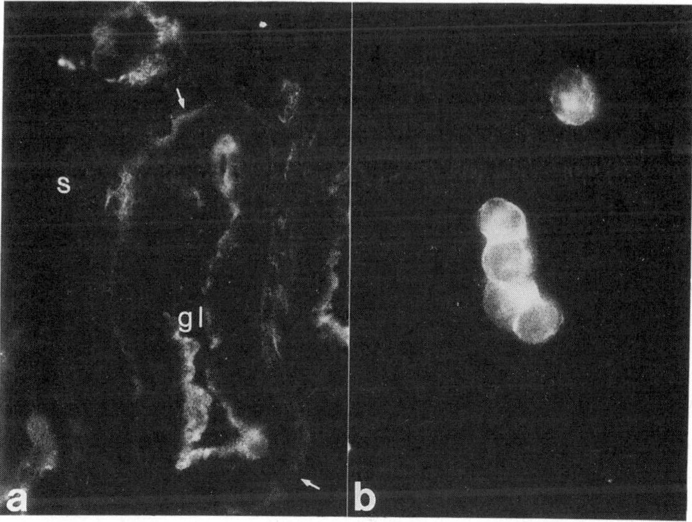

Figure 1. Results of screening of mixed hybridoma cultures. Immunofluorescent staining of secretory phase endometrium (a) shows antigens on the apical and basal (arrows) surfaces of glandular epithelial cells and in some stromal sites. Staining is weak on the lateral surfaces of epithelial cells. gl, gland lumen; s, stroma. Epithelial cell cytospins (b) show antigens in all surface domains (X430).

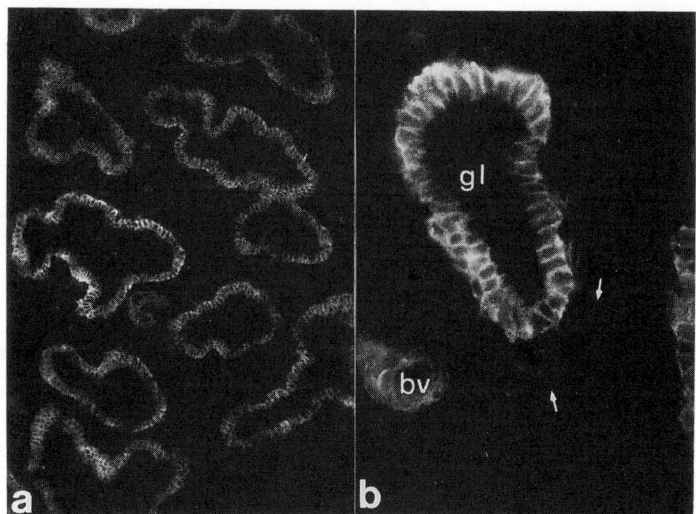

Figure 2.   Immunofluorescent staining of early to mid secretory phase
endometrium with mcAb CC25. (a) All glandular epithelium is uniformly stained
(X130). (b) Examination at a higher magnification (X440) shows that the staining
is associated with the basolateral surface of glandular epithelium; no staining is
detected on the apical surfaces lining the gland lumens (gl). Glancing sections of
glands show staining surrounding each epithelial cell. Blood vessels (bv) are also
stained in the smooth muscle layer and weak expression occurs on a few stromal
cells (arrows).

One hundred microliters of first antibody was added to each specimen and
incubated for 1 hour at room temperature in a humid chamber following which
surplus antibody was removed by washing in three changes of PBS each for 15
minutes.   Excess solution was wiped from around the sections and the second
antibody (fluorescein-conjugated rabbit anti-mouse; Dako) was applied at 1:100
(v/v) dilution. After further washing in PBS, sections were mounted in a non-fade
mountant (Johnson and Nogeira Araujo, 1981).

When dual immunolocalization was required, the above procedure was
conduct sequentially using: (i) the first monoclonal antibody; (ii)
tetramethylrhodamine-conjugated rabbit anti-mouse (Dako) at 1:50 (v/v); (iii) the
second monoclonal antibody; (iv) fluorescein-conjugated rabbit anti-mouse
(Dako) 1:100 (v/v). Each step was followed by an appropriate washing protocol.
Controls were included which demonstrated that the first monoclonal antibody
used was singly but not doubly labeled using the above concentrations of
fluorescent antibody conjugates, and that no inter-channel leakage occurred under
the conditions used.

RESULTS

Polyclonal antibody, either antiserum or mixed hybridoma culture
supernatant, reacted predominantly with the apical and basal surfaces of

endometrial epithelial cells in the secretory phase (Figure 1a). Mixed hybridoma culture supernatant reacted strongly with the cell surface of endometrial epithelial cytospins (Figure 1b). On cloning, the spatial and temporal distribution of various cell surface components could be defined. None of the epitopes described was detected in embryonic mouse, rat or chick (M.W.J. Ferguson, personal communication), nor in endometrium of sexually mature mice at any stage of the reproductive cycle (S.J. Kimber, personal communication), nor in the endometria of pregnant and non-pregnant horse (including the formed equid endometrial cup), donkey, sheep, or pig (A. Whyte, personal communication). The results, therefore, suggest that the epitopes are specific to human. Among the panel of mcAbs obtained, CC25 and F51 recognize cell surface antigens, the expression of which varies through the menstrual cycle, while DH71 and G71 recognize other cell surface components which lack this cycle dependency of expression.

Cyclic Variations Of Endometrial Cell Surface

McAb CC25 recognizes an epitope that is absent from glandular and luminal epithelial cell surfaces in the proliferative phase, appearing after ovulation in the early secretory phase (Figure 2a). The epitope is located at the basolateral surfaces of endometrial epithelial cells (Figure 2b) of glandular and uterine lumen.

An immunohistological study (Aplin and Seif, 1987) of the epitope recognized by mcAb CC25 showed a strong menstrual cycle dependency of expression with a sudden appearance in the early secretory phase followed by a gradual reduction in the mid and late phases with a slightly more diffuse distribution, still confined to the basolateral surfaces of epithelial cells. By the time of menstruation, the epitope is lost. In all stages the epitope is absent from apical surface locations; the appearance of the immunofluorescence complex all around the cells in certain glandular domains arises from glancing planes of section. In contrast, vascular smooth muscle layers (Figure 2b) stain throughout the menstrual cycle. This staining persists in first trimester decidua when some scattered staining is also associated with a subpopulation of interstitial cells despite the complete absence of staining from decidual epithelium. During the secretory phase, the CC25 epitope is occasionally found in stromal cell subpopulations present in the vicinity of some glands (Figure 2b). However, these cells express the epitope at an extremely low level.

In contrast, mcAb F51 recognizes an epitope on the apical surface of endometrial epithelial cells during the proliferative phase (Figure 3a). The epitope is exclusively located on the apical surface and absent from lateral and basal surfaces of both glandular (Figure 3c) and luminal epithelia. Very occasionally, isolated cells, located close to the glands or surface epithelium, express the epitope at a lower concentration.

After ovulation, the epithelial expression of the epitope is lost (Figure 3b). The exact timing of this change is yet to be determined; hormonally timed biopsy material is required for this purpose. However, in cases studied in the transitional phase (as assessed by histopathology), the epitope could be detected in some glands which expressed the antigen in a suppressed and non-uniform pattern.

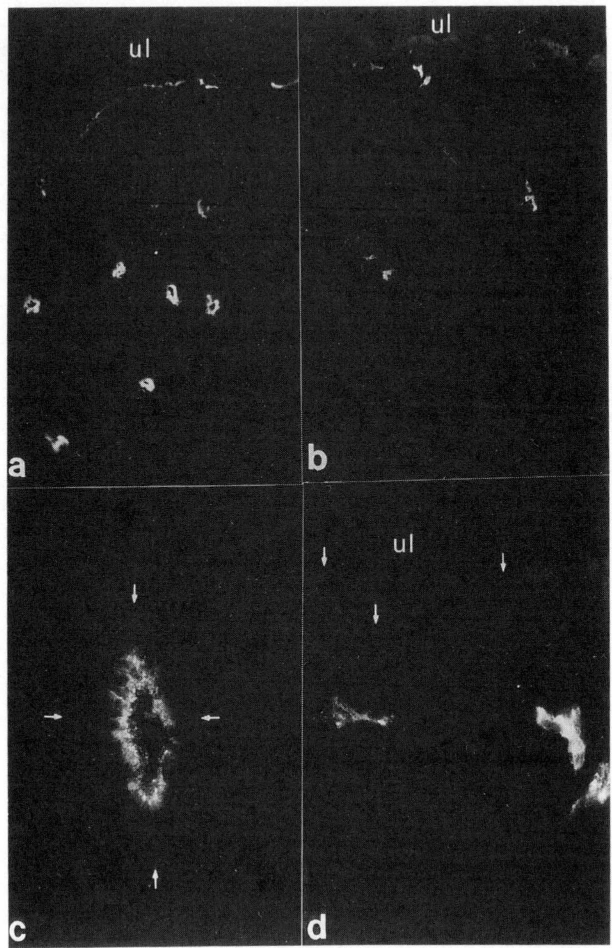

Figure 3. Immunofluorescence showing the distribution of the epitope recognized by mcAb F51 in endometrial tissue. (a), (c): proliferative phase; (b), (d): late secretory phase. In the proliferative phase (a;X90) the epitope is expressed at the surface of uterine luminal (ul) and glandular epithelial cells. (b) In late secretory phase all of the glandular and luminal staining has disappeared, but scattered groups of stromal cells are now stained (X90). (c) A proliferative phase gland showing the exclusively apical distribution of the epitope (X450); the arrows indicate the basal surfaces of the gland cells. (d) A higher magnification (X450) view of late secretory phase tissue. The uterine luminal (ul) epithelium (arrows indicate the apical surface) is, like the glands, negative. F51-positive stromal cell groups are seen near the luminal epithelium as well as in the vicinity of glands.

The disappearance of this epitope from endometrial epithelial cells during the secretory phase is coupled with its appearance on a subpopulation of interstital cells (Figure 3b, d). Isolated cells or groups of linearly arranged cells can be seen in close relation to the glands and below the luminal epithelium (Figure 3d). In

these cells, the polarized pattern of expression is lost, and immunofluorescence is
seen with equal intensity all over the cell surfaces.

In the late secretory phase, this subpopulation of stromal cells is more
evident and is the only carrier of the F51 epitope in endometrial tissue. On the basis
of the timing of expression of the F51 epitope, and the morphology of the F51 epitope-
bearing stromal cells, they do not correspond to pseudo-decidual cells.

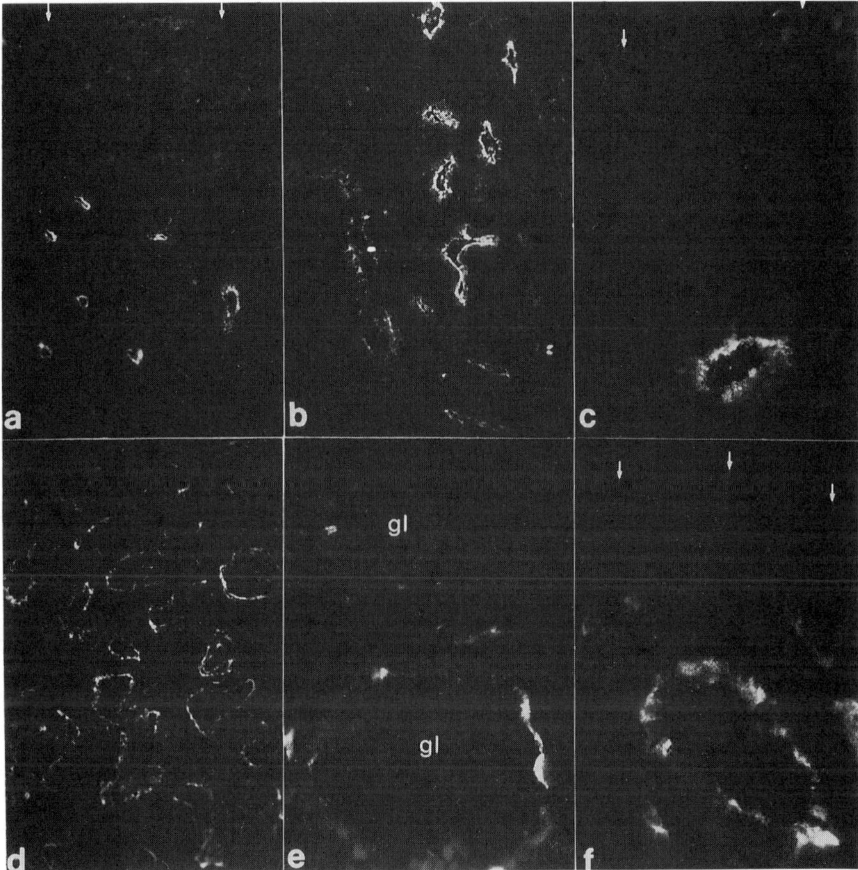

Figure 4. Immunofluorescent staining of endometrial tissue with mcAb DH71. In
the proliferative phase (a, b, c) the antibody binds to apical surfaces of glandular
epithelial cells which appear in cross-section as circular(a) or more complex (b)
patterns of staining. No staining is seen in luminal epithelium (arrows, a, c). (d,
e, f) A similar pattern of binding is observed in the secretory phase. Some
variation in the intensity of staining is observed between different glands (d, e) as
well as within glands (e, f). As in the proliferative phase, no staining is seen in the
luminal epithelium (arrows, f), nor on the basolateral surfaces of glandular cells.
Weakly stained stromal cells are visible in the vicinity of some glands (e, f). (a, b,
d:X60; c, e, f:X290).

Compositional Difference Between Surface And Glandular Epithelium

McAb DH71 stains the apical surface of glandular epithelial cells. In cross sections a ring of apical surface staining is seen (Figure 4a, c), while a continuous thin line of staining is seen in longitudinal sections (Figure 4d). No staining is visible in the secretions, epithelial cell cytoplasms, or the basement membranes (Figure 4c, f). The staining is detectable in glands in proliferative phase (Figure 4a, b, c) as well as all through the secretory phase (Figure 4d, e, f). However, some variation of intensity may occasionally be observed within a gland (Figure 4e, f) and a small proportion of glands lacks detectable staining (Figure 4e). The most prominent feature of the epitope recognized by mcAb DH71 is its absence from surface epithelial cells in all phases of the menstrual cycle (Figure 4c, f). In longitudinal sections where the glandular and luminal epithelia are seen in continuity, staining can be seen to end abruptly at the gland opening.

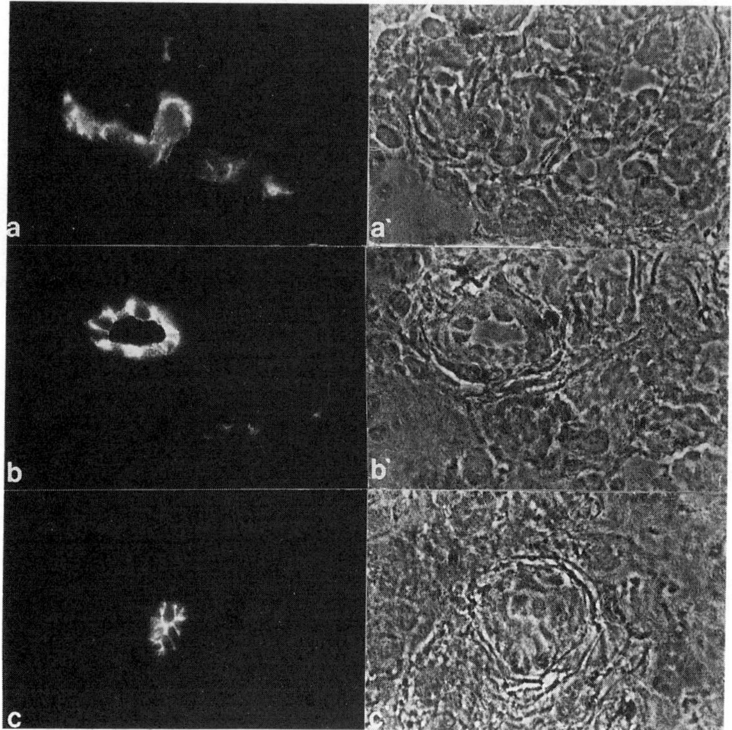

Figure 5. Stromal localization of the F51 epitope by combined immunofluorescence and phase contrast microscopy of endometrial tissue in the late secretory phase. (a) Groups of linearly arranged cells are seen expressing the epitope. (a') These cannot be distinguished from neighboring stromal cells in phase contrast. Occasionally the F51-positive cells are organized in a microfollicular pattern with a patent (b) or obliterated (c) lumen. These structures are evident in phase contrast (b', c') and are surrounded by elongated fibroblasts. Note that the polarized expression of the epitope seen in proliferative phase glands (cf. Figure 3c) is altered in all the structures illustrated (X590).

Extraglandular Expression of Epithelial Cell Surface Epitopes

The switch of the epitope recognized by mcAb F51 from epithelial apical locations in the proliferative phase (Figure 3a, c) to periglandular interstitial sites in the secretory phase (Figure 3b, d) raises the question of the nature of this interstitial cell population. Examination of endometrial sections in the late secretory phase, when the F51-positive interstitial cells are most abundant, revealed that they are detected in close proximity to the glands or uterine luminal epithelium. Most commonly, they are arranged in a linear form (Figure 5a) where each group is formed of 1-6 cells. When the specimen is examined by phase contrast (Figure 5a'), the cells which carry the F51 epitope cannot readily be distinguished from the rest of the stromal cells. Less commonly, these cells are arranged in a microfollicular pattern with a patent (Figure 5b,b') or obliterated (Figure 5c, c') lumen.

In order to investigate the possible epithelial nature of the extraglandular poulation of F51-positive cells, sections of late secretory phase endometrium were stained with a broad-specificity monoclonal antibody to cytokeratin (Figure 6a). The antibody Clone 80 has been shown to recognize an epitope present on several polypeptides of the cytokeratin family and to stain all types of epithelia except myoepithelium and kidney glomerular podocytes. It does not stain any non-epithelial cell population or tissue (van Muijen et al., 1984). The two components of endometrial epithelium, glandular and luminal, are evident, and in addition keratin can be detected in some stromal locations in the vicinity of the glandular epithelium. At higher magnifications (Figure 6b), stromal staining is visible in a subpopulation of cells of variable number and arrangement. Obviously, many of these structures represent glancing sections of glandular epithelium (Figure 6b, c). However, other keratin-positive cells and groups are isolated, elongated, linearly arranged, one cell thick or in other respects lack the characteristics of glands. Single cells or groups of cells can be seen in connection with the adjacent gland (Figure 6d) or completely separated from it (Figure 6b).

The correspondence between cells expressing the F51 epitope and those containing cytokeratin was further investigated by dual immunolocalization. Endometrial tissue sections obtained in the late secretory phase were incubated with mcAb F51. This was followed by labeling with TRITC-conjugated anti-mouse immunoglobulin. The same sections were then incubated with anti-keratin monoclonal antibody which was then labeled with FITC-conjugated anti-mouse immunoglobulin. The sections were examined through the rhodamine channel (Figure 7b) to detect cells which carried the F51 epitope and through the fluorescein channel to detect those which were expressing keratin (Figure 7a).

As well as the keratin-positive glandular epithelium, a variety of stromal sites in the vicinity were keratin-positive. These contained groups of cells in various numbers and arrangements (Figure 7c). Examination through the rhodamine channel showed that the F51 epitope is restricted to the surface of a subpopulation of cells (Figure 7d) within the keratin-positive population in the stroma. In other words, not all of the keratin-containing cells express the F51 epitope. However, all those which express the F51 epitope contain keratin (Figure 7c, d, e, f). F51 does not stain the adjacent glandular epithelium (cf. Figure 3b).

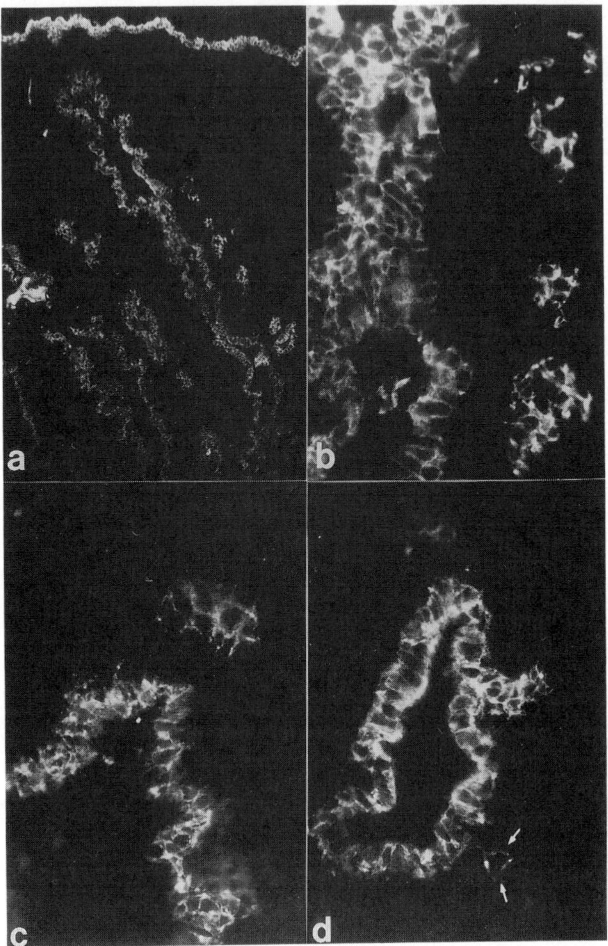

Figure 6. Immunofluorescent localization of the epithelial components of endometrial tissue in late secretory phase using keratin as a marker. (a) The two populations of glandular and luminal epithelium can be readily identified at low magnification. (b) Keratin-containing structures can also be seen in the stroma alongside some glands. These consist of groups of cells varying in number and organisation. Some groups arise as a result of staining of glancing sections through glands (b, c), while in other areas (d) one or two keratin-positive cells (arrows) can be seen in the stroma with a thin connection to an adjacent gland. (a:X90; b, c, d:X450).

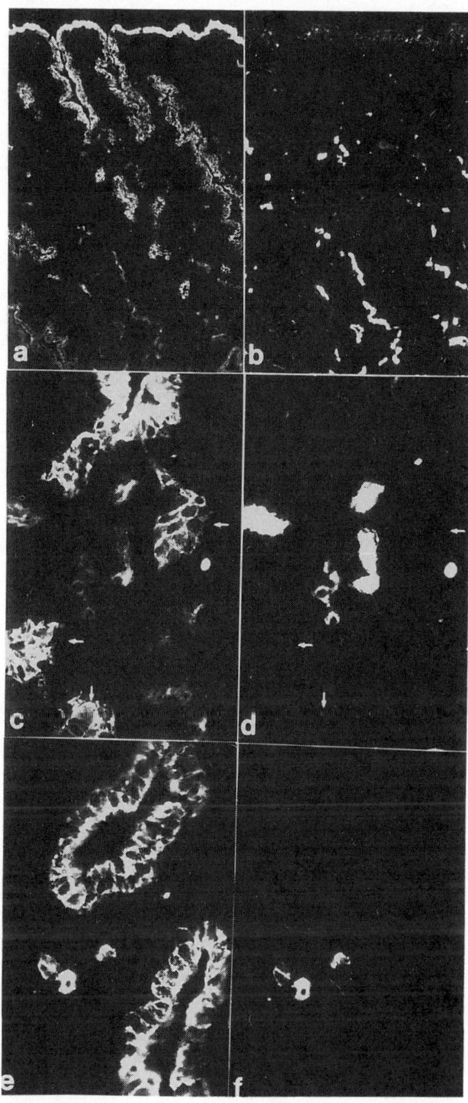

Figure 7.   Coexpression of keratin and the F51 epitope by interstitial cells.
Examination through the fluorescein channel (a, c, e) shows the distribution of
keratin-containing cells in late secretory phase endometrium.   F51-positive cells
are visualised in the tetramethylrhodamine channel (b, d, f).   (a) and (b), (c) and
(d), and (e) and (f) are pairs of identical fields.   Various arrangements of keratin-
containing cells are evident in the stroma (c) and the intensity of the staining
varies.   Some but not all of these express the F51 epitope (d).   Glancing sections of
glands can be identified as keratin-positive structures (arrow, c) and all of these
are F51-negative (arrows, d).   (e) and (f) show clearly that single cells or small
groups in the stroma can be positive for both keratin and F51 and do not correspond
to glancing sections through glands.   The distribution of the F51 epitope is uniform
across the cell surface. (a, b:X60; c, d, e, f:X300).

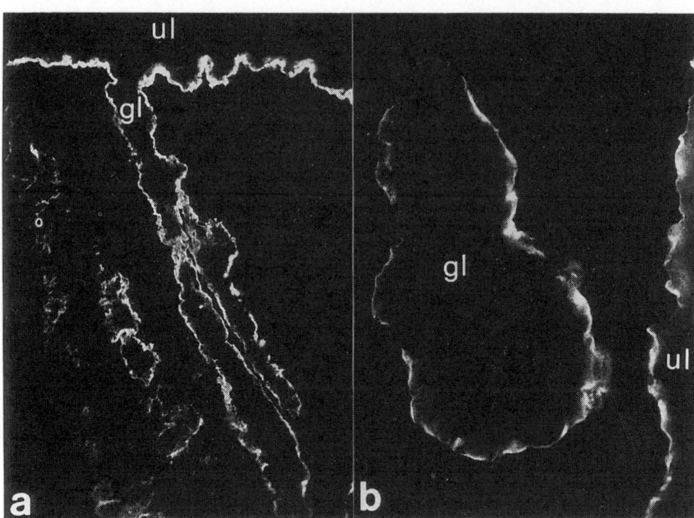

Figure 8. Immunofluorescent staining of late secretory phase endometrium with mcAb G71. The epitope is located in both luminal and glandular epithelia at basal cell surfaces, thus appearing as a continuous line of staining at the epithelial-stromal interface. gl, gland lumen; ul, uterine lumen. (a:X90; b:X400).

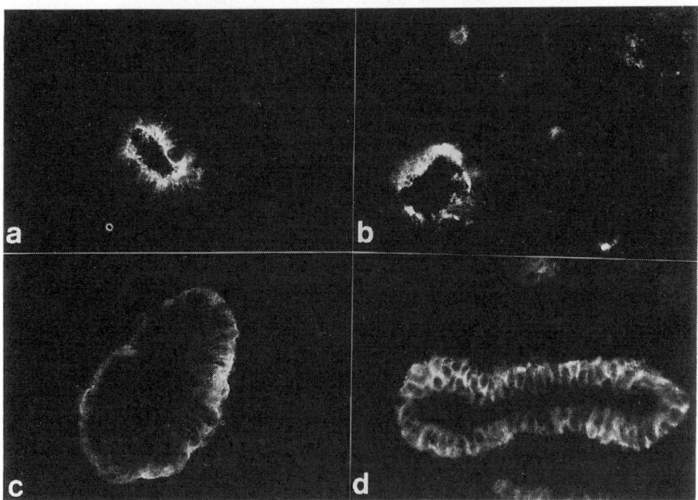

Figure 9. Polarized surface domains in endometrial epithelial cells revealed by various monoclonal antibodies. Different apically expressed epitopes are identified in the proliferative phase by mcAbs F51 (a) and DH71 (b). The former epitope is absent from gland cells in secretory phase as previously illustrated. (c) McAb G71 binds to an epitope located at the basal surface of the glandular cells. (d) McAb CC25 identifies an epitope on the basolateral surface of the cells the expression of which is exclusive to the secretory phase. Thus this panel of mcAbs reflects the polyclonal immune response to endometrial epithelial cells illustrated in Figure 1 (X320).

Subsequently, the F51-positive interstitial cells were examined for expression of other epithelial cell surface epitopes recognized by monoclonal antibodies DH71, CC25, and G71. In each case, this subpopulation of cells expressed the epitope investigated in a rather weak pattern; however, the following observations were consistent. First, the polarity of expression is lost and secondly the epitope-carrying interstitial cells are located in close proximity to endometrial glands (Figure 2b; Figure 4e, f).

Polarity Of Endometrial Epithelial Cells

The initial screening of mixed hybridoma culture supernatants showed a variety of cell surface components in apical, lateral and basal sites (Figure 1). The latter was the most frequently observed pattern in the screened cultures. McAb G71 recognizes an epitope in the area of the basal cell surface of endometrial epithelium. G71 staining is present in both the secretory phase (Figure 8a) and the proliferative phase (Figure 8b). The epitope is located beneath both glandular and luminal epithelial cells (Figure 8b) and also occurs in the basement membrane zone of a variety of other human epithelia (Aplin and Seif, 1985).

Thus, by cloning from a population of hybridoma cells producing antibodies directed to a spectrum of cell surface determinants, monoclonal antibodies to components in each of the three surface domains could be isolated. McAbs F51 and DH71 identify epithelial cell surface epitopes that are located on the apical surface (Figure 9a, b), mcAb CC25 identifies a cell surface epitope located on the basolateral surface of epithelial cells (Figure 9d) and G71 binds to a basally oriented epitope (Figure 9C). In all of the normal endometrial tissue sections examined, the polarity of antigen localization (Figure 9) is clearly maintained. However, in the previously mentioned interstitial cell groups, the polarity is completely lost and each epitope is identified in all areas of the cell surface.

DISCUSSION

A schedule of immunization with secretory phase epithelial cells pooled from several donors gave rise to a panel of monoclonal antibodies that includes antibodies recognizing cell surface components as well as secretory products (Seif et al., 1989). In the present study the distribution of four antigenic markers was characterized whose properties suggest that they are associated with the endometrial epithelial cell surface. None of these epitopes have been detected in endometrial tissue from other species including rodents and domestic animals; this suggests a compositional specificity that may arise from structural and functional properties specific to the human maternofetal interface.

All four epitopes show a polarized distribution: DH71 and F51 epitopes are apically expressed; the G71 epitope is present on basal surfaces; and the CC25 marker is basolaterally distributed. Taken at face value, the results therefore confirm the existence of three surface domains of differing composition in endometrial epithelium.

Of the four markers, only one (DH71) appears to be differentially expressed by glandular as opposed to surface epithelium. It may be that a general similarity in surface composition is to be expected since these two cell types share a common precursor. Despite this, the progesterone-dependent secretory differentiation of the two cell populations has been shown to differ on both morphological and compositional criteria (Ferenczy, 1976a; Fazleabas et al., 1985). Specialized functions associated with preparation for implantation are likely to give rise to a requirement for compositional differences between surface and glandular cells. If indeed the luminal cells are derived from basal gland cells, the DH71 epitope must at some stage be lost from the former. It seems likely that this occurs early in the cycle. DH71 requires further evaluation in this regard, and it should be emphasized that we have not yet undertaken a systematic study of possible variation in the expression of the DH71 epitope in different parts of the endometrium (i.e., isthmic to fundal or anterior vs posterior) or in an implantation site.

Two of the four antigenic markers described (G71 and DH71) show a consistent pattern of expression throughout the menstrual cycle while the other two (CC25 and F51) exhibit interesting stage-specific variations which may render them useful for staging and evaluation of endometrial function. Much more therefore needs to be learned of the mechanism of control of their expression as well as their behavior in an abnormal endocrine environment. Of these two cycle-variable markers, CC25 shows the simpler pattern of behavior. Its absence from proliferative phase epithelium and appearance in early secretory phase (Aplin and Seif, 1987) indicates an association with the development of the secretory phenotype. The observed basolateral distribution prompts speculation that its function may be associated with maternal signaling via stromal-epithelial or vascular-epithelial pathways. The latter is consistent with the presence of the CC25 epitope on the surface of vascular smooth muscle cells, where expression is independent of endocrine status.

The F51 antigen shows an entirely different pattern of behavior, being present in apical locations on proliferative phase epithelial cells and absent in the secretory phase. The loss of the F51 epitope from epithelial cells in early secretory phase is apparently an active alteration in surface composition indicating the onset of differentiation into the secretory phenotype.

The patterns of behavior of the epitopes recognized by antibodies CC25 and F51 confirm that, as in numerous animal species, variations occur in the composition of human endometrial epithelial cell surfaces in response to endocrine signals. It should be stressed that the mechanisms whereby these messages are conveyed may be either direct (i.e., via hormone-receptor interaction within the epithelium) or indirect (e.g., by paracrine signals from differentiating stroma). Compositional changes at the surface have been shown to include both increase and decrease of specific carbohydrate, anionic and glycoprotein components. Thus in rabbits, Anderson et al. (1986) have shown changes in lectin binding to the apical epithelial surface occurring during the first 7 days of pregnancy and arising partly from desialylation of preexisting glycoproteins and partly from the expression of new glycoprotein species in the apical domain. Lectins have also been used by Lee et al. (1983) to show stage-specific changes in mouse endometrium including the disappearance of fucosyl residues recognized by the lectin from Ulex europaeus. Kimber et al. (1988) have shown that binding of a monoclonal antibody to the (fucosylated) blood group H type I oligosaccharide occurs to apical surfaces of most uterine epithelial cells on day 3 of pregnancy, and becomes restricted to groups of 6-20 adjacent cells on days 4 and 5. Morris and Potter (1984) used the interaction between mouse endometrial cells and cationic beads to demonstrate a net reduction in cell surface negative charge occurring at the time of implantation. The current studies demonstrate that changes in cell surface composition occur in association with the development of a secretory phenotype in human endometrial epithelium. Studies in other species have concentrated on the peri-implantation period and have succeeded in demonstrating compositional changes at the luminal surface where the first contact with the embryo is made. It remains to be seen whether analogous changes occur in humans.

The surprising discovery of F51-positive, cytokeratin-containing cells in the periglandular and superficial stroma throughout the secretory phase after the glandular and surface epithelial cells have lost the F51 marker suggests the presence of an interstitial cellular subpopulation that shares some features of the proliferative phase epithelial phenotype. At least two hypotheses can be advanced to account for their presence: the differentiation of a subpopulation of stromal cells at the time of ovulation which subsequently reorganize to form the observed microfollicular structures; or the escape from glands (or luminal epithelium) of epithelial cells which retain elements of the preovulatory phenotype. Evidence for the migration of a class of epithelial cells out of endometrial glands has been adduced by Hopwood and Levison (1976), who associated the phenomenon with apoptotic cell death as a contributory factor in perimenstrual regression. However, it is not clear why this should occur in early secretory phase. Greenburg and Hay (1982; 1986) have shown that epithelial cells from a number of embryonic and adult tissues, when suspended in a three dimensional gel of type I collagen, undergo a transformation into mesenchyme-like cells which cannot be distinguished on morphological criteria from other fibroblasts. The transformed cells show a loss of surface polarity and cease production of certain epithelial-specific polypeptides. The endometrium might provide an in vivo analogue of this phenomenon, but the function of the extraglandular cells remains unknown.

SUMMARY

The tissue distribution is described of epitopes recognized by four monoclonal antibodies to the surface of endometrial epithelial cells. The results demonstrate: (i) the presence of polarized domains of surface composition in glandular and surface epithelium; (ii) compositional differences between the two epithelial cell subpopulations; (iii) changes in cell surface composition during the menstrual cycle including the loss of one epitope and the appearance of another after ovulation; and (iv) the existence in secretory phase of a population of periglandular stromal cells that express cytokeratin as well as an epitope otherwise associated with proliferative phase epithelium.

ACKNOWLEDGEMENTS

We thank Dr. C.H. Buckley for making available routine histopathological reports and clinical colleagues at St. Mary's Hospital for biopsy material.

REFERENCES

Anderson, T.L., Olson, G.E., and Hoffman, L.H. (1986) Stage-specific alterations in the apical membrane glycoproteins of endometrial epithelial cells related to implantation in rabbits. *Biol. Reprod.* 34, 701-720.

Aplin, J.D. (1989) Cellular biochemistry of the endometrium. In: *Biology of the Uterus,* (eds.) R.M.Wynn and W.P.Jollie, Plenum: New York,in press.

Aplin, J.D. and Seif, M.W. (1985) Basally located epithelial cell surface component identified by a monoclonal antibody technique. *Exp. Cell Res.* 160, 550-555.

Aplin, J.D. and Seif, M.W. (1987) A monoclonal antibody to a cell surface determinant in human endometrial epithelium: Stage specific expression in the menstrual cycle. *Am. J. Obstet. Gynecol.* 156, 250-253.

Bell, S.C. (1986) Secretory endometrial and decidual proteins: Studies and clinical significance of a maternally derived group of pregnancy-associated serum proteins. *Human Reprod.* 1, 129-143.

Bychkov, V. and Toto, P.D. (1986) Lectin binding to normal human endometrium. *Gynecol. Obstet. Invest.* 22, 29-33.

Ciocca, D.R., Asch, R.H., Adams, D.J., and McGuire, W.L. (1983) Evidence for modulation of a 24K protein in human endometrium during the menstrual cycle. *J. Clin. Endocrinol. Metab.* 57, 496-499.

Classen-Linke, I., Denker, H.-W., and Winterhager, E. (1987) Apical plasma membrane-bound enzymes of rabbit uterine epithelium. Pattern changes during the periimplantation phase. *Histochem.* 87, 517-529.

Dallenbach-Hellweg, G. (1981) *Histopathology of Endometrium.* Berlin Springer-Verlag.

Denker, H.W. (1982) Proteases of the blastocyst and of the uterus.In: *Proteins and Steroids in Early Pregnancy,* (eds.) H.M. Beier and P. Karlson, Springer, Berlin, pp. 183-208.

Denker, H.W. (1983) Basic aspects of ovoimplantation. *Obstet. Gynecol. Ann.* 12, 15-42.

Dockery, P., Li, T.C., Rogers, A.W., Cooke,I.D.,and Lenton, A.E. (1988) The ultrastructure of the glandular epithelium in the timed endometrial biopsy. *Human Reprod.* 3, 826-834.

Fazleabas, A.T., Bazer, F.W., Hansen, P.J., Geisert, R.D., and Roberts, R.M. (1985) Differential patterns of secretory protein localization within the pig uterine endometrium. *Endocrinol.* 116, 240-245.

Ferenczy, A. (1976a) Studies on the cytodynamics of human endometrial regeneration 1. Scanning electron microscopy. *Am. J. Obstet. Gynecol.* 124, 64-74.

Ferenczy, A. (1976b) Studies on the cytodynamics of human endometriial regenration 11. Trasmission electron microscopy and histochemistry. *Am. J. Obstet. Gynecol.* 124, 582-595.

Ferenczy, A. (1977) Surface ultrastructural response of the human utrine epithelium to hormonal enviroment. A scanning electron microscopic study. *Acta Cytol.* 21, 566-572.

Greenburg, G. and Hay, E.D. (1982) Epithelia suspended in collagen gels can lose polarity and express characteristics of migrating mesenchymal cells. *J. Cell Biol.* 95, 333-339.

Greenburg, G. and Hay, E.D. (1986) Cytodifferentiation and tissue phenotype change during transformation of embryonic lens epithelium to mesenchyme-like cells in vitro. *Dev. Biol.* 115, 363-379.

Hopwood, D. and Levison, D.A. (1976) Atrophy and apoptosis in the cyclical human endometrium. *J. Pathol.* 119, 159-166.

Imakawa, K., Anthony, R.V., Kazemi, M., Marotti, K.R., Polites, H.G., and Roberts, R.M. (1987) Interferon-like sequence of ovine trophoblast protein secreted by embryonic trophectoderm. *Nature* 330, 377-379.

Johnson, G.D. and Nogeira Araujo, G.M.de C. (1981) A simple method of reducing the fading of immunofluorescence during microscopy. *J. Immunol. Meth.* 43, 349-350.

Kimber, S.J., Lindenberg, S., and Lundblad, A. (1988) Distribution of some Galβ1-3(4)GlcNAc related carbohydrate antigens on the mouse uterine epithelium in relation to the peri-implantation period. *J. Reprod. Immunol.* 12, 297-314.

Kirk, D., King, R.J.B., Heyes, J., Peachy, L., Hirsch, P.J., and Taylor, R.W.T. (1978) Normal human endometrium in cell culture.1. Separation and characterisation of epithelial and stromal components in vitro. *In Vitro* 14, 651-662.

Kohler, G. and Milstein, C. (1975) Continuous culture of fused cell secreting antibody of predefined specificity. *Nature* 256, 495-497.

Kosasa, T. (1973) Early detection of implantation using a radioimmunoassay specific for human chorionic gonadotrophin. *J. Clin. Endocrinol. Metab.* 36, 622-624.

Lee, M.C. and Damjanov, I. (1985) Pregnancy-related changes in the human endometrium revealed by lectin histochemistry. *Histochem.* 82, 275-280.

Lee, M.C., Wu, T.C., Wai, Y.J., and Damjanov, I. (1983) Pregnancy-related changes in mouse oviduct and uterus revealed by differential binding of fluoresceinated lectins. *Histochem.* 79, 365-375.

Matlin, K.S. and Simons, K. (1984) Sorting of an apical plasma membrane glycoprotien occurs before it reaches the cell surface in cultured epithelial cell. *J. Cell Sci.* 99, 2131-2139.

Milstein, C. (1981) Monoclonal antibodies from hybrid myelomas. *Proc. Roy. Soc. London* B211, 393-412.

Morton, H., Rolfe, B., Clunie, G.J.A., Anderson, M.J., and Morrison, J. (1977) An early pregnancy factor detected in human serum by the rosette inhibition test. *Lancet* 1, 394-397.

Morris, J.E. and Potter, S.W. (1984) A comparison of developmental changes in surface charge in mouse blastocysts and uterine epithelium using DEAE beads and dextran sulphate in vitro. *Dev. Biol.* 103, 190-199.

Seif, M.W., Aplin, J.D., Foden, L.J., and Tindall, V.R. (1989) A novel approach for monitoring the endometrial cycle and detecting ovulation. *Am. J. Obstet. Gynecol.*, in press.

Smith, R.A., Seif, M.W., Rogers, A.W., Li, T.C., Dockery, P., Cooke, I.D., and Aplin, J.D. (1989) The endometrial cycle: The expression of a secretory component correlated with the luteinising hormone peak. *Human Reprod.* 160, 357-362.

Van Meer, G. and Simons, K. (1986) The function of tight junctions in maintaining differences in lipid composition between apical and basolateral cell surface domains of MDCK cells. *EMBO J.* 5, 1455-1465.

Van Muijen, G.N.P., Ruuter, D.J., Ponec, M., Huiskens-Van Der Mey, C. and Waarnar, S.O. (1984) Monoclonal antibodies with different specificities against cytokeratins. *Am. J. Pathol.* 114, 9-17.

Yen, Y., Lee, M.C., Salzmann, M., and Damjanov, I. (1986) Lectin binding sites on human endocervix: A comparison with secretory and proliferative endometrium. *Anat. Rec.* 215, 262-266.

CHANGES OF INTERMEDIATE FILAMENT PROTEIN LOCALIZATION IN ENDOMETRIAL CELLS DURING EARLY PREGNANCY OF RABBITS

Axel Hochfeld[1], Henning M. Beier[1], and Hans-Werner Denker[2,3]

Institut für Anatomie
Medizinische Fakultät der RWTH Aachen
Melatener Strasse 211
D-5100 Aachen, Federal Republic of Germany

[1]Lehrstuhl für Anatomie und Reproduktionsbiologie
Medizinische Fakultät der RWTH Aachen

[2]Lehr- und Forschungsgebiet Anatomie und Reproduktionsbiologie
Medizinische Fakultät der RWTH Aachen

INTRODUCTION

Implantation is initiated by an interaction of trophoblast with the uterine epithelium via the apical cell poles of both partners. Aspects of changes that must take place in the glycocalyx and in the plasma membrane, to allow this process to be initiated are discussed in other contributions to this volume. There appears to be good reason, however, to bring the cytoskeleton into the picture, since findings obtained in other systems give strong evidence for interactions between it and the cell membranes (Tachi et al., 1970; Jones and Goldman, 1985; Perides et al., 1986a,b; Traub et al., 1987; Lazarides, 1980, for review see: Cowin et al., 1985; Geiger et al., 1985). There are only a few reports on the occurrence of intermediate filaments (IF) in the uterine epithelium. Franke et al. (1986) have studied IF in the proliferative phase of the human endometrial epithelium. Khong et al. (1986) concentrated on the expression of IF in the placenta, amniochorion, and placental bed in humans and Dabbs et al. (1986) used IF as diagnostic tools for histologic differentiation of uterine adenocarcinomas. Viale et al. (1988) described coexpression of vimentin and keratin in endometrial glands.

At present there is no literature concerning the distribution of IF in the uterine epithelium in the phase when attachment is being initiated between it and the trophoblast . In this study, regional distribution and cellular expression of vimentin as a marker for mesenchymal cells, and cytokeratins for epithelial cells, has been investigated during this phase. Also there was interest in the localization characteristics of these two intermediate filament proteins in the corresponding phases of pseudopregnancy. Specifically, the following questions were asked: Are there differences in the distribution between pregnancy and pseudopregnancy? Does the blastocyst influence the maternal cytoskeleton of the uterine epithelium?

[3]Present address: Institut für Anatomie, Lehrstuhl für Anatomie und Entwicklungsbiologie, Universitätsklinikum, Hufelandstrasse 55, D-4300 Essen, Federal Republic of Germany

MATERIALS AND METHODS

Rabbits were caged as described by Fischer and Meuser-Odenkirchen (1988). Does were mated to two fertile bucks. Pseudopregnancy was induced by intravenous injection of 75 I.U. of human chorionic gonadotropin (hCG) (Prolan® Bayer AG, Leverkusen, FRG). The day of mating and of hCG injection was designated day 0 of pregnancy/pseudopregnancy. Two rabbits each were killed by stunning and exsanguination at 3, 4, 5, 6, 7, 8, and 9 days post coitum (p.c.)/post hCG (p.hCG). The uteri were quickly removed, cut into pieces, frozen unfixed in liquid nitrogen and stored at -25°C to -30°C in air tight plastic bags until use.

Uterine ligation was performed under thiobarbital anesthesia (Thiogenal®, Merck, 40 mg/kg body weight given i.v. plus 20 - 40 mg/kg body weight i.p. during surgery) after perphenanzine premedication (Decentan®, Merck, Darmstadt, FRG, 5 mg/rabbit i.m.). Twenty-four hours after mating, one of the uteri was ligated near the uterotubal junction to prevent passage of embryos from the Fallopian Tube into the uterine lumen. Animals were killed 8 d p.c. The uteri were handled as described.

Sections (10 μm) were taken in longitudinal or transverse orientation, on a cryostat (Dittes Duspiva, Heidelberg, FRG) at -25°C, mounted on gelatin coated glass slides, and stored in phosphate buffered saline (PBS: Ca^{++} and Mg^{++} free, Seromed, Berlin, FRG) for about 10 minutes until use.

Immunofluorescence Staining

Unfixed uterine sections were incubated with: rabbit serum diluted 1:20 in PBS with 1.5% bovine serum albumin (BSA, Sigma No. A 7906, Deisenhofen, FRG), followed by first antibody: 1:40 diluted anti-keratin (KL_4, Dianova, Hamburg, FRG, reacting with polypeptides in the molecular range of 50 - 67 kilodaltons) or 1:40 diluted anti-vimentin (Dakopatts, Glostrup, Denmark, reacting with the 57 kilodaltons protein) 2 hours at room temperature. Controls were incubated with mouse serum (1:100) instead. After washing with 4 changes of PBS for 10 minutes each, 1:40 diluted FITC-conjugated rabbit anti-mouse IgG (F 232, Dakopatts, Glostrup) was applied for 1 hour in the dark at room temperature. Sections were again washed in 4 changes of PBS. The sections were then mounted in glycerol:PBS (9:1) and examined under a Zeiss photomicroscope II with an UV-epiilluminator and a filter set for FITC. Samples were photographed immediately after fluorescence-reaction on Kodak Tri X pan 400 film with standardized illumination and identical exposure time.

RESULTS

Distribution Of Vimentin

Myometrial smooth muscle cells and blood vessels were recognized by the vimentin antibody in all sections of uteri, regardless of the state of pregnancy and pseudopregnancy. The stromal cells were also positively stained with anti-vimentin in all cryostat sections. In pregnancy the cells of the endometrial epithelium showed stage-dependent alterations.

Table 1

Intracellular Distribution Of Vimentin In Endometrial Epithelial Cells

	Pseudopregnancy			Pregnancy					
				Interblastocyst Segments			Implantation Chamber		
	Luminal Epithelial Cells		Deepest Parts Of The Crypts	Luminal Epithelial Cells		Deepest parts Of The Crypts	Luminal Epithelial Cells		Deepest Parts Of The Crypts
Day	Apical Region	Basolateral Region		Apical Region	Basolateral Region		Apical Region	Basolateral Region	
3	0	+	0						
4	0	+	0						
5	0	+	0	0	+	0	0	+	0
6	0	+	0	0	+	0	0	+	0
7	0	+	0	0	+	0	+	++	0
8	0	+	0	0	+	0	++	++	0
9	0	+	0	0	+	0	+++	+++	0

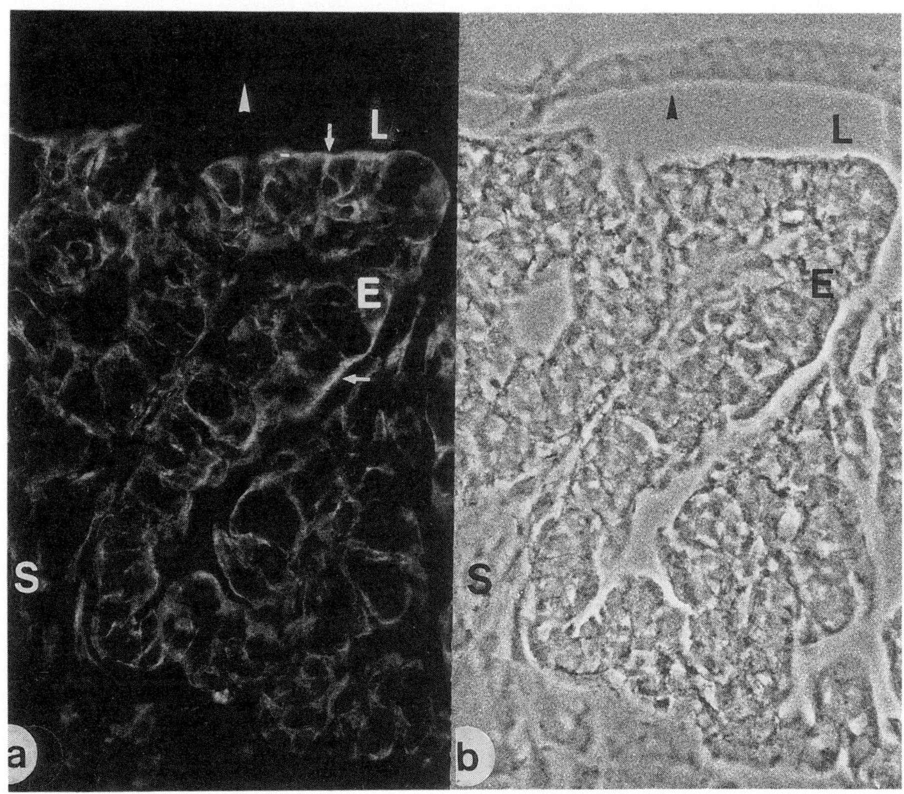

Figure 1. Vimentin immunoreactivity. (a) Implantation chamber 7 days p.c. Epithelium (E) adjacent to the blastocyst shows an intense vimentin reaction in the apical cytoplasm of the luminal cells(arrows). Stromal components were also significantly stained (S). There is no reaction of the blastocyst coverings (arrowhead). (b) Phase contrast, X370.

At 3 days p.c. to 5 days p.c., there was only a weak vimentin reaction in the luminal epithelium, restricted to the basolateral regions of the cells. There was no reaction in the deepest part of the crypts (Table 1).

A difference from these early stages first became detectable at 6 days p.c. The epithelium surrounding the blastocyst in the implantation chamber was more intensely stained, but always mainly restricted to the basolateral regions of the epithelial cells.

At 7 days p.c. the anti-vimentin reaction in the implantation chamber became much more pronounced. The luminal epithelium was significantly more intensely stained in particular at the antimesometrial side. This anti-vimentin reaction was no longer restricted to the basolateral regions of the cells but rather vimentin-positive structures in the apical cytoplasm of the luminal cells were noted (Figure 1). In the deepest parts of the crypts, near the myometrium, still no positive vimentin reaction was visible (Table 1).

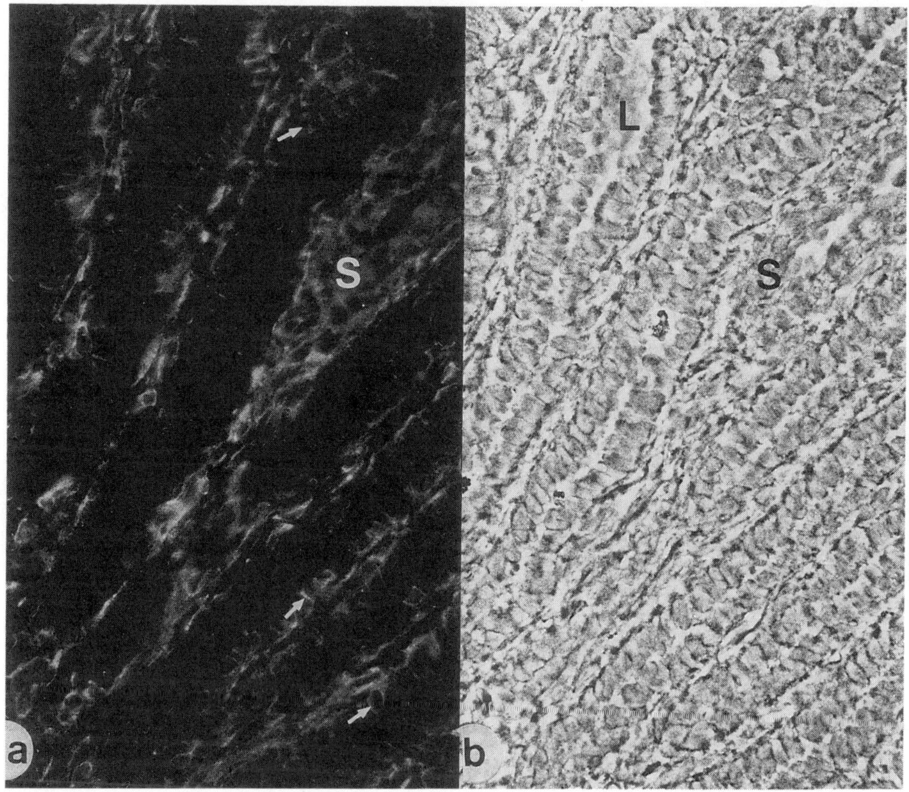

Figure 2. Vimentin immunoreactivity. (a) Interblastocyst segments 7 days p.c. There is only a delicate vimentin reaction in the luminal and middle parts of the endometrial epithelium (arrows). Stromal components (S) are stained with anti-vimentin. (b) Phase contrast, X370.

At 8 days p.c. the epithelial cells in the implantation chamber were strongly stained for vimentin in the apical cytoplasm. A moderate apical reaction could also be detected in the middle parts of the crypts. Again there was no reaction in the deepest parts of the crypts (Table 1).

Nine days p.c. the fluorescence reached a maximum in the epithelial cells surrounding the blastocyst. A massive band of fluorescence could be detected in the apical and basal cytoplasm of the luminal epithelium, clearly marked off from the uterine stromal cells (Figure 3). Decidual cells located near small blood vessels were positively stained with anti-vimentin.

In all states of pseudopregnancy from day 3 to day 9 p. hCG, only a delicate vimentin reaction in the luminal and middle parts of the endometrial epithelium was noted, corresponding with the observations in pregnancy from day 3 p.c. to day 5 p.c. As in pregnancy there was never any positive reaction in the deepest parts of the crypts (Figures 5 and 6).

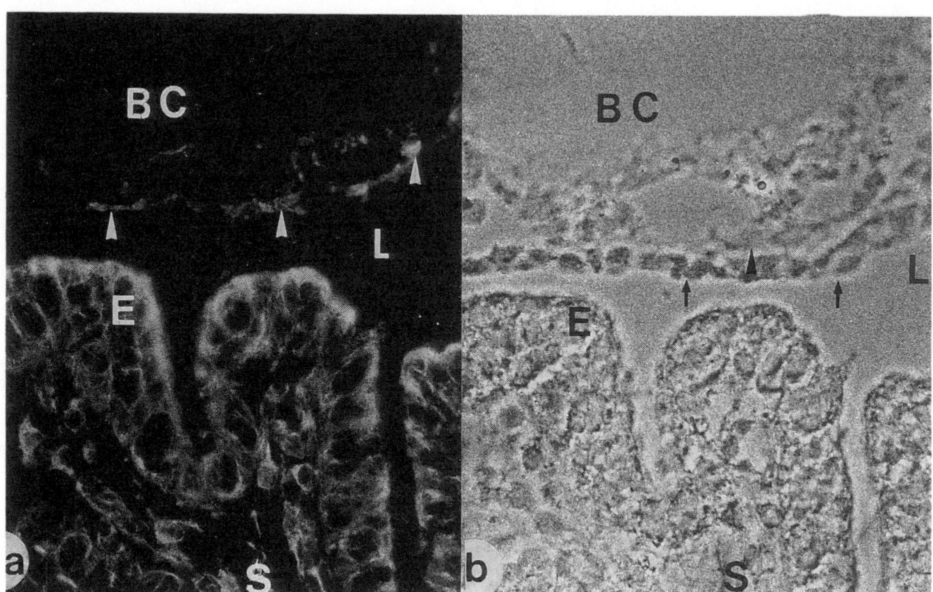

Figure 3. Vimentin immunoreactivity. (a) Implantation chamber 9 days p.c.. Reaction reached a maximum in the apical and basolateral cytoplasm of the epithelial cells surrounding the blastocyst. There is a delicate immunofluorescence in mesoderm (arrowheads). Arrows = trophoblast; BC = blastocyst cavity; L = uterine lumen; S = stroma. (b) Phase contrast, X370.

Vimentin distribution in the interblastocyst segments of pregnant uteri was from day 6 p.c. to day 9 p.c. quite similar to the results in all stages of pseudopregnancy (Figures 2 and 4). After tubal ligation, the epithelium of the ligated side (no blastocyst present) showed a vimentin distribution which was identical with that of interblastocyst segments or of pseudopregnancy. At the contralateral side serving as a control (blastocysts present), the typical distribution pattern of vimentin with increased apical reaction of the epithelium of the implantation chamber was seen.

In the blastocyst there was no distinct vimentin reaction at 6 days p.c. and 7 days p.c., either in fragments of blastocyst coverings or in the trophoblast cells or endoderm. At 8 days p.c. and 9 days p.c. a delicate immunofluorescence was detected in extraembryonic cells other than trophoblast, probably extraembryonic mesoderm. Vimentin was not detected in trophoblast cells (Figure 3).

Distribution Of Cytokeratin

In pregnancy and pseudopregnancy an intense reaction for cytokeratin was noted in the endometrial epithelium, with a higher concentration in the subapical region. There were no significant differences in staining patterns between the various stages; also the deepest parts of the crypts showed the same cytokeratin reaction (Figures 7 and 8; Table 2). Other parts of the uterus including stromal cells or myometrium did not react with the cytokeratin antibody, except for the peritoneal epithelium.

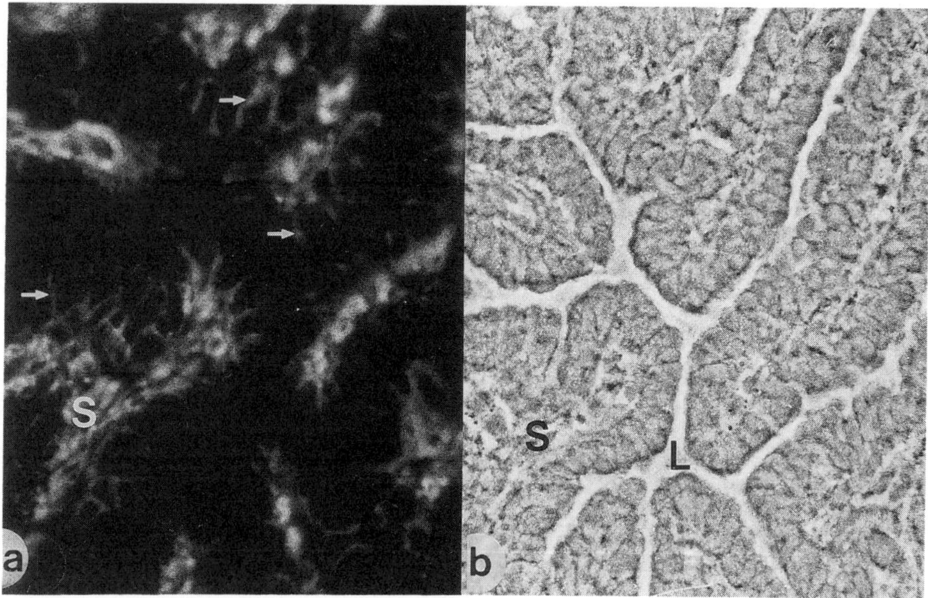

Figure 4. Vimentin immunoreactivity. (a) Interblastocyst segments 9 days p.c.. There is again only a weak reaction in the basolateral regions of the epithelial cells (arrows) and a positive staining of the stromal components (S). (b) Phase contrast, X370.

Distribution Of Cytokeratin In The Blastocyst

In the blastocyst from 6 days p.c. on a faint cytokeratin reaction was visible only in trophoblast cells. Fragments of the zona pellucida showed no staining (Figure 8).

DISCUSSION

According to earlier investigations, the physiological and histological transformation of the rabbit endometrium seems to be very similar in pregnancy and pseudopregnancy (Beier and Kühnel, 1973; Busch, 1982; Winterhager, 1985) except for the implantation chamber (from 6 1/2 - 7 days p.c. on). Biochemical investigations, however, have revealed time-dependent quantitative differences in protein patterns between pregnancy and pseudopregnancy. These differences are obviously due to the presence or absence, respectively, of blastocysts (Beier et al., 1974, 1979).

From embryo transfer experiments it appears that the pseudopregnant uterus acquires characteristics of a pregnant uterus when blastocysts are surgically transferred at any stage between 3 1/2 and 6 days p.c. into such pseudopregnant milieu (Fischer, 1988). It is, however, not yet clear whether success of embryo transfer is only a matter of "appropriateness" of the pseudopregnant endometrium or whether this condition is triggered by the transferred blastocysts.

Table 2

Intracellular Distribution Of Cytokeratin In Endometrial Epithelial Cells

| Day | Pseudopregnancy | | Pregnancy | | | |
| | | | Interblastocyst Segments | | Implantation Chamber | |
	Apical Region	Basolateral Region	Apical Region	Basolateral Region	Apical Region	Basolateral Region
3	++	+	++	+	++	+
4	++	+	++	+	++	+
5	++	+	++	+	++	+
6	++	+	++	+	++	+
7	++	+	++	+	++	+
8	++	+	++	+	++	+
9	++	+	++	+	++	+

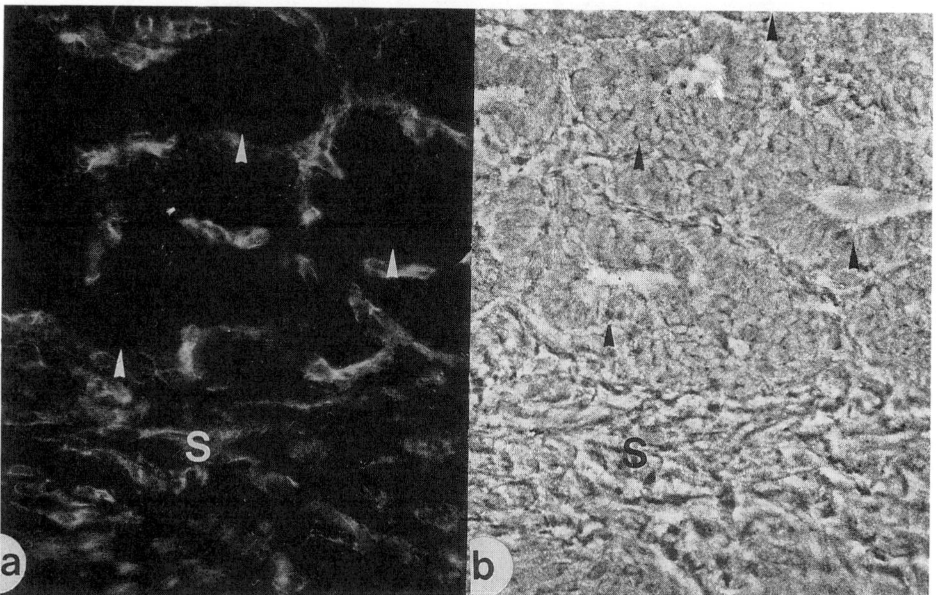

Figure 5. Vimentin immunoreactivity. (a) Interblastocyst segments 9 days p.c..
In pregnancy and pseudopregnancy (not shown) no positive vimentin reaction was
found in the deepest parts of the crypts (arrowheads). (b) Phase contrast, X370.

The possibility that embryonic signals trigger reactions of the
endometrium has been discussed by Beier (1968), when protein patterns of rabbit
uterine secretions were analyzed and compared. These investigations
demonstrated that normal pregnancy patterns differed clearly from
pseudopregnant patterns. Those differences suggested that embryonic signals act
as conditioning factors for an appropriate preparation of the intrauterine protein
milieu for preimplantation development and implantation (Beier, 1968).

The earliest differences between pseudopregnancy and pregnancy in the
rabbit model have been found in the immediate vicinity of blastocysts from 6 d to 6 d
16 h p.c. when membrane-bound enzymes or gap junction proteins were
investigated (van Hoorn and Denker, 1975; Winterhager et al., 1988). The results
documented in the present communication for vimentin are quite consistent with
these findings. During the preimplantation phase the histochemical findings are
quite similar for pseudopregnancy and for pregnancy until 5 d p.c./p. hCG. The
reaction is weak but clearly positive in the basolateral region of all epithelial cells.
In both physiological states, the reaction remains negative in the deepest parts of
crypt epithelia. This shows that the uterine epithelium consists of cell populations
that clearly differ in their content of vimentin intermediate filaments, while
appearing similar with respect to cytokeratin content.

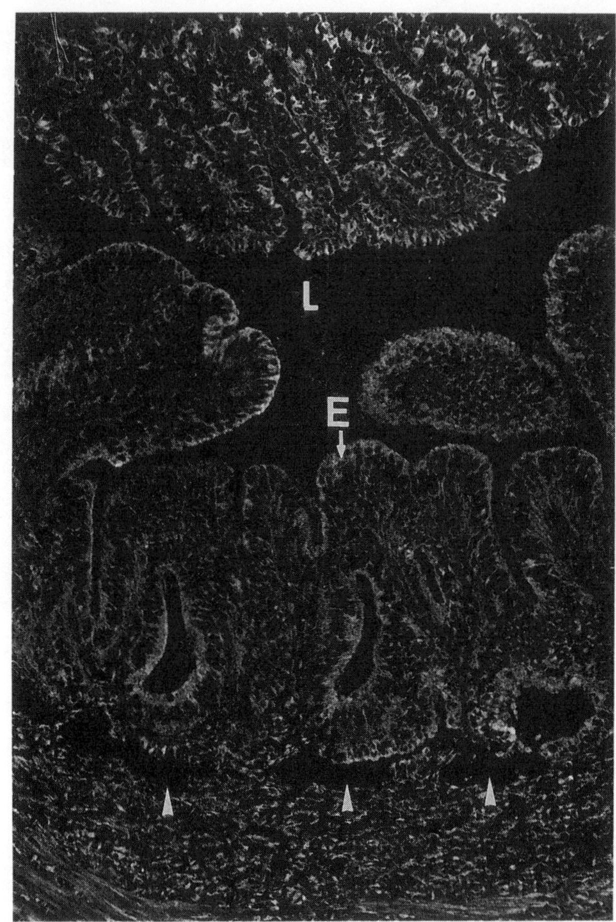

Figure 6. Vimentin immunoreactivity. Overview over the implantation chamber, 8 days p.c. This photograph is from the non-ligated uterus. There is abundant vimentin in the apical regions of the luminal epithelium, but no reaction in the deepest parts of the crypts (arrowheads). (L = uterine lumen; E = luminal epithelial cells). X92.

     There are other reports consistent with the finding that the uterine epithelium of the rabbit represents a combination of different populations of cells. Conti et al. (1984) have described a differential reaction of various cell types of the rabbit uterine epithelium to stimulation by ovarian steroid hormones. A differential response is also suggested to play a role for the outcome of in vitro culture experiments (Gerschenson et al., 1979; Mulholland et al., 1988). Histochemical investigations of in vivo material have revealed that deep parts of crypts react in a different manner as compared with the middle and upper parts of crypts and the surface epithelium of the uterus (phosphorylase: Denker, 1971;

uteroglobin: Hegele-Hartung and Beier, 1985; aminopeptidase M, alkaline phosphatase and dipeptidyl peptidase IV: Classen-Linke et al., 1987).

In contrast to vimentin, cytokeratin immunoreactivity does not show any differences between the uterine luminal (surface) epithelium and any parts of the endometrial crypts including the deepest parts which are all likewise positive. In all investigated stages, cytokeratin immunoreactivity is dominant in the apical part of the cytoplasm of uterine epithelial cells. Cytokeratin filaments are here obviously attached to desmosomes (Franke et al., 1986; Lazarides, 1980; Gounon et al., 1987).

On the basis of these findings it appears that a combination of cytochemical tests for vimentin and cytokeratins is very useful for the identification of various populations of cells within the uterine epithelium of the rabbit. Interestingly, the lack of vimentin immunoreactivity seen in the deepest parts of cryptal epithelium remained constant throughout all investigated stages. This could be useful for the identification, in in vitro culture, of various endometrial cells since exclusively vimentin positivity is found in stroma cells; epithelial cells of middle and upper parts of crypts and of surface epithelium are positive for both vimentin and cytokeratin, whereas epithelial cells of deep crypts are negative for vimentin but positive for cytokeratin. Whether these characteristics remain stable in culture will have to be investigated further.

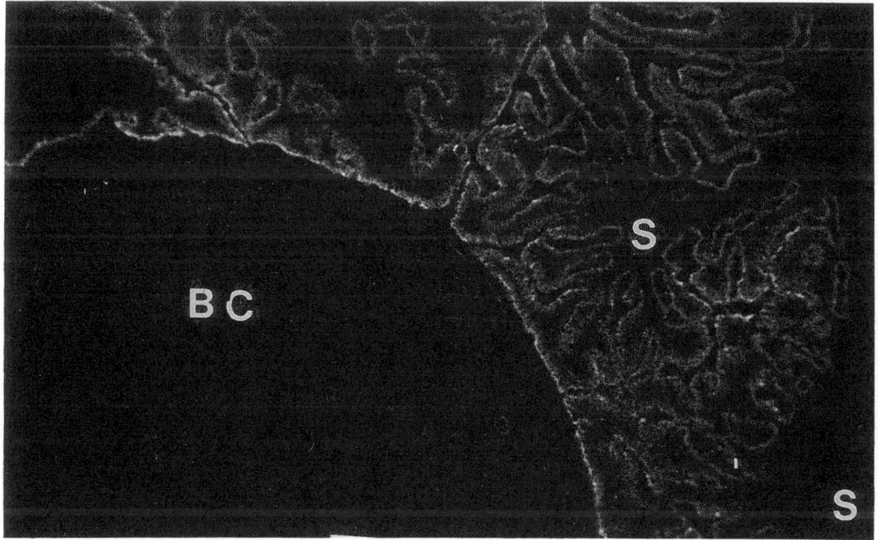

Figure 7. Cytokeratin immunoreactivity, 6 days p.c.. Intense reaction only in the uterine epithelium including the deepest parts of the crypts.(BC = blastocyst cavity; S =stroma). X92

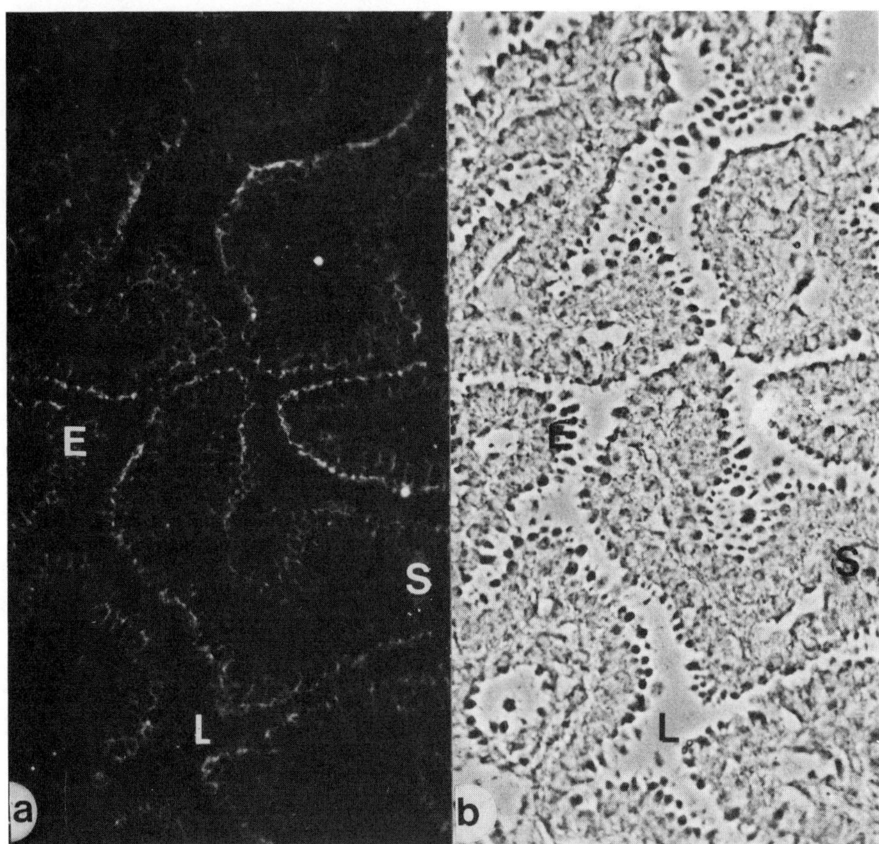

Figure 8. Cytokeratin immunoreactivity.(a) Interblastocyst segments, 6 days p.c..
In pregnancy and pseudopregnancy (not shown) a high cytokeratin reaction is
noted in the subapical cytoplasm of the epithelial cells. (L = uterine lumen; S =
stroma; E = epithelium). (b) Phase contrast, X370.

        Changes of immunoreactivity of intermediate filament proteins in the
rabbit uterus during early pregnancy and pseudopregnancy have not been
described before. The presence of intermediate filaments in the uterine epithelium
has been reported for a number of species, but data on any influence that the
blastocyst might exert have been lacking (Franke et al., 1986; Khong et al., 1986).
Franke et al. reported a case of well differentiated adenocarcinoma in the human
where vimentin was found in both an apical and basal location within the cells.
Glasser et al. (1986, 1987) reported an increase of intermediate filaments, namely
vimentin and desmin, in rat uterine stromal cells undergoing decidualization in
vivo and in vitro. They demonstrated that vimentin is common in all uterine
stromal cells but increased during in vivo and in vitro decidualization. The
increase in vimentin concentration per cell was found to be proportional to the
increase in total cell protein. Desmin was selectively expressed in decidualizing
stroma and may serve as a marker for it.

In the implantation chamber of the rabbit, i.e., in the immediate vicinity of the blastocyst, uterine epithelial vimentin is observed to show a change in the intensity of expression as well as in intracellular distribution from 6 days p.c. on, i.e., before attachment of the trophoblast to the uterine epithelium which occurs at 7 days p.c. Vimentin immunoreactivity increases in the basal parts of the epithelial cytoplasm, but, interestingly, also starts to become positive at the apical cell pole. At the end of the observation period, i.e., 9 days p.c., vimentin is abundant both apically and basally. The apical reaction is most impressive in the surface epithelium adjacent to the blastocyst but reaches also down into the middle parts of crypts. The deepest parts of crypts, however, remain negative even in these regions of the implantation chamber.

The described changes of vimentin reactivity in middle and upper parts of crypts and in the surface epithelium are not seen in inter-blastocyst segments of the uterus. This points to signals that are locally produced by the blastocyst and influence the state of the adjacent uterine epithelium.

Tubal ligation experiments were performed in order to demonstrate the importance of such locally acting signals from the blastocyst more clearly, within the same animal. The increase in vimentin reactivity, and in particular, the apical localization of vimentin was seen only in the implantation chambers of the uterus of the non-ligated side, whereas the blastocyst-free uterus at the ligated side showed the same picture as in pseudopregnancy. Thus the changes in vimentin expression and intracellular distribution depend clearly on the presence of blastocysts. The distribution of cytokeratins does not appear to change in the presence of an embryo.

Signals that the blastocyst might emit just before attaching to the uterine epithelium initiating implantation have often been postulated but have not been identified chemically (cf. Beier, 1984; Kennedy, 1983). Dickmann et al. (1976) have postulated that such signals are identical with steroid hormones produced by the blastocyst. However, although the blastocyst of various species has been shown to possess steroid metabolizing enzymes, actual steroid production by the blastocyst has been shown unequivocally only in the pig (Perry et al., 1973; Heap et al., 1979).

Alternatively, it is known that prostaglandins can exert various effects on the endometrium. It is very probable that cAMP acts as a second messenger for various reactions on the endometrium (Kasamo et al., 1986; Dey et al., 1980). Lazarides (1980) and Inagaki (1987) discussed that cAMP-dependent kinases may be involved in the regulation of cytoskeletal structures by affecting the phosphorylation of intermediate filament proteins, especially vimentin.

As far as the general function of intermediate filaments in uterine epithelial cells is concerned, their role for maintaining the mechanical stability of cells should be considered. Vimentin immunoreactivity was maximal in the large symplasms of uterine epithelium formed in the rabbit implantation chamber. After the lateral membranes have been lost due to cell fusion, intermediate filaments probably have important functions in stabilizing the multinuclear cytoplasmic masses formed in this way. Interestingly, the smaller symplasms formed during pseudopregnancy (Winterhager, 1985; Busch, 1982) do not show any change in vimentin expression and its intracellular distribution, supporting the view that a different type of symplasm is found in the implantation chamber.

An interesting question appears to be to what extent the changes in the intermediate filament proteins described here may be related to membrane changes taking place in uterine epithelial cells (discussed in other contributions to this volume, cf. Classen-Linke et al., 1990; Winterhager and Denker, 1990; Denker, 1986; Lampelo et al., 1985). According to Traub et al. (1987) neutral lipids which are part of all cell membranes may be associated with vimentin filaments. In this way vimentin filaments could influence the fluidity of membranes and the lateral diffusion of integral membrane proteins. The polarity of epithelial cell membranes could easily be influenced. Thus, the current results on changes in the intracellular distribution of vimentin in the uterine epithelium in the implantation chamber should be of relevance for the hypothesis that fundamental changes in the composition of the apical plasma membrane of the uterine epithelium are a prerequisite for trophoblast attachment, and that this depends on a general change of apico-basal polarity characteristics of uterine epithelial cells during this "receptive" phase (Denker, 1986).

SUMMARY

The localization of intermediate filament proteins was analyzed immunohistochemically in the uterus and blastocyst during early pregnancy of the rabbit. Monoclonal antibodies to vimentin (reacting with the 57 kilodalton protein) and cytokeratin (reacting with polypeptides in the molecular range of 50 - 67 kilodaltons) were used. Uteri were investigated from 3 to 9 days post coitum (days p.c.). These uteri were compared with specimens obtained in hCG-induced pseudopregnancy (p. hCG) to demonstrate changes of intermediate filament localization in endometrial cells which might be induced locally by blastocyst-derived signals.

In all stages of pseudopregnancy from day 3 to day 9 p.hCG we noted a delicate vimentin reaction in the basolateral compartments of uterine epithelial cells which were situated mainly in the upper and luminal parts of the endometrial crypts. An identical distribution was found in pregnancy at 3 to 5 days p.c. In the deepest parts of the crypts no positive vimentin reaction was found at any stage of pregnancy or pseudopregnancy.

However, at 7 days p.c. the anti-vimentin reaction in the epithelium surrounding the blastocyst in the implantation chamber became much more pronounced. This immunfluorescence reaction was no longer restricted to the basolateral regions of the cells: vimentin-positive structures in the apical cytoplasm of luminal cells, which reached a maximum at 9 days p.c., were also noted.

An intensely positive reaction of cytokeratin was seen in all uterine epithelial cells. There were no significant differences in staining patterns among any stages of pregnancy or pseudopregnancy.

In the blastocyst, a faint cytokeratin reaction in trophoblast cells was recognized. Vimentin reacted positively in cells which likely respresented the mesoderm.

The current results suggest the following: (1) The uterine epithelium of the adult rabbit (except for the deepest parts of endometrial crypts) coexpresses both types of intermediate filament proteins, cytokeratin, and vimentin; (2) the characteristic intracellular distribution as well as the total amount of vimentin change considerably in the uterine epithelium of the implantation chamber; (3) these changes are assumed to be the result of signals released from the implanting blastocyst; (4) vimentin seems to react to these postulated signals, whereas in contrast the class of cytokeratins investigated here does not show any change of intracellular distribution.

ACKNOWLEDGEMENTS

We wish to thank Gerda Helm for excellent technical assistance, and Gisela Mathieu, Barbara Witte, and G. Bock for secretarial help and for darkroom work. This investigation was supported by the DFG-Schwerpunktprogramm Biologie und Klinik der Reproduktion (Be 524/7-9 and Ho 1059/1-7).

REFERENCES

Beier, H. M. (1968) Uteroglobin: A hormone-sensitive endometrial protein involved in blastocyst development. *Biochim. Biophys. Acta* 160, 289-291.

Beier, H.M. (1974) Oviductal and uterine fluids. *J. Reprod. Fert.* 37, 221-237.

Beier, H.M. and Mootz, U. (1979) Significance of maternal uterine proteins in the establishment of early pregnancy. In: *Maternal Recognition Of Pregnancy*, Ciba Foundation Series 64. Elsevier Exerpta Medica North Holland, Amsterdam, New York, pp. 111-132

Beier, H. M. (1984) Embryonale Signale zu Beginn der Schwangerschaft. Verhandlungen der Gesellschaft Deutscher Naturforscher und Ärzte, 113. *Versammlung*, 419-436.

Beier, H. M. and Kühnel, W. (1973) Pseudopregnancy in the rabbit after stimulation by human chorion gonadotropin. *Hormone Res.* 4, 1-27.

Busch, L. C. (1982) Ultrastrukturelle Untersuchungen zur Transformation des Endometriums. *Habilitationsschrift, Medizinische Fakultät, RWTH Aachen.*

Classen-Linke, I., Denker, H.-W., and Winterhager, E. (1987) Apical plasma membrane-bound enzymes of rabbit uterine epithelium. *Histochem.* 87, 517-529.

Classen-Linke, I. and Denker, H-W. (1990) Preparation of rabbit uterine epithelieum for trophoblast attachment: Histochemical changes in the apical and lateral membrane compart. *Tropho. Res.* 4, 307-322.

Conti, C. J., Gimenez-Conti, I. B., Conner, E. A., Lehman, J. M., and Gerschenson, L. E. (1984) Estrogen and progesterone regulation of proliferation, migration and loss in different target cells of rabbit uterine epithelium. *Endocrinol.* 114, 345-351.

Cowin, P., Franke, W.W., Grund, Chr., Kapprell, H.-P., and Kartenbeck, J. (1985) The desmosome-intermediate filament complex. In: *The Cell In Contact: Adhesion And Junctions As Morphogenetic Determinants*, Ch. 20, (eds.), G.M. Edelmann, and J.-P. Thiery, John Wiley and Sons: New York, Chichester, Brisbane, pp. 427-460.

Dabbs, D. J., Kim, M.D., Geisinger, R., and Norris, H. Th. (1986) Intermediate filaments in endometrial and endocervical carcinomas. *Am. J. Surg. Pathol.* 10(8), 568-576.

Denker, H.-W. (1971) Enzym-Topochemie von Frühentwicklung und Implantation des Kaninchens I. Glykogenstoffwechsel. *Histochemie* 25, 256-267.

Denker, H.-W. (1986) Epithel-Epithel-Interaktionen bei der Embryo-Implantation: Ansätze zur Lösung eines zellbiologischen Paradoxons. *Verh. Anat. Ges.* 80, *Anat. Anz. Suppl.* 160, 93-114.

Dey, S. K., Chen, M.-Ch., Cox, C. L., and Crist, R. D. (1980) Prostaglandin synthesis in the rabbit blastocyst. *Prostaglandins* 19, 449-453.

Dickman, Z., Dey, S. K., and Sen Gupta, J. (1976) A new concept: Control of early pregnancy by steroid hormones originating in the preimplantation embryo. *Vitam. Horm.* 34, 215-242.

Fischer, B. and Meuser-Odenkirchen, G. (1988) 2 year follow up of effects of biotechniques on reproduction in the domestic rabbit, Oryctolaus cuniculus. *Lab. Anim.* 22, 5-15.

Fischer, B. (1988) Charakterisierung entwicklungsspezifischer Strukturen und Lebensäusserungen von Präimplantationsembryonen in vitro und in vivo. *Habilitationsschrift, Medizinische Fakultät, RWTH Aachen.*

Franke, W. W., Moll, R., Achstaetter, Th., and Kuhn, C. (1986) Cell typing of epithelia and carcinomas of the female genital tract using cytoskeletal proteins as markers. Banbury Report 21. Cold Spring Harbor Labr. *Viral Etiology of Cervical Cancer*, 121-148.

Geiger, B., Avnur, Z., Volberg, T., and Volk, T. (1985) Molecular domains of adherens junctions, Ch. 21. In: *The Cell In Contact: Adhesion And Junctions As Morphogenetic Determinants*, (eds.) G.M. Edelmann, and J.-P. Thiery, John Wiley and Sons: New York, Chichester, Brisbane, pp. 461-489.

Gerschenson, L.E., Conner, E.A., and Anderson, J.Y. (1979) Hormonal regulation and proliferation in two populations of rabbit endometrial cells in culture. *Life Sci.* 24, 1337-1344.

Glasser, S. and Julian, J. (1986) Intermediate filament protein as a marker of uterine stromal cell decidualization. *Biol. Reprod.* 35, 463-474.

Glasser, S., Lampelo, S., Munir, M.J., and Julian, J.A. (1987) Expression of desmin, laminin, and fibronectin during in situ differantiation (decidualization) of rat uterine stromal cells. *Differentiation* 35, 132-142.

Gounon, P., Laine, M. Ch., and Sandoz, D. (1987) Cytokeratin filament organization in ciliated cells of the quail oviduct. *Eur. J. Cell Biol.* 44, 229-237.

Heap, R.B., Flint, A.P., Gadby, J.E., and Rice, C. (1979) Hormones, the early embryo and uterine enviroment. *J. Reprod. Fert.* 55, 267-275

Hegele-Hartung, Ch. and Beier, H. M. (1985) Immunocytochemical localization of uteroglobin in the rabbit endometrium. *Anat. Embryol.* 172, 295-301.

Inagaki, M., Nishi, Y., Nishizawa, K., Matsuyama, M., and Sato, Ch. (1987) Site specific phosphorylation induces dissassembly of vimentin filaments in vitro. *Nature* 328, 649-652.

Jones, I. C. R. and Goldman, R.D. (1985) Intermediate filaments and the initiation of desmosome assembly. *J. Cell Biol.* 101, 506-517.

Kasamo, M., Ishikawa, M., Yamashita, K., Sengoku, K., and Shimizu, T. (1986) Possible role of prostaglandin F in blastocyst implantation. *Prostaglandins* 31, 321-336.

Kennedy, T. G. (1983) Embryonic signals and the initiation of blastocyst implantation. *Aust. J. Biol. Sci.* 36, 531-543.

Khong, T. Y., Lane, E.B., and Robertson, W. B. (1986) An immunocytochemical study of fetal cells at the maternal-placental interface using monoclonal antibodies to keratins, vimentin and desmin. *Cell Tiss. Res.* 246, 189-195.

Lampelo, S. A., Ricketts, A. P., and Bullock, D. W. (1985) Purification of rabbit endometrial plasma membranes from receptive uteri. *J. Reprod. Fert.* 75, 475-480.

Lazarides, E. (1980) Intermediate filaments as mechanical integrators of cellular space. *Nature*, 283, 249-256.

Mulholland, J. Winterhager, E., and Beier, H.M. (1988) Changes in proteins synthesized by rabbit endometrial epithelial cells following primary culture. *Cell Tissue Res.* 252, 123-132.

Perides, G., Scherbarth, A., Kuhn, S., and Traub, P. (1986a) An electron microscopic study of the interaction in vitro of vimentin intermediate filaments with vesicles prepared from Ehrlich ascites tumor cell lipids. *Eur. J. Cell Biol.* 41, 313-325.

Perides, G., Scherbarth, A., and Traub, P. (1986b) Influence of phospholipids on the formation and stability of vimentin type intermediate filaments. *Eur. J. Cell Biol.* 42, 268-280.

Perry, J.S., Heap, R.B., and Amoroso, E.C. (1973) Steroid hormone production by pig blastocysts. *Nature* (London) 245, 45-47.

Tachi, S., Tachi, C., and Lindner, H.R. (1970) Ultrastructural features of blastocyst attachment and trophoblastic invasion in the rat. *J. Reprod. Fert.* 21, 37-51.

Traub, P., Perides, G., Kuhn, S., and Scherbarth, A. (1987) Efficient interaction of non polar lipids with intermediate filaments of the vimentin type. *Eur. J. Cell Biol.* 43, 55-64.

van Hoorn, G. and Denker, H.-W. (1975) Effect of the blastocyst on a uterine amino acid arylamidase in the rabbit. *J. Reprod. Fert.* 45, 359-362.

Viale, G., Gambacorta, M., Dell'Orto, P., and Coggi, G. (1988) Co-expression of cytokeratins and vimentin in common epithelial tumors of the ovary: an immunocytochemical study of eighty-three cases. *Virch. Arch. A Pathol. Anat.* 413, 91-101.

Winterhager, E. (1985) Dynamik der Zellmembran: Modellstudien während der Implantationsreaktion beim Kaninchen. *Habilitationsschrift, Medizinische Fakultät, RWTH Aachen.*

Winterhager, E., Brümmer, F., Dermietzel, R., Hülser, D.F., and Denker, H.-W. (1988) Gap junction formation in rabbit uterine epithelium in response to embryo recognition. *Develop. Biol.* 126, 203-211.

Winterhager, E. and Denker, H.-W. (1990) Changes in lipid organization of uterine epithelial cell membranes at implantation in the rabbit. *Tropho. Res.* 4, 323-338, 1990.

Basement Membranes and
Endometrial Stroma

BIOCHEMICAL AND STRUCTURAL CHANGES IN UTERINE ENDOMETRIAL CELL TYPES FOLLOWING NATURAL OR ARTIFICIAL DECIDUOGENIC STIMULI

- A Review -

Stanley R. Glasser

Department of Cell Biology
Baylor College of Medicine
Houston, Texas 77030 USA

Attachment Of The Mammalian Embryo

Attachment of the mammalian embryo to the maternal uterus is a signal event in the highly regulated program of synchronized, apparently independent but interdependent events which define unique interactions between two organisms of different genetic derivation. The focus of this essay will be the role of the individual endometrial cell types, vis-a-vis the blastocyst, in these implantation related events. This paper will suggest that at the present time, only a very generalized understanding of the regulatory mechanisms underlying blastocyst attachment has been established. Present models are, therefore, of limited predictive use in resolving problems in human infertility. In an effort to recognize elements with which to formulate a more functional model particular emphasis will be placed on the responses to endocrine, paracrine, and autocrine mechanisms which come into play following the attachment of trophoblast to the uterine epithelium, i.e., the adhesive and post-implantation phases of implantation.

No single animal model explains all the diverse mechanisms which regulate the varied strategies used by different mammals to direct blastocyst attachment and placental development. However, comparative reproductive biology can be used productively. Thus, while the rat and mouse are imperfect analogies they do offer reasonable homology for the study of human implantation. Attempts to identify and define controls which modify embryo and endometrial cell type development and, depending on species, the long and short range interactions between them are influenced by an anthropocentric interest in interstitial implantation. To facilitate the study of blastocyst development the process of implantation has been divided into five phases: (a) hatching, shedding the zona pellucida, (b) apposition, intrauterine spacing and orientation of the blastocyst, (c) adhesion, attachment of transiently adhesive trophoblast and apical uterine epithelial surfaces, (d) invasion, initiated by focal compromise of apical tight junctions may be followed by removal of depolarized uterine epithelial cells at the implantation site. This is related to alteration of cell-substratum relationships and subsequent passage of trophoblast through the basement membrane into the differentiating stromal compartment, and (e) post-implantation, the reciprocal interactions between the differentiating trophoblast and the stromal compartment, including decidualized cells and the structurally and biochemically altered extracellular matrix. By means of these processes the definitive placenta is established (Glasser and McCormack, 1979b).

Defined in terms of complementary changes of the uterine endometrium, which presumably reflect hormonally regulated alterations in the profile of proteins and glycoconjugates expressed on the apical surface of uterine epithelial cells (Glasser et al., 1988; Carson et al., 1988) and by uterine stromal cells, the role of the uterus in implantation would be described as being sequentially (a) non-receptive and hostile to a non-synchronously transferred blastocyst, (b) non-receptive to attachment of a blastocyst but, in an experimental model (Meyers, 1970; Psychoyos, 1973a, b; Glasser and McCormack, 1979a), can be induced to decidualize in response to an artificial stimulus, after requisite exposure to progesterone, (c) receptive, the definitive response of the progesterone dominated uterus to nidatory estrogen, i.e., attachment of the blastocyst, and (d) refractory i.e., non-receptive to transferred blastocysts. In the rodent the refractory state occurs if implantation fails to take place within 24-30 hours following exposure to nidatory estrogen. Refractoriness is not a null position but requires active hormonal support (Psychoyos, 1973b; Glasser and McCormack, 1979a, 1980). Hatching and apposition occur in the non-receptive uterus, adhesion and its sequelae occur in the receptive and subsequently decidualized uterus.

Decidualization

The complexities of human embryo transfer have emphasized the importance of tandemly related uterine receptivity and decidualization in the establishment and maintenance of pregnancy (Webb and Glasser, 1984; Glasser, 1986). The idea that decidual tissue is important in the control of trophoblast invasion was noted in early studies (Turner, 1876; Bryce and Teacher, 1908) which developed the relationship between decidualization and the extent of hemorrhage and necrosis at the implantation site.

Although an extensive literature on the histology (Krehbiel, 1937), ultrastructure (O'Shea et al., 1983), and endocrine physiology (Psychoyos 1973a, b; Glasser, 1972; Glasser and Clark, 1975; Glasser and McCormack, 1979b, 1980) of the decidual cell reaction has been developed, it is striking how little is understood concerning the cellular and molecular differentiation of the uterine stromal cell and the manner by which progesterone (P) and/or estrogen (E) regulate the biochemical mechanisms which control decidualization (Glasser and McCormack, 1979a, 1980; Bell, 1985). The study of decidualization is still constrained by technical problems involving cell heterogeneity, biochemical sampling, and in vitro methodology.

The histology of decidualization has been well described (Krehbiel, 1937). Endometrial stromal cells of the hormonally sensitized uterus undergo extensive morphological and physiological alterations. In laboratory rodents, these changes are in response to the attaching blastocyst (Psychoyos, 1973a, b; Glasser and Clark, 1975; Glasser and McCormack, 1979b, 1980a) whereas in the human predecidual changes in the stroma can be observed during the late secretory phase of the menstrual cycle independent of blastocyst attachment (Daly et al., 1983b). Not all animals experience decidualization to the same degree (DeFeo, 1967). The process is extensive in the rat but minimal in the rabbit. Experimentally, stromal cell differentiation can be mimicked in pseudopregnant and in some ovariectomized laboratory animals by applying physical or chemical stimuli to the hormonally sensitized endometrium (Glasser and McCormack, 1979b, 1980a).

Presently decidual cell histology provisionally provides the most reliable if not the most useful indices of the process. Unlike other hormonally regulated cell differentiation systems (McKnight and Palmiter, 1979; Swaneck, et al., 1979) the hormones which sensitize the endometrium (Psychoyos, 1973a,b; Glasser and Clark, 1975; Glasser and McCormack, 1979b; 1980a) do not serve as the inductive stimulus. In response to an undefined deciduogenic stimulus a population of hormonally sensitized stromal fibroblasts undergoes transformation (Leroy and Galand, 1970; Leroy et al., 1974). Decidualization begins focally in the subepithelial anti-mesometrial zone at the point of stimulus (attachment). Proliferation, growth and differentiation spread downward and laterally until (approximately 96 hours later) stromal cell decidualization and organization of the metrial gland can be noted on the mesometrial side of the uterus (Kleinfeld and O'Shea, 1983; O'Shea et al., 1983; Bell, 1983). Regional progression of cytodifferentiation constitutes a major impediment to biochemical analysis of decidualization because the cell separation methods which can presently be applied to the stromal compartment do not adequately separate individual stromal cell types or classes. The composition of a sample at any given experimental period is characterized by a profile of heterogeneous cell types at different stages of differentiation. This situation demands a conservative interpretation be applied to biochemical data with respect to the analysis of cause and effect. Recent developments in cell separation techniques, in in vitro culture systems and the application of immunocytochemical and in situ hybridization methods promise new insights into the cellular and molecular mechanisms which underlie differentiation of the cells in the stromal compartment.

Functional Role of Decidual Tissue

The functional role of decidual tissue remains unresolved. Additional functions have been assigned to the decidua since Turner (1876) related it to the control of trophoblast invasion. These include: (a) provision of a cleavage zone for separation of the placenta at parturition, (b) a source of embryonic nutrition, and (c) isolation of individual fetuses of polytoccus species. The importance of these assignments is either equivocal or unproven. Recent investigations have assigned two additional functions, both of potential significance to the establishment and maintenance of pregnancy, to decidual cells.

First, a luteotrophic action has been identified and attributed to human predecidual (Daly et al., 1983a) and decidualizing stromal cells (Basuray and Gibori, 1980; Daly et al., 1983b). Second, the migratory cells, of possible bone marrow origin, have been assigned, via their interaction with trophoblast and decidual cells, critical immunoregulatory and immunotrophic roles in the pregnant animal (Gill, 1988; Wegman, 1988). Even when considered in terms of cell-cell or cell-substratum communication the question of how the acquisition of secretory and immunoregulatory phenotypes serves the purpose of the decidua is open to speculation. There is also a growing appreciation that differential gene expression of cells in the stromal compartments can be modulated by the changing pattern of extracellular matrix proteins, (Hay, 1983).

Decidualization has been studied almost entirely *in vivo*, with morphological (Schlafke and Enders, 1975; Nilsson, 1982) and physiological methods (Glasser, 1972; Glasser and Clark, 1975, 1973b; Psychoyos, 1973a, b). There has been an overdependence on decidual weight as an endpoint. Awareness of the limits of this non-specific type of method to study the regulatory cell and molecular biology of stromal cell decidualization has stimulated an increase in the application of biochemistry (Glasser and McCormack, 1979a, b, 1982; Lejeune et al., 1982; McDonald et al., 1983) and immunocytochemistry (Wewer et al, 1985, 1986; Glasser and Julian, 1986; Glasser et al., 1987). In conjunction with improvements in *in vitro* systems and application of recombinant DNA technologies one is better prepared to inquire into the biochemical mechanisms which regulate cell-specific and stage-specific events determining the direction of decidual cell expression.

Cell Types Associated With The Decidualizing Stromal Cell Compartment

A variety of sources have been used to tabulate (Table 1) the cells that have been associated with decidualization (Bell, 1985; Glasser and McCormack 1980a; Padykula, 1981). Predecidual (human) and decidual cells (human, laboratory rodents) have been the subjects of continuing descriptive studies (Noyes et al., 1950; Tekelioghu-Uysal et al., 1975; Krehbiel, 1937; O'Shea et al., 1983). A number of regulatory roles have provisionally been assigned to the changes in secretory phenotype that accompany the differentiation of the decidual cell which appears to derive from a single cell type, i.e., hormone and growth factor secretion, secretion of immunotrophic factors, multiple functions related to trophoblast invasion including remodeling their extracellular matrix.

Relative to a continuing interaction between trophoblast and maternal host are those cells in the decidual compartment that have been assigned an immunoregulatory and/or immunotrophic function. Their roles are not clearly defined or confirmed. This list includes: (a) endometrial granulocytes, believed to be analagous to rodent granulated metrial gland cells (Stewart and Peel, 1978). This class of cells purportedly derives from either a bone marrow origin (Peel et al., 1983) or differentiates from a lymphocyte precursor (Stewart and Peel, 1978); (b) Fc receptor bearing cells which are not specific to decidua, per se, but appear in high numbers in early pregnancy. These cells possess macrophage-like properties, i.e., ectoenzyme secretion, macrophage surface molecules, phagocytotic capability, and (c) subclasses of leukocytes as represented by the Ia$^+$ cells described below. These non-macrophage cells, resident in the post-implantation decidual compartment (Head and Gaede, 1986), are competent antigen presenters and may play some role in establishing the level of suppressor activity required to sustain pregnancy (Head and Billingham, 1986) and in immunotrophic activity through the release of lymphokines (Wegmann, 1988).

Table 1

Cell Types Of Uterine Stromal Compartment (Based On Observations In Rats, Mice, Oppossum, and Human [H])[1]

Cell Type	Origin	Region
RESIDENT CELLS		
Fixed		
Fibroblasts	Parental Mesenchymal Cell	
Stem Cells	?	
Lymphocytes (T,B)	Marrow Derived	
Mast Cells		
TRANSIENT CELLS		
Constitutive		
Decidual Cell	Stromal Fibroblast	Zona Compacta (H) Late Secretory Phase Pregnancy
Antimesometrium		Antimesometrial Decidua
Mesometrium		Mesometrial Decidua
Endometrial Granular Cell	Bone Marrow Derived Precursor	Zona Compacta (H) Late Secretory Phase Pregnancy
Granulated Metrial	Bone Marrow Derived	Metrial Gland
Gland Cell	(Lymphocyte Precursor)	Mesometrial Decidua
Ameboid		
Macrophage-Monocyte	Infiltrating Migratory Cell	Zona Compacta (H) Late Secretory Pregnancy Mesometrium (Decidua) ?
Fc Receptor Bearing (Small Round Cells)	Infiltrating Migratory Cell	Mesometrial Decidua ?
Fc Receptor Bearing Metrial Gland Cells	Fibroblasts (?)	Metrial Gland
Ia+	Dendritic Non-Macrophage Cells From Lymphoid Tissue	Antigen Presenting Cells Distrubtion Within Stroma Depends On E/P Ratio Expressed In Glandular Epithelium
Lymphocytes (T, B)	Marrow Derived	
Eosinophils		
Heterophils (Neutrophils)		
Plasma Cells		

[1] Data Based On Studies Of Padykula (1981), Feyrtrer (1963), and Bell (1983).

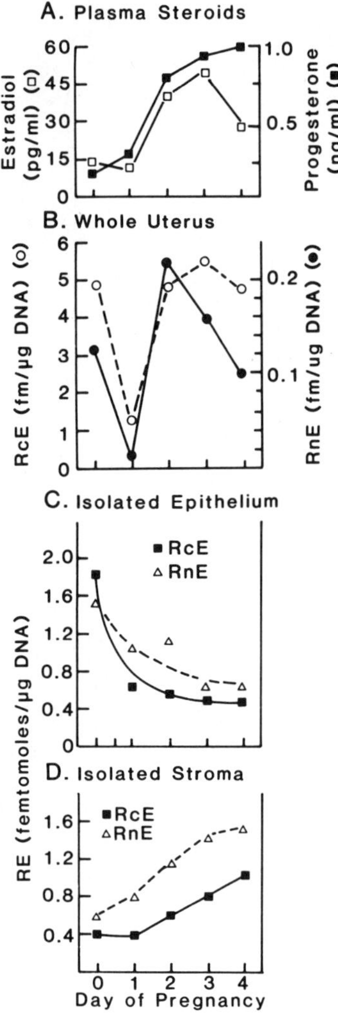

Figure 1. Plasma titers of estradiol and progesterone (A) are used to monitor concentrations of cytoplasmic (R_cE) and nuclear (R_nE) estrogen receptor in the whole uterus of the pregnant rat (B). Differences in the profiles of R_cE and R_nE are evident when their respective concentrations in isolated uterine epithelial (C) and stromal (D) cells are compared with values obtained from analysis of the whole rat uterus (B) on each day of pregnancy.

Steroid Hormone Receptors

Only a few studies, dealing mainly with the mitogenic action of E, have noted that uterine epithelial (UE) and uterine stromal (US) cells did not appear to respond similarly to E in the presence of P (Finn and Martin, 1969; Tachi et al., 1972). The uterine response to changing ratios of P and/or E, reported in most papers, is the net response of all cells without regard to the specific or proportional response of each cell type. The contribution of individual endometrial cell types to the hormonal responses of the uterus has effectively been disregarded. However, many hormonal responses that are implicated in blastocyst attachment appear to be regulated by the dynamics of the cytoplasmic and nuclear steroid hormone receptors of each individual endometrial cell type, i.e., UE and US. Assessing the influence of steroids on implantation and decidualization, in terms of their influence on changing receptors in the entire uterus, can be biased by shifts in the absolute and relative numbers of cell types which comprise the UE and US compartments. Analysis of the whole uterus also includes the myometrial cells whose bulk (~85%) can dilute or mask changes in the endometrium.

Proof of this thesis derives from the demonstration that endometrial UE and US cells of the pregnant rat responded independently and differentially to the endogenous hormonal environment which characterizes the preimplantation period (Glasser and McCormack, 1981). As plasma P and E titers rise on days 1-3 (Figure 1A) the cytoplasmic (R_cE) and nuclear (R_nE) estrogen receptor concentration falls in the post-mitotic UE cells. At the time of implantation (day 4) they have reached castrate levels in the UE cells and increased significantly in the US cell (Figures 1C, D). Analysis on the basis of whole uterus (Figure 1B) yields entirely different data and interpretations. The same types of differential responses of UE and US cells have also been observed in mature ovariectomized rats treated with estradiol, nafoxidine, or clomiphene (Markaverich et al., 1981).

Failure to interpret the endocrinology of implantation in terms of the differential responses of UE and US cells underlies the equivocal utility of previous data. In the case of receptor mediated events changes in R_cE and R_nE in the whole pre-implantation uterus are represented by the summed accumulation of receptors in both UE and US compartments in response to rising titers of plasma P and E (day 2, Figures 1A, B). Following the release of nidatory E there appears to be an inexplicable decrease in R_nE that is even greater than R_cE. Heterogeneity in cell type and response contribute to the controversy (Sartor et al., 1980) surrounding the observation (Martel and Psychoyos, 1980) that P binds selectively and at higher concentrations to implantation sites (vs inter-sites) while RP values for the uterus as a whole are low during the peri-implantation period.

When uterine changes are interpreted in terms of the UE cell and its mitogenic response to E it appears, that after exposure to P for at least 48 hours, P antagonizes the action of E. Tachi and associates (1972) advanced the idea that P redirects (vis-a-vis antagonizes) the action of E to a secondary target, i.e, the US cell. Independent of the action of P, the decrease in RE in post-mitotic daughter UE cells is consonant with the ideas advanced by Smith and Martin (1974). While the absence of a net increase in stromal R_cE through day 2 (Figure 1D) might be anticipated the doubling of R_nE by day 4 is unexpected (Figure 1D). Although this suggests a change in the hormonal responsiveness of the peri-implantation US cell the biological relevance of this increase remains unclear. It is not known which cells in the stromal compartment contribute to this change or where (mesometrial, antimesometrial) they are located.

Physiologically it is the action of nidatory E on the P dominated uterus that transforms it from non-receptive to receptive (Glasser and McCormack, 1979b, 1980a). The question arises, if the concentration of RE in peri-implantation UE cells is depressed to castrate levels, is the increase in sensitivity to blastocyst attachment really a receptor mediated response to nidatory E? The RE response to nidatory E was measured in UE and US cells of ovariectomized rats treated, after 14 days, with three daily injections of P (2 mg/day). At that point the uteri are fully sensitized to a deciduogenic stimulus but are not receptive to the blastocyst (Psychoyos, 1973 a, b; Glasser and McCormack, 1979a, b). A single dose of estradiol-17β (0.2 μg) will transform this uterus to one that is ovoreceptive. The data depicted in Figure 2 show that both cells types respond coordinately to the single nidatory dose of E_2. There are no notable changes in R_cE in either UE or US cells. However R_nE is elevated significantly ($p < .05$) in both UE and US cells at 1 and 4 hours after E_2 injection before returning to baseline values at 16 hours. Attachment of transferred blastocysts occurs during the 24 hours period following E_2, after which the uterus becomes refractory.

These data suggest that in order to achieve the final steps which lead to receptivity the previously divergently responsive cell types (UE,US) become linked in their response (Figures 2A,B). Whether the response to nidatory E by the UE cell is direct, or indirect via the US cells, it occurs via a receptor mediated mechanism which may be related to qualitative gene expression (Cunha, 1976; Cunha et al., 1985; Bigsby et al., 1986; Glasser and McCormack, 1979a, b; 1982). This provides the integration necessary to transform the non-receptive UE apical surface to one that is receptive to the blastocyst and competent to transduce a deciduogenic stimulus. Even better methods for isolating, culturing and analyzing individual uterine cell types may be necessary to extend our understanding how P and E regulate uterine receptivity and its sequelae.

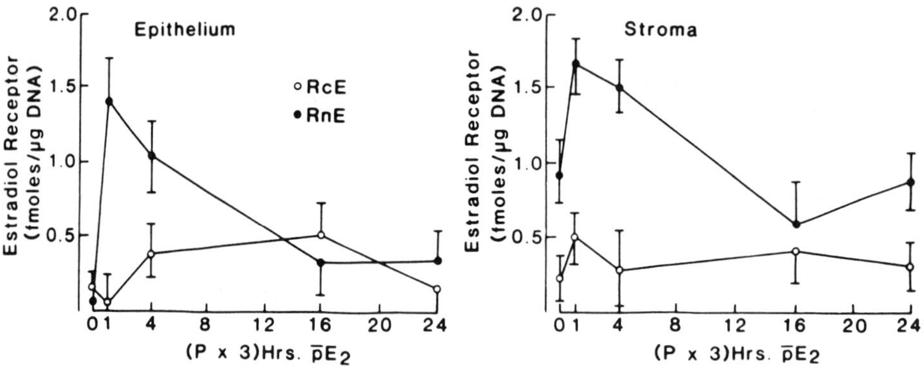

Figure 2. Two weeks after cyclic rats were bilaterally ovariectomized they were injected once daily for three days with 2.0 mg progesterone (PX3). Each animal received a single 0.1 μg injection of estradiol-17β (E_2) with the third P injection. At intervals following the injection of E_2 epithelial and stromal cells were isolated from the uterus and the concentration of cytoplasmic (R_cE) and nuclear (R_nE) estrogen receptor was assayed in each cell fraction. Px3 uteri are not ovoreceptive but are sensitized to a deciduogenic stimulus. Blastocysts transferred after E_2 injection (zero hours) will attach to the uterus within the following 24 hour period.

Invasive Phase

Current research on decidualization (differentiation within the stromal compartment) highlights the importance of the invasive phase of implantation in establishing pregnancy. Legal and ethical constraints which attend such studies with human subjects have placed a marked dependence on research using laboratory rodents. That rats, mice, and humans all exhibit interstitial implantation and some variant of hemochorial placentation is fortuitous. The unique expression of the post-attachment invasion process displayed in these species has skewed an understanding of embryo-uterine interactions. Thus, one does not fully comprehend the strategies of attachment and non-invasive placental development used by other species or how to use them to advance the analysis of human implantation.

The invasive phase occurs, in most of laboratory rodents studied, over a 5-6 day span following blastocyst attachment and thus encompasses the post-implantation period. Except for descriptions of species related patterns of invasiveness which characterize the varying degrees of intimacy between the trophoblast and the maternal vasculature (Amoroso, 1952; Schlafke and Enders, 1975) there is relatively little definitive information regarding the functional aspects of these cell systems. Historically, the chore of providing some type of anchor to place the embryo in proper position to acquire nutrients from the mother has been assigned to the invasive phase.

Fascination with complexities related to the mechanisms by which UE cells become receptive to blastocyst attachment have diverted attention from the role of the US cell in nidation and post-attachment events. It is presumed, but not resolved with certainty, that differentiating uterine stromal cells (resident stromal cells including fibroblasts that decidualize, migratory cells of bone marrow orgin) exert a determinative role on post-attachment events beyond any influence that the mesenchymally derived undifferentiated uterine stromal cell may exercise in regulating UE cell receptivity. New data suggest that cellular components of the decidualizing stromal compartment modulate relationships between trophoblast and the UE cell as well as between the UE cell and its basement membrane and also play an immediate and continuing role in the structural and functional remodeling of the extracellular matrix of the decidual cells and their vasculature (Wewer et al., 1985, 1986; Aplin et al., 1988). The expressed endocrine, paracrine, and autocrine mechanisms of the decidual cell also participate in the control of the immunoregulatory and immunotrophic principles which guarantee programmatic trophoblast invasion (Gill et al., 1987; Gill, 1988).

While there is no question that the rodent trophoblast cell, as a sequel to its attachment, transposes to a migratory or invasive status (Sherman and Wudl, 1976), it is not clear whether these cells are themselves cytolytic with respect to removing the uterine epithelial cell barrier. An emerging consensus (Colwell, 1972; El-Shershaby and Hinchcliffe, 1975; Glass et al., 1983) supports the idea that degeneration of UE cell is intrinsically programmed (Parr et al., 1987) rather than a consequence of trophoblast directed cytolysis. The signal that initiates cell death is not known but could possibly be related to disruption of UE cell-cell and cell-substratum relationships (loss of polarity) which attend blastocyst attachment.

Invasiveness of Trophoblast

Blastocyst attachment to the apical surface of receptive UE cells initiates a cascade of events that describe the beginning of trophoblast invasion. From attachment to the time the differentiating trophoblast pauses at the basement membrane preparatory to its entry into the decidualizing stromal compartment, depolarization, apoptosis, and autolysis of uterine epithelial cells and phagocytosis of dead and living uterine epithelial cells by trophoblast have occurred (Welsh and Enders, 1985; Parr et al., 1987). The mechanisms which govern these initial steps have yet to be identified with certainty (Denker, 1977; Glasser and McCormack, 1982, 1987b; Glasser et al., 1988).

In Utero vs. Ectopic Invasion

Comparison of the invasiveness of trophoblast and tumor cells in utero and at extra-uterine sites (Sherman and Wudl, 1976) classifies trophoblast cells amongst the most invasive (Solomon, 1966). They are exceeded in aggressiveness only by trophoblast cells developing from ectoplacental cone (Kirby, 1965). It has been alleged that trophoblast was more destructive at ectopic sites than the same cells were in utero (Kirby 1965). This suggests that the decidual cell, absent at ectopic loci, plays a definitive role in modulating and directing trophoblast invasion.

Trophoblast cells, following a species specific schedule, advance within the decidualizing stromal compartment. In the interval between attachment and establishing contact with the maternal vasculature these cells exhibit a highly regulated program of differential gene expression, i.e., endoreduplication, cytokeratin expression, synthesis, and secretion of steroids, peptide hormones, and ectoenzymes (Denker, 1977; Glasser and McCormack, 1976; Glasser and Julian, 1986; McCormack and Glasser, 1980b; Soares et al., 1985). It is debatable to what degree cytolysis contributes to trophoblast invasion (Salomon and Sherman, 1975; Glass et al., 1984; Tickle et al., 1978; Schor, 1980). Most recent analyses suggest that cytolysis is secondary to some primary mechanism which could include contact inhibition (Glass et al., 1984) and/or matrix hydration and remodeling (Aplin et al., 1988).

Loss of Invasiveness

Non-neoplastic trophoblast of gestation, although initially invasive, normally loses this aggressiveness. While non-malignant trophoblast cells are deported via the blood stream throughout most of pregnancy, these cells effectively lose their invasiveness coincident with establishment of a definitive placenta.

Only the differentiated forms of trophectoderm or cytotrophoblast appear to possess invasive characteristics. Invasiveness persists longer at ectopic sites than in utero. In general, trophoblast is more invasive in the absence of decidual cells than in their presence (Kirby and Cowell, 1968; Billington, 1971). Thus, experimental evidence suggests that trophoblast is sensitive, to some degree, to control by the decidual cells. Trophoblast is best able to invade through the central necrotic area of terminally differentiated decidua. Trophoblast cells do not display this aggressive behavior on making contact with normal healthy decidual cells. Differential gene expression of developing trophoblast cells (syncytio-trophoblast, trophoblast giant cells) has been implicated in exercising regulatory influence on the acquisition and loss of invasive characteristics. Other than the finite life span of the trophoblast cell changes in the secretory phenotype associated with terminal differentiation could modulate the invasiveness of the trophoblast cell. Such changes could be demonstrated as an altered secretory profile of proteases and collagenases, the secretion of the mature forms of placental hormones (Sherman and Wudl, 1976; Glasser and McCormack, 1979b; McCormack and Glasser, 1980b; Glasser et al., 1987b) and growth factors (Nissley and Rechler, 1984). It has been suggested that trophoblast cells become increasingly receptive to the cytokines and lymphokines that play an immunoregulatory role in pregnancy (Wegmann, 1988).

Functional Analysis

Constraints on Biochemical Analysis

The interactions between mammalian embryo and maternal host which characterize their roles in blastocyst attachment, stromal compartment differentiation and/or invasion, and the establishment of the placenta have been defined mainly in morphological terms. Current research now seeks to elucidate the cellular and molecular processes which regulate blastocyst/ uterine endometrial interrelationships. There is not yet sufficient data to formulate structure/function correlates that might serve to resolve a regulatory pathway.

A major focus of cellular and molecular analyses derives from a pivotal review by Cunha (1976) which reframed the principles of stromal-epithelial cell communication in the form which currently dominates the current thinking. This concept which recognizes that hormonally regulated mesenchymal cells produce signals that instructively or permissively regulate uterine epithelial cells has been extended. Data have been obtained which recognize reciprocal signals, transduced via the UE cell, which initiate certain stromal cell responses (Psychoyos 1973b; Lejeune and Leroy, 1980; Glasser and McCormack, 1979b, 1980b, 1982). This research characterizes the secretory phenotypes of both UE and US cells and suggests the autocrine and paracrine mechanisms which appear to govern their respective structural and functional differentiation.

In terms of analyzing the regulatory processes underlying the responses of the maternal components of implantation and placentation data relating to stromal-epithelial cell reciprocity have not realized their potential. Only marginal research has been devoted to characterizing the secretory phenotype of the UE cell and even less effort has been applied to like studies of the mesenchymally derived stromal cell. In general, these studies are complicated by regionally associated cell heterogeneity of the endometrium (Bell, 1985), which introduces sampling artifacts. While this problem may be resolved in part by methods which allow rapid, reproducible separation of endometrial cell types (McCormack and Glasser, 1980b) biochemical studies of epithelial cells continue to be impeded by the lack of a polarized cell culture system that retains hormone responsiveness and continues to express the specialized functions characteristic of the uterine epithelial cell (proliferation, growth, morphological, and functional differentiation). The majority of in vitro models have failed to appreciate the role of the extracellular matrix in modulating the cell surface and the intracellular environment.

Cultured under non-optimal conditions UE cells remain viable. However, the secretory phenotype of primary endometrial cell type cultures, principally UE cells, has proven unstable. While individual cell types may retain compliments of steroid hormone receptors they have not continued to express their cell type specific functions. Cultured UE cells fail to respond appropriately to mitogens and cease to express some components characteristic of their cell surfaces (Campbell et al., 1988). To achieve fully functional differentiation cells in culture must maintain their proper relationship with their ECM and, particularly with respect to primary simple epithelial cells, a correct, three-dimensional polarization. Nor should the relationship between stromal and epithelial cells be disregarded. The difficulty in achieving these conditions has extended our dependence on organ cultures in which, for 24-48 hours, it appears possible to maintain reasonable positional and cell-cell interrelationships. Although rigorous proof of cell viability and stage of differentiation are often lacking, short term studies using this heterogeneous system are presently accepted as valid.

Recent developments make it possible to perform long term studies on viable, hormonally responsive homogeneous populations of UE cells. These experiments can be done under conditions which make it possible to impose on these in vitro cultures many of the regulatory constraints which control these cells in utero. Primary UE cells cultured on a basement membrane derived, matrix-impregnated semi-permeable filter proliferate to confluence and achieve morphological and functional polarity (Glasser et al., 1988; Carson et al., 1988). The model provides analytical and experimental access to both the apical and basal plasma membrane domains and permits the construction of UE/US cell recombinants. Comparison of the protein (Glasser et al., 1988) and glycoconjugate (Carson et al., 1988) profiles of the apical and basolateral secretory compartments demonstrates a differential polarized response of the cell to steroid hormones. Cell-cell recombinations (UE/US) are facilitated by the efficient separation of cells in the stromal compartment from UE cells (Glasser and Julian, 1986) and a reasonable division of cell types into fibroblastic and bone marrow derived elements can be made (Feyrter, 1963; Padykula, 1981; Bell, 1985). However, while US fibroblastic elements are very stable in culture there is at present no in vitro culture system which consistently preserves their competence to respond to steroid hormones (Glasser and Julian, 1986; Wewer et al., 1986).

The limitations of the present US culture systems emphasizes the significant contribution that can be made by collateral analysis of morphology, immunochemistry, and immunocytochemistry to the localization and indentification of relevant markers (Aplin et al., 1988; Wewer et al., 1985, 1986; Glasser et al., 1987). However, present cell separation methods allow the construction of carefully selected cell type specific cDNA libraries and probes and cRNA probes which will further enhance the productivity of projected biochemical studies.

Intracellular Responses of Stroma

If it is true that in sexually mature animals, as in the fetus and neonate, the hormonal response of UE cells is not direct but is instructively or permissively modulated by factors emanating from the stromal cells (Cunha, 1976; Bigsby et al., 1986), it is important to identify and analyze those stromal factors. If the signal that initiates stromal cell differentiation is transduced via the hormonally sensitized UE cell, it is important to know the nature of the signal, and how it is received and interpreted. It is important to characterize cell responses specific to the decidualizing stroma in order to identify stage specific decidual cell markers. This baseline will allow the resolution of fundamental questions pertaining not only to the regulation of stromal cell differentiation but to interactions between the differentiating stromal cell and the polarized UE cell and later with the trophoblast.

Only rudimentary information is possessed with respect to intracellular changes experienced by a stromal cell exposed to progesterone (P) and/or estrogen (E). Present bias is that hormonally directed responses are stimulatory but the possibility cannot be disallowed that the stromal cell response is the result of derepression or of a repressive hormone action (Glasser and McCormack, 1979a). The resolution of some of these problems is now more feasible because of the availability of an experimental UE cell system which maintains viable hormonally responsive cells that express specific functions (Glasser et al., 1988; Carson et al., 1988).

Markers Of Decidualization

Stromal cell differentiation is most pronounced in species in which the blastocyst implants interstitially. Decidualization occurs in response to the implanting blastocyst or, in pseudopregnant and hormonally sensitized rats and mice, in response to artificial stimuli (DeFeo, 1967; Glasser, 1972; Psychoyos, 1973b). Predecidual cells are found in the stroma of the human uterus during the late secretory phase of the menstrual cycle; differentiation is initiated in the absence of a blastocyst-like signal (Noyes et al., 1950; Wewer et al., 1985). The morphological features which characterize in vivo decidualization (Krehbiel, 1937) have also been described for endometrial cells undergoing a similar process of differentiation during in vitro culture (Vladimirsky et al., 1977; Sananes et al., 1978; Glasser and Julian, 1986; Glasser et al., 1987).

Recent inquiries have attempted to demonstrate biochemical correlates of the morphological markers of stromal cell differentiation (Glasser and Julian, 1986; Glasser et al, 1987; Bell, 1985). Identification of such correlates would facilitate the study of the regulatory mechanisms which underlie decidualization. Proteins unique to decidual cells have been identified in rats (Lejeune et al, 1982), hamsters (MacDonald et al, 1983), and humans (Bell, 1985) by two-dimensional gel electrophoresis. However, in no case was the appearance of any decidual cell associated protein exclusively related to a particular morphological characteristic of the differentiating stromal cell.

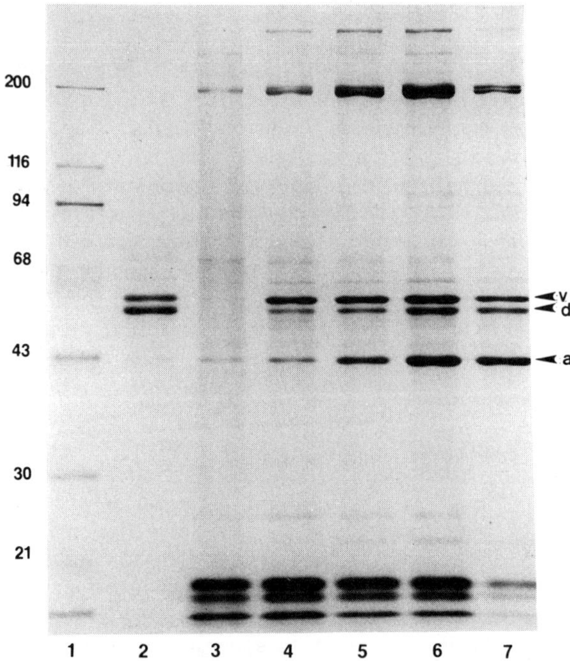

Figure 3. One dimensional polyacrylamide gel eletrophoresis (1-D PAGE) of intermediate filament (IF) enriched fractions isolated from artificially induced decidual tissues after 4 days in vivo (lane 2). IF fractions from control dy 4 stromal cells (lane 3), from day 4 stroma + 24 hours in culture (lane 4), Day 4 stroma + 48 hour in culture (lane 5). Day 4 stroma +72 hours in culutre (lane 6) and dy 4 stroma +96 hours (lane 7). The positions of vimentin (v), desmin (d), and actin (a) are indicated on the right. Ten µg IF protein was applied to each lane except lane 7 (5 µg). Molecular weight standards in lane 1 are myosin, β-galactosidase, phosphorylase B, bovine serum albumin, ovalbumin, carbonic anhydrase, soybean trypsin inhibitor, and lysozyme. Molecular weight, in 1000s, is indicated by the numbers to the left. Gel is stained by Coomassie Blue R-250 (from Glasser and Julian, 1986).

Intermediate Filament Proteins

The accumulation of cytoskeletal proteins identified as intermediate filaments is one of the prominent morphological features of decidual cells, either in vivo (Tachi et al., 1972) or in vitro (Vladimirsky et al., 1977; Sananes et al., 1978). The expression of additional and different intermediate filament proteins has been reported to accompany differentiation in a variety of cells (Jackson et al., 1981; Franke et al., 1982). The expression of these proteins was examined during the process of decidualization in vivo and in vitro (Glasser and Julian, 1986; Glasser et al., 1987) to determine if they might serve as cell-specific and/or stage-specific decidualization markers. Stromal cells, cultured in the absence of epithelial cells (Glasser and Julian, 1986) display the same morphological

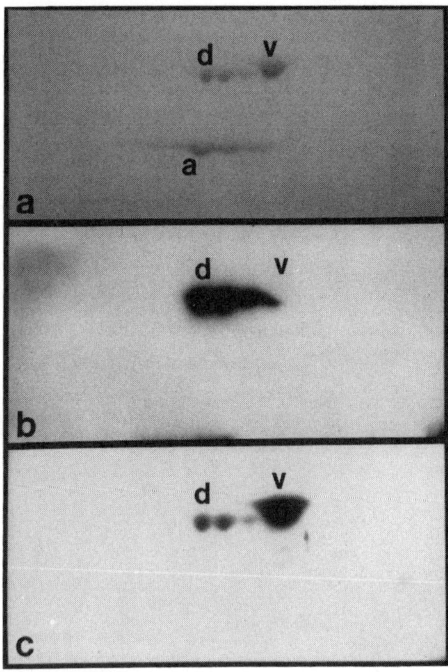

Figure 4. 2-D PAGE of IF-enriched fractions isolated from day 4 + 96 hours cultures of isolated stromal cells. This figure is limited to the area from 35,000 to 65,000 molecular weight, pH 5 to 6. (a) Gel stained by Coomassie R-250 prior to transfer. (b) autoradiogram of nitrocellulose transfer of the gel shown in (a) probed with mouse antidesmin, rabbit antimouse IgG and [^{125}I]-protein A. (c) Autoradiogram of transfer (b) reprobed with rabbit antivimentin and [^{125}I]-protein A. Kodak XRP film and an intensifying screen were used. Exposure times were 72 hours for (b) and 5 hours for (c). The positions of desmin, vimentin, and actin are represented by d, v and a respectively (from Glasser and Julian, 1986).

features, including the accumulation of intermediate filaments, that characterize decidualization in vivo (Tachi et al., 1972) and during in vitro culture of endometrial cells (Vladimirsky et al., 1970; Sananes et al., 1978). Two dimensional gel electrophoresis and immunological criteria (Western blots, indirect immunofluorescent staining) identified the intermediate filament proteins of decidualizing cells as vimentin and desmin (Glasser and Julian, 1986; Glasser et al., 1987; Figures 3, 4, and 5).

Decidualization, in vivo and in vitro, is characterized by an increase in vimentin per cell that is proportional to the increase in total cell protein (Glasser and Julian, 1986). Compared to baseline values (isolated day 4 stromal cells after 24 hours in culture) vimentin and total cell protein accumulate proportionally at approximately 12.5-15% per day (Figure 6). While vimentin appears to be common to both decidualized and non-decidualized stroma desmin makes its major contribution during decidualization (Figures 3, 5) when its accumulation per cell exceeds that of total cell protein (Figure 6). Compared to day 4 baseline values the

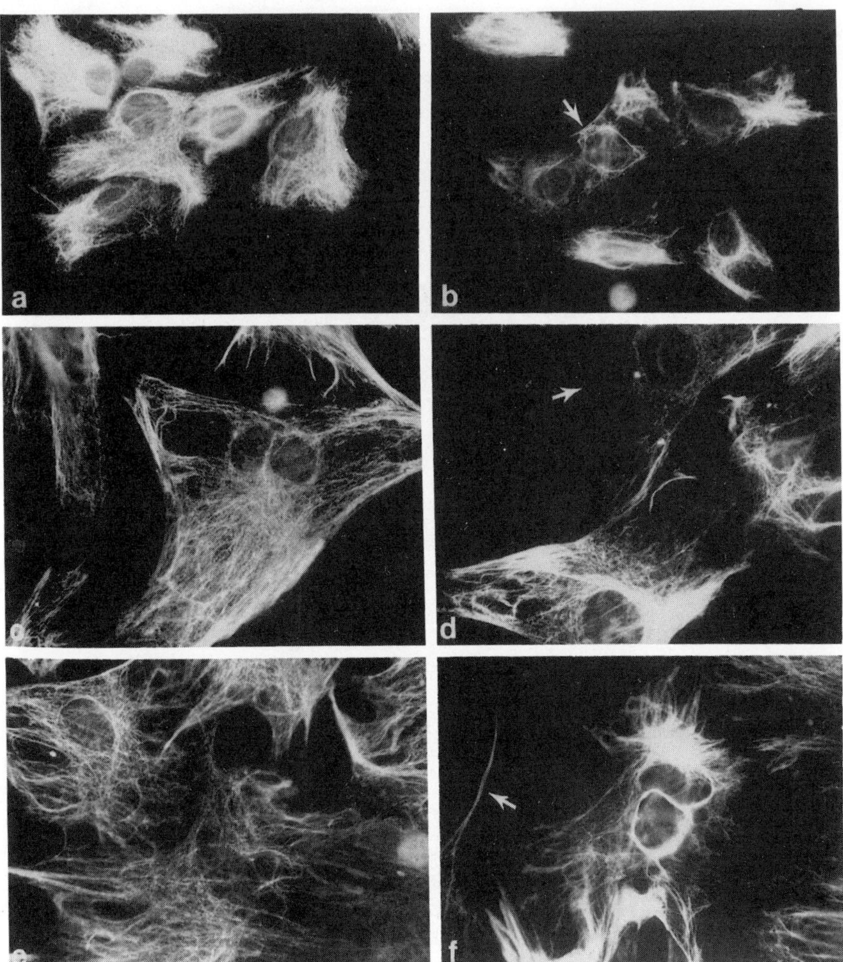

Figure 5. Indirect immunofluorescent micrography of cultured day 4 stromal cells. After isolation from the uterus stromal cells were cultured for 24 hours (a, b), 48 hours (c, d) or 96 hours (e, f). At these time periods cells were incubated with either rabbit antivimentin followed by FITC-goat antirabbit IgG (a, c, e) or with mouse antidesmin followed by FITC-rabbit antimouse IgG (b, d, f). Final magnification for all cells is X275 (from Glasser and Julian, 1986).

desmin concentration per cell, after 96 hours in culture, has increased almost 250%. The daily rate of desmin accumulation, like that of vimentin, is markedly depressed, by more than 50-60%, when decidual cell growth plateaus between 96 and 120 hours.

Desmin is barely detectable in hormonally sensitized but non-decidualized stroma. This may represent its occurrence in a restricted population of cells or desmin may be present in all stromal fibroblasts at marginal levels. In the latter case the marked increase in desmin relative to changes in vimentin and

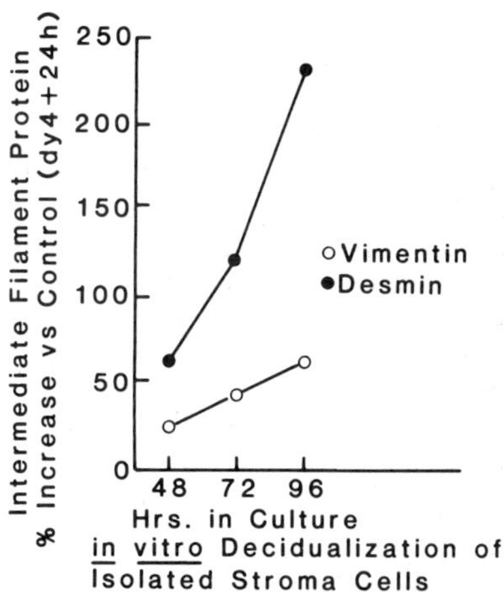

Figure 6. The incremental increases in the intermediate filament proteins vimentin (o) and desmin (●) were measured during the course of in vitro decidualization of rat stromal cells isolated from the uterus on the day of blastocyst attachment (day 4). The changes at 48, 72, and 96 hours are expressed in relation to values assayed 24 hours after the cells are placed in culture. The change in vimentin is proportional to the increase in total cell protein (from Glasser and Julian, 1986).

protein per cell recommends that the increase in desmin is intimately related to the process of decidualization but not to the obligatory hormonal sensitization preceding differentiation.

Increased expression of other intermediate filament proteins relative to vimentin has been functionally related to differentiation of myotubes (Fishman and Danto, 1986) and astroglial cells (Mason et al., 1985). Thus, the presence of desmin in decidual cells may not be considered exclusive in the sense that vimentin and perhaps desmin are constitutive to the ancestral mesenchymal cell and its derived stromal fibroblast. The fact desmin is selectively enhanced in decidualized stroma, but absent or only marginally present in its precursor, suggests that desmin is a useful marker of stromal cell differentiation. This nomination is strengthened by a literature that reports that intermediate filaments are related to the special functions of differentiated cells (Venetanier et al., 1983; Jackson et al., 1981; Franke et al., 1982) rather than the basic functions necessary for growth and proliferation.

Endocrine Role of Decidual Cells

Identification of an endocrine role of decidual cells is a most provocative finding deserving further study. Rat decidual tissue produces a decidual luteotropin, immunologically dissimilar from PRL, which exerts a marked effect on luteal cell function in the absence of PRL (Basuray et al., 1983; Gibori et al., 1984). In the human, decidual cells secrete a prolactin-like hormone which appears indistinguishable from pituitary prolactin (PRL) (Riddick et al., 1982).

Human decidual cells are an extra-pituitary source of this polypeptide which, on the basis of its immunoreactivity, chromatographic behavior, electrophoretic mobility, receptor binding properties and bioassay appears to be identical to pituitary PRL (Tomita et al., 1982; Golander et al., 1978; Riddick et al., 1978). In the absence of trophoblast the secretory endometrium from non-pregnant cycles (Maslar and Riddick, 1979) or decidual tissue obtained from normal or ectopic pregnancies (Maslar et al., 1980) will synthesize and secrete this hormone. In the human decidual tissue has been implicated as the source of amniotic fluid PRL (Riddick and Maslar, 1981). In vitro the ability to induce decidualization of proliferative endometrium is regulated by P (down regulated by the presence of E) and is correlated with histological and functional (PRL secretion) markers of differentiation (Daly et al., 1983b). Once decidualized the stromal cell can only maintain its differentiated secretory phenotype in the presence of P (Riddick and Daly, 1982).

Immunocytochemically localized to decidualized cells (Markoff et al., 1983b; Braverman et al., 1984) regulation of the synthesis and secretion of decidual PRL appears to be different and more complex than pituitary PRL (Golander et al., 1979). Neither bromocriptine, an effective inhibitor at the pituitary level, nor TRH (thyrotropin releasing hormone), a potent stimulator of pituitary PRL production, appear to exercise any influence on decidual cells. Thus, there is no information, at this time, regarding the regulatory biology of human decidual endocrine function or its putative paracrine or endocrine role in regulating luteal cells during the critical period between fertilization and the onset of trophoblast secretion of dimeric hCG (Glasser et al., 1987b).

A decidual cell polypeptide with luteotropic activity is detectable between days 6-11 of pregnancy or pseudopregnancy in the rat (Basuray and Gibori, 1980). This decidual hormone exercises a remarkable influence on luteal cell function in the absence of PRL (Basuray et al., 1983; Gibori et al., 1984). Radioreceptor assays indicate that tissue and blood from rats bearing decidua or deciduoma display high titers of PRL-like activity although the decidual hormone is not immunochemically similar to PRL. The decidual hormone possesses an apparent Mr similar to PRL and competes with it for PRL receptors on the ovary and mammary gland (see Gibori et al., 1984). The ability of decidual luteotropin to sustain luteal P production depends on the continued presence of the pituitary. This proves to be a requirement for LH which suggests that the endocrine repertoire of the decidual cell does not include an LH-like hormone (Jayatilik et al., 1984).

For these reasons it has been suggested that in the rat, which has no hCG-like hormone, the decidual luteotropin may be related to rPL-1 (rat placental lactogen-1). The rPL-1 is one of the two lactogenic hormones produced by the trophoblast giant cell, peaking during days 9-11 of gestation (Soares et al., 1985; Robertson et al., 1982; MacCormack and Glasser et al., 1980b). The rPL-1 possesses luteotropic properties. No information exists regarding the functional homology between human decidual PRL and the rat decidual luteotropin. It is notable that both decidual cell hormones are expressed at the same period relative to the perimplantation period.

Other Decidualization Associated Proteins

The evidence is compelling that both non-decidualized (Cunha, 1976; Cunha et al., 1985) and decidualized (Glasser and McCormack, 1982; Glasser and Julian, 1986; Glasser et al., 1987a; Bell, 1985; Wewer et al., 1986; Aplin et al., 1988) stromal cells are active secretors. The synthesis of extracellular matrix proteins and polypeptide hormones accounts for only a fraction of their secretory activity. Stromal cell secretions have been implicated in the mediation of stage-specific functions of the pregnant uterus (Bell, 1985). Therefore identification of these secretory products could facilitate definition of the role of cell types associated with the stromal compartment in blastocyst attachment and the maintenance of pregnancy (Table 1). Early emphasis on the secretory phenotype of US cells focused on the proliferative and early secretory phases of the menstrual cycle. Attention has shifted to periods of the reproductive cycle which encompass decidualization, i.e., the sensitized or receptive uterus of the rat (Glasser and McCormack, 1979a; Glasser et al., 1987a) or the late secretory phase of the human (Bell, 1985). A catalog of polypeptides has been assembled from studies of the endometrium during the first trimester of pregnancy (Bell, 1985). Comparative analysis of the two studies, for the purpose of understanding the induction and initial stages of stromal cell differentiation, is very difficult. However, careful selection of marker proteins from the lists of pregnancy and decidualization associated macromolecules may be useful in establishing their interactive role via putative paracrine relationships with UE or trophoblast cells. These actions could be mediated through their cell surface receptors, with stromal immunobiological elements responsible for continued recognition of the trophoblast (Gill and Wegmann, 1987; Wegmann, 1988), as well as participation in remodeling of its ECM relative to trophoblast invasion.

The significance of such efforts is validated by the identification of two proteins, insulin-like growth factor or somatomedin binding proteins (IGF-BP) (Nissley and Rechler, 1984) which are proving to be important regulators of pregnancy. Placental protein 12 (PP12) synthesized by endometrial and decidual cells (Rutanen et al., 1986) and pregnancy associated endometrial α_1-globulin (α_1-PEG), synthesized and secreted by decidua (Bell and Keyte, 1988) have been characterized as IGF binding proteins. These two IGF-BP species have a similar molecular weight (29-34 Kd) and immunochemical identity but they differ as to their intracellular localization and N-terminal amino acid sequences. These data suggest that PP-12 and α_1-PEG represent distinct IGF-BP species, specific to different cells and committed to opposite regulatory actions. Although it is accepted that the IGFs exert regulatory actions on growth and development the relationship between IGF-I and II and their relative predominant actions in the fetus (IGF-II) or in the neonate or adult (IGF-I), awaits clarification. Specific roles for each cell type specific factor and binding protein need to be determined at each developmental stage.

These early experiments imply that PP-12 and α_1-PEG could exercise determinative roles in the events related to the peri-implantation period. PP-12 competes with IGF-I receptors for IGF-I. Increased production of PP-12, localized to the glandular epithelium during the luteal phase, would result in inhibition of IGF-I (Bell et al., 1988). Whatever the role of IGF-I in the special or housekeeping functions of endometrial cell types the data suggest a mechanism for the down-regulation of IGF-I action and, by inference, for other growth factors, is in place.

Local synthesis of IGF-BPs implies they can act in an autocrine or paracrine manner to regulate microenvironmental IGF concentrations. If the species of IGF-BP synthesized is PP-12-like then increased production by endometrial cells could inhibit the interactive response of cells which possess IGF-I receptors, i.e, trophoblast cells in the decidual compartment (Bell et al., 1988). If the species of IGF-BP produced is of the type which augments IGF-I action then the recruitment of stromal cells into the decidual cell pool (increased α_1-PEG) could play an important role in selecting developmental options that are regulated by IGF.

The ethical and legal issues which constrain the types of experiments required to define the cell-type specific cause and effect relationships which initiate decidualization and placentation in the human strengthen dependence on studies of comparative reproductive biology. Experiments with rodents are favored, even in the face of certain dissimilarities, because cyclic endometrial changes in the human uterus resemble rodents more than those in closely related primates, (Ramsey et al., 1976). Two caveats must be considered. First, while there are advantages in rodent studies because they are productive and reasonably faithful experimental models the limitations of models should not be disregarded (Glasser, 1985). Second, there is a great deal of important data to be obtained in species, other than rat, mouse, and human, which use other programs for blastocyst-uterine interactions. In each case the continued application of interdisciplinary techniques to these studies will advance our understanding of the control of fertility. This is supported in the preliminary studies of IGF-BPs, now identified as regulatory factors secreted by specific endometrial cell types.

Immunology Of The Fetal Placenta And Decidua

For well over a quarter century reproductive and developmental biologist believed that the general principle that survival of an allograft was compromised by antigenic differences between graft and host did not apply to the relationships between the embryo and mother. In fact, no correlation of this sort has been proved between antibody response to the rat fetus and the success of pregnancy. The basic mechanism by which components of the fetal placenta avoid rejection is related to the role of broadly shared class I antigens, vis-a-vis classical class I transplantation antigens, in the immune response to the paternal contributions to the fetal placenta (Gill, 1988). The idea advanced from these conclusions is that the antibody response is not important for the normal progression of all the phases of implantation. In fact, accumulating evidence suggests that the fundamental criterion for the production of a normal embryo and its successful implantation is genetic (Gill and Wegmann, 1987). Under other circumstances, i.e., embryonic dysgenesis, the immune response may modulate the blastocyst-endometrial interrelationships and allow implantation of an embryo that would otherwise be deleted (Gill, 1988). This salvage response is not without cost since the flawed embryo could develop into an offspring exhibiting developmental anomalies. Thus, the general principle that has influenced our thinking for over a quarter of a century is being rephrased. There is, in fact, growing evidence that MHC disparity improves fetal survival. Not only does paternal cell immunization prevent fetal loss (Gill and Wegmann, 1987; Gill, 1988) but it leads to a modest but significant increase in placental size and a dramatic increase in the size of the fetus (Chaouat et al., 1985). Thus, the components of the fetal-placental immunoregulatory system appear to have a continuing role in post-implantation survival and development.

Based on these data an interesting new decidual cell oriented hypothesis has been proposed that advances the idea that immunotrophism, as opposed to immunosuppression, defines fetal-maternal relationships that sequentially follow the initiation of trophoblast invasiveness (Wegmann, 1988; Gill and Wegmann, 1987; Clarke and Kirby, 1966). This unconfirmed hypothesis proposes that maternal T cell recognition of foreign alloantigens, expressed on invading trophoblast cells at the fetal-maternal interface, promotes lymphokine release from these T cells (Athanassakis et al., 1987; Chaouat et al., 1988). The structural and functional properties of the decidual cell compartment would be an important component of this scheme which defines an autocrine-paracrine pathway for lymphokines and cytokines, derived from either T cells (Mogil and Wegmann, 1988) or from placental components (Hunzicker and Wegmann, 1987). In the latter, case the T cell would play the role of an amplifier. The benefits to placental and fetal growth and function derive from these lymphokines (lymphokines IL3, GM-CSF, cytokine CSF-1) which must be considered as instruments of paracrine regulation in the stromal compartment.

Remodeling The Extracellular Matrix Of The Stromal Compartment

Maintenance of pregnancy is not simply dependent on progesterone titers. Rather it is related to the programmatic differentiation of the various embryonic and maternal cell types which interact, mainly but not necessarily entirely under the direction of progesterone, within the uterine stromal compartment. The components of the stroma do not behave as idiots savants in these events. As part of post-attachment regulatory biology, the secretory phenotype of cells in the stromal compartment is continuously being reexpressed (hormones, growth factors, lymphokines, cytokines, ectoenzymes, matrical proteins, proteoglycans). These functional changes are linked with altered structural correlates, i.e., a change in the appearance of the basal lamina (Sherman and Wadl, 1976) and in the extracellular matrix (ECM) of the stromal compartment (Wewer et al., 1985, 1986; Aplin et al., 1988). Antibodies raised against laminin, entactin, and collagen IV localize such proteins in the basement membrane surrounding the basal surface of the UE cell and outlining the blood vessels of the stromal compartment (Wewer et al., 1985; Aplin et al., 1988). There is no apparent immunoreactivity in the stromal ECM which, in the non-pregnant animal, is a fibrillar network composed mainly of cross-linked fibronectin and collagen I. Within 24 hours after they are placed in culture stromal cells react strongly with the antisera raised against these non-fibrillar basement membrane proteins. Progressively, individual decidual cells are surrounded by a dense pericellular basement membrane-like matrix network which appears amorphous and moderately electron dense in the electron microscope. The changing stromal matrix acts reciprocally to redirect differentiation of cells in the stromal compartment (Hay, 1983). Remodeling of the stromal compartment is requisite to the survival and development of the mammalian embryo.

The extracellular matrix (ECM) of the stromal compartment of the non-pregnant uterus is fibrillar being composed mainly of fibronectin interlinked with the interstitial collagens I, III, V; mainly collagen I (Aplin et al., 1988). The ECM of the stromal compartment differs from that associated with the epithelial cells and the blood vessels. That matrix is of basal lamina composition, i.e., laminin, entactin and collagen IV (Wewer et al., 1985, 1986).

Collagen

The major role in production of collagenous components of the stromal ECM has been assigned to the stromal fibroblast. In human and mouse these cells, which exhibit ultrastructural features representative of a secretory phenotype, produce fibrills that display the 64 nm cross-banding associated with collagen (Zorn et al., 1986). While it is likely that endothelial cells and the vascular smooth muscle associated with the stromal compartment produce their distinctive matrices the contribution of the epithelial cells to the basal lamina and stromal ECM must also be clarified (Aplin et al., 1985).

From the midproliferative through the early secretory phase, human stromal cells produce an increasingly denser and thicker fibrillar network. Immunocytochemically this reticulum is characterized by the presence of collagens I, III, V, and VI (Aplin et al., 1988). The epitopes for collagen V are masked during the proliferative phase; perhaps because these immunoreactive sites are involved in lateral associations with other collagens, i.e., type I. It is this rigid fibrillar ECM of the non-pregnant stromal compartment that progressively and increasingly altered during pregnancy. Immediate to blastocyst attachment (day 21 in the human) the fibrillar collagenous components appear less formally organized. The lattice fibrillar network of the stromal compartment appears diminished. The stroma can be described as edematous.

The mechanisms underlying stromal ECM reorganization are not known. The events, which are defined mainly descriptive terms, suggest that the relationship between synthesis, secretion, and extracellular organization of fibrillar elements is shifted. The apparent change in cross-linking between interstitial collagens and binding to fibronectin is striking. It is not known if the absolute synthesis of collagen, and of fibronectin, is actually down-regulated. These changes appear accented because they occur in conjunction with the enhanced production of extensively hydrated glycosaminoglycans (GAGs) and proteoglycans (PGs).

That collagenolysis is limited and collagen synthesis is not completely arrested is supported by the immunocytochemical detection of collagens I, III, and V in the stroma during the secretory phase of the menstrual cycle. While the immunoreactivity of fibrillar collagens appears marginal in predecidual stroma it is not absent. The progressive unmasking of collagen V may represent a decrease in interchain associations (Aplin et al., 1988). Teleologically structural rearrangement of whatever collagen continues to be synthesized would still contribute to the maintenance of the structural integrity of the remodeled matrix. This would be necessary to support the components of a stromal compartment being enlarged and channelized by the enhanced deposition of hydratable GAGs, PGs, laminin and entactin as a consequence of pregnancy.

Fibronectin

Fibronectin is a secretory product of endometrial US but not UE cells. The dense fibrillar matrix of the proliferative phase ECM is rich in fibronectin. This glycoprotein has been implicated in ECM organization because it interacts with glycosaminoglycans and is capable of cross-linking with interstitial collagens (Owens and Baralle, 1986). Fibronectin is a component of the dense fibrillar network that characterizes the non-pregnant uterus.

As decidualizing fibroblasts round and enlarge the distribution of fibronectin on their surface becomes patchy. Following implantation (24-48 hours), as monitored by indirect immunofluorescent staining, fibronectin seemed to disappear in areas proximal to the implanting rat blastocyst (Grinnell et al., 1982). The loss of fibronectin did not occur uniformly through the stromal compartment since it was retained in areas surrounding the stromal vasculature.

Patched fibronectin is characteristic of fibroblasts that have a rounded morphology as a result of active mitosis (Stenman et al., 1977) or malignant transformation (Chen et al., 1976). Grinnell et al., (1982) have suggested that the regional redistribution of fibronectin endows decidual tissue with a barrier function relative to invasion (Cowell, 1972; Kirby, 1965).

Reorganization of fibronectin from a fibrillar to a patchy distribution on the surface of decidualizing cells has been confirmed but its actual disappearance from the stromal compartment of the pregnant uterus (Grinnell et al., 1982) has been called into question. Fibronectin has been detected as late as day 11 on both antimesometrial and mesometrial decidua. It appears as patches of short fibrils. The impression of its disappearance seems most marked in antimesometrial stroma where the cells are larger (Glasser et al., 1987a).

As is the case with the collagens it is not known if fibronectin synthesis and secretion are different in non-pregnant and pregnant uteri. It is possible that fibronectin continues to be produced at non-pregnant levels but its proportional representation in the stromal matrix is decreased because of the enhanced production of laminin, entactin, and collagen IV. Reorganization of ECM also suggests that the interactions with collagen are interrupted. Uterine stroma is rich in type III collagen (Aplin et al., 1988) for which fibronectin has high affinity. Changes related to decidualization suggests that the adhesive interactions between fibronectin and type III are broken, concommitant with the unmasking of type V collagen, as decidualization progresses. While rearrangement of fibronectin binding sites may alter its interactions with other molecules and specific cell surfaces the structural integrity of the ECM could still be maintained. This property is required to support the a stromal compartment composed of increasing levels of non-fibrillar collagen IV and the hydratable GAGs, PGs, entactin, and laminin.

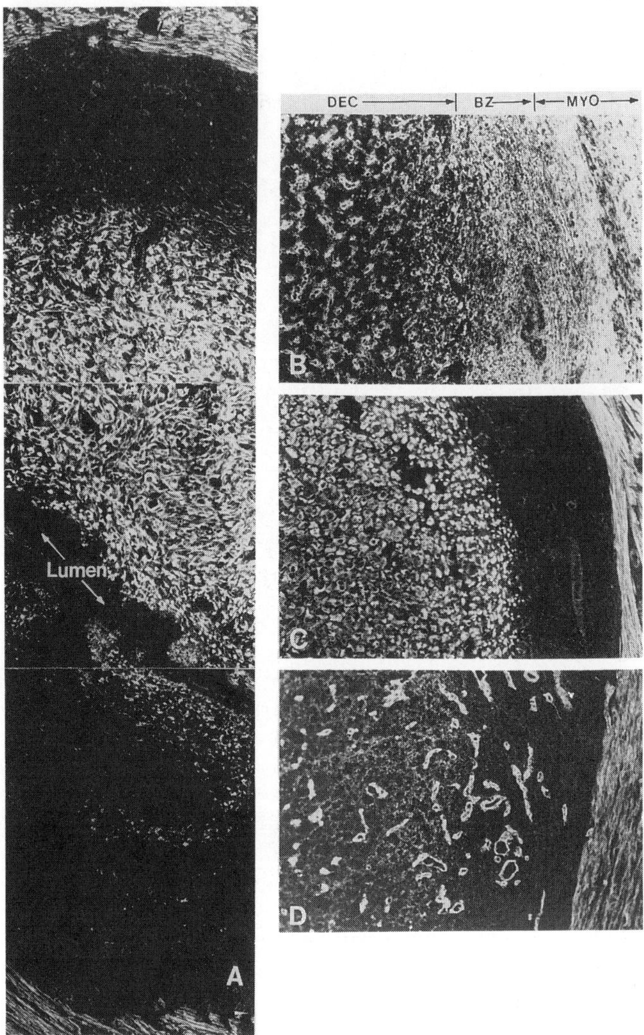

Figure 7. Regional expression of desmin (A, C) relative to fibronectin (B) and laminin (D) during development of deciduoma. Sections were obtained from the decidualized horn of P+E treated castrate (A) or pseudo-pregnant (B, C, D) rat uteri removed 96 hours after the deciduogenic stimulus. Sections were treated with a mouse monoclonal antibody to chicken gizzard desmin (A,C), goat anti-rat fibronectin (B) or rabbit anti-laminin (D). Section (A) is taken through the entire endometrium with the antimesometrial side at the top. The desmin- positive cells at the top and bottom edges of (A) and to the right of (C) are in the inner myometrial layer. (B-D) Only the anti-mesometrial side is pictured. Deciduoma is on the left. M is myometrium, BZ is basal zone and D is deciduoma. Magnification is X75 (from Glasser et al., 1987a).

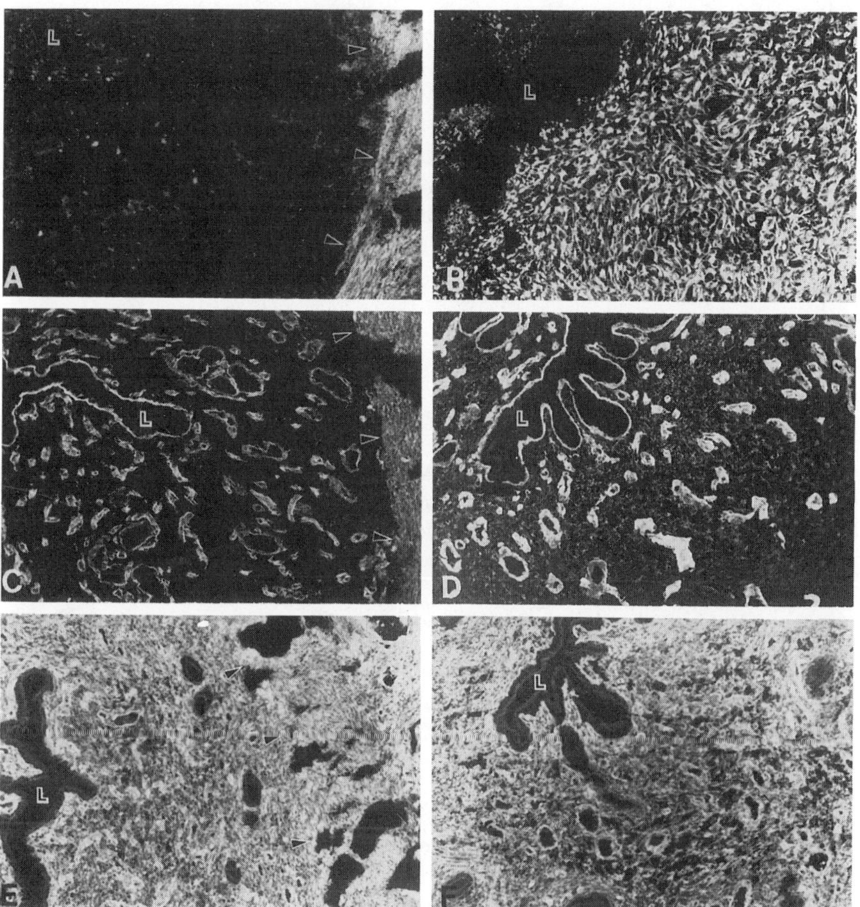

Figure 8. Expression of desmin (A, B), laminin (C, D) and fibronectin (E, F) during the course of deciduoma formation (B, D, F) compared to changes in non-decidualized control horns (A, C, E). Sections were removed from P+E treated castrates 96 hours after decidual stimulation of one uterine horn and reacted for indirect immunofluorescent micrography with mouse monoclonal against desmin (A, B), anti-laminin (C, D) or anti-fibronectin (E, F). The sections in in A, C, and E are the antimesometrial sides of the unstimulated horns contralateral to the deciduoma (B, D, F). M is myometrium, L is lumen. Magnification is X75 (from Glasser et al., 1987a).

Laminin and Entactin

While ECM of the stromal compartment of the non-pregnant uterus is composed mainly of interlinked fibronectin and collagen I. Immunocytochemical analysis of cryostat sections of decidua and deciduoma formed in response to blastocysts or artificial stimulation of uteri, sensitized by endogenous or exogenous hormone regimens, desmonstrated changes in the composition of stromal ECM components to those characteristic of the basal lamina. These alterations were coordinated with the elevated expression of desmin in decidualizing stromal cells (Glasser et al., 1987a). In parallel with the pattern of regional decidualization, monitored by desmin expression, laminin accumulated in the ECM of the decidual cell (Figure 7). This was accompanied by a decrease in fibronectin (Grinnell et al., 1982), more apparent than real, associated with altered organization at the decidual cell surface (Figures 7, 8; Glasser and Julian, 1986). After seven days in culture decidual cells differentiating from stromal cells from non-pregnant uteri or cells cultured as decidua were individually surrounded by a dense meshwork of ECM which ultrastructurally appeared amorphous and of moderate electron density (Wewer et al., 1985, 1986).

The data resolve a number of issues related to the validation of decidual cell markers, (a) desmin and laminin appear to be constitutively expressed in sensitized, non-decidualized stroma at marginally detectable levels; (b) desmin and laminin are produced by decidualizing cells at a rate greater than the increase in cell protein. Therefore, because they are expressed similarly in the decidualized cells, irrespective of different experimental protocols for uterine sensitization and stimulation, laminin and desmin are valid markers of stromal cell differentiation and (c) additional validation is provided by the expression of laminin and desmin along the same path of regional progression described in morphological and biochemical studies of decidualization in the rodent uterus (Kleinfeld and O'Shea, 1983; Glasser et al., 1987a).

The complete profile of matrix proteins produced by cultured decidual cells or by stromal cells from non-pregnant uteri decidualizing in culture (Wewer et al., 1985, 1986) include: (1) both the A (400 KD) and B (200-220 KD, 150 KD) chains of laminin, (2) entactin (150 KD) which anti-laminin coprecipitates with laminin. Further electrophoretic analysis using both reducing and non-reducing conditions and immunoprecipitation with a specific anti-entactin antibody exclude the possibility that this protein is a subspecies of laminin. (3) collagen IV, α, and α_2 (180-190 KD), (4) fibronectin (240 KD) and (5) four species of sulfated glycoproteins. These include a high MW product with a glycosaminoglycan composition of 42% heparan sulfate and 53% chondroitin sulfate. Immunoprecipitation with anti-laminin antibody brings down a 370 KD product which yields a 200/180 KD doublet. This result suggests the the absence of the laminin A chain from the precipitated product (Klein et al., 1988).

The level of efficiency of immunoprecipitations and the susceptibility of the 400 KD laminin chain and enactin to proteolysis, even in the presence of inhibitors, complicates attempts at quantifying changes in these patterns. Qualitatively there are not striking differences between decidualizing stromal cells from non-pregnant uteri and decidual cells after either 1 or 7 days in culture (Wewer et al., 1986).

If stromal fibroblasts are properly sensitized and stimulated they initiate decidualization (Psychoyos, 1973a, b, Glasser, 1972; Glasser and Clark, 1975; Glasser and McCormack, 1979a, b) and alter their phenotypic expression, presumably by selective gene activation. Decidual cells become vigorous producers of basel laminal. However, it must be emphasized that the synthesis and secretion of basement lamina components is not a feature unique to the decidual cell. The cells are unique because they differentiate from cells which do not produce, biochemically or immunocytochemically, detectable amounts of basel laminal components. Experiments benefit from the fact that decidual cells, both in vivo and in vitro (Tachi et al., 1972; Sananes et al., 1978; Vladimirsky et al., 1977; Glasser and Julian, 1986) synthesize and secrete common basement lamina components (Laurie et al., 1983; Wewer et al., 1985, 1986; Aplin and Seif, 1985; Aplin et al., 1988). In those cases where they have been compared, mouse decidual cells closely resemble human cells with respect to decidualization related alterations in the composition of the matrix of the stromal compartment (Wewer et al., 1985).

As decidualization proceeds entactin appears to be preferentially produced relative to laminin so that even qualitatively the amount of entactin that accumulates in decidual cells is impressive. It is not known what importance these shifts in the synthesis and secretion of laminin, entactin, or other basement lamina or extracellular matrix components represent, either in terms of stromal cell differentiation or of trophoblast-decidual cell interactions. It is recognized that the same hormonal signals that stimulate these synthetic processes also influence the activity of lysosomes and both endogenous and exogeneous proteolysis all of which can play roles in remodeling the apical UE cell surface (Schlafke et al., 1985; Morris et al., 1987), the implantation chamber (Welch and Enders, 1985) as well as the extracellular architecture of the stromal compartment (Wewer et al., 1985, 1986: Aplin et al., 1988).

The Remodeled Stromal Compartment and Trophoblast Invasion

The decidual tissue that develops in the pregnant (decidua) or pseudopregnant (deciduoma) rodent uterus has been at times regarded as a modular tumor-like growth of stromal cells (Wewer et al., 1986). While decidual cells differ from tumors because they are not malignant and have a finite life span decidual cells, like certain tumor cells, produce large amounts of basel laminal components. For these and for additional reasons it has been proposed that extrapolation of data gathered during analysis of the more aggressively studied subject of tumor invasion (Liotta et al., 1986) might provide insight and direction to studies of trophoblast interactions in the stromal compartment. The interrelationships between the trophoblast cell and the stromal extracellular matrix (ECM) are compared with the progression of tumor cells through the host ECM in Table 2.

Tumor Invasion Model

ECM components play an important determinative roles in cell morphology, proliferation, and differentiation which characterize mammalian embryonic and fetal development (Bissell et al., 1982). These ECM components provide structural and biophysical parameters which influence selective filtration of soluble macromolecules through the matrix. Regulatory signals are transmitted from ECM components through specific cell surface receptors. In turn autocrine and paracrine cues modulate cell shape, migration, and morphological and functional cell polarity.

Table 2

A Study Model For Trophoblast Invasion: Invasion Is Related To Tumor Cell Interactions With The Extracellular Matrix At Multiple Stages In The Metastatic Cascade (Liotta et al., 1986)

TUMOR CELL*	TROPHOBLAST CELL
1. Tumor can modify ECM in various ways	1. Trophoblast can modify ECMin various ways
a. Loss of basement membrane	a. Loss of basement membrane • specific structural and functional reorganization of the basolateral plasma membrane domain of the receptive UE cell • putative US cell single
b. Disorganization of stromal elements	b. Disorganization of stromal elements • initiated by deciduogenic signal transduced via UE cells at attachment • sustained by secondary signals (?) emanating from differentiating trophoblast (mesometrial decidual response in rodents)
c. Degradiation of matrix components associated with focal tumor invasion	c. Altered synthesis and focal degradation of fibrillar ECM components associated with continuing differentiation of trophoblast
2. Stimulation of accumulation of matrix components by host cells in response to the local presence of tumor cells Hypothesis: tumor cells produce soluble factors that elicit various responses by the host fibroblasts	2. Secondary to changes initiated by the transduced deciduogenic signal the stromal cell accumulates specific matrix (basement membrane-like) components in response to the differentiating trophoblast cells Not studied. Possible directives may be elicited by trophoblast steroids, placental lactogens, growth factors including cytokines and lymphokines.
3. Synthesis of matrix components by tumor cells. New material is generally of the same type produced by normal cells	3. Not adequately studied. Some preliminary evidence supports idea that the profile of proteins, glycoconjugates and their respecitve receptors disposed on trophectoderm surface changes with development, i.e., advance through the stromal compartment. Obversely, ECM of the stromal compartment and its endothelial vasculature is remodeled from a fibrillar ECM to a hydrated basement membrane-like matrix. Specific biologic function of these changes, which includes mobilization of immunocytes, relative to trophoblast invasion is not known.

*For any one type of tumor any one of these three types of matrix modification may characterize tumor invasion. Alternatively, all three motifs may be expressed stimulataneously in different regions of the same tumor. Extrapolation to trophoblast must necessarily be conservative because of the rigorous regulation that applies to its differentation.

In a similar manner profound alterations in the composition and distribution of the adjacent ECM may accompany the development of a malignant neoplasm (Table 2). The nature and progression of events which characterize tumor cell invasion suggests that, perhaps in a more limited sense, similar types of changes in the uterine stromal ECM and basement lamina underlie the expression of the very specialized functions that characterize the sequence of processes that define blastocyst attachment and trophoblast invasion.

Rephrasing the metastatic cascade (Liotta et al., 1986) in terms of implantation related events suggests that the purpose of this regulated ECM remodeling (Table 2) appears to be to render the flexible but insoluble and relatively impermeable (large proteins) basel lamina sensitive to focal penetration by the differentiating trophoblast. Since the stromal compartment does not contain pathways for neoplastic or for trophoblast cell movement remodeling serves to make the ECM permissive to cell migration; perhaps as it occurs in wound healing, inflammation, and neoplasia. Remodeling is a complex sequence of biochemical events (Liotta et al., 1986) which may first include attachment of invading cells to basement membrane laminin and heparan sulfate proteoglycan (HSPG) and subsequently to stromal fibronectin and later to the laminin, entactin, and proteoglycans of the decidual cells and their remodeled ECM.

The advancing tumor or trophoblast cell secretes factors which may induce the decidualizing stromal cells to secrete and/or induce trophoblast enzymes that locally degrade the matrix. The migrating cell (trophoblast, tumor) can enter (invade) into the remodeled region (Table 2). All the components associated with tumor invasion, i.e., laminin, laminin receptor, type IV collagen and collagenase, HSPG, HSPG receptor, HSPG endoglycosidase, fibronectin and fibronectin receptor have all been implicated in trophoblast invasion. Although, the specific role of each component, its regulation and expression is even less well understood in the case of the trophoblast than in the analysis of tumor invasion the tumor invasion model remains a productive paradigm for the study of certain types of placentation.

SUMMARY

The substance of research with blastocyst attachment during the last three decades has produced a directory of practical criteria. These end-points, i.e., tissue weight and gross composition, cytology and ultrastructure, analysis of steroid hormone titers and receptor concentrations, vascular flux, enzyme activities, the measurement of secretory proteins, no longer serve usefully to understand the principles which regulate the establishment and maintenance of mammalian pregnancy. These established experimental models offer few new insights that can be used experimentally or clinically to resolve problems of human or animal reproductive and developmental function.

The complexity of the interdependent processes which define the interactions between embyro and maternal host comes as no surprise. That these processes proceed on many levels of biological organization also is not a novel discovery. However, while these conditions have long been recognized they have not yet been critically addressed. Thus, the interdependence of development programs expressed by the blastocyst and endometrial cell types is probably coordinated by a series of long and short range signals. However, attempts to identify stage and cell specific signals have been unrewarding, largely because appropriate sensitive endpoints are yet to be defined which embody the concept that signal recognition is intimately related to the specific process it initiates. Only recently has the feasiblity of reexamining blastocyst attachment and its sequelae with the purpose of understanding these complexities been extended by the integration of techniques and analytical strategies adapted from studies of cell and molecular biology.

This essay has attempted to discuss, to varying degrees, some of the directions developed within the last decade using these adapted methods. The new data indicate the exciting opportunities open for the analysis of "implantation". Appreciation of the immunobiology operational in the stromal as well as in the luminal compartment, the biochemistry of cytoskeletal, adhesive surface and matrix macromolecules and their receptors, the intricacies of epithelial → stromal and matrix → cellular as well as stromal → epithelial communication can now be factored into practical research. Developed within the context of paracrine and autocrine regulatory mechanisms these gains promise that future research may fuse objective vision with subjective feeling. One may reasonably expect to reveal some basic tenets regarding the reciprocity between the genotypic and phenotypic expression which dictate the success or failure of embryo-uterine interactions.

ACKNOWLEDGEMENTS

Support for the studies from this laboratory was provided by grants from the National Institutes of Health (HD-22785, HD-07495). Particular thanks are due my associates JoAnne Julian, Dr. Daniel Carson, and Jy-Ping Tang. The author thanks his colleagues Drs. Saara Lampelo, M. Idrees Munir, Shaila Mani, Joy Mulholland, and Michael Soares for many helpful discussions. The contributions of Linda Cooper and Georgietta Brown (wordprocessing), Debbie Delmore (photography), and David Scarff (illustration) are most appreciated.

REFERENCES

Amoroso, E.C. (1952) Placentation. In: *Marshall's Physiology Of Reproduction*, Vol. 2., (ed.), A.S. Parkes, Longmans, London, pp. 119-224.

Aplin, J.D. and Seif, M.W. (1985) Basally located epithelial cell surface component identified by a novel monoclonal antibody technique. *Exp. Cell. Res.* 160, 550-555.

Aplin, J.D., Charlton, A.K., and Ayad, S., (1988) An immunohistochemical study of human endometrial extracellular matrix during the menstrual cycle and the first trimester of pregnancy. *Cell Tissue Res.* 253(1), 231-240.

Athanassakis, I., Bleakley, R.C., Paetkau, V., and Wegmann, T.G. (1987) The immunostimulatory effect of T cells and T cell lymphokines on murine fetally-derived placental cells. *J. Immunol.* 138, 37-44.

Barsuray, R. and Gibori, G. (1980) Luteotropic action of decidual tissue in the pregnant rat. *Biol. Reprod.* 23, 507-512.

Basuray, R., Jaffee, R.C., and Gibori, G. (1983) Role of decidual luteotropin and prolactin in the control of luteal cell receptors for estradiol. *Biol. Reprod.* 28, 551-556.

Bell, S.C. (1983) Decidualization: Regional differentiation and associated function. *Oxford Rev. Reprod. Biol.* 5, 220-271.

Bell, S.C. (1985) Comparative aspects of decidualization in rodents and human: Cell types, secreted products and associated function. In: *Implantation Of The Human Embryo*, (eds.), R.G. Edwards, J.M. Purdy, and P.C. Steptoe, Academic Press, London, pp. 71-122.

Bell, S.C. and Keyte, J.W. (1988) N-terminal amino acid sequence of human pregnancy-associated endometrial α_1 globulin, an endometrial insulin-like growth factor (IGF) binding protein. Evidence for two small molecular weight IGF bindng proteins. *Endocrinol.* 123, 1202-1204.

Bell, S.C., Patel, S.R., Jackons, J.A., and Waites, G.T., (1988) Major secretory protein of human decidualized endometrium in pregnancy is an insulin-like growth factor-binding protein. *J. Endocrinol.* 118, 317-328.

Bigsby, R.M. and Cunha, G.R., (1986) Estrogen stimulation of deoxyribonucleic acid synthesis in uterine epithelial cells which lack estrogen receptors. *Endocrinol.* 119, 390-396.

Billington, W.D. (1971) Biology of the trophoblast. *Adv. Reprod. Physiol.* 5, 27-66.

Bissell, M.J., Hall, H.G., and Parry, G. (1982) How does the extracellular matrix direct gene expression? *J. Theor. Biol.* 99, 31-68.

Braverman, M.B., Bagni, A., de Ziegler, D., Den, T., and Gurpide, E. (1984) Isolation of prolactin producing cells from first and second trimester decidua. *J. Clin. Endocrin. Metab.* 58, 521-525.

Bryce, T.H. and Teacher, J.H. (1908) Contribution to the study of early development and imbedding of the human ovum. *MacLeHose, Glasgow*.

Campbell, S., Seif, M.W., Aplin, J.D., Richmond, S.J., Haynes, P., and Allen, T.D., (1988) Expression of a secretory product by microvillous and cilated cells of human endometrial epithelium, in vitro and in vivo. *Hum. Reprod.* 3, 927-934.

Carson, D.D., Tang, J.-Y., Julian, J., and Glasser, S.R. (1988) Vectorial secretion of proteoglycans by polarized rat uterine epithelial cells. *J. Cell Biol.* 107, 2425-2435.

Chaouat, G., Kolb, J.-P., Kiger, N., Stanislawski, M., and Wegmann, T.G. (1985) Immunologic consequences of vaccination against aborton in mice. *J. Immunol.* 134, 1594-1598.

Chaouat, G., Menu E., Athanassakis, I., and Wegmann, T.G. (1988) Maternal T cells regulate placental size and fetal survival. *Regional Immunol.* 1, 143-148.

Chen, L.B., Gallimore, P.H., and McDougall, J.K. (1976) Correlation between tumor induction and the large external transformation sensitive on the cell surface. *Proc. Natl. Acad. Sci. USA* 73, 3570-3574.

Clarke, B. and Kirby, D.R.S. (1966) Maintenance of histocompatability polymorphism. *Nature* (London) 211, 999-1000.

Cowell, T.P. (1972) Control of epithelial invasion by connective tissue during embedding of the mouse ovum. In: *Tissue Interactions During Carcinogenesis*, (ed.), D. Tarin, D., Academic Press, London, pp. 435-463.

Cunha, G.R. (1976) Epithelial-stromal interactions in development of the urogenital tract. *Int. Rev. Cytol.* 47, 137-194.

Cunha, G.R., Bigsby, R.M., Cooke, P.S., and Sugimura, Y. (1985) Stromal-epithelial interactions in adult organs. *Cell Different.* 17, 137-148.

Daly, D.C., Maslar, I.A., and Riddick, D.H., (1983a) Term decidua response to estradiol and progesterone. *Am. J. Obstet. Gynecol.* 145, 679-683.

Daly, D.C., Maslar, I.A., and Riddick, D.H. (1983b). Prolactin production during in vitro decidualization of proliferative endometrium. *Am. J. Obstet. Gynecol.* 145, 672-678.

DeFeo, V.J. (1967) Decidualization, In: *Biology Of The Uterus*, (ed.), R.M. Wynn, Appleton Century Crofts, New York, pp. 192-290.

Denker, H.-W. (1977) Implantation. The role of proteinases and blockage of implantation by proteinase inhibitors. *Adv. Embryol. Cell Biol.* 53, Part 5, Springer-Verlag, Berlin.

El-Shershaby, A.M. and Hinchliffe, J.R. (1975) Epithelial autolysis during implantation of the mouse blastocyst: an ultrastructural study. *J. Embryol. Exp. Morph.* 33, 1067-1080.

Feyrter, F. (1963) Origin and particular morphology of stromal cells. In: *The Normal Endometrium*, (ed.), J. Schmidt-Matthiessen, McGraw-Hill, New York, pp. 65-85.

Finn, C.A. and Martin, L. (1967) Patterns of cell division in the mouse uterus during early prengancy. *J. Endocrinol.* 39, 593-597.

Fischman, D.A. and Danto, S.I. (1985) Monoclonal antibodies to desmin: Evidence for stage-dependent intermediate filament immunoreactivity during cardiac and skeletal muscle development. *Ann. N.Y. Acad. Sci.* 455, 167-184.

Franke, W.W., Schmid, E., Schiller, D.L., Winter, S., Jarasch, E.D., and Moll, R. (1982) Differentiation-related patterns of expression of proteins of intermediate size filaments in tissues and cultured cells. *Cold Spring Harbor Sym. Quant. Biol.* Vol. XLV1, 431-453.

Gibori, G., Kalison, B., Basuray, R., Rao, M.C., and Hunzicker-Dunn, M. (1984) Endocrine role of decidual tissue: Decidual luteotropin regulation of luteal adenyl cyclase activity, luteinizing hormone receptors and steroidogenesis. *Endocrinol.* 115, 1157-1163.

Gill, T.J. (1988) Immunological and genetic factors influencing pregnancy. In: *Physiology Of Reproduction*, (eds.), E. Knobil and J. Neill, J., Raven Press, New York, pp. 2023-2042.

Gill, T.J.and Wegmann, T.G. (1987) Immune recognition, genetic regulation and the life of the fetus. In: *Immunoregulation And Fetal Survival*, (eds.), T.J. Gill and T.G. Wegmann, Oxford Univ. Press, N.Y., pp. 3-11.

Gill, T.J., MacPherson, T.A., Ho, H.N., Kunz, H.W., Cortese-Hassett, A., Stranick, J.S., and Locker, J. (1987) Immunological and genetic factors affecting implantation and development in the rat and human: A unique trophoblast antigen and genes regulating development. In: *Immunoregulation And Fetal Survival*, (eds.), T.J. Gill and T.G. Wegmann, Oxford Univ. Press, NY, pp. 137-155.

Glass, R.H., Haggeler, J. Spindle, A., Pedersen, R.A., and Werb, Z. (1983) Degradation of extracellular matrix by mouse trophoblast outgrowths: A model for implantation. *J. Cell Biol.* 96, 1108-1116.

Glasser, S.R. (1972) The uterine environment in implantation and decidualization In: *Reproductive Biology*, (eds.), H.A. Balin. and S.R. Glasser, Excerpta Medica, Amsterdam, pp. 776-833.

Glasser, S.R. (1985) Laboratory models of implantation. In: *Reproductive Toxicology*, (ed.), R.L. Dixon, Raven Press, New York, pp. 219-238.

Glasser, S.R. (1986) Current concepts of implantation and decidualization. In: *The Physiology And Biochemistry Of The Uterus In Pregnancy And Labor*, (ed.), G. Huszar, CRC Press Boca Raton, FL, pp. 127-154.

Glasser, S.R. and Clark, J.H. (1975) A determinant role for progesterone in the development of uterine sensitivity to decidualization and ovoimplantation. In: *Developmental Biology Of Reproduction*, (eds.), C.L. Markert, and J. Papaeonstantinou, Academic Press, New York, pp. 311-345.

Glasser, S.R. and McCormack, S.A. (1979a). Estrogen modulated uterine gene transcription in relation to decidualization. *Endocrinol.* 104, 1112-1118.

Glasser, S.R. and McCormack, S.A. (1979b) Functional development of rat trophoblast and decidual cells during establishment of the hemochorial placenta. *Adv. Biosci.* 25, 165-197.

Glasser, S.R. and McCormack, S.A. (1980a) Role of stromal cell function diversity in altering endometrial cell responses. *Prog. Reprod. Biol.* 7, 102-114.

Glasser, S.R. and McCormack, S.A. (1980b) Analysis of hormonal responses of the rat endometrium by the use of separated uterine cell types. In: *The Endometrium*, (ed.), F.A. Kimball, Spectrum Publications, New York, pp. 173-192.

Glasser, S.R. and McCormack, S.A. (1981) Separated cell types as analytical tools in the study of decidualization and implantation. In: *Cellular And Molecular Aspects Of Implantation*, (eds.), S.R. Glasser and D.W. Bullock, Plenum Press, N.Y., pp. 217-239.

Glasser, S.R. and McCormack, S.A. (1982) Cellular and molecular aspects of decidualization and implantation. In: *Proteins And Steroids In Early Mammalian Development*, (eds.), H. Beier and P. Karlson, Springer-Verlag, Berlin, pp. 245-310.

Glasser, S.R. and Julian, J.A. (1986) Intermediate filament protein as a marker for uterine stromal cell decidualization. *Biol. Reprod.* 35, 463-474.

Glasser, S.R., Lampelo, S., Munir, M.I., and Julian, J.A. (1987a) Expression of desmin, laminin and fibronectin during in situ differentiation (decidualization) of rat uterine stromal cells. *Different.* 35, 132-142.

Glasser, S.R., Julian, J.A., Munir, M.I., and Soares, M.J. (1987b) Biological markers during early pregnancy: Trophoblastic signals of the peri-implantation period. *Environ. Health. Persp.* 74, 129-147.

Glasser, S.R., Julian, J.A., Decker, G.L., Tang, J.-Y., and Carson, D.D. (1988) Development of morphological and functional polarity in primary cultures of immature rat uterine epithelial cells. *J. Cell Biol.* 107, 2409-2423.

Golander, A., Hurley, T., Barrett, J., Hizi, A., and Handwerger, S. (1978) Prolactin synthesis by human chorion-decidual tissue: A possible source of amniotic fluid prolactin. *Sci.* 202, 311-313.

Golander, A., Hurley, T., Barrett, J., Hizi, A., and Handwerger, S., (1979) Failure of bromocryptine, dopamine and thyrotropin-releasing hormone to affect prolactin secretion by human decidual tissue in vitro. *J. Clin. Endocrinol. Metab.* 49, 787-789.

Grinnell, F., Head, J.R., and Hoffpauir, J. (1982) Fibronectin and cell shape in vivo: studies on the endometrium during pregnancy. *J. Cell Biol.* 94, 597-606.

Hay, E.D. (1983) Cell and extracellular matrix: Their organization and mutual dependence. *Mod. Cell Biol.* 2, 509-548.

Head, J.R. and Billingham, R.E. (1986) Concerning the immunology of the uterus *Am. J. Reprod. Immunol. Microbiol.* 10, 76-81.

Head, J.R. and Gaede, S.D. (1986) Ia antigen expression in the rat uterus. *J. Reprod. Immunol.* 9, 137-153.

Hunzicker, R.D. and Wegmann, T.G. (1987) Placental immunoregulation *C.R.C. Crit. Rev. Immunol.* 6, 245-285.

Jackson, B.W., Grund, C., Winter, S., Franke, W.W., and Ilmensee, K. (1981) Formation of cytoskeletal elements during mouse embryogenesis. II. Epithelial differentiation and intermediate sized filaments in early post-implantation embryos. *Different.* 20, 203-16.

Jayatilak, P.G., Glaser, L.A., Warshaw, M.L., Herz, Z., Gruber, J.R., and Gibori, G. (1984) Relationship between luteinizing hormone and decidual luteotropin in the maintenance of luteal steroidogenesis. *Biol. Reprod.* 31, 556-564.

Kirby, D.R.S. (1965) The "invasiveness" of trophoblast. In: *The Early Conceptus, Normal And Abnormal*, (eds.), W.W Park, Univ. St. Andrews Press, Edinburgh, pp. 68-74.

Kirby, D.R.S. and Cowell, T.P (1968) Trophoblast-host interactions. In: *Epithelial-Mesenchymal Interactions*, (eds.), R. Fleishmajer and R.E. Billingham, Williams and Wilkins, Baltimore, pp. 64-77.

Klein, G., Langegger, M., Timpl, R., and Ekblom, P. (1988). Role of the laminin A chain in the development of epithelial cell polarity. *Cell* 55, 331-341.

Kleinfeld, R.G. and O'Shea, J.D. (1983) Spatial and temporal patterns of deoxyribonucleic acid synthesis and mitosis in the endometrial stroma during decidualization in the pseudopregnant rat. *Biol. Reprod.* 28, 691-702.

Krehbiel, R.H (1937) Cytological studies of the decidual reaction in the rat during early pregnancy and in the production of deciduomata. *Physiol. Zool.* 10, 212-233.

Laurie, G.W., Leblond, C.R., and Martin, G.R. (1983) Light microscopic immunolocalization of type IV collagen, laminin, heparan sulfate proteoglycan and fibronectin in the basement membranes of a variety of organs. *Am. J. Anat.* 167, 71-82.

Lejeune, B. and Leroy, F., (1980) Role of uterine epithelium in inducing the decidual cell reaction. *Prog. Reprod. Biol.* 7, 92-101.

Lejeune, B., Lecoq, R., Lamy, F., and Leroy, F. (1982) Changes in the pattern of endometrial protein synthesis during decidualization in the rat. *J. Reprod. Fertil.* 66, 519-523.

Leroy F. and Galand, P. (1970) Activité du DNA dans l'endomètre de la rate pseudogestante pendant la phase de réceptivité. In: *Ovo-Implantation, Human Gonadotropins And Prolactin*, (eds.), P.O. Hubinont, F. Leroy, C. Robyn, and P. Leleux, Karger, Basel, pp. 130-148.

Leroy, F., Bogaert, C., Van Hoeck, J., and Delacroix, C., (1974) Cytophotometric and autoradiographic evaluation of cell kinetics in decidual growth in rats. *J. Reprod. Fert.* 38, 441-449.

Liotta, L.A., Rao, C.N., and Wewer, U.M. (1986) Biochemical interactions of tumor cells with the basement membrane. *Ann. Rev. Biochem.* 55, 1037-1057.

Markaverich, B.M., Upchurch, S., McCormack, S.A., Glasser, S.R., and Clark, J.H (1981) Differential stimulation of uterine cells by nafoxidine and clomiphene: Relationship between nuclear estrogen receptors and type II binding sites and cellular growth. *Biol. Reprod.* 24, 171-181.

Markoff, E., Zeitler, P., Peleg, S., and Handwerger, S. (1983) Characterization of the synthesis and release of prolactin by an enriched fraction of human decidual cells. *J. Clin. Endocrin. Metab.* 56, 962-968,

Martel, D. and Psychoyos, A. (1980) Behavior of uterine steroid receptors at implantation. *Prog. Reprod. Biol.* 7, 216-233.

Maslar, I.A. and Riddick, D.H. (1979) Prolactin release by human endometrium during the normal menstrual cycle. *Am. J. Obstet. Gynecol.* 135, 757-754.

Maslar, I.A., Kaplan, B.M., Luciano, A.A., and Riddick, D.H. (1980) Prolactin production by the endometrium of early human pregnancy. *J. Clin. Endocrin. Metab.* 51, 78-83.

Mason, C.A., Bovolenta, P., and Liem, R.K.H. (1985) Glial filament protein expression and mitotic activity in astroglial cells in vivo. *N.Y. Acad. Sci.* 455, 790-791.

McDonald, R.G., Morency, K.O., and Leavitt, W.W. (1983) Progesterone modulation of specific protein synthesis in the decidualized hamster uterus. *Biol. Reprod.* 28, 753-768.

McCormack, S.A. and Glasser, S.R. (1980a) Differential response of individual uterine cell types from immature rats treated with estradiol. *Endocrinol.* 106, 1634-1649.

McCormack, S.A. and Glasser, S.R. (1980b) Hormone production by rat blastocysts and mid-gestation trophoblast in vitro. In: *The Endometrium*, (ed.), F.A. Kimball, Spectrum Publications, New York, pp. 145-172.

McKnight, G.S. and Palmiter, R.D. (1979) Transcriptional regulation of the ovalbumin and conalbumin gene by steroid hormones in the chick oviduct. *J. Biol. Chem.* 254, 9050-9058.

Morris, J.E., Potter, S.W., and Gaza-Bulesco, G. (1988) Estradiol induces an accumulation of free heparan sulfate glycosaminoglycan chains in uterine epithelium. *Endocrinol.* 122, 242-253.

Meyers, K.P. (1970) Hormonal requirements for the maintenance of oestriol-induced inhibition of uterine sensitivity in the ovariectomized rat. *J. Endocrinol.* 46, 341-346.

Mogil, R.J. and Wegmann, T.J. (1988) Immunoregulatory effects of lymphokine-driven placental cells on cloned T cells. *Regional Immunol.* 1, 69-77.

Nilsson, B.O. (1982) Ultrastructure of trophoblast-epithelium relations during implantation. In: *Proteins And Steroids In Early Pregnancy*, (eds.), H.M. Beier and P. Karlson, Springer-Verlag, Berlin, pp. 5-14.

Nissley, S.P. and Rechler, M.M. (1984) Insulin-like growth factors. Biosynthesis, receptors and carrier proteins. In: *Hormonal Proteins And Peptides*, Vol. X11 (ed.), C.H. Li, Academic Press, New York, p. 127.

Noyes, R.W., Hertig, A.T., and Rock, J. (1950) Dating the endometrial biopsy. *Fertil. Steril.* 1, 3-25.

O'Shea, J.D., Kleinfeld, R.G., and Marrow, H.A. (1983) Ultrastructure of decidualization in the pseudopregnant rat. *Am. J. Anat.* 166, 271-298.

Owens, R.J. and Baralle, F.E. (1986) Exon structure of the collagen binding domain of human fibronectin. *FEBS Lett.* 204, 318-322.

Padykula, H.A. (1981) Shifts in uterine stromal cell populations during pregnancy and regression. In: *Cellular And Molecular Aspects Of Implantation*, (eds.), S.R. Glasser and D.W. Bullock, Plenum Press, New York, pp. 197-216.

Parr, E.L., Tung, H.N., and Parr, M.B. (1987) Apoptosis as the mode of uterine epithelial cell death during embryo implantation in mice and rats. *Biol. Reprod.* 36, 211-225.

Peel, S., Stewart, I.J., and Bulmer, D. (1983) Experimental evidence for the bone marrow origin of granulated metrial gland cells of the mouse uterus. *Cell Tiss. Res.* 233, 647-656.

Pijnenborg, R. Robertson, W.B., Brosens, I., and Dixon, G. (1981) Trophoblast invasion and the establishment of hemochorial placentation in man and laboratory animals. *Placenta* 2, 71-92.

Psychoyos A. (1973a) Hormonal control of ovoimplantation. *Vitam. Horm.* 31, 201-256.

Psychoyos A. (1973b) Endocrine centrol of egg implantation. In: *Handbook Of Physiology*, Section 7, Endocrinology Vol II, Pt II, (eds.), R.O. Greep and E.B. Astwood, Am. Physiol. Soc., Washington, D.C. pp 187-215.

Ramsey, E.M., Houston, M.L., and Harris, J.W.S. (1976) Interactions of the trophoblast and maternal tissues in three closely related primate species. *Am. J. Obstet. Gynecol.* 124, 647-652.

Riddick, D.H., Luciano, A.A., Kusmik, W.F., and Maslar, I.A. (1978) De novo synthesis of prolactin by human decidua. *Life Sci.* 23, 1913.

Riddick, D.H. and Maslar, I.A. (1981) The transport of prolactin by human fetal membranes. *J. Clin. Endocrin. Metab.* 52, 220-224.

Riddick, D.H. and Daly, D.C. (1982) Decidual prolactin production in human gestation. *Sem. Perinat.* 6, 229-237.

Riddick, D.H. Daly D.C., and Walters, C.A. (1983) The uterus as an endocrine compartment. *Clin. Perinatol.* 10, 627-639.

Robertson, M.C., Gillespie, B., and Friesen, H.G. (1982) Characterization of the two forms of rat placental lactogen (rPL): rPL-I and rPL-II. *Endocrinol.* 111, 1862-1866.

Rutanen, E.-M., Koistinin, R., Sjoberg, J., Julkunen, M., Wahlstrom, T., Bohn, A., and Seppala, M. (1986) Synthesis of placental protein 12 by human endometrium. *Endocrinol.* 118, 1067-1071.

Salomon, D.S. and Sherman, M.I. (1975) Implantation and invasiveness of mouse blastocysts on uterine monolayers. *Exp. Cell Res.* 90, 261-268.

Sananes, N., Weiller, S., Baulieu, E.-E., and LeGoascogne, C. (1978) In vitro decidualization of rat endometrial cells. *Endocrinol.* 103, 86-95.

Sartor, P. (1980) Cell proliferation and decidual morphogenesis. *Prog. Reprod. Biol.* 7, 115-124.

Schlafke, S. and Enders, A.C. (1975) Cellular basis of interaction between trophoblast and uterus at implantation. *Biol. Reprod.* 12,41-65.

Schlafke, S., Welch, A.O., and Enders, A.C., (1985) Penetration of the basal lamina of the uterine epithelium during implantation in the rat. *Anat. Rec.* 212, 47-56.

Schor, S. (1980) Cell proliferation and migration on collagen substrata in vitro. *J. Cell Sci.* 41, 159-175.

Sherman, M.I. and Wudl, L.W. (1976) The implanting mouse blastocyst. In: *The Cell Surface In Animal Embryogenesis And Development*, (eds.), G. Poste G.L. Nicolson, Elsevier/North Holland, Amsterdam, pp. 81-125.

Smith, J.A. and Martin, L. (1974) Regulation of cell proliferation. In: *Cell Cycle Controls*, (eds.), G.M. Padilla, I.L. Cameron, and A. Zimmerman, Academic Press, New York pp. 43-60.

Soares, M.J., Julian, J.A., and Glasser, S.R. (1985) Trophoblast giant cell release of placental lactogens: Temporal and regional characteristics. *Dev. Biol.* 107, 520-526.

Solomon, J.B. (1966) Relative growth of trophoblast and tumor cells co-implanted into isogenic mouse testes and the inhibitory action of methotrexate. *Nature* 210, 716-718.

Steinman, R.M., Gutchinov, B., Witmer, M.D., and Nussenzweig, M.C. (1983) Dendritic cells are the principal stimulators of the primary mixed lymphocyte reaction in mice. *J. Exp. Med.* 157, 613-627.

Stenman, S., Wartiovaara, J., and Vaheri, A. (1977) Changes in the distribution of a major filament protein, fibronectin, during mitosis and interphase. *J. Cell Biol.* 74, 453-467.

Stewart, I. and Peel, S. (1978) The differentiation of the decidua and the distribution of metrial gland cells in the pregnant mouse uterus. *Cell Tiss. Res.* 187, 167-179.

Swaneck, G.E., Nordstrom, J.L., Kreuzaler, F., Tsai, M.-J., and O'Malley, B.W. (1979) Effect of estrogen on gene expression in chicken oviduct: Evidence for transcriptional control of ovalbumin gene. *Proc. Natl. Acad. Sci. USA* 76, 1049-1053.

Tachi, S., Tachi, C., and Lindner, H.R. (1972) Modification by progesterone of oestradiol-induced cell proliferation, RNA synthesis and oestradiol distribution in the rat uterus. *J. Reprod. Fertil.* 31, 59-76.

Tekelioghu-Uysal, M., Edwards, R.G., and Kisnisci, H.A. (1975) Ultrastructural relationships between decidua, trophoblasts and lymphocytes at the beginning of human pregnancy. *J. Reprod. Fertil.* 42, 431-438.

Tickle, C., Crowley, A., and Goodman, M. (1978) Mechanism of invasiveness of epithelial tumors. *J. Cell Sci.* 33, 133-155.

Tomita, K., McCoshen, J.A., Fernandez, C.S., and Tyson, J.E. (1982) Immunologic and biologic characteristics of human decidual prolactin. *Am. J. Obstet. Gynecol.* 142, 420-426.

Turner, W. (1876) *Lectures On The Comparative Anatomy Of The Placenta.* First Series, A and C Block, Edinburgh.

Vladimirsky, F., Chen, L., Amsterdam, A., Zor, L.L., and Lindner, H.R. (1977) Differentiation of decidual cells in cultures of rat endometrium. *J. Reprod. Fertil.* 49, 61-68.

Webb, P.D. and Glasser, S.R. (1984) Implantation. In: *Human In Vitro Fertilization And Embryo Transfer*, (eds.), D.P. Wolf and M.M. Quigley, Plenum Press, New York, pp. 341-364.

Wegmann, T.G. (1988) Maternal T cells promote placental growth and prevent spontaneous abortion. *Immunol. Lett.* 17, 297-299.

Welch, A.D. and Enders, A.C. (1985) Light and electron microscopic examination of the mature decidual cells of the mature decidual cells of the rat with emphasis on the antimesometrial decidua and its degeneration. *Am. J. Anat.* 172, 1-29.

Wewer, U.M., Faber, M., Liotta, L.A., and Albrechtsen, R. (1985) Immunochemical and ultrastructural assessment of the nature of the pericellular basement membrane of human decidual cells. *Lab. Invest.* 53, 624-633.

Wewer, U.M., Damjanov, J.A., Weiss, J., Liotta, L.A., and Damjanov. I. (1986) Mouse endometrial stromal cells produce basement membrane conponents. *Different.* 32, 49-58.

Zorn, T.M.T., Bevilacqua, E.M.A.F., and Abrahamson, P.A. (1986) Collagen remodeling during decidualization in the mouse. *Cell Tissue Res.* 244, 443-338.

PENETRATION OF THE BASAL LAMINA BY PROCESSES OF THE UTERINE EPITHELIAL CELLS DURING IMPLANTATION IN THE RABBIT

Michael Marx, Elke Winterhager[1,2], and H.-W. Denker[1]

Institut für Anatomie
RWTH Aachen, Melatener Strasse 211
D-5100 Aachen, Federal Republic of Germany

INTRODUCTION

The basal lamina is a scaffold whose integrity is required for maintenance of ordered tissue structure and the regulation of cell growth within an epithelium. When tissue remodeling is to occur, it must undergo complex changes. The ability of invasive cells including those of malignant tumors to move across a basal lamina is thought to require the acquisition of specific properties (Vracko, 1974; England, 1982; Foltz et al., 1982; Gospodarowicz et al., 1982; Ingber and Jamieson, 1982; Bernfield et al., 1984; Liotta et al., 1983; Liotta et al., 1986; Gehlsen et al., 1988).

Implantation of the mammalian embryo in the uterus provides an example of invasion by a non-malignant type of cell, the trophoblast. The invasive properties of the trophoblast are very impressively shown when blastocysts are transplanted to ectopic sites, e.g., under the kidney capsule (Kirby, 1960). However, in the uterus, the process of trophoblast invasion is limited in time and space, so that the host organ avoids destruction. There is evidence that the process seems to be regulated from both sides, the trophoblast (invasiveness limited in time) and the uterus ("permissiveness" limited). Details of the interplay between both partners are, however, largely unknown.

Penetration of the trophoblast through the basal lamina has been studied in several species by electron microscopy (cf. Schlafke and Enders, 1975). In general, the morphological analysis of successive states has led to the conclusion that the trophoblast pauses, after penetration through the uterine epithelium, at the residual basal lamina before continuing with penetration into the stroma. However, one report gave surprising evidence that, in the rat, it is not the trophoblast but decidual cell processes that are the first to penetrate the basal lamina of the uterine epithelium (Schlafke et al., 1985).

The present communication describes observations on structural changes in the basal lamina of the uterine epithelium that can be observed in implantation chambers of the rabbit. Comparison is made between the changes that occur in the vicinity of the invading trophoblast, in parts of the implantation chamber which the trophoblast has not reached yet, and blastocyst-free segments of the uterus. It is found that considerable alterations of the uterine epithelial basal lamina do occur

[1]Present address: Institut für Anatomie, Universitätsklinikum, Hufelandstr. 55, D-4300 Essen, Federal Republic of Germany
[2]To Whom Correspondence Should Be Addressed.

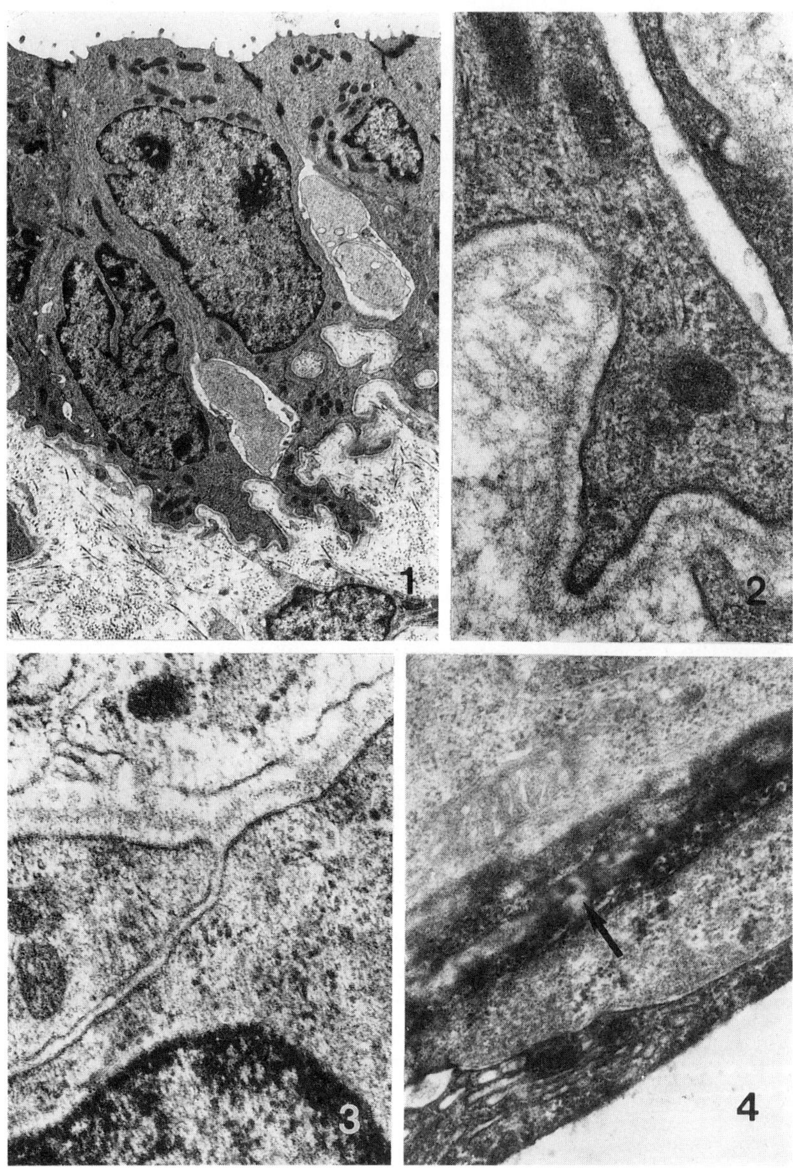

even independent of contact with the invasive trophoblast, and they provide evidence for an active role of the uterine epithelium in remodeling this structure in preparation for embryo implantation.

MATERIALS AND METHODS

Female rabbits of mixed breeds were housed and mated as previously described (Denker, 1977). Two animals of each of the stages 6, 7, and 8 days post coitum (d p.c.; 6 d p.c. = 144 hours p.c.) were included in this investigation. For comparison, 2 animals of the non-pregnant, unestrous stage were also studied, behavioral unestrus being defined as lack of mating response after a couple of trials by two different vigorous bucks.

Perfusion fixation via the abdominal aorta was performed as previously described (Denker, 1977) using 0.1 M cacodylate buffer (pH 7.4) for the pre-rinse followed by a mixture of 2.5% glutaraldehyde and 2% formaldehyde in the same buffer (Karnovsky, 1965), at room temperature for about 10 minutes. Isoosmolarity of the buffer was obtained by addition of sodium chloride. The excised uteri were stored in the same fixative at 3-4°C for 24 hours. In order to detect morphological signs pointing to local influences of the blastocyst on the surrounding part of the endometrium, implantation chamber and blastocyst-free segments of the uteri were cut out in fixative and were investigated separately, from 6 d p.c. onwards. Uterine segments were rinsed in 0.1 M cacodylate buffer (with 5% saccharose), postfixed in 2% osmiumtetroxide and some of them additionally in 4% tannic acid in the same buffer, dehydrated in ethanol, bloc-stained with uranyl acetate, and embedded in araldite epoxy resin.

Ultrathin sections were stained with uranyl acetate and lead citrate and examined in a Zeiss EM 10 transmission electron microscope.

Figure 1. Uterine epithelium of a non-pregnant animal; the cylindric cells form plump foot processes anchoring in the stroma. X5700.

Figure 2. A well structured basal lamina with a lamina lucida and a lamina densa underlies the basal cell processes described in Figure 1. X45000.

Figure 3. Basal lamina of a uterine epithelium 6 d p.c.; the lamina lucida is not as clearly delineated, is reduced in width and electron dense material bridges the lamina lucida in many places. Notice higher magnification than in Figure 2. X60000.

Figure 4. Basal lamina region of the uterine epithelium in the blastocyst-free segments (8 d p.c.). Dark stained amorphous material (arrow) is accumulated in the region of the basal lamina. X38500.

RESULTS

Non Pregnant Animals

The epithelial lining of the rabbit endometrium consists, in the non-pregnant state, of columnar cells decorated with sparse microvilli, and of a few scattered ciliated cells. The epithelial cells are basally anchored in the stroma with plump foot processes (Figure 1). The whole basal surface of the cells including these foot processes is underlain by a well-developed, typically structured basal lamina with a distinct lamina densa (28 nm in diameter) and lamina lucida (30 nm in diameter) (Figure 2). The morphology of this basal lamina is the same in the different parts of the mucosa including luminal and cryptal epithelium. Foot processes, however, are not seen in the deepest parts of the crypts.

The endometrial stroma is relatively rich in collagen and appears to have more collagen per unit volume than in the pregnant state.

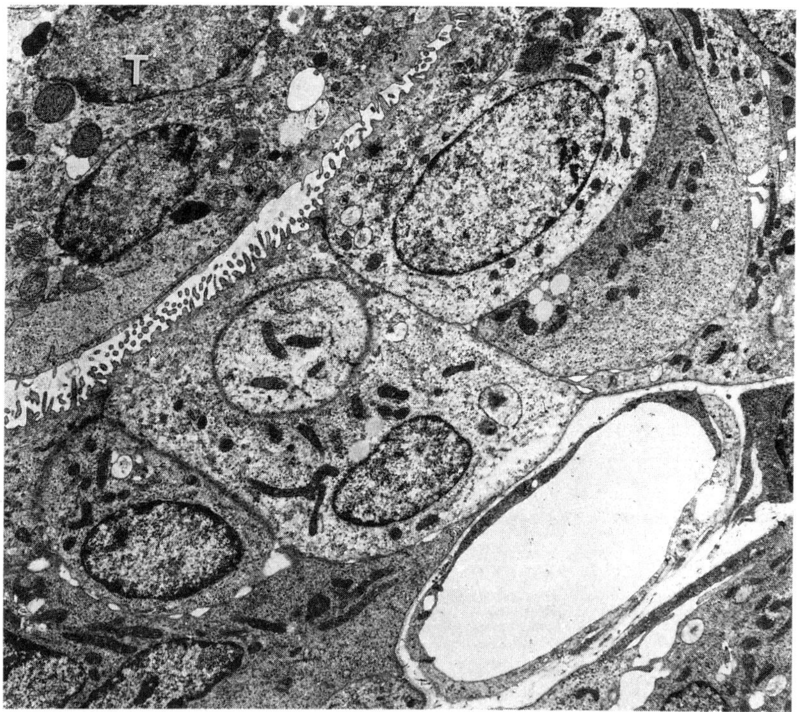

Figure 5. Trophoblastic knob (T) attaching to the antimesometrial uterine epithelial surface 7 d p.c.; in this region the trophoblast has not yet penetrated the uterine epithelium. X4800.

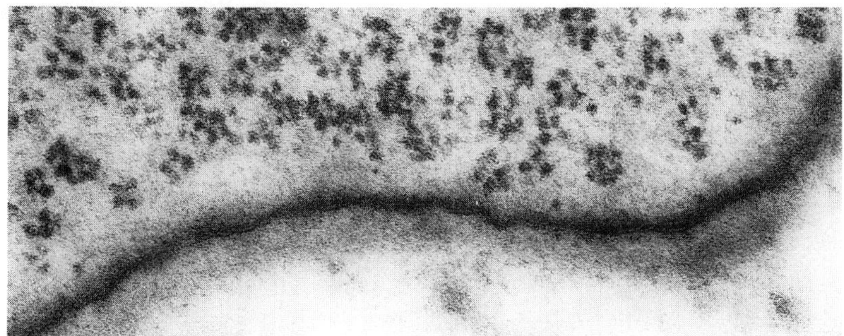

Figure 6. Higher magnification of the basal lamina in a similar region as indicated in Figure 5 reveals a fuzzy lamina densa, whereas a lamina lucida is missing. X172000.

Periimplantation Phase

At 6 d p.c., i.e., one day before implantation starts, the rabbit blastocyst is immobilized in the forming implantation chamber but the trophoblast is still non-attached, the blastocyst coverings being still complete. Here as well as in the blastocyst-free segments most of the luminal cells of the endometrium show conspicuous apical protrusions which contain various cell organelles, and their apical plasma membrane is covered by sparse microvilli. The basal membrane underlying this epithelium follows a fairly smooth contour, since basal cell processes are now lacking in contrast to the non-pregnant state. The fine structure of the basal lamina is slightly changed. The lamina densa is thickened (40 nm) and does not appear as clearly delineated as previously (Figure 3). Electron dense material of the same texture as the lamina densa bridges the lamina lucida in many places. These discrete changes are seen at the uterine epithelium of the implantation chamber as well as in the blastocyst-free segments. Due to epithelial proliferation, the stromal space is now considerably reduced as compared to the non-pregnant state, and collagen fibers are less densely packed.

At 7 d p.c. when implantation starts in the abembryonic-antimesometrial region, more considerable changes in the morphology of the basal lamina are seen in the implantation chamber. In the antimesometrial region, the lamina lucida has totally disappeared in some places (Figure 6), and the remaining lamina densa is here in direct contact with the plasma membrane of the uterine epithelial cells. These remains of the lamina densa appear irregular in texture and outline. The described alterations are most typically seen in those parts of the uterine epithelium where trophoblastic knobs have attached or just fused but not reached the basal lamina yet (Figure 5). In contrast, in those regions where the trophoblastic knobs have already penetrated more deeply into the epithelium so that cytoplasm and nuclei of the trophoblastic type are seen adjacent to the basal lamina, cellular processes have penetrated the lamina densa and are not covered by a comparable structure. Next to them, a lamina densa may be seen that has detached from the epithelium so that a broad and irregular lamina lucida is found here (Figure 7). Further away from the points of attachment of trophoblastic knobs, but still within the implantation chamber, the basal lamina of the uterine epithelium keeps the structure already described for the 6 d p.c. stage. This holds also true for the blastocyst-free segments.

Most striking alterations of basal lamina morphology are observed in the implantation chamber at 8 d p.c. at the uterine luminal epithelium of the placental folds. This is the stage just before attachment of the embryonic pole trophoblast which starts here after 8.5 d p.c. The luminal epithelium consists here, at this stage, of single-nucleated cells as well as of symplasms that are variable in size but contain considerably fewer nuclei than those seen simultaneously in the antimesometrial region. The apical plasma membrane shows numerous microvilli. The lateral plasma membranes are rich in complicated interdigitations. Most interesting are basal cell processes seen in this part of the epithelium. They are different from those found in the non-pregnant state, having a short stem and extending branches that are thin and irregular in shape giving the base of the cell a fringe-like appearance (Figures 9 and 10). These cell processes penetrate the basal lamina (Figure 10). In the basal portion of the epithelial symplasms numerous large vacuoles are found often linearly arranged. This is most typically seen in those cells which clearly show the described processes. The basal lamina consists of a thin and fuzzy lamina densa; the lamina lucida is mostly indistinct and has locally completely disappeared (Figure 11). The described changes in morphology of the basal lamina and penetration by basal processes are seen not only in the luminal epithelium but also in upper and middle parts of the crypts. In the same regions, the stroma is extremely edematous, and collagen fibers are rare. However, fibroblasts are still found here, in particular, close to the epithelium. The epithelium of the deepest parts of the crypts does not show the described changes: It remains without basal cell processes and stays clearly separated from the stroma by a continuous basal lamina as in the other stages.

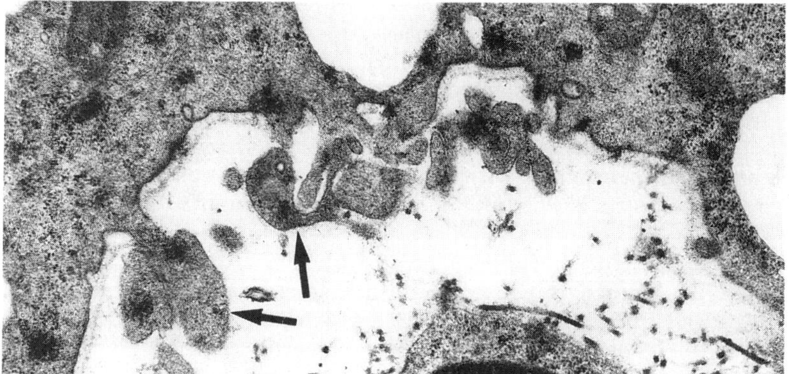

Figure 7. In a region where a trophoblastic knob has fused with the antimesometrial uterine epithelium cytoplasmic processes penetrate the basal lamina (arrows) and extend into the stroma (7 d p.c.); the lamina lucida is relatively narrow and irregular in width. X44000.

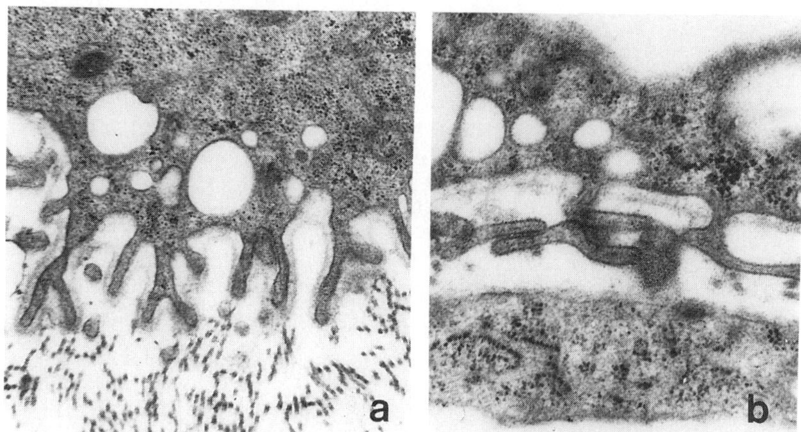

Figures 8a and 8b. At the antimesometrial pole 8 d p.c., the mixed symplasm of uterine epithelial and trophoblastic origin shows a thin but continuous basal lamina which a) underlines numerous thin cell processes or b) is being penetrated by them. X21000.

In the antimesometrial part of the implantation chamber, widespread fusion of uterine epithelial cells with each other and fusion of trophoblastic knobs with uterine epithelium have led to the formation of broad symplasms that already begin to express signs of degeneration. These large symplasms send small finger-like extensions of cytoplasm from their basal parts into the stroma (Figures 8a, 8b). In contrast to the placental folds described above, however, penetration of these cell processes through the basal lamina is seen more rarely here. Some of these processes touch the lamina densa more or less directly, i.e., the lamina lucida is locally absent or irregular in width (Figure 8a). This is similar to phenomena seen next to penetrating trophoblastic knobs at 7 d p.c. However, some of the processes are found to penetrate through the basal lamina in the same way and to branch as described for the placental fold region (Figure 8b).

In the blastocyst-free segments, electron microscopical evaluation of the basal lamina is very difficult at the same stage (8 d p.c.). A typical basal lamina cannot be recognized in any part of the epithelium, since there is an accumulation of amorphous material which is electron densely stained in tannic acid treated specimens. This material fills the whole stromal space underneath the epithelium reaching up to its base (Figure 4). Morphological criteria do not allow to decide whether this is basal lamina material or precipitated protein that possibly obscures the basal lamina structure.

The basal lamina of the epithelial cells of the deep parts of the crypts keeps its morphology throughout the periimplantation phase independent of the distance from the blastocyst.

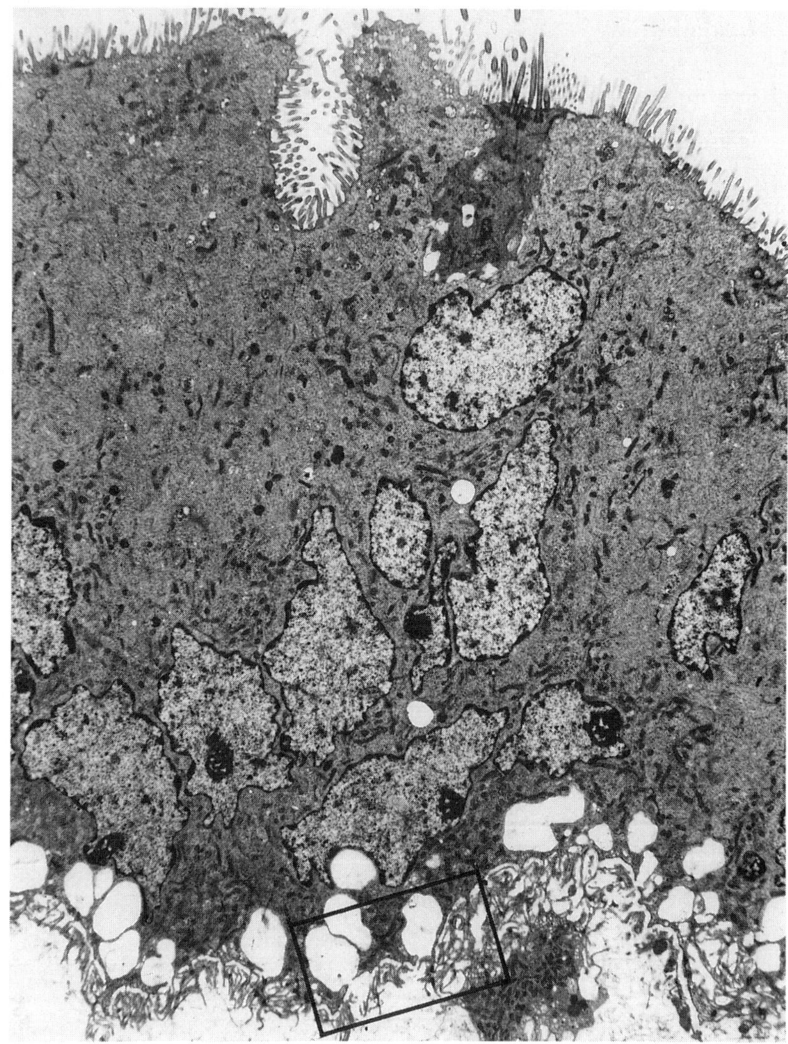

Figure 9. Luminal epithelium of the placental folds 8 d p.c., before trophoblast attachment. Most cells have fused forming a broad symplasm containing a degenerating ciliated cell. Basally, big vacuoles are often continuous with the stromal compartment. The numerous cell processes give a fringe-like appearance. X4200.

DISCUSSION

A number of morphological and biochemical changes in the uterine epithelium that are elicited by maternal steroid hormones during the preimplantation phase have already been described in the rabbit and are thought to represent characteristics of the so-called "receptive phase" during which trophoblast attachment, adhesion, and implantation are possible (Davies and Hoffman, 1973, 1975; Beier and Kühnel, 1973).

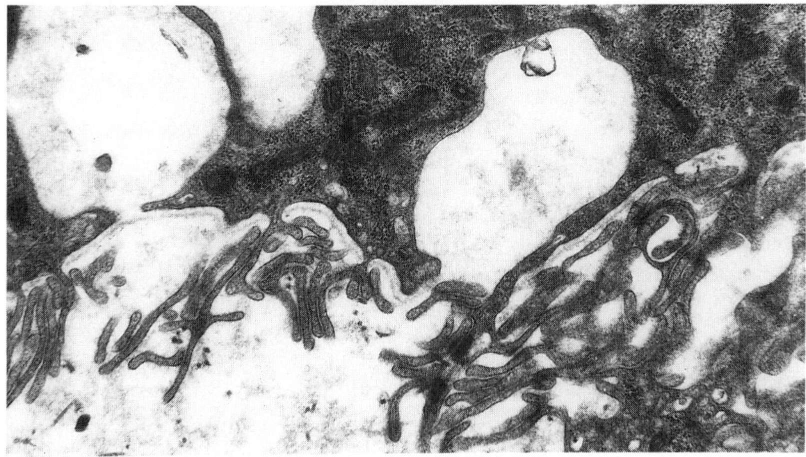

Figure 10. Higher magnification of the part indicated in Figure 9 reveals a thin basal lamina penetrated by stem-like cell processes which branch in the stroma. X17000.

The slight morphological changes in basal lamina structure which were observed during the preimplantation phase until 6 d p.c. and which are maintained in the blastocyst-free segments and in the deepest crypts, may be considered as an additional morphological marker for the pre-conditioning of the endometrium for implantation. The present study provides, in addition, data on more pronounced changes restricted to the implantation chamber. This latter phenomenon obviously needs the presence of a blastocyst since it was not observed outside the implantation chamber. However, it does not require any direct contact with the invasive trophoblast as seen clearly at the placental folds at 8 d p.c., i.e., half a day before the trophoblast attaches here. This suggests that the described changes in the implantation chamber are elicited or stimulated by diffusible signals provided by the blastocyst and which act at a distance, but that the immediate cause for the changes must be in the uterine epithelium.

The structural changes of the basal lamina observed in the implantation chamber are, in a number of details, different in the antimesometrial part of the endometrium (obplacenta, yolk sac placenta) as compared to the mesometrial part (where the chorioallantoic placenta is formed). Differences include the time course as well as the morphological patterns. Antimesometrially, structural changes of the basal lamina are only minor before the trophoblast has attached to the uterine epithelium. After trophoblastic knobs have attached and start to fuse with the uterine epithelium, however, the lamina lucida disappears leaving only remnants of the lamina densa exactly underneath the attachment/fusion site. The most basally located part of the cytoplasm of the trophoblast-uterine epithelial syncytium still maintains structural characteristics of uterine epithelium. Later on, when cytoplasm of trophoblastic type has penetrated more deeply in this syncytium and has reached the basal lamina, cytoplasmic processes penetrate through the vestiges of the lamina densa to reach the stroma. Already one day later, however, at the time when the antimesometrial uterine epithelium has been transformed into broad syncytia containing a mixed population of uterine and trophoblastic nuclei, a typical and nearly complete basal lamina is reestablished. Only rarely do processes of this syncytium penetrate the basal lamina. During the following two days, the synctia will degenerate and will be sloughed off, and non-fused uterine epithelium will grow out of the crypts and regenerate a surface lining. Further investigations will be needed to understand the mechanisms involved in the termination of invasion.

At the placental folds, i.e., in the central part of the mesometrial endometrium, the basal lamina undergoes much more widespread and pronounced changes. Interestingly, these changes are seen here before the trophoblast has reached this scaffold, even before it has contacted the uterine epithelium at all. Penetration of the basal lamina before trophoblast invasion has already been observed in the rat; in this case, however, by decidual cell processes (Schlafke et al., 1985). This, on the other hand, cannot be a mechanism for erosion of the uterine epithelial cell basal lamina at this phase in the rabbit, since in this species decidualization starts about one day later, and not underneath the epithelium but around deeper blood vessels. Both studies are in agreement, however, in showing that in both species, uterine cells modify and even penetrate this barrier first. This suggests that the invasive trophoblast does not play the major role in penetrating or dissolving the basal lamina. This idea is supported by a recent publication (Roberts et al., 1988) where it has been shown that the human uterine epithelial cells including the basal lamina undergo the same morphological changes observed here. In the early secretory phase numerous epithelial cell projections extend through the basal lamina which has become focally diffuse. Interestingly these projections are often in close contact with underlying stromal cells. Since these phenomena are observed in cycling endometria, they seem to be regulated exclusively by maternal steroid hormones.

The mechanisms of change in the uterine epithelial basal lamina can only be speculated on at present. The basal lamina is linked by proteins and proteoglycans to the basal cell membrane (Bernfield et al., 1984) which is directly or indirectly coupled with the cytoskeleton. Therefore, changes in the basal lamina must be linked to profound changes in the cell biology of the uterine epithelium at implantation. A reciprocal relationship appears to exist between epithelial cells and their basal lamina. Maintenance of an ordered basal lamina structure requires physiological integrity of epithelium, in particular a functioning apico-basal polarity (e.g., intracellular sorting and transport processes). Therefore, changes in the physiological state of epithelial cells must result in changes in basal lamina structure. On the other hand, cells respond to basal laminae in various ways, and a changed basal lamina or its loss must be of relevance for the behavior of the uterine epithelium. If the basal lamina is destroyed, cells have been found (in other systems) to lose functional differentiation, and this leads to a disorder in organotypic architecture (Vracko, 1974). For isolated thyroid follicles and for corneal epithelial cells it was shown that cells lose polarity and produce cytoplasmic processes when the basal lamina is removed (Hay, 1977; Greenburg and Hay, 1988). Implications of this concept for understanding the mechanism of endometrial "receptivity" are presently discussed more in details elsewhere in this volume (Denker, 1990).

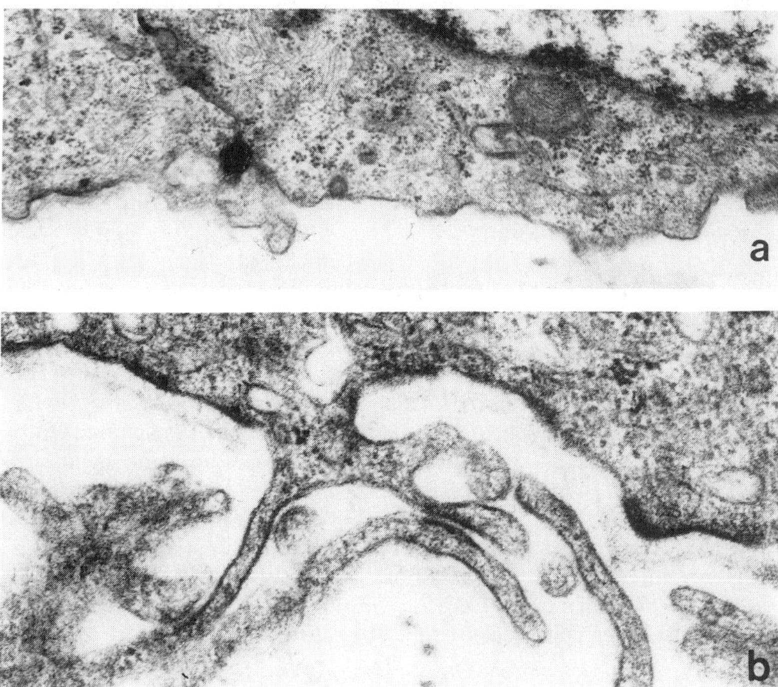

Figures 11a and 11b. Basal lamina of the epithelium of an endometrial crypt 8 d
p.c., placental fold a) The basal lamina is focally missing. b) This vestigial basal
lamina is penetrated by a basal cell projection. X60000.

During tumor invasion, the extracellular matrix is often modified in a
localized region (Liotta et al., 1979, 1986). Studies of the enzymes involved are still
in their infancy. Collagen type IV is degraded by a specific type of collagenase,
while laminin can be degraded by a number of different proteinases (Liotta et al.,
1979, 1986). Various glycosidases and proteinases have been detected in the uterine
secretion as well in implanting blastocysts. As far as their function is concerned,
what has been proven experimentally so far is only a role in the dissolution process
of the blastocyst coverings while effects on components of the endometrium, in
particular during stromal invasion, need to be defined (Denker, 1977, 1983). Even
though uterine epithelial cells appear to destroy their own basal lamina as
suggested by results shown in the present paper, this seems to be under the influence
of the blastocyst since it is observed only in the implantation chamber. It remains
to be seen what components of the basal lamina are being altered during trophoblast
invasion and which have already been altered in the preimplantation phase.

SUMMARY

In order to shed light on the function of the uterine epithelial basal lamina,
its fine structure has been studied during the periimplantation phase (6-8 days post
coitum, d p.c.) in the rabbit. The endometrium of the blastocyst-free segments and

the implantation chamber were investigated seperately and compared to the situation in the non-pregnant animal.

Slight morphological changes are already observed from 6 d p.c. onwards independent of the different parts investigated. The lamina densa ceases to be clearly delineated, expanding partly into the lamina lucida. This phenomenon is enhanced 7 d p.c. in the implantation chamber, at a time when the trophoblastic knobs start to penetrate the antimesometrial uterine epithelium. Here, locally, a lamina lucida is totally missing.

The most obvious changes in basal lamina structure are observed at the mesometrial side 8 d p.c., i.e., before trophoblast invasion has started. At the epithelium of the placental folds the lamina densa appears fuzzy and the lamina lucida is thin or even missing. In some places the basal lamina completely disappears. The uterine epithelium forms numerous cell processes at the basal side which penetrate the residual basal lamina and extend into the stroma. At the same time big symplasms have formed at the antimesometrial side by fusion of the uterine epithelial cells with each other and with the trophoblast. Here, similar basal processes extend into the stroma partly penetrating an otherwise relatively typically structured basal lamina. In contrast, the blastocyst-free segments of these uteri reveal an intact basal lamina structure with a lamina lucida and a lamina densa in all stages up to 7 d p.c.. On 8 d p.c. an accumulation of amorphous material often obscures the basal lamina structure.

In conclusion, remarkable changes are found in the fine structure of the basal lamina of the rabbit uterine epithelium in the implantation chamber starting even before trophoblast invasion. Basal cell processes of uterine epithelium or of symplasms penetrate this scaffold independently of trophoblast attachment, thus facilitating trophoblast invasion. These findings suggest that the uterine epithelial basal lamina, at least in the receptive state of the endometrium, should not be considered as an effective barrier for the invasive trophoblast.

ACKNOWLEDGEMENTS

The authors thank Mrs. Seelis for excellent technical assistance, Mrs. Bock for photographic work as well and Mrs. Mathieu for typing the manuscript. This work was supported by a grant of the DFG (Wi 774/1-1).

REFERENCES

Beier, H.M. and Kühnel, W. (1973) Pseudopregnancy in the rabbit after stimulation by human chorionic gonadotropin. *Hormone Res.* 4, 1-27.

Bernfield, M., Banerjee, S.D., Keda-Alan, J.E., and Rapraeger, C. (1984) Remodeling of the basement membrane as a mechanism of morphogenetic tissue interaction. In: *The Role Of Extracellular Matrix In Development*, (ed.), R.L. Trelstad, Alan R. Liss, Inc., New York, NY, pp. 545-572.

Davies, J. and Hoffman, L.H. (1973) Studies on the progestational endometrium of the rabbit. I. Light microscopy, day 0 to day 13 of gonadotrophin-induced pseudopregnancy. *Am. J. Anat.* 137, 423-446.

Davies, J. and Hoffman, L.H. (1975) Studies on the progestational endometrium of the rabbit. II. Electron microscopy, day 0 to day 13 of gonadotrophin-induced pseudopregnancy. *Am. J. Anat.* 142, 335-366.

Denker, H.-W. (1977) Implantation. The role of proteinases, and blockage of implantation by proteinase inhibitors. Springer-Verlag Berlin, Heidelberg, New York. *Advances In Anatomy, Embryology And Cell Biology,* 53, Part 5.

Denker, H.-W. (1983) Basic aspects of ovoimplantation. In: *Obstetrics and Gynecology Annual,* 12, (ed.), R.M. Wynn, Appleton-Century-Crofts/Norwalk, Connecticut, pp. 15-42.

England, M.E. (1982) Interactions at basement membranes. In: *The Functional Integration Of Cells In Animal Tissues,* (eds.), J.D. Pitts and M.E. Finbow, Cambridge Univ. Press, pp. 209-228.

Foltz, C.M., Russo, R.G., Siegal, G.P., Terranova, V.P., and Liotta, L.A. (1982) Interactions of tumor cells with whole basement membrane in the presence or absence of endothelium. In: *Interaction Of Platelets And Tumor Cells,* (eds.), G.A. Jamieson and A.R. Scipio, Alan R. Liss, Inc., New York, pp. 353-371.

Gehlsen, K.R., Argraves, W.S., Pierschbacher, M.D., and Ruoslahti, E. (1988) Inhibition of in vitro tumor cell invasion by Arg-Gly-Asp-containing synthetic peptides. *J. Cell Biol.* 106, 925-930.

Gospodarowicz, D., Cohen, D., and Fujii, D.K. (1982) Regulation of cell growth by the basal lamina and plasma factors: Relevance to embryonic control of cell proliferation and differentiation. In: *Growth Of Cells In Hormonally Defined Media,* (eds.), A. Book, G.H. Sato, A.B. Pardee, and D.A. Sirbasku, Cold Spring Harbor Conference on Cell Proliferation, 9, pp. 95-124.

Greenburg, G. and Hay, E.D. (1988) Cytoskeleton and thyroglobulin expression change during transformation of thyroid epithelium to mesenchyme-like cells. *Develop.* 102, 605-622.

Hay, E.D. (1977) Interaction between the cell surface and extracellular matrix in corneal development, In: *Cell And Tissue Interaction,* (eds.), J.W. Lash, and M.M. Burger, Raven Press, New York, p. 115.

Ingber, D.E. and Jamieson, J.D. (1982) Tumor formation and malignant invasion: role of basal lamina. In: *Tumor Invasion and Metastasis,* (eds.), L.A. Liotta and I.R. Hart, Martinus Nijhoff Publishers, The Hague, Boston, London, pp. 335-357.

Karnovsky, M.J. (1965) A formaldehyde-glutaraldehyde fixative of high osmolality for use in electron microscopy. *J. Cell Biol.* 27, 137A-138A.

Kirby, D.R.S. (1960) Development of mouse eggs beneath the kidney capsule. *Nature* 187, 707-708.

Liotta, L.A., Rao, C.N., and Barsky, S.H (1983) Tumor invasion and the extracellular matrix. *Lab. Invest.* 49(6), 636-649.

Liotta, L.A., Rao, C.N., and Wewer, U.M. (1986) Biochemical interactions of tumor cells with the basement membrane. *Ann. Rev. Biochem.* 55, 1037-1057.

Roberts, D.K., Walker, N.J., and Lavia, L.A. (1988) Ultrastructural evidence of stromal/epithelial interactions in the human endometrial cycle. *Am. J. Obstet. Gynecol.* 158, 854-861.

Schlafke, S. and Enders, A.C. (1975) Cellular basis of interaction between trophoblast and uterus at implantation. *Biol. Reprod.* 12, 41-65.

Schlafke, S., Welsh, A.O., and Enders, A.C. (1985) Penetration of the basal lamina of the uterine luminal epithelium during implantation in the rat. *Anat. Rec.* 212, 47-56.

Vracko, R. (1974) Basal lamina scaffold-anatomy and significance for maintenance of orderly tissue structure. *Am. J. Pathol.* 77, 313-346.

ENDOMETRIAL LEUKOCYTES IN HUMAN PREGNANCY

Judith N. Bulmer, Anne Ritson, and Denise Pace

Department of Pathology
University of Leeds
Leeds LS2 9JT, United Kingdom

INTRODUCTION

Recent morphological studies of pregnancy hysterectomies have highlighted the extent of invasion of maternal uterine tissues and vessels by fetal trophoblast in normal human pregnancy (Pijnenborg et al., 1980). In early gestation, columns of cytotrophoblast proliferate from the tips of tertiary villi, spreading laterally to form a cytotrophoblast shell. Interstitial trophoblast invades maternal decidua and myometrium and endovascular trophoblast migrates in a retrograde direction along the spiral arteries effecting the so-called 'physiological changes' (Pijnenborg et al., 1980; Robertson et al., 1986). The mechanisms controlling trophoblastic proliferation and invasion are unknown and may involve complex interactions between maternal uterine cells and fetal trophoblast. Failure of invasion into spiral arteries by trophoblast in their myometrial segments, and sometimes throughout their entire length is associated with maternal pregnancy-induced hypertension and intrauterine fetal growth retardation (Khong et al., 1986; Robertson et al., 1986).

Development of monoclonal antibodies (mAbs) has allowed immunohistochemical characterization of maternal and fetal cells in uteroplacental tissues (Bulmer and Johnson, 1985a) and has highlighted the phenotypic heterogeneity of trophoblast populations. Recent attention has focused on materno-fetal cellular relationships in the placental bed. Extravillous trophoblast in the placental bed and amniochorion expresses an unusual class I MHC antigen which is reactive with some markers of monomorphic determinants (Hsi et al., 1984; Redman et al., 1984; Wells et al., 1984) and appears to possess a 40-41 kd heavy chain associated with β2-microglobulin (Ellis et al., 1986). The precise nature of the antigen remains unclear; the possibility of a human Qa analogue has been considered (Ellis et al., 1986; Stern et al., 1986) and a role in induction of immunosuppression has been considered (Stern et al., 1986).

Maternal uterine cells coexist in apparent harmony with class I MHC-bearing semiallogenic fetal trophoblast and attention has recently been directed towards the constituent cells and function of decidualized uterine endometrium.

Morphological Studies Of Human Decidua

Human decidua is a complex tissue, and many cell types cannot be confidently identified at either light or electron microscope level (Tekelioglu-Uysal et al., 1975; Pijnenborg et al., 1980). As well as glycogen-containing large decidualized endometrial stromal cells, other smaller cells are seen; these include a population of granulated cells variously termed endometrial stromal granulocytes (EGs), Körnchenzellen or 'K' cells (Dallenbach-Hellweg, 1981) (Figure 1). EGs are characterized by a hyperchromatic round, oval or indented nucleus and varying numbers of cytoplasmic granules which stain red with the phloxine tartrazine stain (Figure 1A). They can also be detected in imprints prepared from fragments of early pregnancy decidua and stained with May Grünwald Giemsa (Figure 1B). EGs are present in endometrium in the late secretory phase of the menstrual cycle and in the first trimester of pregnancy, thereafter declining in number to be virtually absent at term. Their origin and function has been a source of controversy; it has been suggested that they derive from endometrial stromal cells and secrete relaxin (Dallenbach-Hellweg et al., 1981) but more recent investigations indicate that they are granulated lymphocytes (Bulmer et al., 1987a).

Leukocyte Populations In Human Decidua

Immunohistochemical studies have considerably clarified the numerous cell types present in human decidua in normal pregnancy (see Table 1). Leukocytes can be identified in large numbers within decidual stroma and are aggregated particularly around vessels and adjacent to endometrial glands (Figure 2) (Bulmer and Sunderland, 1983). A major proportion of leukocytes are macrophages which are present in both decidua basalis and decidua parietalis throughout gestation (Figure 3) (Bulmer and Johnson, 1984; Kabawat et al., 1985a; Khong, 1987). Decidual macrophages express the CD14 (Leu-M3) macrophage differentiation antigen but many are CD11b (OKM1/Mac-1) negative. The majority are class II MHC antigen-positive (Bulmer and Johnson, 1984; Kabawat et al., 1985a; Bulmer et al., 1988a) and express the CD11c (p150, 95) antigen. These latter surface antigens and the close association between class II MHC-positive maternal decidual macrophages and class I MHC-bearing extravillous fetal trophoblast in decidua basalis (Figure 3) (Bulmer et al., 1988b) may suggest that decidual macrophages have an immunological role. Antigen presenting capacity has been demonstrated in early pregnancy decidua; the cells responsible have not been fully characterized but may be macrophages or dendritic cells (Oksenberg et al., 1986). Decidual macrophages also contain acid phosphatase and non-specific esterase activity (Nehemiah et al., 1981; Bulmer and Johnson, 1984; Kabawat et al., 1985a) which could indicate a primarily phagocytic function. Macrophages in murine and human decidua have also been implicated as mediators of immunological suppressor activity (see below) (Lala et al., 1986; Tawfik et al., 1986) by secretion of prostaglandin E_2. Despite the possibilities suggested by these in vitro studies, the role of decidual macrophages in normal human decidua remains to be determined.

Macrophages are prominent in decidua basalis and decidua parietalis throughout gestation, but an additional population of lymphoid cells is also

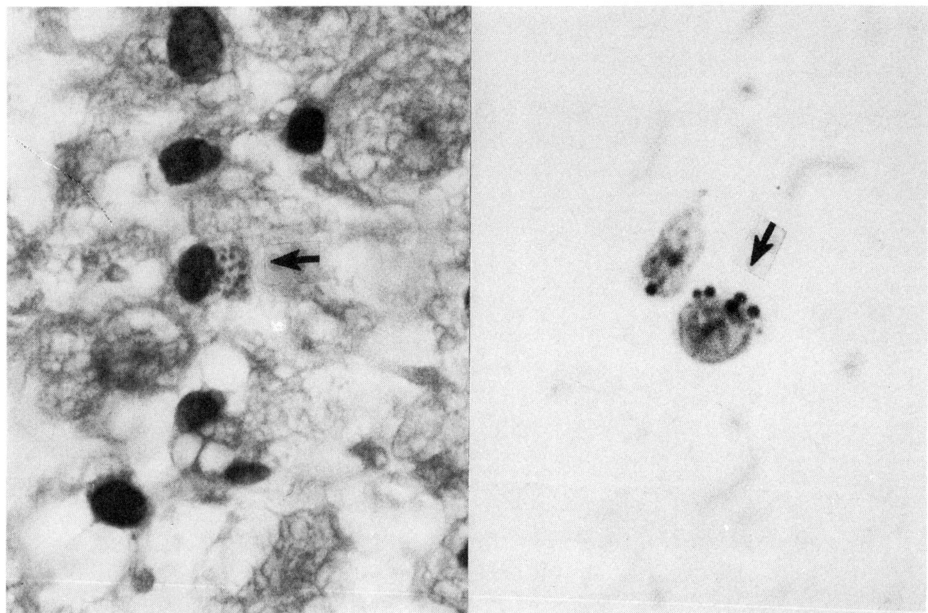

Figure 1. A) Paraffin-embedded section of first trimester decidua stained with phloxine-tartrazine showing a granulated lymphocyte (endometrial stromal granulocyte/Körnchenzellen) (arrowed). B) Imprint prepared from first trimester decidua stained with May Grünwald Giemsa showing a granulated lymphocyte (arrowed). Magnification X900.

abundant in first trimester decidua. These lymphoid cells have an unusual antigenic phenotype: they express the CD2 and CD7 T cell differentiation antigens as well as CD38 (OKT10), but they fail to label for the classical T cell antigens CD3, CD5, CD4, or CD8, nor for the interleukin-2 (IL2) receptor, CD25 (Bulmer and Sunderland, 1984; Bulmer and Johnson, 1986; Bulmer et al., 1987b). The distribution of this unusual lymphoid population within decidual tissue throughout gestation mirrors that of the endometrial stromal granulocytes (EGs). These cells lose their characteristic cytoplasmic granules after freezing but recent studies with mAbs such as MT1 and UCHL1, which are reactive on formalin-fixed paraffin-embedded sections, and with cell suspensions and imprint preparations have provided conclusive evidence that the EGs are granulated lymphocytes (Bulmer et al., 1987a).

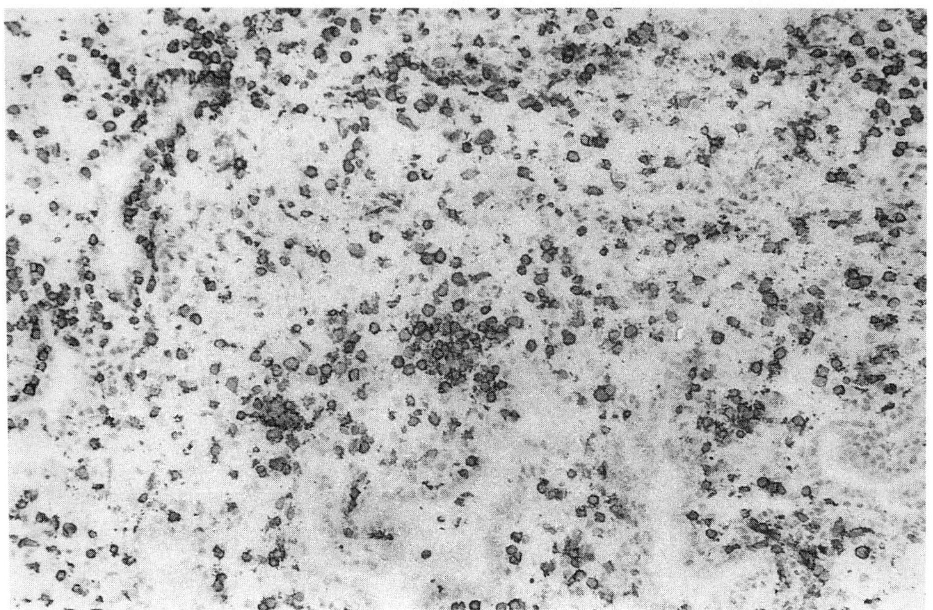

Figure 2. Cryostat section of first trimester decidua stained for leukocyte-common antigen (CD45) using an indirect immunoperoxidase technique. Note numerous leukocytes. Magnification X63.

Table 1

Leucocytes In Human Endometrium

	NON-PREGNANT			PREGNANT		
	Proliferative	Early secretory	Late secretory	1st trimester	2nd trimester	3rd trimester
Macrophage	+++	+++	+++	+++	+++	+++
T lymphocyte CD2+ CD3+	+	+	+	+	+	+
Granulated lymphocyte CD2+ CD3-	-	+	+++	+++	+	+
B lymphocyte	(+)	(+)	(+)	(+)	(+)	(+)

Despite the close morphological resemblance of decidual granulated lymphocytes to natural killer (NK) cells, only rare cells in early pregnancy decidua label with the 'classical' NK cell markers, Leu-7 and Leu-11 (CD16). However, numerous cells in decidua label particularly intensely with NKH-1, a mAb which is reactive with peripheral blood large granulated lymphocytes (LGL) including NK cells (Ritson and Bulmer, 1987a). Double immunoenzymatic labeling studies have suggested that almost all CD2-positive, CD3-negative cells in decidua are NKH-1-positive but an additional population of NKH-1-positive, CD2-negative cells may also be present (Ritson and Bulmer, 1987a; Ritson and Bulmer, unpublished results).

Other minor leukocyte subtypes have been identified within human decidua; these include CD3-positive T lymphocytes which are mainly of the CD8-positive suppressor/cytotoxic subset, B lymphocytes, 'classical' NK cells and scanty polymorphs particularly in the superficial zones (Bulmer and Sunderland, 1984; Bulmer and Johnson, 1984). In contrast to observations in several animal species, immunoglobulin-containing plasma cells are uncommon in non-pregnant and pregnant human endometrium (Bulmer et al., 1986a).

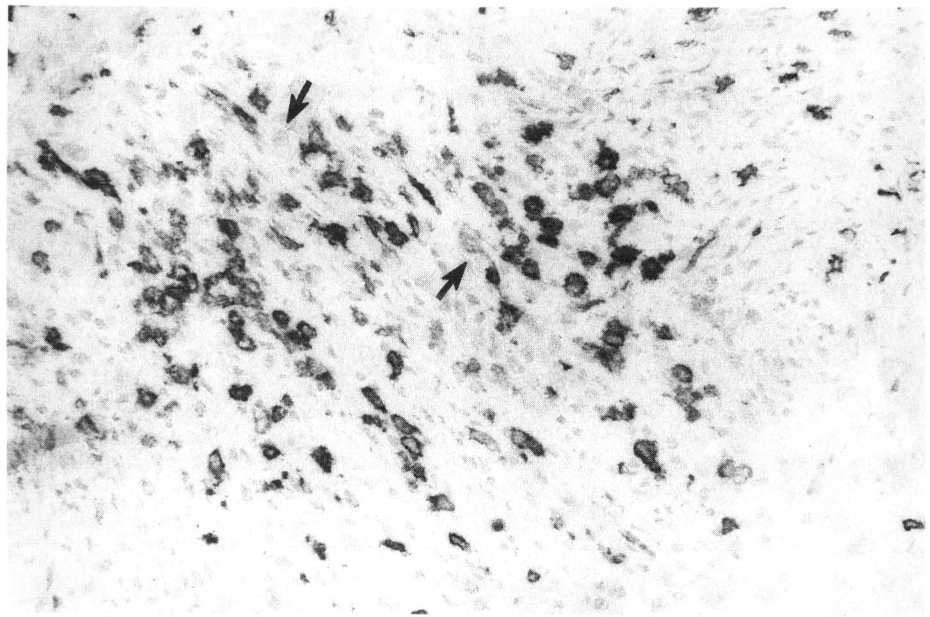

Figure 3. Cryostat sections of first trimester decidua labeled for class II MHC antigens with an indirect immunoperoxidase technique. Note MHC class II + macrophages of varying size and shape, often close to extravillous trophoblast (arrowed). Magnification X160.

Decidual Leukocytes Related To Endometrial Glands

In early pregnancy, endometrial glands appear to decrease their expression of class I MHC antigens (Johnson and Bulmer, 1984) and are associated with an unusual mononuclear cell response (Bulmer and Johnson, 1985b). Large leukocytic aggregates may be identified adjacent to the glands on routine H&E stained slides; these include macrophages and CD2-positive, CD3-negative granulated lymphocytes. In addition, immediately beneath the epithelium there are scattered single cells reactive for the CD1 antigen, which is normally detected on cortical thymocytes and Langerhans cells (Bulmer and Johnson, 1985b). Similar CD1-positive cells have been noted adjacent to glands in non-pregnant endometrium (Kamat and Isaacson, 1987). It has been suggested that only isolated gland remnants can be observed after the first half of pregnancy at light microscope level, thus raising the possibility of an immune-mediated attack on glandular structures. However, numerous glands have been identified in normal third trimester placental bed biopsies using immunohistochemical markers of epithelial cells such as HMFG1 or CAM 5.2, although the gland structures are often distorted or attenuated (Bulmer et al., 1986b).

Kinetics Of Decidual Granulated Lymphocytes

Endometrial stromal granulocytes have been considered to be a fully differentiated cell type derived from endometrial stromal cells. They are scanty in proliferative endometrium but gradually increase in number during the secretory phase, reaching maximal numbers premenstrually, particularly in areas of pseudodecidual change. EGs are prominent in decidua in early pregnancy but after the first trimester they decline in numbers (Hamperl and Hellweg, 1958; Kazzaz, 1972; Dallenbach-Hellweg, 1981). Their source remains unclear, as does the explanation for their rapid decline in numbers. Mitotic figures can be clearly seen in EGs, detected with the phloxine tartrazine stain, in normal non-pregnant endometrium, particularly in the late secretory phase of the cycle. EGs in early pregnancy decidua also show mitotic activity but mitoses are considerably less abundant than in late secretory phase non-pregnant endometrium.

Ki67 is a mAb which is reactive with the nucleus of dividing cells at all stages of the cell cycle, excluding G_o (Gerdes et al., 1983; 1984). As expected, this mAb labels gland epithelial cells and stromal cells in proliferative endometrium but in the secretory phase of the menstrual cycle numerous small stromal cells are Ki67-positive while glands are unreactive (Pace and Bulmer, unpublished). Scattered small Ki67-positive stromal cells have also been observed in smaller numbers in first trimester decidual stroma (Figure 4). Double immunoenzymatic labeling techniques empolying Ki67 with either Dako-T11 (CD2) or NKH-1 have shown that the majority of the Ki67-positive stromal cells in secretory phase endometrium and early pregnancy decidua are also CD2-positive and/or NKH-1-positive and most probably represent EGs.

It is therefore apparent that EGs divide within human endometrium, probably accounting for their dramatic increase in numbers in premenstrual endometrium. Although it is established that EGs are a form af granulated

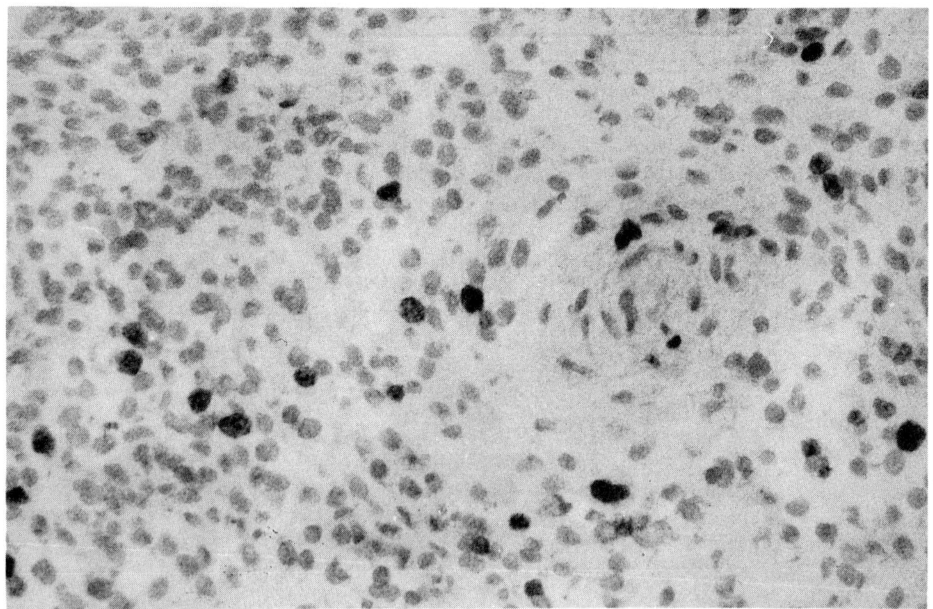

Figure 4.   Cryostat section of first trimester decidua labeled with Ki67 (proliferating cells) using an avidin-biotin technique. Note nuclear reactivity in many lymphoid cells around spiral artery. Magnification X250.

lymphocyte, it is not known whether they differentiate in situ within the endometrium from primitive bone marrow-derived precursor cells. Granulated metrial gland (GMG) cells in rat and mouse have been considered to be analogous to human EGs and evidence from ultrastructural studies suggests that they differentiate within the uterus from lymphocyte-like precursors (Peel and Bulmer, 1977; Stewart and Peel, 1977). In preliminary electron microscope studies of human EGs no comparable sequence was seen (Bulmer, 1985).

Immunohistochemical studies of non-pregnant human endometrium have revealed a population of CD2-positive CD3-negative stromal lymphocytes in the late secretory phase of the menstrual cycle. These cells are antigenically similar to those observed in early pregnancy and are particularly common in areas of pseudodecidual change (Bulmer et al., 1987b). Characterization of endometrial leukocyte populations at various stages of the cycle has not provided sufficient information to deduce whether endometrial granulated lymphocytes differentiate in situ under hormonal or other influences or whether they influx to endometrium in the late secretory phase of the cycle via the bloodstream as fully differentiated cells. These possiblities can be addressed by studies of stromal lymphocytic populations in endometrium undergoing decidualisation in an in vitro system. By this approach it should be possible to determine whether EG precursors are present in proliferative endometrium and whether they can be induced to differentiate during in vitro decidualization.

However, EGs may be detected within blood vessel lumena in late secretory phase endometrium and in early pregnancy decidua using a phloxine tartrazine stain (Bulmer, unpublished). A small subpopulation of NK cells in peripheral blood possess a similar surface antigenic phenotype to endometrial granulated lymphocytes (CD2+, CD16-, NKH1++) (Lanier et al., 1986). Changes in peripheral blood NK cells during pregnancy have been reported (Gregory et al., 1987) but it would be of interest to focus on any changing profile of this latter small specific NK subpopulation in peripheral blood during the menstrual cycle and early pregnancy when they are present in maximal numbers within the endometrium.

Function Of Decidual Granulated Lymphocytes

The distribution of granulated lymphocytes in early pregnancy decidua at a time when many pregnancies fail and in secretory endometrium from the time of implantation implies that they may play an important role in implantation and early placental development. Many previous functional studies of human and murine decidua have been hampered by poor documentation of cells under investigation so that results of different laboratories cannot be adequately correlated. Separation and purification of a particular cell type from a tissue fragment involves initial tissue dispersal. It is worth noting that the proportion of various cell types in cell suspensions depends on the method of disaggregation of decidual tissue (Ritson and Bulmer, 1987b). Mechanical methods of disaggregation yield a larger percentage of true decidualized stromal cells and fewer bone-marrow-derived cells, whereas collagenase digestion appears to select leukocytic populations. Similar results have been reported for murine decidua (Gambel et al., 1985). Initial tissue preparation may thus have a dramatic and important effect on subsequent functional studies. The duration of enzyme digestion and exposure of cells to enzyme may also alter their surface characteristics and lead to difficulties in immunohistochemical characterization of cell types within the suspensions.

Decidua contains many different cell types and pure populations are difficult to achieve. Decidual granulated lymphocytes (endometrial stromal granulocytes) (DGL) have been partially purified using a discontinuous Nycodenz (Nygaard A/S, Norway) gradient. This method yields suspensions enriched up to 80% for CD2-positive, NKH-1-positive cells but purification clearly requires improvement. The large majority of cells in smears are granulated when stained with toluidine blue or May Grünwald Giemsa. Decidual cell suspensions enriched for CD2-positive lymphocytes have been tested in a variety of functional assays.

DGLs fail to proliferate in response to either phytohemagglutinin or concanavalin A, even in the presence of a high dose of interleukin 2. The DGL fraction did, however, show significant cytotoxic activity in a standard K562 chromium release assay at all effector: target ratios tested. Levels of killing were low compared with peripheral blood lymphocytes but DGL consistently showed higher cytotoxicity than unfractionated decidual cell suspensions.

Interest in local decidual suppression and reports of granulated lymphoid suppressor cells in mouse decidua (Clark et al., 1986a; 1986b) led to the suggestion that decidual granulated lymphocytes may be natural suppressor (NS) cells. NS cells are large granular lymphocytes which show non-MHC-dependent immunosuppressive activity (Maier et al., 1986). However, cell suspensions enriched for decidual granulated lymphocytes (DGLs) have failed to show consistent suppression of lymphocyte mitogen responses and mixed lymphocyte reactions. When paired specimens of unfractionated decidual cell suspensions and semi-purified DGLs were compared, DGLs consistently showed lower levels of suppression than the corresponding unfractionated decidual cell suspension. These results would imply that suppressor activity in human decidua cannot be attributed solely to CD2-positive granulated lymphocytes.

Thus, in functional studies in vitro, decidual granulated lymphocytes failed to proliferate in response to mitogens and IL2, failed to exhibit suppressor activity and were poor effectors in a K562 cytotoxicity assay (Ritson and Bulmer, unpublished). Classical natural killer (NK) cells may require induction of IL2 receptors by gamma IFN to induce proliferation, and responses of DGL to gamma IFN and high doses of IL2 are currently being studied. The lack of suppressor activity in DGL was disappointing in view of numerous reports of suppression by cells in both human and murine decidua (see below). It has been suggested that collagenase digestion of tissues may lead to loss of suppressor activity in murine decidua (Clark et al., 1986b) but recent studies of immunosuppression by decidual cell suspensions prepared by digestion with various enzymes have not shown any significant effect on suppressor reactivity of enzymatic dispersal compared with mechanically disaggregated tissue. The only function demonstrated for DGLs was a low level of cytotoxic activity in a standard K562 chromium release assay for NK activity. Lanier et al. (1986) have described a small subset of NK cells in peripheral blood that shows surface antigenic features similar to decidual granulated lymphocytes and in particular is brightly NKH-1-positive. These cells are also poor effectors in a K562 cytotoxicity assay. It is possible that at least a proportion of DGLs is comparable to this small peripheral blood NK subset.

Suppressor Activity In Decidua

There are numerous reports of immunological suppressor activity in murine and human decidua. Supernatants produced by explants of human decidua suppress lymphocyte responses to mitogens and one-way mixed lymphocyte reactions (Golander et al., 1981; Nakayama et al., 1985). Several partially purified cell types have been shown to mediate suppressor activity.

Both human and murine decidualized endometrial stromal cells have been reported to derive ultimately from bone marrow precursors and to show suppressor activity, apparently mediated by prostaglandin E_2 (Lala et al., 1986). However, decidual cells are difficult to characterize with certainty in cell suspensions by morphological or phenotypic criteria and other cell types could be responsible for immunosuppression. Furthermore, the collagenase digestion technique used for initial dispersal of human decidual tissue selects against decidualized endometrial stromal cells and in favor of bone marrow derived cells (Ritson and Bulmer, 1987b).

Decidual macrophages have also been implicated in decidual immunosuppression in both murine and human systems (Lala et al., 1986; Tawfik et al., 1986), also mediated by prostaglandin E_2. Pregnancy loss in mice treated with indomethacin, which blocks effects of prostaglandin E_2, has been proposed as evidence for prostaglandin-mediated suppression of immune responses (Lala et al., 1986). Decidual macrophages in mouse and human may also function as antigen presenting cells (Searle et al., 1986; Oksenberg et al., 1986). The role of decidual macrophages remains unclear: the existence of functionally distinct populations cannot be excluded and more detailed phenotypic characterization of macrophages both in normal and abnormal pregnancy tissues and in functional studies may be valuable.

Immune suppression in early murine decidua has been associated with a small granulated non-T lymphocyte which bears Fc receptors but is distinct from NK cells (Clark et al., 1986a; Slapsys et al., 1986). These cells have been reported to act by suppression of lymphocyte responses to IL2, and it is of interest that administration of IL2 to pregnant mice may cause pregnancy loss (Tezebwala and Johnstone, 1986). Deficiency of granulated lymphoid suppressor cells has been reported in spontaneous pregnancy resorption in Mus musculus/Mus caroli xenogeneic pregnancies and CBA X DBA/2 matings (Clark et al., 1986a). Evidence for a suppressor lymphocyte system in human decidua is scanty. 'Large' and 'small' suppressor cells have been separated from early human decidua but they have not been fully characterized (Daya et al., 1985) and analogy with the murine granulated lymphocytes is speculative. Reports of MHC-independent natural suppressor (NS) cells which have the morphological features of large granulated lymphocytes has raised the possibility that decidual granulated lymphocytes are natural suppressor cells. However, functional studies do not support the proposal that human DGL mediate immunosuppression and higher levels of suppression were noted in supernatants obtained from unfractioned decidual cells and decidual explants than those from semi-purified DGL (Ritson and Bulmer, unpublished).

Endometrial gland epithelial cells separated from non-pregnant endometrium have also been reported to possess immunosuppressive activity, which is maximal in the secretory phase (Johnson et al., 1987). Epithelial cells have not been separated from decidualized endometrium but epithelial cells identified by labeling with HMFG1, an epithelial membrane marker, and with CAM 5.2, directed against low molecular weight cytokeratins are a common contaminant of 'decidual lymphocytes' prepared by density gradient centrifugation over Ficoll-hypaque (Ritson and Bulmer, unpublished).

Uterine Granulated Cells In Other Species

Granulated stromal cells are a prominent feature in decidualized endometrium in early human pregnancy and evidence now indicates that they are granulated lymphocytes. Human endometrial stromal granulocytes have been considered to be analogous to granulated metrial gland (GMG) cells which accumulate in decidua and metrial gland of rat, mouse, and hamster. Initial reports that GMG cells in rat are stromal cells which secrete relaxin were refuted, and there is now convincing evidence that both rat and mouse GMG cells are derived from bone marrow precursors (Peel et al., 1983; Mitchell and Peel, 1984).

The function of GMG cells remains uncertain despite numerous different suggestions. The granules in rat GMG cells have been shown to contain perforins, an observation which has led to the proposal that they are NK cells (Parr and Parr, 1987), although neither in mouse nor rat do GMG cells have the morphology typical of large granular lymphocytes. Stewart (1984) commented on the close relationship between GMG cells and labyrinthine trophoblast in the mouse placenta and recent studies in vitro suggest that mouse GMG cells may be directly cytotoxic to occasional labyrinthine trophoblast cells (Stewart and Mukhtar, 1988). The effects of human DGL on trophoblast cells have not been investigated but it is noteworthy that DGL may show low but significant cytotoxic activity.

Despite their long historical association, the relationship of human decidual granulated lymphocytes to GMG cells in rat and mouse is not clear. Morphologically the cells show a closer resemblance to the population of granulated lymphoid suppressor cells in murine decidua described by Clark et al. (1986a). However, any extrapolation between species must remain purely speculative. Classical NK cells have also been described in murine decidua (Croy et al., 1985). The precise role of decidual NK cells in vivo is a source of some speculation but is worthy of further investigation.

Granulated cells have also been described in species which do not exhibit hemochoral placentation. Among these are the globule leukocytes observed in sheep pregnancy which have recently been characterized as granulated lymphocytes (Lee et al., 1987).

Thus, although granulated cell populations have been described in several species, their inter-relationships are not clear. It is possible that phenotypically distinct populations may represent cells in different stages of differentiation. In vitro studies have provided little information about in vivo function and two phenotypically distinct cell types may perform similar functions both in vitro and in vivo.

Decidual Leukocytes In Abnormal Pregnancy

Spontaneous Abortion

Of all pregnancy disorders, spontaneous abortion most closely resembles graft rejection. Recently the suggestion was made that recurrent spontaneous miscarriage may be due to immunological mechanisms and may be treated by immunization with paternal or allogeneic leukocytes or trophoblast membrane preparations (Taylor and Faulk, 1981; Mowbray et al., 1985; Beer et al., 1987; Johnson et al., 1986). However, there have been few attempts to characterize either trophoblast phenotype or decidual leukocytes in aborted pregnancy tissues. A major problem in any such studies is the intense inflammatory cell infiltrate observed in many specimens at the time of uterine evacuation; leucocytes which may have mediated immune attack on the fetoplacental unit cannot be distinguished from those resulting from secondary inflammation.

Histopathological and functional studies of uteroplacental tissues in spontaneous abortion and pregnancy loss after in vitro fertilization and embryo transfer (IVF-ET) suggest that immunohistological studies may be worthwhile.

Khong et al. (1987) noted absence of trophoblast invasion and associated deficiency of physiological vascular changes in the placental bed in spontaneous abortion tissues. Similar absence of physiological vascular changes is well established in pregnancy-induced hypertension and fetal intrauterine growth retardation; Khong et al. (1987) suggested that there may a continuum of pregnancy failure, sharing a form of defective hemochorial placentation but not necessarily indicating a common pathogenesis.

Characterization of decidual leukocytes in recurrent spontaneous abortion has been neglected. Nebel et al. (1986) noted dense lymphocytic infiltrates in deep decidua and around blood vessels in early clinical abortions occuring 22-30 days following embryo transfer in an IVF-ET program. However, cells were not characterized, nor were functional studies performed. Clark et al. (1987) have recently reported deficiency of mononuclear cells with large phloxinophilic cytoplasmic granules and additional presence of large granular lymphocytes with small cytoplasmic granules in four pregnancies aborting between 8 and 21 week gestation; no criteria were given for making the important distinction between these two cell types.

Spontaneous abortion is a potentially fruitful area for study of decidual leukocytes and their functions but investigations will continue to be hampered by the generally poor quality of tissues available at the time of endometrial evacuation.

Maternal Pregnancy-Induced Hypertension

Several observations have supported the concept that pregnancy-induced hypertension (pre-eclampsia) may be due to a disordered materno-fetal relationship (Redman, 1980). Physiological vascular changes in pregnancy-induced hypertension are restricted to decidual segments of spiral arteries due to absence of endovascular trophoblast migration into myometrial segments (Robertson et al., 1986). More recently, Khong et al. (1986) have suggested that physiological changes are absent throughout the entire length of some spiral arteries. These changes raise the possibility of an abnormal materno-fetal interaction in the placental bed in pre-eclamptic pregnancy but the nature of such a defective interaction is not known. Khong (1987) characterized decidual leukocytes in normal and pre-eclamptic pregnancies and noted no difference in quantity or type of infiltrate. However, all studies were performed on third trimester tissues, whereas any defective interaction causing pregnancy-induced hypertension may arise in the second trimester or earlier, well before the clinical features are manifest.

Several other abnormal pregnancies including intrauterine growth retardation, systemic lupus erythematosus and placenta accreta have been suggested to result from an abnormal interaction between maternal uterine cells and fetal extravillous trophoblast in the placental bed. Decidual leukocytes and function have not been extensively studied in these conditions but such investigations may be worthwhile.

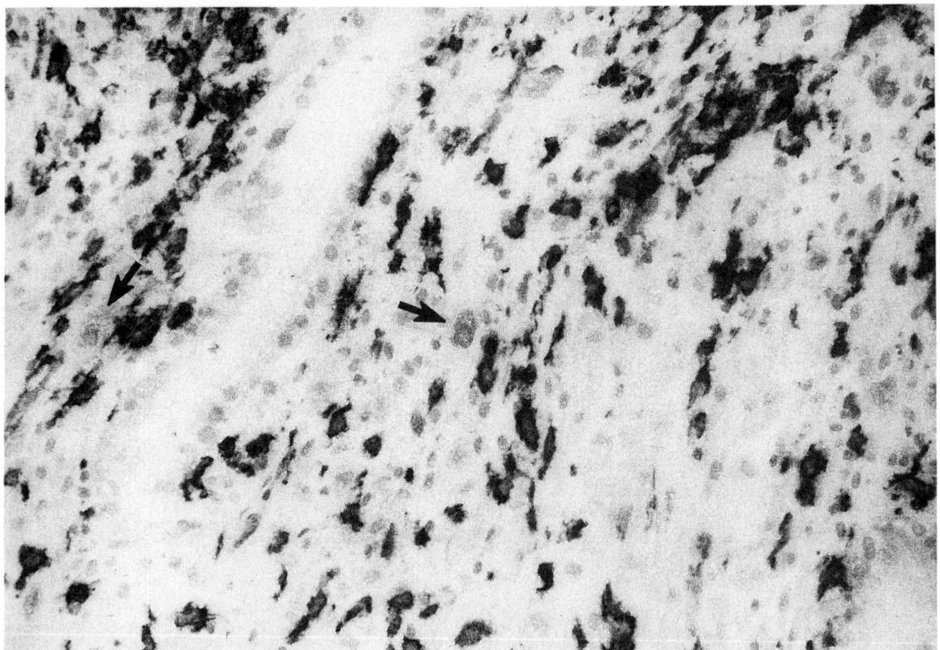

Figure 5. Cryostat section of implantation site from a partial molar pregnancy labeled for MHC class II antigens using an indirect immunoperoxidase technique. Note numerous MHC class II + macrophages closely associated with trophoblast cells (arrowed). Magnification x 250.

Ectopic Pregnancy

Ectopic pregnancy occurs in approximately 1 in 200 conceptions and implantation is usually in the fallopian tube. Placental development proceeds until mechanical complications supervene, usually necessitating surgical removal. Decidualization may occur at the tubal implantation site but is usually focal and patchy and endometrial stromal granulocytes/decidual granulated lymphocytes are detectable only in these areas of decidualization (Bulmer et al., 1987c). The large majority of leukocytes at the ectopic tubal implantation site have been characterized as Leu-M3-positive, class II MHC-positive macrophages: the large population of CD2-positive CD3-negative granulated lymphocytes detectable in decidua in normal pregnancy cannot be detected in the absence of decidualization (Bulmer et al., 1987c; Earl et al., 1987). However, the endometrium undergoes decidualization in ectopic pregnancy and appears to contain leukocyte populations analogous to those in first trimester intrauterine gestation with large populations of macrophages and CD2+CD3-lymphocytes (Bulmer et al., 1987c). Any essential role for decidual granulated lymphocytes in normal control of trophoblast invasion or in vivo immunosuppression would therefore require secretion of a soluble factor which is capable of exerting an effect at a distant site. In contrast, macrophages are prominent at the local implantation site and are often closely associated with trophoblast cells.

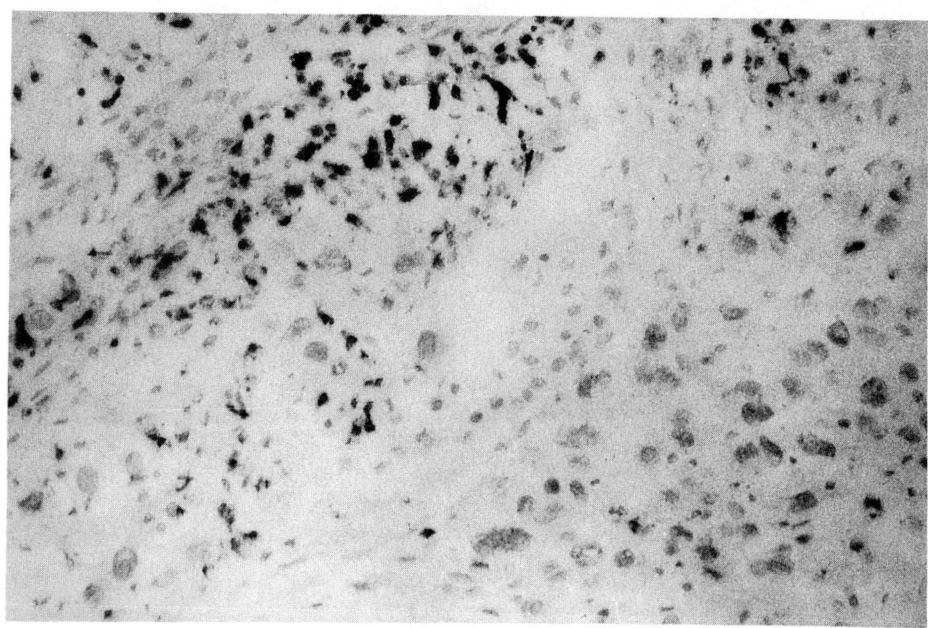

Figure 6. Cryostat section of uterine tissues removed for gestational choriocarcinoma stained with Dako-TII (CD2) by an indirect immunoperoxidase method showing T lymphocytes clustered around the tumor margin. Magnification X250.

Trophoblast Neoplasia

Immunohistochemical studies of complete and partial molar pregnancies have failed to detect any major discrepancies between trophoblast antigens in hydatidiform mole and invasive mole compared with normal pregnancies of comparable gestation. Choriocarcinoma shows trophoblast heterogeneity and the phenotypically distinct trophoblast populations observed in normal pregnancy may all be detected in choriocarcinoma tissues (Bulmer et al., 1988b).

Immunohistological characterization of cells at the molar implantation site have shown a four-fold increase in CD3-positive T lymphocytes, consisting predominantly of the CD8-positive suppressor/cytotoxic subtype (Kabawat et al., 1985b). CD2-postive, CD3-negative lymphocytes have also been reported and correspond to the endometrial stromal granulocytes observed in molar pregnancy decidua (Bulmer et al., 1988b). Macrophages are also prominent at the molar implantation site and they are often closely associated with extravillous trophoblast cells (Figure 5).

Mononuclear cell infiltrates have been noted in association with gestational choriocarcinomas (Elston, 1978) and immunohistological studies have shown that these are composed predominantly of CD3-positive CD2-positive T lymphocytes (Figure 6) and Leu-M3-positive MHC class II-positive macrophages (Bulmer et al., 1988c). Gestational choriocarcinoma arising after normal term pregnancy may be associated with residual decidualization but cell types present have not yet been documented. Any maternal leukocytic populations associated with the rare placental site trophoblastic tumor have also not been characterized.

SUMMARY

Many different leukocytic types can be identified in human pregnancy decidua; the two major leukocyte populations consist of macrophages and granulated lymphocytes. Macrophages are often very closely associated with extravillous trophoblast, although they are also common in decidua parietalis. Their function may be concerned with antigen presentation, immunosuppression or phagocytosis. Granulated lymphocytes (CD2+, CD3-, NKH1++) are prominent in early pregnancy decidua and are analogous with endometrial stromal granulocytes or Körnchenzellen. They appear to divide within decidual tissue itself. In studies of semi-purified suspensions of decidual granulated lymphocytes they failed to respond to mitogens or interleukin-2, they did not appear to mediate immune suppressor activity but they did show low levels of cytotoxicity in a standard K562 chromium release assay. Decidual granulated lymphocytes show a similar antigenic phenotype to a recently described NK cell subset in peripheral blood which are also very poor effectors in standard NK cell assays. Granulated cells with a variety of morphological features and functions have been described in decidua in other species but analogy with human decidual granulated lymphocytes is uncertain. Studies of decdicual leucocytes in abnormal pregnancy to date are limited but more extensive investigation would be worthwhile.

ACKNOWLEDGEMENTS

This work was supported by grants from Yorkshire Regional Health Authority and from Birthright.

REFERENCES

Beer, A.E., Shekar, S.S., Quebbeman, J.F., and Zhu, C. (1987) Paternal and nonpaternal leukocyte immunization in women with recurrent spontaneous abortions: immune responses and subsequent pregnancy outcome. In: *Reproductive Immunology: Materno-Fetal Relationship*, (ed.) G. Chaouat, Paris: Inserm, pp. 161-178.

Bulmer, J.N. and Sunderland, C.A. (1983) Bone marrow origin of endometrial granulocytes in the early human placental bed. *J. Reprod. Immunol.* 5, 383-387.

Bulmer, J.N. and Johnson, P.M. (1984) Macrophage populations in the human placenta and amniochorion. *Clin. Exp. Immunol.* 57, 393-403.

Bulmer, J.N. and Sunderland, C.A. (1984) Immunohistological characterisation of lymphoid cell populations in the early human placental bed. *Immunol.* 52, 349-357.

Bulmer, J.N. (1985) Studies on the immunology of the human placenta in normal and pathological pregnancy. PhD Thesis, University of Bristol.

Bulmer, J.N. and Johnson, P.M. (1985a) Antigen expression by trophoblast populations in the human placenta and their possible immunobiological relevance. *Placenta* 6, 127-140.

Bulmer, J.N. and Johnson, P.M. (1985b) Immunohistological characterization of the decidual leucocytic infiltrate related to endometrial glands in early human pregnancy. *Immunol.* 55, 35-44.

Bulmer, J.N. and Johnson, P.M. (1986) The T-lymphocyte population in first-trimester human decidua does not express the interleukin-2 receptor. *Immunology* 58, 685-687.

Bulmer, J.N., Hagin, S.V., Browne, C.M., and Billington, W.D. (1986a) Localization of immunoglobulin-containing cells in human endometrium in the first trimester of pregnancy and throughout the menstrual cycle. *Eur. J. Obstet. Gynecol. Reprod. Biol.* 23, 31-44.

Bulmer, J.N., Wells, M., Bhabra, K., and Johnson, P.M. (1986b) Immunohistological characterization of endometrial gland epithelium and extravillous fetal trophoblast in third trimester human placental bed tissues. *Br. J. Obstet. Gynaecol.* 93, 823-832.

Bulmer, J.N., Hollings, D., and Ritson, A. (1987a) Immunocytochemical evidence that endometrial stromal granulocytes are granulated lymphocytes. *J. Pathol.* 153, 281-287.

Bulmer, J.N., Johnson, P.M., and Bulmer, D. (1987b) Leukocyte populations in human decidua and endometrium. In: *Immunoregulation and Fetal Survival,* (eds.) T.J. Gill, III and T.G. Wegmann, New York: Oxford University Press, pp. 111-134.

Bulmer, J.N., Ritson, A., Earl, U., and Hollings, D. (1987c) Immunocompetent cells in human decidua. In: *Reproductive Immunology: Materno-Fetal Relationship,* (ed.) G. Chaouat, Paris: Inserm, pp. 89-100.

Bulmer, J.N., Smith, J.C., and Morrison, L. (1988a) Expression of class II MHC gene products by macrophages in human utero-placental tissues. *Immunology* 63, 707-714.

Bulmer, J.N., Johnson, P.M., Sasagawa, M., and Takeuchi, S. (1988b) Immunohistochemical studies of fetal trophoblast and maternal decidua in hydatidiform mole and choriocarcinoma. *Placenta* 9, 183-200.

Bulmer, J.N., Smith, J., Morrison, L., and Wells, M. (1988c) Maternal and fetal cellular relationships in the human placental basal plate. *Placenta* 9, 237-246.

Clark, D.A., Slapsys, R., Chaput, A., Walker, C., Brierley, J., Daya, S., and Rosenthal, K.L. (1986a) Immunoregulatory molecules of trophoblast and decidual suppressor cell origin at the materno-fetal interface. *Am. J. Reprod. Immunol. Microbiol.* 10, 100-104.

Clark, D.A., Brierley, J., Slapsys, R., Daya, S., Damji, N., Chaput, A., and Rosenthal, K. (1986b) Trophoblast-dependent and trophoblast-independent suppressor cells of maternal origin in murine and human decidua. In: *Reproductive Immunology,* (eds.) D.A. Clark and B.A. Croy, Amsterdam: Elsevier Science Publishers B.V, pp. 219-226.

Clark, D.A., Mowbray, J., Underwood, J., and Liddell, H. (1987) Histopathologic alterations in the decidua in human spontaneous abortion: loss of cells with large cytoplasmic granules. *Am. J. Reprod. Immunol. Microbiol.* 13, 19-22.

Croy, B.A., Gambel, P., Rossant, J., and Wegmann, T.G. (1985) Characterization of murine decidual natural killer cells and their relevance to the success of pregnancy. *Cell. Immunol.* 93, 315-326.

Dallenbach-Hellweg, G. (1981) The normal histology of the endometrium. In: *Histopathology of the Endometrium,* Berlin: Springer-Verlag, pp. 22-28.

Daya, S., Clark, D.A., Devlin, C., Jarrell, J., and Chaput, A. (1985) Preliminary characterization of two types of suppressor cells in the human uterus. *Fertil. Steril.* 44, 778-785.

Earl, U., Lunny, D.P., and Bulmer, J.N. (1987) Leucocyte populations in ectopic tubal pregnancy. *J. Clin. Pathol.* 40, 901-905.

Ellis, S.A., Sargent, I.L., Redman, C.W.G., and McMichael, A.J. (1986) Evidence for a novel HLA antigen found on human extravillous trophoblast and a human choriocarcinoma cell line. *Immunology* 59, 595-601.

Elston, C.W. (1978) Trophoblastic tumours of the placenta. In: *Pathology of the Placenta,* (ed.), H. Fox, London: W.B. Saunders, pp. 368-425.

Gambel, P., Rossant, J., Hunziker, R.D., and Wegmann, T.G. (1985) Origin of decidual cells in murine pregnancy and pseudo-pregnancy. *Transplantation* 39, 443-445.

Gerdes, J., Schrowal, U., Lemke, M., and Stein, H. (1983) Production of a mouse monoclonal antibody reactive with a human nuclear antigen associated with cell proliferation. *Int. J. Cancer* 31, 13-20.

Gerdes, J., Lemke, H., Baisch, H., Wacker, H.-H., Schwab, U., and Stein, H. (1984) Cell cycle analysis of a cell proliferation-associated human nuclear antigen defined by the monoclonal antibody, Ki67. *J. Immunol.* 133, 1710-1715.

Golander, G., Zakuth, V., Shechter, Y., and Spirer, Z. (1981) Suppression of lymphocyte reactivity in vitro by a soluble factor secreted by explants of human decidua. *Eur. J. Immunol.* 11, 849-851.

Gregory, C.D., Lee, H., Scott, I.V., and Golding, P.R. (1987) Phenotypic heterogeneity and recycling capacity of natural killer cells in normal human pregnancy. *J. Reprod. Immunol.* 11, 135-145.

Hamperl, H. and Hellweg, G. (1958) Granular endometrial stroma cells. *Obstet. Gynecol.* 11, 379-387.

Hsi, B.-L., Yeh, C.-J.G., and Faulk, W.P. (1984) Class I antigens of the major histocompatibility complex on cytotrophoblast of human chorion laeve. *Immunol.* 52, 621-629.

Johnson, P.M. and Bulmer, J.N. (1984) Uterine gland epithelium in human pregnancy often lacks detectable maternal MHC antigens but does express fetal trophoblast antigens. *J. Immunol.* 132, 1608-1610.

Johnson, P.M., Chia, K.V., and Risk, J.M. (1986) Immunological question marks in recurrent spontaneous abortion. In: *Reproduction Immunology*, (eds.), D.A. Clark, and B.A. Croy, Amsterdam: Elsevier Science Publishers B.V, pp. 239-245.

Johnson, P.M., Risk, J.M., Bulmer, J.N., Niewola, Z., and Kimber, I. (1987) Antigen expression at human maternofetal interfaces. In: *Immunoregulation and Fetal Survival*, (eds.), T.J. Gill, III and T.G. Wegmann, New York: Oxford University Press, pp. 181-196.

Kabawat, S.E., Mostoufi-Zadeh, M., Berkowitz, R.S., Driscoll, S.G. and Bhan, A.K. (1985a) Implantation site in normal pregnancy. A study with monoclonal antibodies. *Am. J. Pathol.* 118, 76-84.

Kabawat, S.E., Mostoufi-Zadeh, M., Berkowitz, R.S., Driscoll, S.G. Goldstein, D.P., and Bhan, A.K. (1985b) Implantation site in complete molar pregnancy: a study of immunologically competent cells with monoclonal antibodies. *Am. J. Obstet. Gynecol.* 152, 97-99.

Kamat, B.R. and Isaacson, P.G. (1987) The immunocytochemical distribution of leukocyte subpopulations in human endometrium. *Am. J. Pathol.* 127, 66-73.

Kazzaz, B.A. (1972) Specific endometrial granular cells. A semi-quantitative study. *Eur. J. Obstet. Gynecol.* 3, 77-84.

Khong, T.Y., DeWolf, F., Robertson, W.B., and Brosens, I. (1986) Inadequate maternal vascular response to placentation in pregnancies complicated by pre-eclampsia and by small-for-gestational age infants. *Br. J. Obstet. Gynaecol.* 93, 1049-1059.

Khong, T.Y. (1987) Immunohistologic study of the leukocytic infiltrate in maternal uterine tissues in normal and pre-eclamptic pregnancies at term. *Am. J. Reprod. Immunol. Microbiol.* 15, 1-8.

Khong, T.Y., Liddell, H.S., and Robertson, W.B. (1987) Defective haemochorial placentation as a cause of miscarriage: a preliminary study. *Br. J. Obstet. Gynaecol.* 94, 649-655.

Lala, P.K., Parhar, R.S., Kearns, M., Johnson, S., and Scodras, J.M. (1986) Immunological aspects of the decidual response. In: *Reproductive Immunology*, (eds.), D.A. Clark and B.A. Croy, Amsterdam: Elsevier Science Publishers B.V, pp. 190-198.

Lanier, L.L., Le, A.M., Civin, C.I., Loken, M.R., and Phillips, J.H. (1986) The relationship of CD16 (leu 11) and leu 19 (NKH1) antigen expression on human peripheral blood NK cells and cytotoxic T lymphocytes. *J. Immunol.* 136, 4480-4486.

Lee, C.S., Gogolin-Ewins, K., and Brandon, M.R. (1987) Identification of a unique lymphocyte subpopulation in the sheep uterus. *Immunol.* 63, 157-164.

Maier, T., Holda, J.H., and Claman, H.N. (1986) Natural suppressor (NS) cells: members of the LGL regulatory family. *Immunol. Today* 7, 312-315.

Mitchell, B.S. and Peel, S. (1984) Identification of cells bearing leucocyte surface antigens in metrial gland tissue from rats of different gestational ages, strains or parities. *Immunol.* 53, 63-68.

Mowbray, J.F., Gibbings, C., Liddell, H., Reginald, P.W., Underwood, J.L., and Beard, R.W. (1985) Controlled trial of treatment of recurrent spontaneous abortion by immunisation with paternal cells. *Lancet* i, 941-943.

Nakayama, E., Asano, S., Kodo, H., and Mirra, S. (1985) Suppression of mixed lymphocyte reaction by cells of human first trimester pregnancy endometrium. *J. Reprod. Immunol.* 8, 25-31.

Nebel, L., Fein, A., Rudak, E., Blank, M., Mashiach, S., Dor, J., Lerran, D., and Goldman, B. (1986) Structural aspects of embryo failure following in-vitro fertilization and embryo transfer: Immune rejection or malimplantation. In: *Reproductive Immunology* , (eds.), D.A. Clark and B.A. Croy, Amsterdam: Elsevier Science Publishers B.V, pp. 227-235.

Nehemiah, J.L., Schnitzer, J.A., Schulman, H., and Novikoff, A.B. (1981) Human chorionic trophoblasts, decidual cells and macrophages; a histochemical and electron microscopic study. *Am. Obstet. Gynecol.* 140, 261-268.

Oksenberg, J.R., Mor Yosef, S., Persitz, E., Schenker, Y., Mozes, E., and Brautbar, C. (1986) Antigen presenting cells in human decidual tissue. *Am. J. Reprod. Immunol. Microbiol.* 11, 82-88.

Parr, E.L. and Parr, M.B. (1987) Localization of perforin in granulated metrial gland cells. *Am. J. Reprod. Immunol. Microbiol.* 14, 9.

Peel, S. and Bulmer, D. (1977) The fine structure of rat metrial gland in relation to the origin of the granulated cells. *J. Anat.* 123, 687-696.

Peel, S., Stewart, I.J., and Bulmer, D. (1983) Experimental evidence for the bone marrow origin of granulated metrial gland cells of the mouse uterus. *Cell Tissue Res.* 233, 647-656.

Pijnenborg, R., Dixon, G., Robertson, W.B., and Brosens, I. (1980) Trophoblast invasion of human decidua from 8-18 weeks of pregnancy. *Placenta* 1, 3-18.

Redman, C.W.G. (1980) Immunological aspects of eclampsia and pre-eclampsia. In: *Immunological Aspects of Reproduction and Fertility Control* (ed.) J.P. Hearn, pp. 83-103. Lancester:MTP Press.

Redman, C.W.G., McMichael, A.J., Stirrat, G.M., Sunderland, C.A., and Ting, A. (1984) Class I major histocompatibility complex antigens on human extravillous trophoblast. *Immunol.* 52, 457-468.

Ritson, A. and Bulmer, J.N. (1987a) Endometrial granulocytes in human dedicua react with a natural-killer (NK) cell marker, NKH1. *Immunol.* 62, 329-331.

Ritson, A. and Bulmer, J.N. (1987b) Extraction of leucocytes from human decidua. A comparison of dispersal techniques. *J. Immunol. Methods* 104, 231-236.

Robertson, W.B., Khong, T.Y., Brosens, I., DeWolf, F., Sheppard, B.L., and Bonnar, J. (1986) The placental bed biopsy: Review from three European centers. *Am. Obstet. Gynecol.* 155, 401-412.

Searle, R.F. (1986) Intrauterine immunization. In: *Reproductive Immunology* 1986 (eds.), D.A. Clark and B.A.Croy, Amsterdam: Elsevier Science Publishers B.V., pp. 211-218.

Slapsys, R.M., Richards, C.D., and Clark, D.A. (1986) Active suppression of host-versus-graft reaction in pregnant mice. VIII. The uterine decidua-associated suppressor cell is distinct from decidual NK cells. *Cell Immunol.* 99, 140-149.

Stern, P.L., Beresford, N., Friedman, C.I., Stevens, V.C., Risk, J.M., and Johnson, P.M. (1986a) Class I-like MHC molecules expressed by baboon placental syncytiotrophoblast. *J. Immunol.* 138, 1088-1091.

Stewart, I.J. and Peel, S. (1977) The structure and differentiation of granulated metrial gland cells of the pregnant mouse uterus. *Cell Tissue Res.* 184, 517-527.

Stewart, I. (1984) A morphological study of granulated metrial gland cells and trophoblast cells in the labyrinthine placenta of the mouse. *J. Anat.* 139, 627-638.

Stewart, I. and Mukhtar, D.D.Y. (1988) The killing of mouse trophoblast cells by granulated metrial gland cells in vitro. *Placenta* 9, 417-425.

Tawfik, O.W., Hunt, J.S., and Wood, G.W. (1986) Implication of prostaglandin E_2 in soluble factor-mediated immune suppression by murine decidual cells. *Am. J. Reprod. Immunol. Microbiol.* 12, 111-117.

Taylor, O.W., Hunt, J.S., and Wood, G.W. (1986) Prevention of recurrent abortions with leukocyte transfusions. *Lancet* ii, 68-70.

Tekelioglu-Uysal, M., Edwards, R.G., and Kisnisci, H.A. (1975) Ultrastructural relationships between decidua, trophoblast and lymphocytes at the beginning of human pregnancy. *J. Reprod. Fert.* 42, 431-438.

Tezabwala, B.U. and Johnstone, A.P. (1986) Effect of administration of interleukin-2 in pregnancy. *J. Reprod. Immunol. Suppl.* 147.

Wells, M., Hsi, B.-L., and Faulk, W.P. (1984) Class I antigens of the major histocompatibility complex on cytotrophoblast of the human placental basal plate. *Am. J. Reprod. Immunol.* 6, 167-174.

LIST OF CONTRIBUTORS

Ted L. Anderson
Department of Obstetrics and
    Gynecology
Vanderbilt University
School of Medicine
Nashville, Tennessee 37232

John D. Aplin
Research Floor
St. Mary's Hospital
Whitworth Park
Manchester M13 0JH United Kingdom

Henning M. Beier
Lehrstuhl für Anatomie and
    Reproduktionsbiologie
Medizinische Fakultät der RWTH
    Aachen
Melatener Strasse 211
D5100 Aachen, Federal Republic of
    Germany

M. Bernfield
Department of Pediatrics
School of Medicine
Stanford University
Stanford, California 94305

A. Bükers
Institute für Anatomie
RWTH Aachen
Melatener Strasse 211
D5100 Aachen, Federal Republic of
    Germany

Judith N. Bulmer
Department of Pathology
University of Leeds
Leeds LS2 9JT, United Kingdom

Patricia Calarco
Department of Anatomy
School of Medicine
University of California
San Francisco, California 94143

Daniel D. Carson
Department of Biochemistry
    and Molecular Biology
The University of Texas
M.D. Anderson Cancer Center
1515 Holcombe Blvd, Box 117
Houston, Texas 77030

Anna K. Charlton
Departments of Obstetrics and
    Gynecology, Biochemistry,
    and Molecular Biology
University of Manchester
Manchester M13 0JH United Kingdom

Daniel J. Chávez
Department of Anatomy
School of Medicine
Southern Illinois University
Carbondale, Illinois 62901

Y. Christiane
Laboratory of Biology
University of Liege
Tower of Pathology (B35)
Sart Tilman
B4020 Liege, Belgium

I. Classen-Linke
Institute für Anatomie
RWTH Aachen
Melatener Strasse 211
D5100 Aachen, Federal Republic of
    Germany

E. Crowley
Periodontology School of Dentistry
University of California
San Francisco, California 94143

C. Damsky
Periodontology School of Dentistry
Department of Anatomy
School of Medicine
University of California
San Francisco, California 94143

Hans-Werner Denker
Institute für Anatomie
Universitätsklinikum
Hufelandstr. 55
D4300 Essen 1, Federal Republic of
Germany

Anuradha Dutt
Department of Biochemistry
    and Molecular Biology
The University of Texas
M.D. Anderson Cancer Center
1515 Holcombe Blvd, Box 117
Houston, Texas 77030

H. Emonard
Laboratoire de Pathologie Cellulaire
CNRS UA 602
Institut Pasteur
F69365 Lyon, France

Susan J. Fisher
HSW 60-4, Box 0512
University of California
    at San Francisco
San Francisco California 94143

J.M. Foidart
Laboratory of Biology
University of Liege
Tower of Pathology (B35)
Sart Tilman
B4020 Liege, Belgium

J. Friedrich
Institute für Anatomie
RWTH Aachen
Melatener Strasse 211
D5100 Aachen, Federal Republic of
    Germany

Stanley R. Glasser
Department of Cell Biology
Baylor College of Medicine

Houston, Texas 77030

Anna Grabowska
Division of Cellular and
    Genetic Pathology
Department of Pathology
University of Cambridge
Cambridge, England

Ruth Grümmer
Institut für Anatomie der
    RWTH Aachen
Melatener Strasse 211
D5100 Aachen, Federal Republic of
    Germany

L. Hartman
Divisions of Oral Biology
School of Dentistry
University of California
San Francisco, California 94143

Mats Hjortberg
Department of Human Anatomy
Biomedical Center, Box 571
S75123 Uppsala, Sweden

Axel Hochfeld
Lehrstuhl für Anatomie and
    Reproduktionsbiologie
Medizinische Fakultät der RWTH
    Aachen
Melatener Strasse 211
D5100 Aachen, Federal Republic of
    Germany

Gary D. Hodgen
The Jones Institute for Reproductive
    Medicine
Laboratory for Cellular and Molecular
    Biology
Department of Obstetrics and
    Gynecology
Eastern Virginia Medical School
Norfolk, Virginia 23507

Loren H. Hoffman
Department of Cell Biology
Vanderbilt University
School of Medicine
Nashville, Tennessee 37232

Hans-Peter Hohn
Institut für Anatomie der
    RWTH Aachen
Melatener Strasse 211
D5100 Aachen, Federal Republic of
    Germany

Ashley King
Division of Cellular and
    Genetic Pathology
Department of Pathology
University of Cambridge
Cambridge, England

Y.W. Loke
Division of Cellular and
    Genetic Pathology
Department of Pathology
University of Cambridge
Cambridge, England

Michael Marx
Institut für Anatomie
RWTH Aachen
Melatener Strasse 211
D5100 Aachen, Federal Republic of
    Germany

John E. Morris
Department of Zoology
Oregon State University
Corvallis, Oregon 97331

L. Moss
Department of Anatomy
School of Medicine
University of California
San Francisco, California 94143

B.P. Nalbach
Institute für Anatomie
RWTH Aachen
Melatener Strasse 211
D5100 Aachen, Federal Republic of
    Germany

B. Ove Nilsson
Department of Human Anatomy

Biomedical Center, Box 571
S75123 Uppsala, Sweden

Gary E. Olson
Department of Cell Biology
Vanderbilt University
School of Medicine
Nashville, Tennessee 37232

Denise Pace
Department of Pathology
University of Leeds
Leeds LS2 9JT, United Kingdom

Robert Pijnenborg
Department of Obstetrics and
    Gynecology
Katholieke Universiteit
Leuven, Belgium

Sandra W. Potter
Department of Zoology
Oregon State University
Corvallis, Oregon 97331

Anne Ritson
Department of Pathology
University of Leeds
Leeds LS2 9JT, United Kingdom

Mourad W. Seif
Departments of Obstetrics, Gynecology,
    Biochemistry and Molecular
    Biology
University of Manchester
Manchester, United Kingdom

James A. Simon
Department of Obstetrics and
    Gynecology
Georgetown University Medical School
Washington, DC 20007

Ann Sutherland
Department of Anatomy
School of Medicine
University of California
San Francisco, California 94143

Oswald F. Wilson
Department of Biochemistry
    and Molecular Biology
The University of Texas
M.D. Anderson Cancer Center
1515 Holcombe Blvd, Box 117
Houston, Texas 77030

Virginia P. Winfrey
Department of Cell Biology
Vanderbilt University
School of Medicine
Nashville, Tennessee 37232

Elke Winterhager
Institute für Anatomie
RWTH Aachen
Melatener Strasse 211
D5100 Aachen, Federal Republic of
    Germany

INDEX

All page numbers listed below represent the first page of each chapter,
where the subject is located.